ROUTLEDGE LIBRARY EDITIONS:
20TH CENTURY SCIENCE

Volume 16

THE COLLECTED PAPERS OF LORD RUTHERFORD OF NELSON

Volume 6

THE COLLECTED PAPERS OF LORD RUTHERFORD OF NELSON

Volume Two: Manchester

ERNEST RUTHERFORD

Edited by
JAMES CHADWICK

Routledge
Taylor & Francis Group

LONDON AND NEW YORK

First published in 1963

This edition first published in 2014
by Routledge
2 Park Square, Milton Park, Abingdon, Oxfordshire OX14 4RN

and by Routledge
711 Third Avenue, New York, NY 10017

First issued in paperback 2016

Routledge is an imprint of the Taylor & Francis Group, an informa business

British Library Cataloguing in Publication Data
A catalogue record for this book is available from the British Library

ISBN: 978-0-415-73519-3 (Set)
ISBN 13: 978-1-138-98893-4 (pbk)
ISBN 13: 978-1-138-01366-7 (hbk)

Publisher's Note
The publisher has gone to great lengths to ensure the quality of this book but points out that some imperfections from the original may be apparent.

Disclaimer
The publisher has made every effort to trace copyright holders and would welcome correspondence from those they have been unable to trace.

THE COLLECTED PAPERS

OF

LORD RUTHERFORD

OF NELSON

O.M., F.R.S.

PUBLISHED UNDER THE SCIENTIFIC DIRECTION OF
SIR JAMES CHADWICK, F.R.S.

VOLUME TWO

MANCHESTER

LONDON
GEORGE ALLEN AND UNWIN LTD

FIRST PUBLISHED IN 1963

This volume © George Allen & Unwin Ltd, 1963

PRINTED IN GREAT BRITAIN
in 10 pt. Times Roman type
BY UNWIN BROTHERS LIMITED
WOKING AND LONDON

FOREWORD

A PROPOSAL to publish the scientific work of the late Lord Rutherford was discussed shortly after the war, but the acute shortage of paper at that time, and, as a corollary, the heavy commitments of those publishers who were interested in the matter, caused it to be abandoned. This earlier proposal was for the publication of a selection of Rutherford's most important papers.

The present venture was, from the outset, conceived on different lines. It was proposed to include every scientific paper which Rutherford had published either alone or with collaborators; and also a number of general articles, formal public lectures, letters to editors, and other communications which seemed worthy of preservation.

This comprehensive and, indeed, ambitious scheme was brought to my attention in the autumn of 1956 by Dr Paul Rosbaud. I readily agreed to give it my full support and to act as scientific editor.

This publication of Rutherford's Collected Works will consist of four volumes, of which the first three will contain the papers published in the usual way in scientific journals.

The first volume includes his work in New Zealand, at the Cavendish Laboratory and in Montreal, covering the years from 1894 to April 1907; the second volume will contain the papers of the Manchester period, 1907 to 1919; and the third volume will cover his period as Cavendish Professor from 1919 to 1937. The fourth volume will include miscellaneous articles, public lectures, letters to editors and, in addition, some obituary notices of Rutherford. It will also include a bibliography which, it is hoped, will be complete.

Each of the four volumes will contain accounts of personal recollections and appreciations by some of his friends and colleagues and also portraits and photographs of historical interest.

Some of Rutherford's papers were published at about the same time in German or in French as well as in English journals. When the two versions are identical, the English version has generally been chosen for publication here. There are, however, occasions when the German or French version contains additional material, and in these cases that version has been adopted.

The primary purpose in this publication of Rutherford's Collected Works is, of course, to set up a visible memorial to one of the greatest figures in the history of science; and, at the same time, to make it readily possible for the succeeding generations of young scientists to see what he did, to follow the development of his ideas, and to get at first hand some idea of the magnitude of his contribution to our knowledge of the physical world. No reader of these volumes can fail to be impressed by the vigour and directness of Rutherford's mind, or fail to become aware that the pursuit of scientific truth was to him an activity of the highest intensity, and also a very personal activity.

It has been the wish and endeavour of all concerned with this project that these volumes should be produced at the lowest price consistent with good printing and reproduction. In furtherance of this aim, the publishers, Messrs George Allen & Unwin, have been most generously helped by large grants towards the very substantial cost of a four-volume publication and the work of preparation and revision which it entailed. It is with deep appreciation and gratitude that such aid is acknowledged from the Government of New Zealand, the National Research Council of Canada and the Leverhulme Trust.

The copyright of Lord Rutherford's publications is held by his grand-children, Dr Peter Fowler, Mrs Elizabeth Rutherford Taylor, Mr Patrick Fowler and Dr Ruth Edwards. We are indebted to them for permission to publish.

Acknowledgments for Volume Two

As stated in the Foreword, this second volume of the Collected Papers contains the scientific papers published by Rutherford from Manchester, where he was Langworthy Professor of Physics from October 1907 to October 1919.

These papers are introduced by a general survey of his work during that period under the title 'Rutherford at Manchester: an epoch in Physics', contributed by Professor N. Feather; and I record here my gratitude to him for this article, in which he sets Rutherford's work against the background of that time.

Also included are some reminiscences of Rutherford and his Manchester laboratory by the late Professor H. Geiger, by Professor E. N. da C. Andrade, and by Dr A. B. Wood. I am grateful for permission to publish these contributions.

It is a pleasure to give thanks to Professor Otto Hahn for the loan of the photograph of Rutherford and Geiger, to Professor Sir Nevill Mott and the Cavendish Laboratory for the reproductions from Rutherford's notes, and to Dr A. B. Wood for the photographs of the laboratory group at Manchester in 1913, and of the Admiralty Physics Board, the latter being reproduced by courtesy of the Editor of the *Journal of the Royal Naval Scientific Service*.

In my acknowledgments at the end of the Foreword to Volume One I referred to my debt to Dr Paul Rosbaud—'but for whose initiative and sustained interest this publication would not have been undertaken'. I record with sorrow that Dr Rosbaud died in January 1963. He had carried out most of the work required to get this second volume ready for publication, but he did not live to see it in a complete state. I repeat with emphasis what I have said above; and I add, in justice to his memory, that any merit which this publication of the Collected Papers may have is almost entirely due to him.

J. CHADWICK

CONTENTS

NOTE *The papers appear here in the chronological order of publication. No attempt has been made to impose uniformity in the use of abbreviations, in the quotation of references, in the consistency of the use of symbols, etc. The sequence of the illustrations and figures in the original papers has been maintained. In consequence, some repetition and some inconsistencies may become apparent when one paper is compared with another. In some cases, however, obvious misprints and errors have been corrected; others may have escaped detection.*

Foreword by Sir James Chadwick   page   7

Rutherford at Manchester: an epoch in physics, by N. Feather   15

1907

The Origin of Radium   34

The Effect of High Temperature on the Activity of the
Products of Radium (J. E. Petavel)   36

Origin of Radium   38

The Production and Origin of Radium   40

The Production and Origin of Radium   42

1908

A Method of Counting the Number of α Particles from
Radioactive Matter   57

Recent Advances in Radio-activity   59

Spectrum of the Radium Emanation (T. Royds)   70

Experiments with the Radium Emanation. I, The Volume
of the Emanation   72

Spectrum of the Radium Emanation (T. Royds)   84

An Electrical Method of counting the Number of α Particles
from Radioactive Substances (H. Geiger)   89

The Charge and Nature of the α Particle (H. Geiger)   109

The Nature and Charge of the α Particles from Radioactive
Substances   121

The Action of the Radium Emanation upon Water (T.
Royds)   128

The Nature of the α Particle (T. Royds)   134

Some Properties of the Radium Emanation   136

The Chemical Nature of the α Particles from Radioactive
Substances   137

A*

1908   *page*

The Discharge of Electricity from Glowing Bodies   147

Der Ursprung des Radiums   150

1909

The Boiling Point of the Radium Emanation   161

The Nature of the α Particle from Radioactive Substances
(T. Royds)   163

Differences in the Decay of the Radium Emanation (Y.
Tuomikoski)   168

Condensation of the Radium Emanation   170

The Action of the α Rays on Glass   176

Production of Helium by Radium (B. B. Boltwood)   177

1910

Action of the α Rays on Glass   178

Properties of Polonium   180

Theory of the Luminosity produced in Certain Substances
by α Rays   182

Radium Standards and Nomenclature   193

The Number of α Particles emitted by Uranium and
Thorium and by Uranium Minerals (H. Geiger)   196

The Probability Variations in the Distribution of α Particles
(H. Geiger. With a note by H. Bateman)   203

1911

The Scattering of the α and β Rays and the Structure of the
Atom   212

Untersuchungen über die Radiumemanation. II, Die
Umwandlungsgeschwindigkeit   214

Die Erzeugung von Helium durch Radium (B. B. Boltwood)   221

The Scattering of α and β Particles by Matter and the
Structure of the Atom   238

Transformation and Nomenclature of the Radioactive
Emanations (H. Geiger)   255

The Transformation of Radium   262

1912

A Balance Method for Comparison of Quantities of Radium
and some of its Applications (J. Chadwick)   271

1912 *page*

The Origin of β and γ Rays from Radioactive Substances   280

Photographic Registration of α Particles (H. Geiger)   288

On the Energy of the Groups of β Rays from Radium   292

Wärmeentwicklung durch Radium and Radiumemanation (H. Robinson)   294

Some Reminiscences of Rutherford during his time in Manchester, by H. Geiger, E. N. da C. Andrade and A. B. Wood   295

1913

Heating Effect of Radium and its Emanation (H. Robinson)   312

A New International Physical Institute   328

The Age of Pleochroic Haloes (J. Joly)   330

The Analysis of the γ Rays from Radium B and Radium C (H. Richardson)   342

Analysis of the γ Rays from Radium D and Radium E (H. Richardson)   353

The Reflection of γ Rays from Crystals (E. N. da C. Andrade)   361

Scattering of α Particles by Gases (J. M. Nuttall)   362

The Analysis of the β Rays from Radium B and Radium C (H. Robinson)   371

Über die Masse und die Geschwindigkeiten der von den radioaktiven Substanzen ausgesendeten α Teilchen (H. Robinson)   383

The British Radium Standard   406

The Structure of the Atom   409

Analysis of the γ Rays of the Thorium and Actinium Products (H. Richardson)   410

1914

The Structure of the Atom   423

The Wavelength of the Soft γ Rays from Radium B (E. N. da C. Andrade)   432

The Structure of the Atom   445

The Spectrum of the Penetrating γ Rays from Radium B and Radium C (E. N. da C. Andrade)   456

Spectrum of the β Rays excited by γ Rays (H. Robinson, W. F. Rawlinson)   466

1914   *page*

The Structure of Atoms and Molecules   471

The Connexion between the β and γ Ray Spectra   473

Radium Constants on the International Standard   486

1915

Origin of the Spectra given by β and γ Rays of Radium   493

Radiations from Exploding Atoms   495

Maximum Frequency of the X Rays from a Coolidge Tube
for Different Voltages (J. Barnes, H. Richardson)   505

Efficiency of Production of X Rays from a Coolidge Tube
(J. Barnes)   524

1916

Long-range α Particles from Thorium (A. B. Wood)   531

1917

Penetrating Power of the X Radiation from a Coolidge Tube   538

1919

Collision of α Particles with Light Atoms

   I Hydrogen   547
  II Velocity of the Hydrogen Atoms   568
 III Nitrogen and Oxygen Atoms   577
 IV An Anomalous Effect in Nitrogen   585

PLATES

Professor Ernest Rutherford, 1909   *frontispiece*

Rutherford and Geiger with their apparatus for counting α particles   *facing page* 112

Rutherford's first rough note on the nuclear theory of atomic structure; written, probably, in the winter of 1910–11   *between pages* 240–241

Another page of these rough notes, in which Rutherford estimated the chance of a large deflection by gold   *between pages* 240–241

Physics Staff and Research Group, Manchester, 1913   *facing page* 304

Admiralty Physics Board at the Mining School, Portsmouth, 1921   528

A page from Rutherford's Laboratory Notebook; recorded 9 November, 1917   584

PROFESSOR ERNEST RUTHERFORD 1909

Rutherford at Manchester: an epoch in physics

by N. FEATHER, F.R.S.

IN the summer of 1907, Rutherford moved from Montreal to Manchester. Twelve years previously he had arrived in Cambridge, a raw student from an outpost of empire; twelve years later he was to return there, as Cavendish professor. The first three years which he spent in Cambridge (1895–98) were the years of his maturing; the last three or four years in Manchester were lean years—they were years of war. Thereafter, as head of the most famous physical laboratory in the world of those days, he was to lose, gradually but inevitably, the opportunity for personal participation in experimental research. The summer of 1907, therefore, provides a natural climacteric: Rutherford's personal achievements during the preceding nine years of his Montreal professorship may be compared on equal terms with his achievements in Manchester—over the first nine years of his tenure of the Langworthy chair, until the privations of war brought fundamental research to a stop in the universities of Europe. On this basis we set the contents of this volume of the *Collected Papers* against the background of Volume I.

But there is one observation, before we embark on our survey. To the papers included in these volumes must be added the books. In his sixth year at Montreal, Rutherford sent the manuscript of his book *Radio-activity* to the Cambridge University Press; at the beginning of his sixth year at Manchester he wrote the preface to *Radioactive Substances and their Radiations* and despatched the completed work to the same publisher. These two substantial monographs remain major classics to this day. Before he left Montreal, Rutherford had already revised the first of them for a new edition—and he had seen his Silliman Memorial Lectures at Yale (1905) published in an impressive volume (*Radioactive Transformations*) to fulfil the conditions of the lectureship—but here we concentrate on the originals, the Cambridge books of 1904 and 1913, whose titles have already been quoted.

In 1903 the theory of spontaneous disintegration had first been formulated explicitly. In 1911 the large-angle scattering of α-particles was 'explained', for the first time satisfactorily, on the basis of the nuclear model of the atom: in that context the atom nucleus was discovered. Universally, after fifty years, these two achievements are regarded as Rutherford's outstanding contributions to physics in Montreal and Manchester, respectively. From this point of time, then, it is interesting to look back to the books, written when these particular contributions were new—and their ideas, to many, strange—in the hope of assessing Rutherford's own estimate of their significance and value.

The reader who makes this enquiry for the first time is likely to be surprised by the outcome. In the preface to *Radio-activity*, written in February 1904, he will find the statement, simple and direct, 'The interpretation of the results has, to a large extent, been based on the disintegration theory . . .'—and in the 382 pages of text which follow he will find this claim amply substantiated: the entire account is informed by the new ideas which Rutherford and Soddy had given to the world, whole and irrefutable, in the paper 'Radioactive Change' in May of the previous year. In the preface to the other book, by contrast, he will look in vain for any reference either to experiments on α-particle scattering, or to the nuclear atom. He will find a paragraph which begins, 'It is of interest to signalise some of the main directions of advance since the publication of the second edition of my *Radio-activity*'. But he will not find any mention of these topics there. Yet this preface was written in October 1912. In the 670 pages of following text, the paper of May 1911, 'The Scattering of α and β Particles by Matter and the Structure of the Atom', is quoted only three times, and the space devoted to the problem of large-angle scattering is no more than three pages in all. The atom nucleus had been discovered in a particular context, but the new concept had not been assimilated into physics generally; radioactivity had not been recognized explicitly as a property of the nucleus ('the transformation theory advanced in explanation of radio-active phenomena has undergone no essential modification', *p. vi*), even the α-particle is still a doubly-charged atom of helium rather than a bare helium nucleus ('it does not seem possible that the α particle can retain any of its constituent electrons in escaping from the radio-active atom', *p. 620*). To some extent, no doubt, it was merely the obtuseness of words and the difficulty of a new vocabulary, to some extent it was Rutherford's natural caution which imposed restraint, but, whatever may have been the reason, the fact is clear: the discovery of May 1911 was accorded no central place in the book of 1913 as the hypothesis of May 1903 had been, in the book of 1904. Perhaps throughout 1911 and the following years the process of re-orientation was unaccountably slow, but let an unbiased reader look through the papers included in this volume, and follow the references, and he will discover that, slow or not, the advance from the first experimental observation of Geiger and Marsden in 1909 to the atom model of Bohr, exhibited in its full potentiality six years later, was a product of Rutherford's inspiration almost exclusively. Independent contributions of others were trivial in comparison; the whole episode was focused in Manchester.

Let us, then, look at the collected papers in more detail, surveying the whole period, and bearing in mind the comparison to which we are pledged. From Montreal Rutherford published seventy papers, twenty-five in collaboration with colleagues; here, in this Manchester volume, there are seventy-two reprinted, thirty-two of joint authorship. The prodigious output of the earlier period was thus maintained in the later; it is only surprising that the proportion of joint papers was so little increased. For at Manchester Rutherford was the administrative head of a big department, with responsibilities for teaching as well as for research; at Montreal, as Macdonald professor, he had been spared such routine duties to a large extent. But ceaseless enquiry of nature was his way of life, and in the full power and authority of his pre-eminent position he was not to be denied the exercise of his genius. At the end of a long day he could say, simply, to a young colleague, 'Robinson, I'm sorry for those fellows who haven't got laboratories to work in'—and the remark came from the heart, and rang true. In the light of it we need enquire no further how it was that the output of publication was maintained.

It was a diverse output, all the time, with interweaving threads of enquiry continually in evidence, but we may see in it, also, the steady development of a planned attack: at first a preponderance of work on the α-emitters, on the nature of the α-particle itself and on the chemical and radioactive properties of radium and radium emanation; then a detailed study of the β- and γ-emitters, involving the first precision spectrometry of these radiations; overlapping both in time, the investigation of α-particle scattering, revealing the nucleus in 1911, and, eight years later, providing the first faint hint of artificial disintegration—'the anomalous effect in nitrogen', as Rutherford described it. For the historian, the nucleus and artificial nuclear transmutation provide the highlights of the period (we have already given pride of place to the former), but the greatness of Rutherford's genius touches all these investigations and we shall do well to consider each in turn.

Broadly, the work on the nature of the α-particle comes first in time—and its origins lay in Montreal rather than Manchester. In November 1902 Rutherford and Soddy had written ('The Cause and Nature of Radioactivity', Part II): 'In light of these results . . . the speculation naturally arises whether the presence of helium in minerals and its invariable association with uranium and thorium may not be connected with their radioactivity'. Then, in August of the following year, Ramsay and Soddy announced the discovery of the presence of helium in the gases obtained from radium bromide, and of its production by radium emanation. Rutherford was in Britain, on holiday at Bettws-y-Coed, at the time. As soon as his copy of *Nature* reached him, he sat down and made an order-of-magnitude calculation. A few months previously the rate of evolution of heat by radium had been determined by Curie and Laborde, and he himself had 'shown that the α or easily absorbed rays from radium consist of a stream of positively charged bodies, of mass about twice that of the hydrogen atom'. Now he assumed that the heating effect was a measure of the dissipation of the kinetic energy of the α-particles in the radioactive material, and that 'the α bodies after expulsion can exist in the gaseous form': on this basis he calculated the rate of evolution of gas. 'The determination of the mass of the α body, taken in conjunction with the experiments on the production of helium by the emanation, supports the view that the α particle is in reality helium', he wrote (*Volume I, p. 610*). This was on August 15th, two days after the communication of Ramsay and Soddy had been published in London. Rutherford's reaction had indeed been swift: within a matter of hours there had crystallized in his mind a precise hypothesis which it took him six years to bring to conclusive test. But, as events were to prove, his intuition was unerring.

The conclusive test belongs to the Manchester period; at Montreal Rutherford could only pave the way for the future. There he was content to establish that the α-particles from all radioactive substances are identical, differing only in velocity of emission. Otto Hahn was his collaborator in some of this work. In July 1906, on a visit to Berkeley, in the intervals of lecturing to an advanced class in the summer session of the University of California, he found time to write up the details in two papers for the *Philosophical Magazine*. Herein was described the work that he had done over the past year, alone and with Hahn, on the magnetic and electric deflection of the α-particles from radium and its products, and from the active deposits of thorium and actinium. Technically, it was a remarkable achievement—and the result left little room for doubt. Within fairly narrow limits of experimental uncertainty, the α-particle was indeed characterized by the same value of e/m whatever its origin. In the longer of the two papers there was a section headed: 'Connexion of the α particle with the helium atom'. In this section Rutherford wrote (*Volume I, p. 896*): 'The value of e/m for the α particle may be explained on the assumptions that the α particle is (1) a *molecule* of hydrogen carrying the ionic charge of hydrogen; (2) a helium atom carrying *twice* the ionic charge of hydrogen; or (3) *one-half* of the helium atom carrying a single ionic charge'. He dismissed the first explanation as being against the evidence; of the other two he preferred (2) for its 'simplicity and probability', but he retained an open mind, admitting the plausibility of (3).

Assumptions (2) and (3) differed primarily in respect of the charge assigned to the α-particle. At Manchester, with Geiger's help, Rutherford determined this charge directly. He had already made a careful examination of the conditions necessary for the successful determination of the total rate of transfer of charge by the α-particles emitted from a strong source of radium, in 1905 in Montreal (*Volume I, p. 816*); now he had merely to refine the experimental procedure and to develop a successful method of determining the rate of emission of α-particles. The papers describing these classic experiments are included in this volume (*pp. 89, 121*), and need not be referred to in detail here. Suffice to say that the chosen method of α-particle counting was the electrical method (the first success of which owed everything to Geiger's persistence and experience)—and to point out that before the investigation was completed Regener's simpler scintillation method was also tested and, to Rutherford's initial surprise, was found to be equally trustworthy. The details of the experiments can be read and appreciated at leisure; our concern is only with the final result—and with Rutherford's interpretation of it. The result, expressed in his own words, was as follows: 'the positive charge E carried by an α-particle from radium C is $9\cdot3 \times 10^{-10}$ E.S. units'; the final conclusion: 'that *an α-particle is a helium atom*, or, to be more precise, *the α-particle, after it has lost its positive charge, is a helium atom*.

Obviously, this is the favoured assumption (2) of the paper of October 1906, justified only if the charge on the α-particle is twice 'the ionic charge of hydrogen'. But, as Rutherford himself admitted, the ionic charge of hydrogen, as originally determined by Townsend, by Thomson and by H. A. Wilson lay between $3 \cdot 0 \times 10^{-10}$ and $3 \cdot 4 \times 10^{-10}$ e.s.u. How, then, could the charge on the α-particle possibly be two, rather than three, units of charge? True, a preliminary report of a new determination by Millikan and Begeman had pointed to a value of $4 \cdot 06 \times 10^{-10}$ e.s.u. for the ionic unit. But Rutherford was never unduly swayed by preliminary reports of unfinished experiments; he relied rather on his own intuition. For five years his intuition had told him that the α-particle was indeed a charged atom of helium, and, as independent evidence, he put forward an ingenious argument (though it was clumsily presented in the paper) to show that the number of ionic charges on the α-particle could be deduced from measurements of the heating effect, the electrostatic deflection of the α-particles, the magnitude of the faraday, the atomic weight of radium and the rate of growth of radium from ionium. This roundabout calculation, gave nothing more decisive than $2 \cdot 2$ ionic charges, with the uncertain data available, but honour was thereby satisfied. The question of probable error was by-passed: the α-particle was a doubly charged atom of helium, as it had to be!

Once this conclusion was accepted, 'the value of e, the charge on a hydrogen atom, becomes $4 \cdot 65 \times 10^{-10}$', so Rutherford wrote. Then, the intriguing aside: 'It is of interest to note that Planck deduced a value of $e = 4 \cdot 69 \times 10^{-10}$ E.S. unit from a general optical theory of the natural temperature radiation'. Now, Planck had utilized the values of Stefan's and Wien's constants, of the universal gas constant, the velocity of light and the faraday, to deduce this result—and in 1908 only a minority of physicists regarded his 'general optical theory' with approval. It is the more intriguing to find Rutherford among these early adherents—and for no other reason than that he knew all along that the α-particle is a doubly charged atom of helium!

On December 11, 1908, Rutherford, having received a Nobel Prize, as the citation said, for 'researches on the disintegration of the elements and the chemistry of radioactive matters', delivered his prize lecture before the Royal Academy of Science at Stockholm. Clearly content, for a day, to be dubbed a chemist, he entitled the lecture, 'The chemical nature of the α-particles from radioactive substances' (*this volume, p. 137*). The chance was too good to be missed, the situation being what it was. Introducing his subject, he said 'during the last six years there has been a persistent attack on this great problem, which has finally yielded to the assault when the resources of the attack seemed almost exhausted'. That was his sober estimate of the state of affairs after the publication of his two papers with Geiger: 'the resources of attack almost exhausted'—in spite of the conviction with which his conclusions had been expressed at the time. α-particles, of mass about four units, and helium, from radium, certainly, but 'It might be argued, for example, that the helium atom* appeared as a result of the disintegration of the radium atom in the same way as the atom of the emanation and had no direct connection with the α-particle'. But, happily, the problem had 'finally yielded to the assault', and Rutherford was able to tell his audience of the success of the experiment which, with Royds, he had completed only four weeks previously. The formal paper describing this work appeared in the *Philosophical Magazine* in February of the following year (*this volume, p. 163*). This was the conclusive test of which we wrote earlier: Mr Baumbach, the departmental glassblower, had succeeded 'after some trials' in blowing very thin-walled glass tubes which proved completely impervious to helium gas under the conditions of the experiment but which were thin enough to allow the α-particles to pass through. In this way the α-particles were separated physically from the emanation in which they originated: so separated, they were shown to build up a sample of helium capable of identification spectroscopically in a capillary discharge. 'The long and arduous path trodden by the experimenter', as Rutherford described it, rather uncharacteristically, in the opening sentences of his Stockholm lecture, had at last come to its appointed end.

The long path had come to its end, or nearly so—for in the years that followed there were the researches of Rutherford and Boltwood on the production of helium by radium (*this volume, p. 221*) and of Rutherford and Robinson on the heating effect (*this volume, p. 312*) and on the value of e/m for the α-particles of radium emanation and of its short-lived products (*this volume, p. 383*). Even Geiger's electrical method of α-particle counting, though it was to be abandoned for twenty years in favour of the scintillation method, received further attention in its own right (*this volume, p. 288*). The electrical method had served its purpose, admirably, in the researches of 1908, and the rough values of e/m, and of the rates of production of helium and heat, then available, had likewise been adequate for the use to which they were put, but the future might require more accurate values (indeed, the remaining sceptics had to be converted, come what may), and it was not Rutherford's intention to leave a job half-done. The conclusive proof of the identity of the α-particle and the doubly-charged atom of helium was completed in 1908, but the consolidation of the secondary evidence in the case continued for another five years, providing challenging problems for the younger members of the Manchester group—and ultimate satisfaction for Rutherford himself, as numerical precision was improved, and the doubts of the most inveterate critic were removed.

The small-angle scattering of α-particles in passing through thin sheets of mica was first observed by Rutherford, in Montreal, in 1906. This was in an experiment in which the magnetic deflection was being studied, and at one

---

* The printed text has 'α-particle', but this is an obvious error undetected in proof-reading.

stage Rutherford covered half of the defining slit with a sheet of mica less than 20 microns thick. The edges of the photographic trace formed by the particles were thereby blurred, indicating a change of direction of some two degrees for a large fraction of the particles in traversing the mica sheet. When the magnetic field was applied, the overall deflection of the particles which passed through the unobstructed portion of the slit was not much greater than this, although the particles travelled several centimetres in the field. Rutherford was not slow to appreciate the significance of the conjunction of these two simple, almost qualitative, observations. To produce the observed deflection in a distance equal to the thickness of the mica it 'would require . . . an average transverse electric field of about 100 million volts per cm.' he wrote (*Volume I, p. 867*), and he concluded: 'Such a result brings out clearly the fact that the atoms of matter must be the seat of very intense electrical forces'. Although he regarded this result as self-evident, once 'the electronic theory of matter' had been accepted, Rutherford also realized, from this single observation, that he possessed in the α-particle a natural probe well suited to the exploration of fields of force of atomic dimensions. It is not, then, surprising that one of the entries in his first list of 'Researches possible', which he drew up in the summer of 1907, on his arrival in Manchester, was 'Scattering of α-rays'. Opportunity offered when he and Geiger had completed their experiments on the rate of emission of α-particles from radium, and had verified the claim that Regener had made for the scintillation method of observation: Geiger at once applied this new method to an investigation of the distribution of intensity in angle in the α-particles emerging from very thin foils of metal on which they had impinged as a well-defined pencil. Preliminary results were published in August 1908, and a full account of the investigation followed in April 1910.

By and large, there was nothing unexpected in the results described in these papers: 'the electronic theory of matter', in the form of the Thomson atom model, appeared adequate to encompass them, if the overall (small-angle) deflection of the α-particle were regarded as the resultant of the individual (even smaller) changes of direction suffered by the particle as it traversed successive atoms in its path. Only in one respect was this description manifestly artificial: it required the volume distribution of positive charge in the Thomson atom to be 'transparent' in relation to the particle traversing it.

But there was one very real skeleton in the cupboard. In the paper of April 1910, Geiger wrote: 'It is also of interest to refer to experiments made by E. Marsden and myself on the diffuse reflection of the α-particles. . . . It does not appear profitable at present to discuss the assumption which might be made to account for [the results of these experiments]'. These results had already been published in July of the previous year. They were entirely anomalous. The statistical theory of 'multiple' scattering seemed to be verified so far as more than 99·9 per cent of the incident particles were concerned, but for the small residual fraction—one particle in ten thousand, perhaps—the theory was wildly inadequate. The particles which Geiger and Marsden had found 'diffusely reflected' should not, according to the theory, have been there at all.

Rutherford had deliberately assigned this search for possible diffuse reflection of α-particles to Marsden, in the spring of 1909, ostensibly as a training exercise, under Geiger. What prompted him to do so will remain unknown. In later years he confessed, openly, to his complete amazement at the positive result of the search: 'It was quite the most incredible event that has ever happened to me in my life'—but one must believe that irrational as well as rational elements enter into the make-up of the experimenter of genius, and in this instance, at least, Rutherford was asking a simple question of nature, outside the realm of the rational, not knowing how pertinent a question it was.

It was more than a year before Rutherford was able to formulate a satisfactory explanation of Marsden's results. Intuitively, it seems, he came close to the beginnings of such an explanation at a very early stage, but for some reason which we do not know the clue was lost for a while, and in April 1910 Geiger found it unprofitable to discuss the problem further. At Winnipeg, in the late summer of the previous year, in his presidential address to Section A of the British Association, Rutherford had almost seized on the idea of 'single' as distinct from 'multiple' scattering in order to explain the anomalous results: 'otherwise it would be impossible to change the direction of the particle in passing over such a minute distance as the diameter of a molecule', he said. But he made no further reference to the problem in public during the closing months of that year, and throughout 1910 he was equally reticent. Then, on March 7, 1911, he made a preliminary announcement at a meeting of the Manchester Literary and Philosophical Society (*this volume, p. 212*), and a month or so later he sent off a full-length paper for publication in the *Philosophical Magazine*. This paper, 'The scattering of α and β Particles by Matter and the Structure of the Atom' (*this volume, p. 238*) marks the discovery of the nucleus.

There is some reason to suppose that the order of treatment in this classical paper—first large-angle, single, scattering, then multiple (or 'compound') scattering—is the reverse of the order in which Rutherford explored the properties of his nuclear model by detailed calculation: he would naturally realize that any new model must at least do what the old Thomson model could do. It may even be that it was the apparent difficulty of assimilating the two aspects of the phenomenon in a single explanatory scheme which impeded progress and delayed for so long any public discussion of the issue at stake. It was not in Rutherford's character to play into the hands of others—even his colleagues—by the premature announcement of a half-baked scheme. However this may be, it is certain that in the published paper the sections on multiple scattering are the least convincing. There is the implicit claim, nowhere quite explicit, that the evidence from small-angle scattering in thick foils itself favours the nuclear model as against the

Thomson model. Twenty years later (*Radiations from Radioactive Substances, 1930, p. 209*), any such claim was specifically disowned: 'The experiments on multiple scattering have indeed led to no definite conclusions about atomic structure . . . the best that can be done is to show that the results are in accord with the nuclear theory'.

The paper of May 1911 ends with a section of 'General Considerations'. Here is an open-minded survey in which Rutherford admits that an atom model involving several massive scattering centres might be as satisfactory as a single-centre model, except that the total charge on the scattering centres would have to increase as their number increased—or a model with a negatively charged central mass be as satisfactory as one with a positively charged centre. It was obvious that the model which had been explored mathematically was merely the simplest possible model: the next step was to submit its predictions to the test of experiment.

We have said that the paper of May 1911 marks the discovery of the nucleus. In fact, the word 'nucleus' is not used in the paper. In Rutherford's published writings on the subject it occurs first in the short account of the scattering problem which is given in *Radioactive Substances and their Radiations*, written probably in the following year. Here (p. 184) we find the statement: 'This indicates that the atom must contain a highly concentrated charged nucleus. . . .', but in the next sentence the writer slips back into his previous usage, 'central charge'; thereafter the word 'nucleus' does not recur throughout the rest of the volume.

At Manchester, Geiger and Marsden were already well advanced with experiments designed to provide a thorough-going test of Rutherford's scattering formula when the paper of May 1911 was published. Indeed, Geiger had been able to present some preliminary results which were decidedly favourable towards the nuclear model, when Rutherford first presented it in public at the meeting of the Literary and Philosophical Society on March 7th. For nearly two years these experiments continued, then in April 1913 a full account of them appeared in the *Philosophical Magazine*. Within the experimental limitations of the time, the test was exhaustive; within those limits the predictions of the theory were verified in every particular. Nowhere else in the world had any other physicist taken up the challenge which the scattering formula presented: only in Manchester, therefore, in these two years, had it been possible for anyone to sense the growing conviction that the bold hypothesis of 1911 was destined to become the central tenet of belief of the physicist of the future. Those around Rutherford had this experience; in particular, Niels Bohr shared it during the four months which he spent in the laboratory in the spring and early summer of 1912.

Bohr returned to Copenhagen towards the end of July. On March 6, 1913, he sent Rutherford the first draft of 'the first chapter of my paper on the constitution of atoms', adding, 'I hope that the next chapters shall follow in a few weeks.' After an exchange of letters, Rutherford agreed to send the first chapter, duly amended, for immediate publication. It appeared in the *Philosophical Magazine* of July 1913. The first sentence of the published paper clearly identifies the source of its inspiration, the fifth reflects its author's enthusiasm for the task he had undertaken: 'In order to explain the results of experiments on scattering of α rays by matter Prof. Rutherford has given a theory of the structure of atoms. . . . Great interest is to be attributed to this atom-model . . .'.

Rutherford received the other chapters in due course. They were published serially, in September and November 1913, in March 1914 and September 1915, respectively. We have already briefly assessed this series of papers by Bohr; here it is unnecessary further to apostrophize the theoretical genius of the young Dane who first attempted 'to show that the application of [Planck's] ideas to Rutherford's atom-model affords a basis for a theory of the constitution of atoms'. From that time onwards, the nuclear model of the atom has commonly, and with justice, been referred to as the model of Rutherford and Bohr. Nothing more need be said.

In the summer of 1913 the scientific world, generally, became suddenly aware of the nucleus. At the British Association meeting in Birmingham, at the Solvay Conference in Brussels and at an international congress in Vienna, the new ideas provoked the greatest interest. A little later, in London, on March 19, 1914, the Royal Society held a special discussion on the subject. Except in Vienna, Rutherford was the central figure on each occasion. In the end, there could be no permanent opposition, for there was no plausible alternative to the new theory, and its quantitative successes appeared utterly convincing. But at Brussels, Thomson did not acquiesce, unreservedly. In his original paper, Rutherford had not thought it necessary to question the validity of the coulombian law even when distances of the order of 10^{-12} cm. were involved. Thomson took the opposite point of view: 'Now we have no direct evidence', he said, 'as to what is the law of force between electrified bodies when the distance separating them is as small as this . . .'—and, later, he continued, 'This seems to me to indicate that the large deflections . . . of the α particles are not produced by forces . . . due to electrical charges . . . which would act upon a charged corpuscle as well as upon a charged α particle. They are in my opinion more likely to be due to special forces which come into play when two α particles are within less than a certain distance of each other. That in fact when two α particles come into collision inside an atom the forces between them are not merely [electrostatic forces]'. To Rutherford, understandably, this appeared as an obtuse opinion: had not the experimental results of Geiger and Marsden confirmed, at the same time, both the essential correctness of the nuclear model and also the validity of the coulombian law at nuclear distances? Obviously they had—for all practical purposes: Thomson's was indeed an obtuse opinion, in so far as the collisions of 8 MeV α-particles with heavy atoms were concerned, but, from the vantage point of today, we might ask whether it would have been equally obtuse in relation to the collisions of

8 MeV α-particles with free α-particles (or with helium atoms). Luck—or the very nature of things—was on Rutherford's side!

It is no coincidence, perhaps, that, when the special meeting of the Royal Society took place on March 19, 1914, the then current issue of the *Philosophical Magazine* carried a paper by Rutherford surveying the position as he saw it at the time (*this volume, p. 423*). He had not written specifically on the nuclear model, at least for general publication, since his first formal paper on the subject nearly three years previously. Now he was able to 'deal with certain points in connexion with the "nucleus" theory of the atom which were purposely omitted in my first communication' and to give 'a brief account . . . of the later investigations which have been made to test the theory'. Among the matters 'purposely omitted', we may identify, first of all, the old problem of the validity of the coulombian law. Rutherford was now able to report that Darwin had proved conclusively (and the proof was in an accompanying paper in the same journal) that no other power-law would serve to explain the results of experiment. And we may also identify the problem of the scattering law as it applies to collisions with the lightest atoms, for example, hydrogen and helium (for this problem was referred to, and definitely held over for later discussion, in the original paper). Again, Darwin had worked out the necessary modification of the theory—and Rutherford and Nuttall had already made some tests of its validity (*this volume, p. 362*). But the 'later investigations' provide the real content of the paper: the unexpected confirmation of the assumption of single scattering in the first photographs of α-particle cloud tracks taken by C. T. R. Wilson in Cambridge; the fitting together of van den Broek's simple suggestion regarding the magnitude of the nuclear charge, both with Bohr's dictum that α- and β-particles alike originate in the nucleus, and with the empirical 'displacement law' of Fajans, Russell and Soddy; the brilliant experiments of Moseley ordering the lighter elements, at least, uniquely in terms of the frequencies of the characteristic X-radiations. Finally, there was Bohr's own massive contribution, as it existed at that time in the three papers which had then been published.

We may note with amusement that Rutherford mentioned Bohr's papers last of all in his survey—almost as an afterthought. 'While there may be much difference of opinion as to the validity and of the underlying physical meaning of the assumptions made by Bohr', he was prepared to applaud his intentions! But he had already said of Moseley, with no such reservations, 'he has shown that the variation of wave-length can be simply explained by supposing that the charge on the nucleus increases from element to element by exactly one unit'. The mere onlooker may be forgiven for thinking that, at that stage, the two issues were very much entangled.

The paper of March 1914 marks the beginning of the assimilation of the nucleus into the general world-picture of the physicist; it is also remarkable in that it indicates the lines of Rutherford's thinking on the problem of nuclear constitution. The signature of the chemical atom was the number of unit charges on the nucleus, but the nucleus itself was a system of some considerable complexity, its mass almost certainly less than the sum of the masses of its constituent particles—hydrogen nuclei and electrons, with helium nuclei (α-particles) as sub-units of 'very stable configuration'. Obviously, Rutherford had done a lot of thinking in the three years which had elapsed since the original announcement. And he had done some adventurous experimenting, too: 'In conjunction with Mr Robinson, I have examined whether any charged atoms are expelled from radioactive matter except helium atoms . . . if such particles are expelled, their number is certainly less than 1 in 10,000 of the number of helium atoms'. So, let us leave the nucleus for a while, noting only that Marsden was already counting the scintillations of hydrogen nuclei projected forwards 'elastically' with roughly four times the residual range of the α-particles (according to Darwin's prediction), and consider another line of work. It is a natural breaking-point: we have followed the history of the nucleus to the outbreak of war.

During much of the period that we have just been describing, Rutherford was heavily committed with work on the β- and γ-radiations. As first discoverer—indeed, almost as 'onlie begetter'—of the α-particle, he had at times been somewhat neglectful of the claim of the more penetrating radiations to serious study. 'I have often pointed out what an important part the α particle plays in radioactive transformation. In comparison, the β and γ rays play quite a secondary rôle', he wrote in July 1906 (*Volume I, p. 895*). Five years later the position had changed. von Baeyer, Hahn and Meitner had discovered the magnetic line spectrum of the electrons from meso-thorium 2 and other β-emitters, and Danysz, confirming their results, had already revealed the great complexity of this spectrum in certain cases. Obviously, there was much detailed information to be gathered, precise energies to be determined—and used to develop and test an interpretative scheme.

Rutherford did not at once enter into competition with the workers in Berlin and Paris in the field of magnetic spectroscopy. His first reaction was to assign to Moseley the problem of determining the mean number of electrons emitted in the successive β-disintegrations of radium B and radium C. This was merely one of the many assignments which, during his three years at Manchester, Moseley tackled and fulfilled with unflagging energy and consummate skill. Then, with Robinson, Rutherford re-determined the heating effect of radium and its products, paying particular attention to the heating due to the more penetrating radiations, the β-particles and the γ-rays (*this volume, p. 312*). These measurements completed, he wrote up his own provisional views on the subject in a paper for the *Philosophical Magazine* (*this volume, p. 280*). Some of his views were soon to be discarded, in the light of the displacement law and the other evidence which we have already discussed: 'The instability of the atom which leads to its disintegration may be conveniently considered to be due to two causes . . . the instability of the central nucleus [which 'leads to the expulsion of an α particle'] and the instability of the electronic distribution [which leads 'to the appearance of β and γ rays']'. None survived unchanged, save the basic view that there must exist a close connection between the appearance of the electrons of the line spectrum and the emission of the γ-rays: 'It may prove significant that only those products which emit well-defined groups of β rays emit also a strong γ radiation'.

Rutherford naturally followed the clue which this last quotation suggested; the fact that 'as far as observation has gone, the β rays from uranium X and radium E give a continuous spectrum in a magnetic field', was awkward, but he shrugged it off, for the time being, as an effect which a young man in a hurry was unlikely to make sense of.

At this stage Rutherford, himself, was really in a hurry to obtain further information regarding the spectra of the β- and γ-radiations of as many products as possible. As providing a background to more detailed studies, a general survey of the γ-radiations was instituted, using the standard method of absorption analysis. Much of this work was done by H. Richardson, under Rutherford's personal direction (*this volume, pp. 342, 353, 410*), but Moseley, Fajans and Makower also contributed—and, somewhat outside the main line of interest, Russell and Chadwick detected weak radiations of the γ-ray type with the α-emitters radium, ionium and polonium, and examined them by the same method. As regards the more detailed studies, Robinson was chosen to be chiefly responsible for the magnetic spectroscopy: to develop an instrument with which the work of Danysz could be checked—and overtaken. Andrade was given an even more exacting assignment: no one had yet applied the new method of crystal diffraction, which Moseley was using to such good effect in the domain of the soft X-rays, to the more penetrating γ-rays. Andrade was to do just this. Let us look at the results of these detailed studies, very briefly, in turn.

Before the experiments of Rutherford and Robinson had produced much of significance, Danysz published some further results (*Le Radium*, January 1913), in obtaining which a new experimental technique had been employed. In Manchester, the main technical problem had already been broadly identified: 'to devise a method of bringing out the presence of groups of β rays, the total energy [intensity] of which might be only a small fraction of that distributed in the more intense groups'—and it appeared that Danysz's new arrangement met this requirement. Rutherford and Robinson were quick to take advantage of the new technique, and to improve on it. Later, they wrote laconically (*this volume, p. 371*), 'we have used a special method which appears to be very similar in principle to that employed by Danysz in his latest investigation'. So was the method of semi-circular focusing brought to bear on the problem of the moment. Surprisingly, the principle of the method was not presented in the published paper (*Philosophical Magazine*, October 1913) with the clarity and directness of which Rutherford was undoubtedly capable, but it was applied with real insight and the results to which it led were remarkable in their scope and complexity.

The sources which Rutherford and Robinson had used were 'line' sources—thin-walled emanation tubes and wires coated with the active deposit of radium, or with radium C. The β-radiation from these sources was spread out into a momentum spectrum in the magnetic field: the γ-radiation went free. Rutherford, with good reason, suspected a close connection between the two. The whole situation cried out for someone to ask Nature the next question—with his hands. The question when it came— and there was no unseemly delay—came in the authentic Rutherford manner. A thin sheath of lead was slipped over the line source in the spectrograph; then the γ-radiation did not all go free, some of it produced 'secondary' electrons in the lead, and the momentum spectrum of 'the β rays excited by γ rays' was observed for the first time. As Rutherford expected, the general structure of the 'excited' spectrum—in so far as it could be observed, for the experiment was technically a difficult one—was very similar indeed to that of the 'natural' line spectrum observed with the unsheathed source. Very definitely there was a close connection between the electrons of the natural line spectrum and the γ-rays.

The experiments that we have just been describing were reported by Rutherford, Robinson and Rawlinson in the *Philosophical Magazine* for August 1914 (*this volume, p. 466*). The last paragraph of this paper begins: 'Experiments on this subject will be continued by Robinson and Rawlinson'. That was to reckon without the war-lords of Europe.

Rutherford and Andrade published the results of their experiments by the crystal diffraction method in two papers in May and August 1914 (this *volume, pp. 432, 456*). The first paper described a straightforward experiment, using the technique of Bragg reflection from the cleavage face of a crystal. It required intense sources and long exposures. Qualitatively, it demonstrated without doubt that the least penetrating component of the radiation was L X-radiation, according to the nomenclature of Barkla. Technically, the achievement was considerable (the spectrum 'lines' were faithful images of the emanation tube used as source, showing bright edges and dark centres, as would be expected from the distribution of active deposit over the inner walls of the tube), but it is doubtful whether the identification of the characteristic X-radiation helped, at that stage, to clarify the situation generally. It directed attention away from the nucleus as the emitting body. And, for once, by some strange circumstance, Rutherford must have allowed his own convictions to colour his estimate of the merit of an experimental determination. According to the displacement law, the atomic number of radium B had to be that of lead; according to Moseley, the atomic number of lead was 82. The authors of the paper of May 1914 expressed the conviction that they had a large factor of safety in hand in claiming that the L X-rays in the spectrum of radium B were demonstrably those of an atom of atomic number 82, not 81 or 83. It all seemed to be so obviously straightforward. Yet, something had been at fault. More than ten years later Rutherford and Wooster repeated the experiment in Cambridge. The

L X-rays in the radiation spectrum of radium B were shown to be those of an atom of atomic number 83. Which is not to say that radium B is not an isotope of lead!

The second paper of Rutherford and Andrade is noteworthy for the remarkably simple and effective arrangement, therein described, by which sharp spectrum lines were obtained, in a pattern of four-fold symmetry, by 'focal isolation' of the diffracted ('reflected') radiation transmitted through a rock-salt crystal. The source was a short 'line' source of radium emanation, directed along the normal to a crystal face, the isolating 'stop' a circular hole in a lead screen placed at an equal distance on the other side of the crystal. With this arrangement, exposure-times were reduced and background fogging was almost eliminated. But the elegance of the method could not alone ensure that the results which were achieved added anything of significance to the general theoretical picture. Rutherford and Andrade convinced themselves—and rightly so—that they had observed for the first time the K series lines due to radium B, but preoccupation with ideas of characteristic X-radiations led them to regard some γ-rays of shorter wavelength which they also observed as belonging to 'the "H" series, for no doubt evidence of a similar radiation will be found in other elements when bombarded by high speed cathode rays' (so they wrote). They were pioneering an entirely new field of experimental research; it should not be held to their discredit that, in discussing the assignment of the homogeneous component radiations as between the two constituents of the active deposit, they were entirely unaware that more than 95 per cent of the γ-radiations of radium C in fact lay outside the range of analysis of the method they were using.

For nearly two years, as we have seen, much of the effort of the Manchester laboratory had been devoted to a concerted attack on the problem of β- and γ-ray changes. On June 30, 1914, having directed this attack, Rutherford completed his second paper reviewing the general situation. Essentially, it had the same title as the first, and it appeared in the September number of the *Philosophical Magazine* (*this volume, p. 473*). Seen from this distance in time, it scarcely appears to advance our understanding of the problem, in any notable particular. Experimentally, much information had been gathered, but a coherent scheme of interpretation still eluded the Manchester group. Bohr was still in Copenhagen. Lip-service was paid to the displacement law: 'Suppose . . . that the disintegration of the atom leads to the expulsion of a high speed β particle from or near the nucleus'. But the basic simplicity of the atomic model which provided the only available explanation of Moseley's results—the only explanation which made them intelligible, and utterly fundamental, in the scheme of things—was not seized upon firmly; instead, localized 'vibrating systems' were postulated within the atom, responsible each for the emission of a characteristic radiation when suitably excited. The exciting and entirely novel result which Chadwick had obtained in a few months' work with Geiger in Berlin was recorded, and accepted at its face value, but the attempt to specify the conditions under which the electrons of the line spectrum should appear was artificial in the extreme. The line spectrum was missing, so Rutherford said, when the disintegration electron never succeeded in exciting the 'vibrating systems' on its way out of the atom—and he made much play with hypothetical directional preferences as a basis for such possibility. He accepted the electrons of the continuous spectrum as the disintegration electrons—he had no other alternative in the light of Chadwick's results—and in one sentence he reached an entirely luminous conclusion, 'The present theory supposes that the homogeneous groups of β rays arise from the conversion of the energy of the γ rays into the β ray form', but the confusion between γ-rays and characteristic X-rays remained and clouded the whole picture. How soon a way would have been found through the complexities of the problem, had not war intervened, is a matter of speculation. In the upshot, the issue remained unresolved for eight years, until Ellis, in Rutherford's laboratory in Cambridge, produced compelling evidence for the modern view that the γ-rays, like the α-particles and the β-particles, originate in the nucleus. Even then, the difficulty of the continuous spectrum was in no way abated, and another ten years or more were required to bring it into focus in a symbol, Pauli's neutrino. On reflection, surely, our strictures have been too severe: in 1914 the time was not ripe for an understanding of the problem of the β- and γ-ray changes.

There is only one postscript to be added to this opinion here. In the first year of the war Rutherford was able to organize some work in Manchester, with Barnes and Richardson, on the X-radiation obtainable with the then newly developed Coolidge-type tube. Two papers were published describing the experiments in the *Philosophical Magazine* in September 1915 (*this volume, pp. 505, 524*). Thereafter, he managed to continue these experiments, sporadically, with such help as he could enlist, and a further paper appeared in the same journal exactly two years later (*this volume, p. 538*). Throughout all this work, amid all the distractions and duties of the time, quite obviously Rutherford had kept the problem of the penetrating γ-rays not far out of mind. In the end he had come round by a devious route to a most important conclusion. Let it be expressed in his own words: 'In our present ignorance ... it is only possible to estimate the actual wave-length of the most penetrating gamma rays. It is clear, however, that ... they correspond to waves generated by voltages between 600,000 and 2,000,000 ... that the gamma rays from radium C ... are of considerably shorter wave-length than any so far observed in an X-ray tube, with the highest voltages at our disposal ... the β rays from radium C consist mainly of groups lying between 500,000 and 2,000,000 volts ... It would thus appear probable that the observed groups of β rays are due to the conversion of the energy, $E = h\nu$, of a wave of frequency ν into electronic form, and that consequently the energy of the β ray groups may be utilized by the quantum relation to determine the wave-lengths of the penetrating gamma rays'. It should now be clear how it was that Ellis was able to start off on the right foot, when he began research under Rutherford in Cambridge, after the war!

Rutherford finished writing the paper, from which we have just quoted, on May 12th. In less than a week he was in Paris, where a joint Anglo-French mission was assembling, for the journey to Washington. They were going to acquaint the naval authorities of the United States with the situation in respect of anti-submarine research and development as it then was. This had been Rutherford's main concern for nearly two years: it had taken him on frequent journeys to Rosyth, and Harwich, and London, and it had involved much hard work and improvisation in a field of experiment which was new to him. The visit to America provided a welcome break, and marked the end of his active involvement in the British effort, though it was arduous in itself. It provided opportunity for renewal, also: Rutherford saw Boltwood at Yale, and received an honorary degree from the university, and he made the journey to Montreal, as well. He was back in Manchester by the end of July.

Perhaps his visit to Washington reminded Rutherford of the last occasion on which he had been there—in April 1914, when he had delivered the first course of William Ellery Hale lectures before the National Academy of Sciences. His title had been: 'The constitution of matter and the evolution of the elements', and he had been bold enough to discourse in public on the possibilities of bringing to pass in the laboratory some of the transmutations —some of the stages of the 'inorganic evolution of the elements'—concerning which Lockyer had speculated so many years previously. 'It is possible', he had said, 'that the nucleus of an atom may be altered either by direct collision of the nucleus with very swift electrons or atoms of helium such as are ejected from radioactive matter'. We have mentioned, already, that early in 1914 Marsden was at work investigating the long-range hydrogen nuclei projected forwards by α-particles passing through hydrogen gas. When Rutherford returned from Washington in May of that year, he surely had in mind to continue these investigations using gases other than hydrogen. Indeed, some few observations were made, though not deliberately, of the effect in air. But, by that time, the country was at war, and the matter could not be pursued to its conclusion. Had it been so pursued, with Rutherford's active participation, it is unlikely that the effect would have been missed; as it was, Marsden and Lantsberry merely reported their observations. They had observed what appeared to be long-range hydrogen nuclei with an α-particle source situated in air—but they concluded (*Philosophical Magazine, August 1915*) that the hydrogen nuclei were probably emitted by the source. In this they were doubtless mistaken, but they had at least convinced themselves that the magnitude of the effect was too great to be explained in terms of water vapour, or other hydrogenous material, condensed on the source—and in that conviction they were probably correct.

When Rutherford returned from Washington in July 1917, he took up, on his own, the work which Marsden had begun. He was without assistance, save for William Kay, the laboratory steward. But Kay was a prince among assistants, and the work went well. It started in earnest on Saturday, September 8th, and within three days the 'unexpected' scintillations of long-range particles had definitely been observed, with air in the tube. By September 28th, similar experiments were being made with nitrogen, and oxygen, and carbon dioxide, in turn. Some days earlier, helium had been used. Already there was fairly definite evidence of an effect specific to nitrogen. But Marsden had thought that there were hydrogen nuclei of high energy emitted from the source. Rutherford, therefore, spent laborious days throughout October investigating the scintillations observed when the α-particles were allowed to pass into an absorbing foil in an otherwise evacuated tube. He found nothing to substantiate Marsden's suggestion. Early in November he returned to the experiments with air. Careful drying, rigorous elimination of dust: nothing that he could do by way of purification produced any diminution in the number of scintillations due to the long-range particles. Under otherwise identical conditions, he compared the effect in chemically prepared nitrogen with the effect in air. The ratio was $5:4$, as nearly as his statistical accuracy could be trusted. He was now convinced that the effect arose in α-particle collisions with nitrogen nuclei. The next step, as he recorded in his laboratory notebook on November 9, 1917, was: 'To settle whether these scintillations are N, He, H or Li'.

Admittedly, the four possibilities symbolized in this brief exhortation were not equally plausible. That unchanged nitrogen nuclei, or α-particles, should have acquired an abnormally large increment of energy in mutual interaction, when carbon nuclei and α-particles, or oxygen nuclei and α-particles, did not interact in this (unclassical) way, was most unlikely. But these possibilities were at least entertained, although the other two were the less outrageous. If the long-range particles were hydrogen nuclei, the suggestion was, in effect, that a 'chip' had been knocked off a nitrogen nucleus in close collision with an α-particle; if the long-range particles were lithium nuclei, then the collision had split the nitrogen nucleus in two. Rutherford set to work to examine these possibilities—and on January 10, 1918, he added another to his list: the long-range particles might be deuterons. He did not, of course, use that word (the word 'proton' was not yet in the vocabulary of physics): he wrote 'atom charge $+ e$ and mass $M = 2$ called x'.

It was a brave exhortation—this exhortation to take each possible identification in turn, devise experimental tests, and sift the evidence. Naturally, it did not work out in that way. The scintillations that Marsden had observed with α-particles in hydrogen were certainly those of hydrogen nuclei. Rutherford knew what they looked like, and he was prepared to bet that the 'unknown' scintillations were due to hydrogen nuclei, also. So he planned a long series of experiments in which he compared objectively the two radiations: the hydrogen nuclei projected by elastic collision from hydrogenous materials, and the particles of long range produced in nitrogen. It required all his ingenuity—and the greater part of his faith in the rightness of his judgment. But the demonstrable results of the comparisons were definite enough to give colour to his view: 'It is difficult to avoid the conclusion that the long-range atoms arising from collision of α-particles with nitrogen are not nitrogen atoms but probably atoms of hydrogen, or atoms of mass 2'.

The quotation is from the last paper that Rutherford wrote during the Manchester period. Under the title: 'An anomalous effect in nitrogen', it constituted Part IV of a sequence of papers published in the *Philosophical Magazine* for June 1919. The general title of the sequence (*this volume*, p. 547) is: 'Collisions of α-particles with light atoms', and its length fifty pages of original text. Part IV occupies a mere six pages of this total. In this compass one of the most momentous experiments in the history of science is presented unostentatiously, without emphasis, or any sensational claim.

Single-handed, in the last dark days of war, Rutherford had been adventuring into the unknown. In happier times, much of this adventure would have fallen to the lot of the young men who thronged his laboratory and knew the inspiration of his genius. He had not failed them: forty-four of the fifty pages of the published account of the work that he had done in their absence was of work such as they might have achieved under his guidance— honest work, and worth while. Only the kernel of it, the residue which did not belong to any previous category of thought, was outside the range of their powers. That residue, those six pages of print, were to startle the world.

The Origin of Radium

From *Nature*, 76, 1907, p. 126

IN a previous letter to NATURE (January 17)* I gave an account of some experiments which I had made upon the growth of radium in preparations of actinium. The results obtained were in substantial agreement with the earlier observations of Boltwood in this Journal (November 15, 1906), but it was pointed out that there was no definite evidence that actinium itself was the true parent of radium. The experimental results could be equally well explained by supposing that the parent substance of radium was ordinarily separated from radio-active ores with the actinium, but had no direct radio-active connection with the latter.

Observations have been continued upon the growth of radium in the actinium solution prepared in the manner indicated in my first letter. The rate of growth was found to be uniform over a period of 120 days, and to agree closely with the rate of growth observed in the solid preparation of actinium which had been set aside for a period of two and a half years. Another sample of actinium was then taken and successively precipitated with ammonium sulphide in order to remove the radium from the solution. In this way a solution of actinium was obtained initially almost entirely free from radium. By examination of the α-ray activity, it was found that the actinium after this chemical treatment contained an excess of radio-actinium. This was shown by the rise of the activity to twice its initial value in about twenty days, and then a gradual decay to a steady value. Special care was taken to measure accurately the rate of growth of radium in the solution at short intervals in order to see whether it depended in any way upon the variation of the activity. No such connection was observed, for the radium was produced at a constant rate over the whole period of examination, viz. 111 days.

For equal quantities of actinium, the rate of growth of radium observed in this solution was $1 \cdot 5$ times greater than the normal. This indicated that only a portion of the actinium had been precipitated, while the radium-producing substance had been precipitated with the actinium in excess of the normal amount. This conclusion was confirmed by an examination of the filtrates, which were found to contain more than half the actinium. After suitable chemical treatment, a small precipitate of actinium was again obtained which was about one hundred times as active, weight for weight, as the original preparation. This actinium precipitate was dissolved in hydrochloric acid, and observations of the amount of radium in it were made at regular intervals. *No appreciable growth of radium was observed over a period*

* *Vol. I, p. 907.*

of eighty days. If there were any growth at all, it was certainly less than one two-hundredth part of that normally to be expected. In order to make certain that the absence of apparent growth of radium in this solution could not be ascribed to the precipitation of the radium in some non-emanating form, the solution was again chemically treated. The actinium was precipitated with ammonia and re-dissolved in hydrochloric acid. Again no growth was observed over the period of examination, viz. twenty days. The solution in its present state contains a just measurable quantity of radium, viz. about 2×10^{-12} gram.

From these observations I think we may safely conclude that, in the ordinary commercial preparations of actinium, there exists a new substance which is slowly transformed into radium. This immediate parent of radium is chemically quite distinct from actinium and radium and their known products, and is capable of complete separation from them.

It is not possible at present to decide definitely whether this parent substance is a final product of the transformation of actinium or not. It is not improbable that it may prove to be the long-looked-for intermediate product of slow transformation between uranium X and radium, but with no direct radio-active connection with actinium. If this be the case, the position of actinium in the radio-active series still remains unsettled.

It is intended to continue observations on the growth of radium in the solutions described above. Experiments are also in progress to isolate this new substance in order to examine its chemical and radio-active properties.

E. RUTHERFORD.

Manchester
May 30

The Effect of High Temperature on the Activity of the Products of Radium

by PROFESSOR E. RUTHERFORD, F.R.S.

and J. E. PETAVEL, F.R.S.

Abstract of the British Association *Report*, 1907, pp. 456–7

BRONSON has shown that the activity of the products of radium is not appreciably altered by exposure to a temperature of 1600° C. On the other hand, Makower, working with the active deposit of radium, found that there was a small decrease of its activity, measured by the β and γ rays, when exposed for some time to a temperature of about 1100° C. The experiments of Schuster and of Eve have shown that the highest obtainable pressures have no influence on the activity of radium.

In the present experiments the emanation from about four milligrams of radium bromide was momentarily exposed to the influence of the very high temperature produced by the explosion of cordite in a closed steel bomb. The bomb used in these experiments was constructed by Mr Petavel, and had been used by him in previous experiments on the pressures developed during explosions. The bomb was a complete sphere of mild steel, about 4 inches internal diameter and about 2 inches thick. About forty-six grains of cordite were placed in the bomb, and after exhaustion the emanation was introduced. About four hours later the emanation is in equilibrium with its products, and the activity due to the γ rays, which passed through the bomb, was observed by means of an electroscope placed outside the bomb. The cordite was fired electrically, and observations were made of any change of activity. By running the electroscope *during* the explosion, it was found that no sudden burst of activity occurred, showing conclusively that the normal rate of disintegration of the product, radium C, was not much altered by this process. Three experiments were made with equal weights of cordite, but of different diameter, in order to vary the suddenness of the explosion. In every case the activity measured by the γ rays was found to have decreased about 9 per cent. after the explosion. The activity gradually rose again, reaching nearly the equilibrium value after three hours. A special experiment showed that the rate of change of the emanation itself was not altered by the explosion.

The maximum pressure of the gases during the explosion was about 1200 atmospheres, and the maximum temperature certainly not lower than 2500° C.

The change of activity produced by the explosion may be due either to a sudden alteration of the distribution of the active deposit or to a change in the amount or period of the products, radium B and radium C. Since the active deposit of radium is volatilised at about 1200° C, it would be rendered gaseous by the high temperature of the explosion, and redeposited when it cooled. Since the bomb was exactly spherical, a change of distribution of the active deposit does not appear very probable. In one experiment two electroscopes were used, one by the side of the bomb and the other underneath it. Both showed about an equal decrease of activity.

The experiments recorded here are preliminary, and it is intended to examine still further whether there is a real change of activity of radium products by the action of the high temperature.

Origin of Radium

From *Nature*, **76**, 1907, p. 661

IN a letter to NATURE (June 6)* I gave the experimental evidence which led me to conclude that in ordinary actinium preparations a new substance was present which was slowly transformed into radium. By a chemical method this substance was separated from actinium, and a solution of the latter was obtained which showed no appreciable growth of radium over a period of eighty days. Observations on this solution have been continued over a total period of 240 days, and there is still no detectable increase in the quantity of radium. The growth of radium, if it occurs at all, is certainly less than 1/500 of that observed in other experiments.

In two recent letters to NATURE (September 26 and October 10) Dr Boltwood has given the results of his later experiments in this direction. He has confirmed my conclusions, and has, in addition, been successful in devising a satisfactory method of separating this new substance from actinium, and has examined its radio-active and chemical properties. He suggests that the name 'ionium' be given to this new body, which is probably the immediate parent of radium. Dr Boltwood is to be congratulated for his admirable work on this very difficult problem, for, apart from the chemical operations, the radio-active analysis required for correct deduction is unusually complicated and difficult.

Dr Boltwood has not been able to separate the parent of radium from actinium by the reagent employed by me, viz. ammonium sulphide, but has found the use of sodium thiosulphate effective. In explanation of this discrepancy, he suggests that I employed old ammonium sulphide. As a matter of fact, I did not use the ordinary laboratory solution of ammonium sulphide, but added ammonia to the actinium solution, and then saturated it with sulphuretted hydrogen. The complete separation effected in my experiment was, I think, probably due to an accidental production of finely divided sulphur in the solution.

In a letter to NATURE of last week, Mr N. R. Campbell raised objections to the name 'ionium' given by Dr Boltwood to the new body, from the point of view that every radio-active substance should be given a name to indicate its position in the scheme of radio-active changes. This system is very excellent in theory, but I have found it extremely difficult to carry out in practice. The continual discovery of new products in very awkward positions in the radio-active series has made any simple permanent system of nomenclature impossible. Besides uranium and thorium, twenty-four distinct radio-active

* *This vol., p. 34.*

substances are now known to exist in radio-active minerals. The number of products still to be discovered is, I think, nearly exhausted. When there is a general consensus of opinion that this is the case, I feel it will be very desirable for physicists and chemists to meet together in order to revise the whole system of nomenclature. There is not much to be gained in doing so immediately, as the discovery of a new product in the midst of a series would entail the alteration of the names of a possible half-dozen others which follow it. At the same time, I think it will be desirable to retain a distinctive name for those radio-active substances which, like radium, have a long enough life to be separated in sufficient quantity for an examination of properties by the ordinary chemical and physical methods. It is probable that the parent of radium fulfils these conditions, and should thus have a distinctive name like radium.

Personally, I do not much like the name 'ionium,' but for similar reasons neither do I care for the name 'actinium.' It is not easy to suggest a name that is at once simple and explanatory. I have for some time thought that possibly 'paradium' or 'picradium' might be suitable for the new substance. The former name suggests that it is the parent of radium, but I recognise that a possible play on words may make it unsuitable. The name uranium A, suggested by Mr Campbell, in itself innocuous, is open to the objection that in the case of radium, thorium, and actinium the suffix A is applied to the first product of the disintegration of the respective emanations, while no such emanation has been observed in the initial series of changes of uranium.

E. RUTHERFORD

University of Manchester
October 27

The Production and Origin of Radium

by PROFESSOR E. RUTHERFORD, D.SC., F.R.S.

From the *Proceedings* of the Manchester Literary and Philosophical Society
Abstract of a paper read before the Society on October 29, 1907

AN account was given of the historical development of our ideas in regard to radium. On the disintegration theory, radium is regarded as a substance undergoing slow spontaneous transformation with a period of about 2,000 years. In order to account for the existence of radium in minerals of great age, it is necessary to suppose that radium is produced from another substance of long period of transformation. There is an undoubted genetic connection between uranium and radium, for investigation has shown that the amount of radium in minerals is in all cases proportional to their content of uranium. If this be the case, radium should gradually appear in a preparation of uranium, initially freed from radium. No such growth of radium has been observed over a period of several years although a very minute growth of radium can be easily detected. This is not necessarily inconsistent with the disintegration theory for if one or more products of slow transformation exist between uranium and radium, no appreciable growth of the latter is to be expected in a short interval. A search for this intermediate product has recently proved successful. Boltwood found that a preparation of actinium, initially freed from radium, grew radium at a constant and rapid rate. Boltwood at first considered that actinium was this intermediate product and that actinium changed directly into radium. The growth of radium in actinium solutions was confirmed by the writer, who had commenced experiments in that direction three years before. The experiments showed, however, that actinium did not, as Boltwood supposed, change directly into radium. By a special method, a preparation of actinium was obtained by the writer which showed no appreciable growth of radium over a period of 240 days. The growth of radium, if it occurred at all, was certainly less than 1/500th of that ordinarily observed.

In another case, a solution of actinium was obtained which produced radium faster than the normal.

These results are completely explained by supposing that a new substance of slow transformation is present with actinium, and this substance is transformed directly into radium. This parent of radium has distinct chemical properties, which allow it to be separated from both actinium and radium. The absence of growth of radium observed in the actinium solution mentioned above is due to the fact that, by the special method, the parent of radium had been completely separated from the actinium.

In recent letters to *Nature*, Boltwood has confirmed the results of the writer, and has devised a satisfactory method of separating the radium parent from actinium. He has shown that this new body, which he proposes to call 'ionium,' gives out α and β rays, and has the chemical properties of thorium.

The Royal Society recently loaned the writer the actinium residues from about a ton of pitchblende. These residues contain the parent of radium, and experiments are in progress to isolate and concentrate both the actinium and ionium in these residues.

B*

The Production and Origin of Radium

by E. RUTHERFORD, F.R.S.

Professor of Physics, University of Manchester

From the *Philosophical Magazine* for December 1907, ser. 6, XIV, pp. 733–49
Read before the British Association, Leicester, August 1907. Previous accounts of the
results were given in letters to *Nature*, January 17 and June 6, 1907 (*Vol. I, p.* 907; *this
Vol. p. 34*)

§ 1

THE present point of view of regarding radium as a substance which is
undergoing slow transformation was first put forward definitely by Rutherford
and Soddy in the paper entitled 'Radioactive Change' (*Phil. Mag.*, May
1903, p. 590), in the following terms: 'In the case of radium, however, the
same amount, (viz. about 1 milligram), must be changing per gram per year.
The "life" of the radium cannot in consequence be more than a few thousand
years on this minimum estimate, based on the assumption that each particle
produces one ray at each change. . . . So that it appears certain that the
radium present in a mineral has not been in existence as long as the mineral
itself, but is being continuously produced by radioactive change.' (*Vol. I,
p.* 607.)

On this theory, the parent substance which produces radium must always
be present in minerals containing radium. Uranium from the first appeared
to be the most probable parent, since it possessed a life long compared
with radium and was always found associated with it. There were two
obvious methods of attack to throw light upon this question, one direct
and the other indirect. The first consisted in an examination to see whether
in course of time radium appeared in a solution of uranium initially freed
from radium. The second depended upon an examination of the relative
amount of radium and uranium in radioactive minerals. According to
theory, if uranium is the parent of radium, the ratio of the amount of radium
in any mineral to that of uranium should be constant. The constancy of this
ratio has been completely substantiated by the independent work of
Boltwood*, Strutt†, and McCoy‡; and there can be no doubt that uranium
and radium are genetically connected. Rutherford and Boltwood§ have

---

* Boltwood, *Phil. Mag.*, April 1905.
† Strutt, *Proc. Roy. Soc.*, March 2, 1905.
‡ McCoy, *Ber. d. D. Chem. Ges.*, No. 11, p. 2641, 1905.
§ Rutherford and Boltwood, *Amer. Journ. Sci.*, July 1906. (*Vol. I, p. 856*).

found that for every gram of uranium in a mineral, there is present $3 \cdot 8 \times 10^{-7}$ gram of radium.

The question of the growth of radium in a uranium solution was first attacked by Soddy[*] and, later, by Boltwood.[†] Without entering into the details of these important investigations, it suffices to say that, in carefully purified uranium solutions, no growth of radium has been observed, over the space of the few years that observations have been in progress. If radium is produced at all, it is certainly produced at less than one-thousandth of the rate to be expected theoretically. This result is not necessarily inconsistent with the view that radium is a transformation product of uranium, for the absence of observable growth of radium in a limited time is to be expected, if one or more products of slow transformation exist between uranium and radium.

In the meantime, Boltwood[‡] had approached the problem from a different direction. By a special method, the actinium was separated from a kilogram of carnotite. A solution of this actinium, initially containing very little radium, was placed aside and examined 120 days afterwards. A notable increase in the amount of radium was observed. In addition, the rate of growth in this interval was about that to be expected if radium were half transformed in about 2,000 years—a result in conformity with calculations of the probable life of radium. The work of Boltwood marks a definite and important stage in the attack on this problem, for it clearly shows that radium, as theory predicted, is produced from another substance and that this parent substance is normally present with actinium.

Boltwood concluded that actinium was the direct parent of radium and was itself an intermediate product between uranium and radium. This conclusion was strongly supported by his observation that the amount of actinium in minerals, like the amount of radium, was proportional to the amount of uranium. Since actinium has probably a life comparable with that of radium, such a conclusion is consistent with the observed absence of growth of radium in uranium solutions, for the uranium must first form a considerable quantity of actinium before the transformation product of the latter, viz. radium, could be detected in the solution. This question will be discussed later in the paper after the consideration of further experimental results. It will be seen that the problem is more complicated than at first appeared.

§ 2. *Old Experiments*

It may be of interest to give a brief account of some experiments commenced by myself in 1904 to determine whether radium was continuously produced from actinium. A preliminary account of this work was given in the Bakerian

---

[*] Soddy, *Phil. Mag.*, June 1905; August 1907.
[†] Boltwood, *Amer. Journ. Sci.*, September 1905.
[‡] Boltwood, *Nature*, November 15, 1906.

Lecture (*Phil. Trans.*, A, p. 218, 1904).* Two grams of an active preparation, of activity about 250 times that of uranium, obtained from Giesel, were taken and dissolved in acid. The initial content of radium was determined by the emanation method, and the greater part of it then removed by successive precipitations in the solution of small quantities of barium as sulphate. Measurements were then made of the amount of radium in this solution at intervals over a space of three months, but with no certain evidence of the growth of radium. The amount of radium was estimated by the emanation method. The radium emanation, which was allowed to collect in the solution for a known interval, was removed into a large electroscope by aspirating a considerable amount of air through the solution. Later work of Boltwood has shown that this aspiration method is unreliable for an accurate determination of the amount of radium present, but it no doubt serves for comparative measurements under identical conditions.

In the light of later knowledge, the method employed for the separation of the radium present initially in the solution was very unsuitable for several reasons. A trace of sulphuric acid remaining in the solution after the removal of the barium might possibly precipitate the radium as sulphate—a form in which it would be very unlikely to release all its emanation by aspiration of the solution.

After three months' observations, this solution was put aside with the intention of testing its radium content at intervals; but the pressure of other work and the recognition of the danger of contaminating the solution in a laboratory in which a large quantity of radium was in use, led to a postponement of further tests for a period of over two and a half years. On the appearance of Boltwood's paper I immediately examined this solution to see whether there had been a growth of radium in this long interval. A preliminary test showed that there had been a considerable increase in the content of radium, but in making a more accurate determination, the solution was unfortunately contaminated with radium, probably by the use of some tap grease for a stopcock. This accident brought home to me the danger of making experiments of this character in a laboratory contaminated with radium, so that most of the experimental work recorded in this paper was carried out in the Chemical Laboratory, in which no radioactive matter had been introduced.

At the same time that the actinium solution had been prepared, a quantity of a solid actinium preparation weighing $0 \cdot 32$ gr. of activity about 250 times uranium, had been set aside in a closed glass tube. The radium content of a gram of the same sample had been determined in 1904 by comparison with a standard radium solution prepared at that time. Corrected in terms of the recently prepared radium standards of Rutherford and Boltwood, the amount of radium per gram of actinium in 1904 was $1 \cdot 16 \times 10^{-8}$ gr. Assuming, as was probable, that the content of radium was equally distributed throughout the whole mass of the actinium, the amount of radium

* *Vol. I, p. 721.*

in the 0.32 gr. was 3.7×10^{-9} gr. After an interval of 2.7 years this actinium was removed and dissolved in hydrochloric acid, and the amount of radium present was found by the emanation method, described later, to be 1.05×10^{-8} gr. The growth of radium in the interval of 2.7 years between the two tests was thus 6.9×10^{-9} gr., or assuming the rate of growth constant 2.9×10^{-9} gr. per year. While not much weight can be attached to this result by itself, on account of the imperfect aspiration method employed in the initial determination of the radium constant, yet the rate of growth observed will be seen to be in good agreement with that determined later for a similar preparation.

§ 3. *Experimental Methods*

Before considering further experiments, a brief description will be given of the experimental methods employed to determine accurately the quantity of radium in the various preparations. The preparation was obtained in the form of a solution and placed in a glass flask. The solution was then boiled to expel completely all the emanation, and the exit-tube sealed before the flask had cooled. After a definite time-interval, the flask was opened and the air, mixed with emanation, was expelled by vigorous boiling and collected over water. Boltwood has shown that boiling is the only satisfactory method of expelling all the radium emanation. The air plus emanation was then introduced through a drying-tube of phosphorous pentoxide into an exhausted electroscope of capacity greater than the volume of air to be introduced. Air was then let in to fill the electroscope to atmospheric pressure and the electroscope closed. The electroscope employed in most of these experiments consisted of an Erlenmeyer flask of about one litre capacity silvered on the inside, in which the gold-leaf system was insulated after Wilson's method by a sulphur bead. The motion of the gold-leaf was read through openings in the silvering by a microscope with a scale in the eyepiece in the usual manner. The natural leak of the electroscope was small and corresponded to 0.14 division per minute of the scale in the eyepiece. Since the emanation from 10^{-9} gr. of radium in equilibrium produced a movement of 11.5 divisions per minute, the natural leak was equivalent to that produced by 1.2×10^{-11} gr. of radium. As is generally observed, the natural leak increased gradually for several days on standing, probably due to a small trace of radium present, but always came back to the same value if the electroscope was exhausted and refilled two hours before observations were begun. Since the vessel was always exhausted to introduce fresh emanation, this peculiarity of the electroscope was no disadvantage. The natural leak tested in this way was remarkably steady, and never varied more than 10 per cent over the course of several months.

Readings of the movement of the gold-leaf were commenced three hours after the introduction of the emanation. At this time, the active deposit is nearly in equilibrium with the emanation, and there is only a very slight

change of the rate of movement for several hours. The electroscope was calibrated and standardized by means of the emanation from a standard solution
of radium bromide, prepared by Rutherford and Boltwood. For the above
electroscope the emanation from 10^{-9} gr. of radium gave a movement of
the gold-leaf of $11 \cdot 5$ divisions per minute.

An amount of emanation which increased the natural leak by 10 per
cent could be detected with certainty, so that the electroscope was capable
of showing the presence of 10^{-12} of a gram of radium in a solution. Ten
times this quantity could be measured with a probable error not more than
a few per cent.

In the experiments to be described later, it will be shown that there was
a constant rate of growth of radium in most of the solutions under examination. Since the amount of emanation in the various solutions was determined
at irregular intervals, it is necessary to consider how the electroscope readings
are connected with the amount of radium existing in the solution at the
moment of expulsion of the emanation.

Let q_0 = amount of radium present initially.

Let q = rate of growth of radium.

Then after the solution has stood for a time t, the amount of radium
present is $q_0 + q.t$.

Suppose that the emanation is completely removed after a time t_1 since
the preparation of the solution, and is tested for the amount of emanation
after a further interval t_2. If a constant quantity of radium is allowed to
produce emanation for a time t, it is well known that the fraction of the
equilibrium quantity of emanation produced is $1 - e^{-\lambda t}$, where λ is the
constant of decay of the radium emanation.

Consequently the amount of emanation present after a time of collection
t_1 is proportional to:

$$(q_0 + qt_1)(1 - e^{-\lambda t_2}) + q \int_0^{t_2} (1 - e^{-\lambda \overline{t_2 - t}})dt.$$

The left-hand side of the expression is proportional to the amount of
emanation due to the radium present in the solution at the time t_1, while
the integral is proportional to the emanation produced by the quantity of
radium formed in the interval t_2.

After reduction, the amount of emanation is seen to be proportional to

$$\left[q_0 + q\left(t_1 - \frac{1}{\lambda}\right)\right](1 - e^{-\lambda t_2}) + qt_2. \qquad . \quad . \quad . \quad . \quad (1)$$

This expression is proportional to the observed rate of movement of the
gold-leaf, so that knowing q_0, t_1, t_2 and λ, the value of q may be expressed
in terms of divisions per minute of the electroscope.

In all the experiments to be discussed, the value of q was found to be
constant over the whole time of observation. The value q_0—the initial amount
of radium present—is best determined from the first observation assuming

that the rate of growth of radium during the first few days is the same as that found later. As an example of the method of calculation, let us consider the solution labelled actinium II (*see* § 7). The average rate of growth of radium per week corresponded to $0 \cdot 26$ divisions per minute of the electroscope; i.e., the equilibrium quantity of the emanation from the amount of radium produced per week would give that rate of movement of the electroscope.

The first measurement was made after an interval of four days, and the emanation from the solution gave $0 \cdot 050$ division per minute of the electroscope. Since $t_1 = 0$, the equation given above reduces to:

$$\left(q_0 - \frac{q}{\lambda}\right)(1 - e^{-\lambda t_2}) + qt_2 = 0 \cdot 050.$$

For convenience, we shall take a week as the unit of time. Since the emanation is half transformed in $3 \cdot 8$ days,

$$\lambda = 1 \cdot 28 \text{ (week)}^{-1}; \quad q = 0 \cdot 26 \text{ and } t_2 = \tfrac{4}{7} \text{ week}.$$

Substituting these values, $q_0 = 0 \cdot 02$; i.e., the emanation from the amount of radium initially present in the solution would give a rate of movement of the electroscope of only $0 \cdot 02$ division per minute—a just detectable quantity.

§ 4. *New Experiments*

A part of the contaminated actinium solution, previously mentioned, was chemically treated to free it from radium. For this purpose, ammonium sulphide was added; this precipitated the actinium and left the radium in solution. By two successive precipitations the greater part of the radium was removed. The precipitate was dissolved in hydrochloric acid, and the radium content of the solution tested at intervals. The quantity of radium initially present in the solution (called actinium I) was found from the first observation by the method already described. The results are given in the following table. Column I gives the interval since the preparation of the solution; column II the time of collection of the emanation; column III the observed movement of the gold-leaf in scale divisions per minute due to the emanation in the solution. In column IV is given the value of q, the average quantity of radium produced per week, calculated from equation (1) and expressed in terms of divisions per minute of the electroscope. The value of q is calculated from each observation on the assumption that the rate of growth has been constant since the preparation of the solution. In column V is given the value of qt, the amount of radium present in the solution at the time of testing.

The amount of radium initially present corresponded to $3 \cdot 95$ divisions per minute.

The results are shown graphically in Fig. 1 in the curve marked

Actinium I. The ordinates represent the amount of radium present in scale divisions per minute (taken from column V), and the abscissae time in days. It will be seen from the curve and also from column IV that the rate of growth of radium is constant within the limit of experimental error over

Growth of Radium in Actinium I

I Time in days	II Time of collec- tion in days	III Observed movement of electroscope	IV q	V qt
4	4	2·13	0·56	0·32
11	7	3·33	0·58	0·91
18	7	3·74	0·58	1·49
25	7	4·12	0·56	2·00
32	7	4·53	0·56	2·56
38	6	4·65	0·60	3·26
53	15	7·25	0·55	4·16
82	29	10·1	0·56	6·56
121	7	9·36	0·54	9·33

Mean value .... 0·566

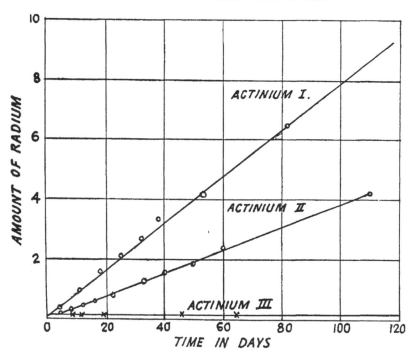

Fig. 1

the time of observation, viz. 121 days. In that interval, the amount of radium in the solution has increased 2·36 times the initial value. The equilibrium amount of emanation from a standard solution of 10^{-9} gr. of radium gave $11 \cdot 5$ divisions per minute in the electroscope. The rate of growth of radium in the solution thus was $4 \cdot 9 \times 10^{-11}$ gr. per week, and assuming the rate of growth constant, $2 \cdot 55 \times 10^{-9}$ gr. per year.

§ 5. *Activity Measurements*

In order to follow the results of the chemical operations, the activity due to a definite fraction of the solution was examined over a long interval. 1/2,000 of the solution was taken and evaporated to dryness on a watch-glass. This gave an extremely thin film of active matter from which the α rays escaped with little absorption. The α ray activity of this film was tested in an α ray electroscope. The variation of activity is shown in Fig. 2, where

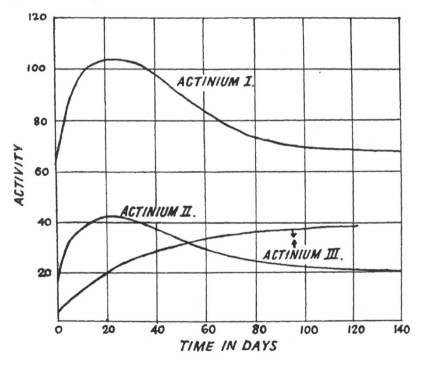

Fig. 2

the ordinates represent the activity in arbitrary units, and the abscissae time in days. It will be seen that the activity at first rises, passes through a maximum in a little over 20 days, and then gradually decays to a constant value about equal to that initially observed. This decrease of the activity after rising to a maximum shows clearly that some of the actinium had not been precipitated by the ammonium sulphide. In addition, the rise of activity to a maximum in about 20 days shows that an excess of the normal amount of radioactinium was removed with the part of the actinium precipitated. Most of the actinium was left behind in the filtrate. The radioactinium in the precipitate produced fresh actinium X and the density rose. The gradual fall of the activity to a constant value is due to the decay of the excess of the radioactinium together with its transformation products. The activity curve will be seen to be very similar to the curves given by Hahn (*Phil. Mag.*, January 1907) for cases where radioactinium is initially in excess. It will be seen that the activity curve is of great value for determining the effect of the chemical operation in removing the various products associated with actinium.

§ 6. *Further Experiments with Actinium I*

After an interval of 121 days, the solution was removed, precipitated with ammonia, and the growth of radium in the precipitated actinium again observed. The object of this experiment was to test whether this reagent was as effective as ammonium sulphide in removing radium from actinium, and also whether the substance that produced radium was precipitated completely with the actinium. Most of the radium was removed by a single precipitation, while the rate of growth of radium in the actinium was unaltered by the process. Such a result is of importance in showing that while radium is soluble in ammonia, the substance that produces it is not, but is precipitated with the actinium. The observation of the growth of radium in the solution has been tested over a further period of 184 days or 305 days in all. There has been no certain change in the rate of growth of radium in this interval.

§ 7. *Experiments with Actinium II*

The experiments with actinium I show that radium is produced at a constant rate in a normal actinium solution, but do not show whether radium is produced from actinium itself or from another substance ordinarily separated with the actinium. The products of actinium in order of sequence are radioactinium, actinium X, emanation, actinium A and actinium B, with periods of half transformation of $19 \cdot 5$ days, 10 days, $3 \cdot 9$ sec., 34 min. and $1 \cdot 5$ min., respectively. Is radium the final product of actinium, i.e. is radium formed from actinium B? There are two methods of attacking this question:

(1) To examine whether the active deposit of radium (composed of actinium A and B) produces radium; or
(2) To test whether the rate of growth in actinium is initially altered by the removal from it of actinium X or radioactinium.

Experiments using the method (1) are described later in § 9 of this paper, but we shall here only consider the second method. If radium is produced directly from actinium B, the rate of production of radium should be nearly proportional to the amount of actinium X, since after a few hours the rapidly transformed products actinium A and B are in equilibrium with it. Consequently, if actinium X is removed from an actinium solution in equilibrium, the rate of growth of radium in the latter should at first be very small, but should gradually increase as fresh actinium X is formed. Since the half period of actinium X is 10 days, the rate of growth of radium should reach half value in 10 days and be within less than 2 per cent of the final constant value in about 60 days.

In order to test this point accurately, it is necessary that the solution under examination should initially contain an amount of radium small compared with the growth per week. In the case of actinium I, the initial content of radium was too large to be quite certain that the rate of growth in the first few days was identical with that determined later.

A part of the actinium preparation of weight 0·32 gr., discussed in § 2, was used for this purpose. It was dissolved in hydrochloric acid and twice precipitated with ammonium sulphide and finally with ammonia, in order to be sure that the actinium X as well as the radium was almost completely removed. These operations were successful, for a solution of actinium was obtained whose initial content of radium corresponded to only 0·02 division per minute in the electroscope, while the rate of growth per week was 0·26 division. A careful examination of the radium content was made at successive intervals of four days until it was clear that the rate of growth was constant. The results are shown in the following table, and are arranged in the same way as the results in the table for actinium I.

I Time of preparation	II Time of collection	III Observed movement of electroscope	IV q in divs. per min.	V qt in divs. per min.
4 days	4 days	0·05 div. per min.	0·26	0·15
8 ,,	4 ,,	0·120 ,,	0·23	0·26
12 ,,	4 ,,	0·194 ,,	0·24	0·41
16 ,,	4 ,,	0·265 ,,	0·24	0·55
22 ,,	6 ,,	0·443 ,,	0·23	0·72
33 ,,	11 ,,	0·965 ,,	0·27	1·27
40 ,,	7 ,,	1·06 ,,	0·275	1·57
50 ,,	10 ,,	1·44 ,,	0·255	1·82
60 ,,	80 ,,	1·85 ,,	0·27	2·32
111 ,,	51 ,,	3·93 ,,	0·259	4·11

The amount of radium at different times is shown graphically in Fig. 1 (curve actinium II), drawn on the same scale as the curve for actinium I. It will be seen from column IV that the rate of production of radium per week, within the limits of experimental error, is constant over the whole interval of 111 days. If radium had been produced directly from actinium B, the rate of growth observed at an interval of 8 days should have been about 0·11, or less than half that actually observed. We may thus conclude that (1) radium is not produced directly from actinium B, and (2) that if radium is produced directly from actinium, a product of slow transformation must exist between actinium B and radium.

On account of my departure from Montreal, the experiments were stopped after 111 days. The electroscope was removed and set up in Manchester and the solution tested at intervals. The results indicate that the rate of growth is still the same after a period of 246 days.

§ 8. *Experiments with Actinium III*

The variation of the α ray activity of a thin film of the solution actinium II is shown in Fig. 2 on the same scale as that of actinium I. It will be seen that the variation of the activity is very similar to that observed for actinium I, and is to be explained in a similar manner. As in the first case, only a part of the actinium was precipitated by the addition of ammonium sulphide. The residue of the actinium remained in the filtrates. After suitable treatment of the latter, ammonia was added in order to precipitate the remaining actinium. A very small precipitate was obtained which was not more than one hundredth of the weight of that initially obtained using ammonium sulphide. This small precipitate contained more than three quarters of the actinium in the original preparation, showing that, under the experimental conditions, a considerable concentration of the actinium had been effected. This small precipitate (called actinium III) was dissolved in hydrochloric acid and its activity examined. The variation of its activity is shown in Fig. 2 (curve actinium III). The ammonia removed most of the actinium X, while most of the radioactinium had been separated with actinium II. The activity consequently rapidly increased, due to the fresh production of radioactinium, and was still rising after an interval of 120 days. This curve is very similar in shape to that given by Hahn (*Phil. Mag.*, June 1907) for the rise of α ray activity of actinium freed from all its products.

The solution actinium III was then tested to see if there were any growth of radium in it. The observations are shown in the following table:

Actinium III

I Time in days	II Divisions per minute of electroscope	III Divisions per minute in equilibrium
4	0·071	0·140
8	0·074	0·148
12	0·077	0·154
19	1·07	0·151
46	1·39	0·139
64	1·30	0·135

Mean 0·144

Column II gives the observed rate of leak of the electroscope due to the emanation; column III the calculated rate of leak of the equilibrium amount of emanation, supposing there was no growth of radium in the solution. Considering the very small rate of leak to be measured, the differences between the numbers in column III are not greater than the experimental errors. The results thus clearly show that there is no certain growth of radium in the solution. If there is any growth of radium at all, it is certainly less than 0·02 of a division per minute over a period of 64 days. The growth of the radium per week, in the solution actinium II, which contains only about half of the actinium in the solution III, corresponds to 0·26 division or to 2·38 divisions in 64 days. For equal quantities of actinium, the growth of radium in solution III, is certainly less than 1/200 of that observed in solution II.

In order to make certain that the observed absence of growth of the radium cannot be ascribed to some chemical action, the solution, after 64 days, was removed and again treated with ammonia. The precipitate was dissolved in hydrochloric acid and again tested for growth of radium. By this treatment the initial content of radium was reduced from 0·14 to 0·04. No growth of radium was observed over a period of 20 days. The solution was then removed to Manchester and tested at intervals over a further period of 136 days. The growth in the total interval of 220 days is certainly less than 0·03 or not greater than 1/500 of the normal growth to be expected.

§ 9. *Experiments with the Active Deposit*

We have seen in § 7 that the observed constant growth of radium in a solution freed from actinium X shows that actinium B is not directly transformed into radium. This result has been confirmed by testing directly whether there was any growth of radium in the active deposit of actinium. The active deposit (actinium A and B) was concentrated on a platinum plate by keeping it negatively charged in the presence of the emanation from an active solid preparation of actinium. Four different experiments were tried in which a platinum plate was exposed for 4, 7, 7, and 14 days respectively. After exposure, the platinum plate was placed in a solution of hydrochloric acid to dissolve off the deposited matter, and the solution with the platinum plate *in situ* tested for radium. The first two experiments showed a just measurable quantity of radium, but with still greater precautions against radium contamination, the last two experiments showed no measurable amount. It may be of interest to consider briefly a method of calculating the amount of radium theoretically to be expected, if actinium B changes directly into radium. Suppose as the basis of calculation that the growth of radium in the solution actinium I (§ 4) is normal, and compare the amounts of actinium B in this solution and on the platinum plate. The α ray activity of the actinium and its products in the solution spread in a thin film corresponded to 9,900 divisions per minute in the α ray electroscope, while the activity of both sides of the platinum plate, tested immediately after removal from the actinium emanation, was 3,800 divisions. Now the actinium in equilibrium contains 4 α ray products whose ranges are 4·8, 6·55, 5·8, and 5·5 cm. respectively, while the active deposit contains only one range of 5·5 cm. Taking as a first approximation that the ionization due to an α particle from each product is proportional to its range in air, the solution contains an amount of active deposit corresponding to an activity of 2,400 divisions. Consequently, the amount of radium to be expected on the platinum plate is 1·6 of the growth of radium in the solution in the time of exposure of the platinum plate to the constant supply of actinium emanation. Now this solution grew per week an amount of radium corresponding to 0·56 division per minute of the electroscope. Consequently, on this hypothesis, the platinum plate exposed for 2 weeks should contain an amount of radium corresponding to 1·8 divisions per minute. The actual amount observed was not more than 0·01 division, or less than 1/180 of the theoretical amount. Such a result conclusively shows that actinium B does not change directly into radium. It is not inconsistent, however, with the possibility that a slowly changing product exists between actinium B and radium. In such a case radium, in the course of time, should appear in the solution containing the platinum plate.

§ 10. *Discussion of Results*

In the following table are given the collected results of the experiments on the growth of radium in the various solutions. In column II is given the total activity of the preparations when in radioactive equilibrium, expressed in divisions per minute of the α ray electroscope. This activity was determined in each case by taking a definite small fraction of the preparation in solution and evaporating it to dryness on a watch-glass. Column III gives the observed rate of growth per week expressed in grams of radium. Column IV gives the rate of growth per year, on the assumption that it is constant

I	II	III	IV	V
Preparation	Total constant activity of preparation	Rate of growth of radium per week	Rate of growth of radium per year	Relative growth of radium
Actinium I	9,900	$4·9 \times 10^{-11}$ gr.	$2·55 \times 10^{-9}$ gr.	1
Actinium II	3,000	$2·3 \times 10^{-11}$ gr.	$1·2 \times 10^{-9}$ gr.	1·55
Actinium III	6,000	not measurable	—	not greater than 0·002
0·32 gr. of actinium tested over a period of 2·7 years	12,900	$5·6 \times 10^{-11}$ gr.	$2·9 \times 10^{-9}$ gr.	0·97

over that interval. Column V shows the relative growth of radium for equal quantities of actinium, taking the rate of growth in the solution actinium I as unity.

It is seen from the above table that for equal amounts of actinium, the growth in actinium I is very nearly the same as for the 0·32 gr., examined over a period of 2·7 years. The closeness of the agreement is no doubt accidental on account of the uncertainty (*see* § 2) in regard to the initial content of radium in the 0·32 gr. of actinium preparation. The table brings out clearly the differences in the growth of radium in the three solutions for equal quantities of actinium.

The solution actinium II grows radium at 1·5 times the rate of actinium I, while solution III, if it grows radium at all, certainly does so extremely slowly.

These experiments can all be readily explained on the simple hypothesis that in ordinary actinium preparations there exists a new substance of slow change which is directly transformed into radium. This new substance is separated with the actinium from the mineral by the methods ordinarily employed for the removal of the actinium. It differs, however, in chemical properties both from radium and actinium, and by special chemical methods can be separated from them both. For example, the preparation actinium I may be considered as possessing the normal quantity of the radium producing substance; while the preparation II contains an excess over the normal. In the case of the solution II, the treatment with ammonium sulphide precipitated all this new substance, but only a fraction of the actinium. The filtrate consequently contained actinium, but no trace of the parent of radium. Under such conditions, there was relatively a large growth of radium in the solution II, but none in the solution III.

As far as the investigations have gone, there is no definite evidence whether this new substance is itself produced by actinium, or whether it is merely associated with the actinium in the same sense that barium always appears with the radium. In the former case, the solution III is gradually producing this new substance, and in the course of time there will be an appreciable growth of radium in it. In the latter case, the solution III will never show any growth of radium comparable with that ordinarily observed. It is not unlikely that this new substance is in reality an intermediate product in the direct line between uranium and radium, and has no direct genetic connexion with the actinium.

The observed constant growth of radium in the solution shows that the parent of radium has a slow rate of change. At a minimum estimate, its period cannot be less than several years and may be much longer.

Summary of Results

(1) Over the time of observation (305 days) radium is produced in actinium preparations at a constant rate.

(2) By suitable chemical treatment actinium preparations can be obtained which grow radium extremely slowly.

(3) The active deposit of actinium does not change directly into radium.

(4) The results indicate that in the ordinary actinium preparations there exists a new substance which is slowly transformed into radium. This direct parent of radium can be chemically separated both from actinium and radium.

(5) Observations have not extended over sufficient time to settle whether this direct parent of radium has any direct genetic connexion with actinium or not.

Experiments are in progress to devise more definite methods for separation and isolation of this new substance in order to examine its physical and chemical properties, and to determine its position in the long series of transformations of uranium.

Manchester
September 20, 1907

A Method of Counting the Number of α Particles from Radio-active Matter

by PROFESSOR E. RUTHERFORD, F.R.S., *and* H. GEIGER, PH.D.

From the *Memoirs* of the Manchester Literary and Philosophical Society, 1908,
Vol. lii, No. 9, pp. 1–3
Received and read February 11, 1908

THE total number of α particles expelled per second from one gram. of radium has been estimated (Rutherford *Phil. Mag.*, Aug. 1905)* by measuring experimentally the total positive charge carried by the α rays from a thin film of radium, on the assumption that each α particle has the same charge as an ion produced in gases. If the α particle is an atom of helium it is necessary to assume that each α particle carries twice the ordinary ionic charge. The need of a method of directly counting the number of α particles shot out from radio-active matter has long been felt in order to determine with the minimum of assumption the charge carried by the α particle and also the magnitude of other radio-active quantities.

It can be calculated that an α particle expelled from radium produces about 80,000 ions in a gas before its ionizing power is lost. With very sensitive apparatus, it should be just possible to detect the ionization produced by a single α particle by electrical methods. The effect, however, would be small and difficult to measure with accuracy. In order to overcome this difficulty, we have employed a method which automatically increases the ionization produced by an α particle several thousand times and so makes the electrical effect easily observable with an ordinary electrometer. This is done by making use of the property discovered by Townsend, that an ion moving in a strong electric field in a gas at low pressure, produces a number of fresh ions by collision with the gas molecules. If the electric field is adjusted nearly to the value required for the passage of the spark, a single ion generated in the gas by external agencies, produces in this way several thousand fresh ions by collision. In the experimental arrangement, the testing vessel consists of a brass tube 60 cms. in length, along the axis of which passes a thin insulated wire attached to the electrometer. With a gas pressure of about 2 cms. a potential difference of about 1,000 volts between the brass tube and the wire is required. The α particles are fired down the tube through a small hole at the end of the tube about 2 mms. in diameter covered with a thin plate of mica. In order to use a narrow pencil of α rays, the active matter in the form of a

* *Vol. I, p. 816.*

thin film on a surface about one square cm. in area is placed in an exhausted tube which is a prolongation of the testing vessel. The distance of the active matter from the hole is usually between 50 and 75 cms. and the amount of active matter adjusted so that from six to ten α particles are fired through the hole per minute. The effect of the α particle entering the testing vessel is shown by a sudden throw of the electrometer needle. Under good conditions this throw is about 50 divisions using an electrometer which has a sensibility of 300 divisions per volt. By observing the number of throws of the electrometer needle, we can count the average number of α particles shot through the opening per minute. The total number fired out by the active matter can be calculated from the known area of the opening and the distance of the latter from the active matter. Preliminary observations show that the number of α particles counted by this method is of the same order as the calculated number, but special experiments are in progress to determine with accuracy the value of this important constant. By counting at intervals the number of α particles expelled per minute, we have been able to obtain the curves of decay of activity of a plate coated with radium C or actinium B.

The α particles from a constant source are shot out at irregular intervals. The time interval between the entrance of successive α particles has been observed over a long interval, and the results show that the distribution curve with time is similar in general shape to the probability curve of distribution of the velocity of molecules in a gas. Further observations, however, are in progress to determine the distribution curve with the accuracy required for comparison with the mathematical theory.

Recent Advances in Radio-activity

From *Nature*, **77**, 1908, pp. 422–6
A discourse delivered at the Royal Institution on Friday, January 31, 1908

IN 1904 I had the honour of giving an address at the Royal Institution on the subject of radio-activity. In the interval steady and rapid progress has been made in unravelling the tangled skein of radio-active phenomena. In the present lecture I shall endeavour to review very shortly some of the more important advances made in the last few years, but as I cannot hope to mention, even briefly, the whole additions to our knowledge in the various branches of the subject, I shall confine my attention to a few of the more salient facts in the development of which I have taken some small share.

In my previous lecture I based the explanation of radio-active phenomena on the disintegration theory put forward in 1903 by Rutherford and Soddy, which supposes that the atoms of the radio-active bodies are unstable systems which break up with explosive violence. This theory has stood the test of time, and has been invaluable in guiding the experimenter through the maze of radio-active complications. In its simplest form, the theory supposes that every second a certain fraction (usually very small) of the atoms present become unstable and explode with great violence, expelling in many cases a small portion of the disrupted atom at a high speed. The residue of the atom forms a new atomic system of less atomic weight, and possessing physical and chemical properties which markedly distinguish it from the parent atom. The atoms composing the new substance formed by the disintegration of the parent matter are also unstable, and break up in turn. The process of degradation of the atom, once started, proceeds through a number of distinct stages. These new products formed by the successive disintegrations of the parent matter are in most cases present in such extremely minute quantity that they cannot be investigated by ordinary chemical methods. The radiations from these substances, however, afford a very delicate method of qualitative and quantitative analysis, so that we can obtain some idea of the physical and chemical properties of substances existing in an amount which is far below the limit of detection of the balance or spectroscope.

The law that governs the breaking up of atoms is very simple and universal in its application. For any simple substance, the average number of atoms breaking up per second is proportional at any time to the number present. In consequence, the amount of radio-active matter decreases in a geometrical progression with the time. The 'period' of any radio-active product, *i.e.* the time for half the matter to be transformed, is a definite and characteristic property of the product which is uninfluenced by any of the laboratory agents at our command. In fact, the period of any radio-active product, for example, the radium emanation, if determined with sufficient accuracy, might well be taken as a definite standard of time, independent of all terrestrial influences.

The law of radio-active transformation can be very simply and aptly illustrated by an hydraulic analogy. Suppose we take a vertical cylinder filled with water, with an opening near the base through which the water escapes through a high resistance.* When the discharge is started the amount of water escaping per second is proportional to the height of water above the zero level of the cylinder. The height of water decreases in a geometrical progression with the time in exactly the same way as the amount of radio-active matter decreases. We can consequently take the height of the column of water as representing the amount of radio-active matter A present at any time. The quantity of water escaping per second is a measure of the rate of disintegration of A and also of the amount of the new substance B formed per second by the disintegration of A. The 'period' of the substance is controlled by the amount of resistance in the discharge circuit. A high resistance gives a small flow of water and a long period of transformation, and *vice versa*. By a suitable arrangement we can readily trace out the decay curve for such a case. A cork carrying a light vertical glass rod is floated on the water in the cylinder. A light camel's hair brush is attached at right angles, and moves over the surface of a smoked-glass plate. A vertical line drawn on the glass through the point of contact of the brush gives the axis of ordinates, while a horizontal line drawn through the brush when the water has reached its lowest level gives the axis of abscissæ. If the glass plate is moved with uniform velocity from the moment of starting the discharge a curve is traced on the glass which is identical in shape with the curve of decay of a radio-active product, where the ordinates at any time represent the relative amount of active matter present, and the abscissæ time. With such an apparatus we can illustrate in a simple way the increase with time of radio-active matter B, which is supplied by the transformation of a substance A. This will correspond, for example, to the growth of the radium emanation with time in a quantity of radium initially freed from emanation. Let us for convenience suppose that A has a much longer period than B. In the hydraulic analogy A is represented by a high head of water discharging at its base through a circuit of high resistance into the top of another cylinder representing the matter B. The water from the cylinder B escapes at its base through a lower resistance. Suppose that initially only A is present. In this case the water in the cylinder B stands at zero level. On opening the stop-cock connecting with A, water flows into B. The rise of water with time in the cylinder B is traced out in the same way as before by moving the glass plate at a constant rate across the tracing brush. If the period of A is very long compared with that of B the water is supplied to B at a constant rate, and the water in B reaches a constant maximum height when the rate of supply to B equals the rate of escape from the latter. The curve traced out in that case is identical in shape with the 'recovery curve' of

* A short glass tube in which is placed a plug of glass wool is very suitable.

a radio-active product supplied at a nearly constant rate. The quantity of matter reaches a maximum when the rate of supply equals its own rate of transformation. The relative height of the columns of water in A and B represents at any time the relative amounts of these substances present.

If the period is comparable with that of B, the height of water in B after reaching a maximum falls again, since as the height of A diminishes the supply to B decreases. Ultimately, the height of B will decrease in a geometrical progression with the time at a rate corresponding to the longer period of the two. This is an exact illustration of the way the amount of a radio-active substance B varies when initially only the parent substance A is present. By using a number of cylinders in series, each with a suitable resistance, we can in a similar way illustrate in a quantitative manner the variation in amount with time of a number of products arising from successive disintegrations of a primary substance. By suitably adjusting the amount of resistance in the discharge circuits of the various cylinders, the curves could be drawn to scale to imitate approximately the variation in amount of the various products with time when the initial conditions are given.

During the last few years a very large amount of work has been done in tracing the remarkable succession of transformations that occur in the various radio-active substances. The known products of radium, thorium, actinium, and uranium are shown graphically below, together with the periods of the products and the character of the radiations they emit. It will be seen that a large list of these unstable bodies are now known. It is probable, however, that not many more remain to be discovered. The main uncertainty lies in the possibility of overlooking a product of rapid transformation following or succeeding one with a very slow period. In tracing out the succession of changes, the emanations or radio-active gases continuously evolved by radium, thorium, and actinium have marked a very definite and important stage, for these emanations can be easily removed from the radio-active body and their further transformations studied quite apart from the parent element. The analysis of the transformation of the radium emanation has yielded results of great importance and interest. After passing through three stages, radium A, B, and C, of short period, a substance, radium D, of long period, makes its appearance. This is transformed through two stages E and F of short period into radium G, of period 140 days. Meyer and Schweidler have conclusively shown that radium D is the primary constituent of the radio-active substance separated by Hofmann and called by him radio-lead. Radium G is identical with the first radio-active substance separated from pitchblende by Madame Curie, viz. polonium. We are thus sure that these bodies are transformation products of radium. It will be seen that I have added another product of period 4·5 days between radium D and polonium. The presence of such a product has been shown by Meyer and Schweidler.

In the case of thorium, a very long list of products is now known. For several years thorium X was thought to be the first product of thorium, but Hahn has recently shown that at least two other products of slow transformation intervene, which he has called mesothorium and radiothorium. The radiothorium emits α rays, and has a period of more than 800 days. Mesothorium apparently emits β rays, and has a still longer period of transformation, the exact value of which has not yet been accurately determined. Since thorium is used commercially on a large scale, there is every prospect that we shall soon be able to obtain considerable quantities of very active preparations of mesothorium and radiothorium. The separation of these bodies from thorium does not in any way alter its commercial value. It is to be hoped that if these active preparations are separated in quantity, the physicist and chemist may be able to obtain a supply of very active material at a reasonable cost, and that there will not be an attempt to compete with the ridiculously high prices charged for radium.

Succession of Substances produced by the transformation of radium, thorium, actinium and uranium. The period of transformation of each substance is added below.

From the radio-active point of view, the radio-elements are only distinguished from their families of products by their comparatively long period of transformation. Now we have reason to believe that radium itself is transformed according to the laws of other radio-active products with a period of about 2000 years. If this be the case, in order to keep up its supply in a mineral, radium must be produced from another substance of relatively long period of transformation. The search for this elusive parent of radium has been one of almost dramatic interest, and illustrates the great importance of the theory as a guide to the experimenter. The view that radium was a substance in continuous transformation was put forward by Rutherford and Soddy in 1903. The most probable parent of radium appeared to be uranium, which has a period of transformation of the order of 1000 million years. If this were the case, uranium, initially freed from radium, should in the course of time grow radium, *i.e.* radium should again appear in the uranium. This has been tested independently by Soddy and Boltwood, and both have shown that in carefully prepared uranium solutions there is no appreciable growth of radium in the course of several years. The rate of production of radium, if it occurs at all, is certainly less than 1/1000 of the amount to be expected from theory. This would appear at first sight to put out of count the view that uranium is the parent of radium. This, however, is by no means the case, for such a result could be very easily explained if one or more substances of very slow period of transformation appeared between uranium and radium. It is obvious that the necessity of forming such an intermediate product would greatly lengthen the time required before an appreciable amount of radium appeared.

There is, however, another indirect but very simple method of attack to settle the parentage of radium. If radium is derived from the transformation of uranium, however many unknown products intervene, the ratio between the amount of radium and uranium in old minerals should be a definite constant. This is obviously the case, provided sufficient time has elapsed for the amount of radium to have reached its equilibrium value. The constancy of this relation has been completely substantiated by the independent work of Boltwood, Strutt, and McCoy. It has been shown that the quantity of radium corresponding to 1 gram of radium is $3 \cdot 8 \times 10^{-7}$ gram, and is the same for minerals obtained from all parts of the world. Since the radium is always distributed throughout the mass of uranium, we cannot expect to find nuggets of radium like nuggets of gold, unless by some chance the radium has been dissolved out of radio-active minerals and re-deposited within the last few thousands of years. To those who had faith in the disintegration theory, this unique constant relation between the amounts of two elements was a satisfactory proof that radium stood in a genetic relation with uranium. A search was then made for the unknown intervening product which, if isolated, must grow radium at a rapid rate. A year or so ago Boltwood observed that a preparation of actinium separated from a uranium mineral did grow radium at a constant but rapid rate. It thus appeared as if actinium were the long-looked-for parent of radium, and that actinium and its long family of products intervened between uranium X and radium. I was, however, able to show that actinium itself was not responsible for the growth of radium, but another unknown substance separated with it. These results were confirmed by Boltwood, who finally succeeded in isolating a new substance from uranium minerals, which was slowly transformed into radium. This substance, which he termed 'ionium,' has apparently chemical properties similar to those of thorium, and emits α rays of penetrating power less than those of uranium.

The main previsions of the theory have thus been experimentally verified. Radium is a changing substance the amount of which is kept up by the disintegration of another element, ionium. In order to complete the chain of evidence, we require to show that uranium grows ionium, and it is probable that evidence in this direction will soon be forthcoming. We thus see that we are able to link uranium, ionium, radium, and its long line of descendants, into one family, with uranium as its first parent. As uranium has a period of transformation of more than one thousand million years, it will not be profitable at the moment to try and trace back the family further.

It appears almost certain that, from the radio-active point of view, uranium and thorium must be considered as two independent elements. The case of actinium is different, for Boltwood has shown that the amount of actinium in minerals, like the amount of radium, is proportional to the amount of uranium. This indicates that actinium stands in a genetic relation with uranium. Unless our experimental evidence is at fault, it does not appear probable that actinium belongs to the main line of descent of uranium, for the activity of actinium separated from a mineral compared with radium is only about one-quarter of what we should expect under such conditions. I think that a suggestion which I put forward some time ago may account for the obvious connection of actinium with uranium, and at the same time for the anomaly observed. This supposes that actinium is a branch descent from some member of the uranium family. It does not appear improbable that at one stage of the disintegration two distinct substances may be produced, one in greater quantity than the other. After the expulsion of an α particle, it may happen that there are two possible arrangements of temporary stability of the residual atom. The great majority of the atoms may fall into one arrangement, and the remainder into the other. Actinium in this case would correspond to the substance in lesser quantity. It would act as a distinct element, and would break up in a different way from the main amount. It is probable that a large amount of accurate work will be required before the position of actinium in the scheme of changes can be fixed with certainty. It is a matter of remark how closely actinium resembles thorium in its series of transformations. It would appear that the atom of actinium has many points in common with thorium, or rather with its product, mesothorium.

The recent observations on the growth of radium offer a very simple and straightforward method of determining experimentally the period of radium. Suppose that we take a uranium mineral and determine by the emanation method the quantity of radium contained in it. If the immediate parent of radium (*i.e.* ionium) is next completely separated from the uranium and radium, it will begin to grow radium at a constant rate. Now the rate of growth of radium observed is a measure of the rate of breaking up of the radium parent in the mineral, since before separation the rate of production was equal to the rate of breaking up. Now the growth of radium observed for a short interval, for example, a year, divided by the quantity present in the mineral, gives the fraction of the radium breaking up per year. Proceeding in this way, Boltwood found that the fraction breaking up per year is about 1/3000, and that the period of radium is about 2000 years—a value which lies between the most probable values deduced from quite distinct data.

From an inspection of the radio-active families, it will be seen that out of twenty-six radio-active substances that have been identified, seventeen give out α rays or α and β rays, four give out only β rays, and five emit no rays at all. The rayless and β-ray products are transformed according to the same law as the α-ray products, and there is the same sudden change of physical and chemical properties as the result of the transformation. In the case of the substances which throw off atoms of matter in the form of α particles, there are obvious reasons for anticipating a change in properties of the substance, but this is not the case for the rayless or β-ray products. We must either suppose that the mass of the atom is not appreciably changed by the transformation, which consists in an internal rearrangement of the parts of the atom, or that the atom expels a particle at too low a velocity to be appreciated by the electrical methods. Unfortunately, it is very difficult to study the rayless products with care, as in practically every case they are succeeded by a ray product of comparatively rapid transformation. The rayless products are of great interest as indicating the possibility of transformations which can occur without any detectable radiation.

In the course of the analysis of radio-active changes, special methods have been developed for the separation of the various products from each other. It is only in a few cases, however, that we can hope to obtain a sufficient quantity of the substance to examine by means of the balance. It should be possible to obtain workable quantities of actinium, radium D (radio-lead), and radium G (polonium), but the isolation of these substances in any quantity has not yet been effected. Sir William Ramsay and Mr. Cameron have made a number of important investigations of the properties and volume of the radium emanation, freed so far as possible from any traces of known gases. The remarkable initial contraction of the volume due to the emanation shows that there is still much to be done to obtain a clear understanding of the behaviour of this intensely radio-active gas when obtained in a pure state.

Simultaneously with the work on the analysis of radio-active changes, a large number of investigations have been made on the laws of absorption by matter of the three primary types of radiation from active matter, viz. the α, β, and γ rays, and the secondary radiations to which they give rise. It has generally been accepted for some years that the γ rays are a type of penetrating X-rays. The latter are supposed to consist of electromagnetic pulses in the ether, set up by the impact or escape of electrons from matter, and akin in many respects to very short waves of ultra-violet light. Recently, however, Bragg has challenged this view, and has suggested that the γ rays (and probably also the X-rays) are mainly corpuscular in character, and consist of uncharged particles or 'neutral pairs,' as he terms them, projected at a high velocity. Such a view serves to explain most of the experimental observations

C

equally well as the pulse theory; Bragg has recently brought forward additional evidence, based on the direction of the secondary radiation from the γ rays, which he considers to be inexplicable by the pulse theory. We must await further data before this important question can be settled definitely, but the theory of Bragg, which carries many important consequences in its train, certainly deserves very careful examination.

From the radio-active point of view, the α rays are by far the most important type of radiation emitted by active matter, although their power of penetration is insignificant compared with the β or γ rays. They consist of veritable atoms of matter projected at a speed, on an average, of 6000 miles per second. It is the great energy of motion of these swiftly expelled masses that gives rise to the heating effect of radium. In addition, they are responsible for the greater part of the ionisation observed near an uncovered radio-active substance. On account of their importance in radio-active phenomena, I shall devote some little attention to the behaviour of these rays. The work of Bragg and Kleeman, of Adelaide, first gave us a clear idea of the nature of the absorption of these rays by matter. The α particles from a very thin film of any simple kind of radio-active matter are all projected at an identical speed, and lose their power of ionising the gas or of producing phosphorescence or photographic action after they have traversed exactly the same distance, which may conveniently be called the 'range' of the α particle. Now every product emits α particles at an identical speed among themselves, but different from every other product. For example, the swiftest α particle from the radium family, viz. that from radium C, travels 7 cm. in air under ordinary conditions before it is stopped, while that from radium itself is projected at a slower speed, travelling only 3·5 cm. We may regard the α particle as a projectile travelling so swiftly that it plunges through every molecule in its path, producing positively and negatively charged ions in the process. On an average, an α particle before its career of violence is stopped breaks up about 100,000 molecules. So great is the kinetic energy of the α projectile that its collisions with matter do not sensibly deflect it, and in this respect it differs markedly from the β particle, which is apparently easily deflected by its passage through matter. At the same time, there is undoubted evidence that the direction of motion of some of the α particles is slightly changed by their passage through matter.

The sudden cessation of the ionising power produced by the α particle after traversing a definite distance of air has been shown by Bragg to be a powerful method of analysis of the number of α-ray products present in a substance. For example, suppose the amount of ionisation in the gas produced by a narrow pencil of α rays is examined at varying distances from the radium. At a distance of 7 cm. there is a sudden increase in the amount of ionisation, for at this distance the α particles from radium C enter the testing vessel. There are again sudden changes in the ionisation at distances of 4·8 cm., 4·3 cm., and 3·5 cm. These are due to the rays from the radium A, the emanation and radium itself respectively entering the testing vessel. The α-ray analysis thus discloses four types of α rays present in radium in equilibrium—a result in conformity with the more direct analysis. This method allows us to settle at once whether more than one α-ray product is present in a given radio-active material. For example, an analysis by Hahn by this method of the radiation from the active deposit of thorium has disclosed the existence of two α-ray products instead of one as previously supposed. We can consequently gain information on the complexity of radio-active material, even though no chemical methods have been found to separate the products concerned. The range of the α particle from each product is a definite constant which is characteristic of each product.

The α particle decreases in velocity as it passes through matter. This result is clearly brought out by photographs showing the deflection of a homogeneous pencil of α rays in a magnetic field before and after passing through an absorbing screen. The greater divergence of the trace of the α rays on the plate, after passing through the screen, shows that their velocity is reduced, while the sharpness of the band shows that the α particles still move at an identical speed.

In order to make an accurate determination of the constants of the α particles, it is necessary to work with homogeneous rays, and we consequently require to use a thin layer of matter of one kind. For experiments of this character, a wire coated with a thin film of radium C by exposure to the radium emanation is very suitable. The velocity of the α particle and the value e/m, the ratio of the charge carried by the α particle to its mass, can be deduced by observing the deflections of a pencil of α rays exposed in a magnetic and in an electric field of known strengths. The deflection of a pencil of α rays in an electric field is small under normal conditions, and special care is needed to determine it with accuracy.

In this way I have calculated the velocity and value of e/m for a number of α-ray products. The velocity of expulsion varies for different products, but is connected by a simple relation with the range of the α particle in air. The value of e/m has been determined for selected products of radium, thorium, and actinium, and in each case the same value has been found. This shows that the α particles expelled from radio-active substances in general are identical in constitution. They have all the same mass, but differ from one another in the initia velocity of their projection. Although we are sure that the α particles, from whatever source, are identical atoms of matter, we are still unable to settle definitely the true nature of the α particle. The value of e/m found by experiment is nearly 5×10^3. Now the value of e/m for the hydrogen atom in the electrolysis of water is 10^4. If the charge carried by the α particle and the hydrogen atom is the same, the mass of the α particle is twice that of the hydrogen atom, *i.e.* a mass equal to the hydrogen molecule But we are not certain that they do carry the same charge. Here we are, unfortunately, confronted by a number of possibilities, for the magnitude of m for the α particle is conditioned by the value assumed for e. If the charge of the α particle is assumed to be twice the value of the hydrogen atom, the mass comes out four times the hydrogen atom—the value found for the helium atom. The weight of evidence still supports the view that the α particle is in some way connected with the helium atom. If the α particle is a helium atom with twice the ionic charge, we must regard the helium produced by radio-active bodies as actually the collected α particles the charges of which have been neutralised. This at once offers a reasonable explanation of the production of helium by actinium as well as by radium. In addition, Strutt has recently contributed strong evidence that helium is a product of thorium. Such results are only to be expected on the above view, since the α particle is the only common product of these elements.

The determination of the true character of the α particle is one of the most pressing unsolved problems in radio-activity, for a number of important consequences follow from its solution. Unfortunately, a direct experimental proof of its true character appears to be very difficult unless a new method of attack is found. We have seen that if the charge carried by the α particle could be experimentally determined, the actual value of m could be determined in terms of the hydrogen atom, since the value of the charge carried by the latter is known. This could be done if we could devise a method of detecting the emission of a single α particle, and thus counting the number of particles expelled from a known quantity of a radio-active substance, for example, from radium. In considering a possible method of attack of this question, the remarkable property of the α particles of producing scintillations in zinc sulphide at once suggests itself. Apart from the difficulty of counting the scintillations, it is very doubtful whether more than a small fraction of the α particles which strike the screen produce the scintillations. Viewed from the electrical side, a simple calculation from the data at our disposal shows that the ionisation produced in a gas by a single α particle should be detectable. The electrometer or electroscope used for measurement would, however, require to be extremely sensitive, and under such conditions it is known that small electrical disturbances are very difficult to avoid.

In order to obtain a reasonably large effect, we require some method of magnifying the ionisation produced by the α particle. In conjunction with Dr. Hans Geiger, I have recently developed a method whereby the electrical effect produced by the α particle can be magnified several thousand times. From the work of Townsend it is known that if a strong electric field acts on gas at low pressure, any ions generated in the gas by an external agency are set in motion by the electric field, and under the proper conditions produce fresh ions by collision with the gas molecules. The negative ion is the most effective ioniser in weak fields, but when the voltage is increased near the point at which a discharge passes, the positive ion also produces fresh ions by collision. In the experimental arrangement the α particle from the active matter is fired through a small opening about 2 mm. in diameter, covered with a thin layer of mica, into a cylinder 60 cm. long and 2·5 cm. in diameter, in which the gas pressure is about 3 cm. of mercury. A thin insulated wire connected to the electrometer is fixed centrally in the cylinder. If the outside cylinder is charged negatively, for a difference of potential of about 1000 volts any ionisation produced in the cylinder is increased about 2000 times by collision. This can be simply illustrated by using the γ rays of radium as a source of ionisation. When a difference of potential is applied to the cylinder, the ionisation produced by the γ rays only causes a slight movement of the electrometer needle. By applying, however, a voltage nearly equal to that required for a discharge through the gas there is a very rapid movement of the needle. On removing the radium there is no appreciable current through the gas. On placing a source of α rays near the small opening in the cylinder so that some of the α particles can be fired along the axis of the cylinder, the electrometer needle does not move uniformly, but with a succession of rapid throws with a considerable interval in between. Each of these throws is due to the discharge produced by a single α particle entering the cylinder, increased several thousand times by the intermediary of the strong electric field. If a sheet of paper which stops the α rays is placed before the opening, the electrometer needle at once comes to rest. The interval of time between the throws is not uniform. This is exactly what we should expect if the number of α particles entering such a small opening is governed by the law of probability. On the average, a certain number of α particles are fired through the opening per minute, but in some cases the interval is less than the average, in others much greater. In fact, by observing the intervals between the entrance of a large number of α particles, we should be able to determine accurately the 'probability' curve of distribution of the α particles with time. For purposes of measurements, the active material, in the form of a thin film covering a small area, is placed in an exhausted tube connected in series with the ionisation cylinder, and at a considerable distance from the hole. The number of α particles entering the opening per minute is counted, and from this the total number expelled can be calculated. Preliminary measurements show that the number of α particles expelled from a known weight of radium is of the same order as the calculated value. When the measurements are completed it should be possible to determine the charge carried by each α particle, since the total charge carried by the α particles from 1 gram of radium is known. In this way it may be possible to settle whether the α particle is a helium atom or not. In any case, it is a matter of some interest to be able to detect by its electrical effect a single atom of matter, and so to determine directly with a minimum of assumption the magnitude of some of the most important quantities in radio-active phenomena.

The same paper was also published in *Chemical News*, 1909, **99**, 171/4 and 181/3.

Spectrum of the Radium Emanation

From *Nature*, **78**, 1908, pp. 220–1

A FEW months ago, through the generosity of the Academy of Sciences of Vienna, one of us was loaned a radium preparation containing about 250 mg. of radium. Observations were at once begun to purify the emanation produced by it, and to determine its volume. An account of these investigations was read before the Academy of Sciences of Vienna on July 2. It was found that the maximum volume of the emanation per gram of radium was in good accord with that to be expected from calculation (about $0 \cdot 6$ cubic mm.), and the initial volume was about one-tenth of that determined by Ramsay and Cameron (Journ. Chem. Soc., p. 1266, 1907). In the course of this work we have had occasion to test the purity of the emanation by the spectroscope, passing an electric discharge in the capillary in which the volume was measured. We have on four different occasions during the last two months determined the spectrum of the radium emanation by visual observations, using a direct-reading Hilger spectroscope, leaving a more accurate determination of its spectrum until the measurements of the volume had been completed. We have now photographed the emanation spectrum, using a prism of 2 inches base. Pure emanation, corresponding to the equilibrium amount from 130 mg. of radium, was condensed by liquid air in an exhausted spectrum tube of about 50 cubic millimetres capacity, provided with thin platinum electrodes. Two photographs were immediately taken, one giving about thirty of the more intense lines, and the other, with much longer exposure, showing more than one hundred lines. For a comparison spectrum a helium tube was used. The colour of the discharge in the tube was bluish. Visual observations of the spectrum were made during the exposure of the photographs.

When the emanation was condensed in a side tube by means of liquid air, the great majority of the lines vanished at the moment of condensation, which was readily noted by the phosphorescence of the glass. The colour of the discharge then completely changed, and became of a pale rose colour. At the instant of volatilisation, the emanation lines flashed out again. The hydrogen lines were visible in the spectrum, and these became much more brilliant when the emanation was condensed. In the electrodeless discharge of previous experiments, the hydrogen lines were never observed. Their occurrence in the present experiment was probably due to the platinum electrodes. By observations of the intensity of the phosphorescence when the emanation was condensed, it was noted that the amount of pure emanation in the tube gradually diminished with increase of time of discharge. The spectrum of the emanation, however, persisted until practically all the emanation had been driven into the walls of the tube. The phosphorescence on the walls of the tube showed that the occluded emanation was fairly uniformly distributed. This effect has been observed by us on several occasions.

The first determination of the spectrum of the emanation was made in 1904 by Ramsay and Collie, who determined the wave-lengths of about eleven lines by visual observations. As shown by them, the spectrum of the emanation is a bright line spectrum with sharply defined lines. We observed also visually a weak band spectrum in the yellow, which slightly decreased in intensity when the emanation was condensed. This, however, may not be connected with the emanation itself. The wave-lengths of the lines of the photographic plate were accurately measured, using a Kayser's measuring machine. The accuracy obtained is indicated by the agreement of the wave-lengths of some of the hydrogen lines with their known values. In most cases, for well marked lines, the error is not more than half an Ångström unit. The following table gives the wave-lengths of the more prominent lines. The wave-lengths of the lines initially determined by Ramsay and Collie (marked R. and C.) are added for comparison. Visual observations of three of the more prominent lines in the yellow and green are also given:—

Intensity	Observed λ	Remarks	Intensity	Observed λ	Remarks
5	5721	(Visual) R. & C. 5725	15	4350·3	
8	5589	(Visual)	7	4340·9	H = 4340·66
3	5393	,,	4	4225·8	
4	5084·5		10	4203·7	
4	4979·0	R. & C. 4985	7	4188·2	
10	4861·3	H = 4861·49	20	4166·6	
4	4817·2		10	4114·9	
5	4721·5		2	4102·2	H = 4101·85
10	4681·1	R. & C. 4690	4	4045·4	
10	4644·7	R. & C. 4650	15	4018·0	
8	4625·8	R. & C. 4630	12	3982·0	
7	4609·9		7	3957·5	
4	4604·7		4	3917·5	
7	4578·7		—	3888·9	H = 3889·15
9	4509·0		6	3867·6	
10	4466·0		10	3753·6	
8	4435·7		7	3739·9	
6	4391·8		10	3664·6	
4	4372·1		5	3622·2	

A more detailed list of lines will be published later. We understand that Sir William Ramsay showed a photograph of the spectrum of the emanation at the meeting of the Royal Society on June 25. It will be of interest to compare the two spectra.

E. Rutherford
T. Royds

University, Manchester
July 4

Experiments with the Radium Emanation
I. The Volume of the Emanation

E. RUTHERFORD, F.R.S.,
Professor of Physics, University of Manchester

From the *Philosophical Magazine* for August 1908, ser. 6, XVI, pp. 300–12
(Read before the Academy of Sciences of Vienna, July 2, 1908)

THE amount of radium emanation to be obtained from 1 gr. of radium in equilibrium is a definite quantity, and is equal to q/λ where q is the rate of production of emanation per second, and λ is the radioactive constant of the emanation. Taking the half period of the emanation as 3·75 days, the value of λ is 1/468,000. I have on different occasions* calculated the volume of the emanation (at normal pressure and temperature) to be expected from 1 gr. of radium from the radioactive data at our disposal. As the simplest and most probable assumption, it is supposed that one atom of radium in breaking up emits one α particle and then becomes an atom of the emanation. On the assumption that each α particle carries the ordinary ionic charge e of $3·4 \times 10^{-10}$ electrostatic unit, it was calculated that the volume† of the emanation from 1 gr. of radium should be 0·8 cu. mm. Later work† indicated the probability that the α particle carried the charge $2e$. This reduces the calculated volume of the emanation to one-half of the above value. Recently, in conjunction with Dr Geiger, the number of α particles expelled per second from 1 gr. of radium has been accurately determined, and also the charge carried by each α particle. From these data, we have calculated that the volume of the emanation is 0·57 cu. mm.—a value about intermediate between the other two values.‡

The first experiments to measure directly the volume of the emanation were made by Ramsay and Soddy.§ The emanation after suitable treatment was condensed in a glass tube surrounded by liquid air. The residual gases were pumped off, and the emanation after volatilization was forced by raising the mercury into a capillary tube where its volume was measured. From the volume of the collected gas observed after two days, they concluded that the volume of the emanation was about 1·2 cu. mm. Later a number

* *Radioactivity*, 2nd edition, p. 288.
† Rutherford, *Phil. Mag.*, October 1906. (*Vol. I, p. 880.*)
‡ An account of this work was given to the Royal Society, June 18, 1908. (*This vol., pp. 89, 109.*)
§ *Proc. Roy. Soc.*, lxxiii, p. 346 (1904).

of systematic observations of the volume of the emanation by a similar method have been made by Ramsay and Cameron.*

They conclude that the volume of the emanation is about 7·07 cu. mm., and suggest that the smaller value initially obtained by Ramsay and Soddy was due to the greater part of the emanation being pumped off during the experiment. The volume of the emanation (7·07 cu. mm.) obtained in their experiments is of quite a different order from the calculated volume (0·57 cu. mm.). It is of importance to determine the cause of this wide discrepancy between theory and experiment. If the experimental value proves correct, it would indicate that much of the radioactive data and of the theory on which the calculations are based is seriously in error. Apart from the interest attaching to the comparison of theory with experiment, the separation of the radium emanation in a pure state is now of the highest practical importance. Not only is pure emanation required in order to study carefully the physical and chemical properties of this remarkable gas, but it is also required in the experiments similar to those initiated by Ramsay and Cameron,† where the radium emanation is added to different solutions and the resulting products determined.

By the generosity of the Academy of Sciences of Vienna, I was recently loaned a preparation of radium containing about 250 milligrams of radium. Experiments were immediately begun in order to purify the emanation produced by it, and to determine its volume. In all, a large number of experiments have been made, but for brevity I shall here only indicate the general results obtained in these investigations.

Separation of the Emanation

There are two general methods of obtaining the emanation from preparations of radium, viz., by heat or by solution. Both of these methods have been used. In the earlier experiments, the radium preparation was placed in a thin quartz tube which was enclosed in a larger quartz tube. The latter was heated to the desired temperature by means of an electric furnace. The emanation is practically all released at the temperature of fusion (about 830°C)‡ of barium-radium chloride. After the preliminary heating, a very small quantity of intensely radioactive gas was released, the volume of which was so small that it was found necessary to add a small quantity of hydrogen or oxygen in order to pump off the emanation completely. In later experiments, the radium preparation was in solution in a quartz tube. The emanation was pumped off together with a large quantity of hydrogen and oxygen formed in solution. After the method employed by Ramsay and Cameron, this was sparked down and a small quantity of oxygen added in order to reduce the excess of hydrogen. The emanation, mixed with about 1 c.c. of

* *Journ. Chem. Soc.*, p. 1266 (1907).
† *Journ. Chem. Soc.*, p. 1593 (1907).
‡ See paper by L. Kolowrat, *Le Radium*, September 1907.

C*

hydrogen, was then collected over mercury in a small burette in which was placed a piece of caustic potash in order to absorb any carbon dioxide present. The apparatus for purification of the emanation and measurement of its volume is shown in Fig. 1. The measuring apparatus consisted essentially of a Macleod gauge. A capillary tube, 15 cm. long and 0·58 mm. diameter, was attached to a long glass tube of volume about 25 c.c. By raising the reservoir with the stopcock A closed, the gas in the tube E was forced into the capillary tube F and its volume measured. A mercury trap R was used to avoid the entrance of any gas which crept along the surface of the glass. The general method of purification of the emanation is best seen from a description of an experiment.*

The whole apparatus was first exhausted to a low vacuum by means of a mercury pump. The emanation, conveyed by means of 1 c.c. of gas, was transferred over mercury into the reservoir C. The stopcocks A and B were closed, and the emanation was forced by raising the mercury reservoir through the stopcock H along the tube D, coated with caustic potash, into the U-tube T. The U-tube, of volume about 1·5 c.c. was surrounded by liquid air in order to condense the emanation. The whole emanation was condensed by successively raising and lowering the mercury in the reservoir. The stopcock B was then opened into the pump and the uncondensed gases completely pumped off. The mercury was then lowered in the tube D to the position of the dotted line in the figure. The liquid air was then removed. The emanation after volatilization was left some hours in contact with the caustic potash in the tube D to remove the last trace of carbon dioxide. The U-tube was then surrounded by a vessel filled with pentane, which was cooled down by liquid air to a temperature between the temperature of condensation of the emanation ($-150°C$) and the temperature of liquid air ($-186°C$). The whole apparatus was then completely exhausted again by the mercury pump, a portion of the emanation being volatilized during the process and removed by the pump. Finally, when a very low vacuum was obtained, the liquid air was removed, the stopcock B closed, and the emanation after volatilization was allowed to expand into the tube E. Since the volume of the U-tube was small compared with the tube E, the greater part of the emanation after volatilization expands into the tube E. The experiments were made in a darkened room, so that the moment of almost complete volatilization of the emanation could be observed by the sudden phosphorescence of the tube E due to the entrance of the emanation. The stopcock A was then closed and the emanation forced into the capillary, where its volume was measured at regular intervals.

Activity Measurements

In this type of experiment, it is of importance to determine accurately the amount of emanation in the capillary where the volume is measured. This

* In this work, I have found the methods developed by Ramsay and others for manipulating small quantities of gases of great assistance. *See* Travers, 'Study of Gases'.

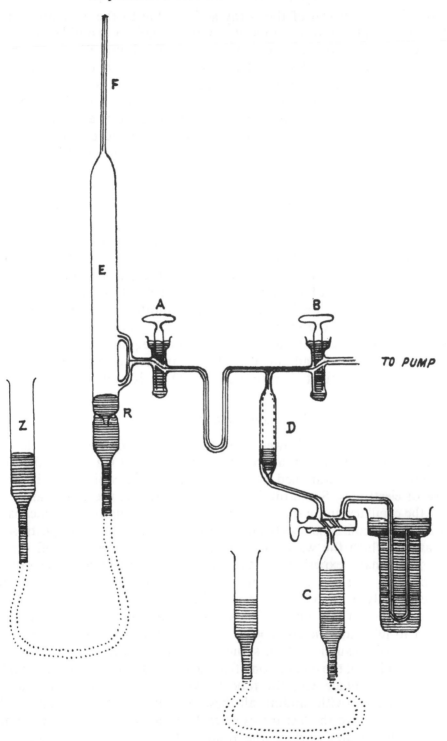

Fig. 1

was done by comparison of the γ ray activity due to the emanation in the capillary with that due to a standard sample of radium bromide which gave a heating effect of 110 gram-calories per gram per hour. When the emanation had been forced into the capillary, the residual emanation in the U-tube and connections was completely pumped out and all sources of γ rays removed to a distance. A small closed lead electroscope was placed at a distance of 76 cm. from the emanation in the capillary, and the rate of discharge observed about three hours after the introduction of the emanation. The γ ray activity compared with a standard is then a measure of the amount of emanation expressed in terms of the equilibrium quantity present in one milligram of pure radium. Measurements were made daily of the γ ray activity of the emanation. Experiment showed that the highly concentrated emanation compressed into a small volume had the usual rate of decay, viz. half period in 3·75 days. Measurements of this character are essential if any accuracy is required. It is not sufficient to assume that all the emanation formed in a certain interval of days is pumped off from the radium solution. Moreover a small part of the emanation is left behind in every operation of transferring the emanation from one vessel to another. It may be mentioned that in most of the experiments the amount of emanation whose volume and activity were measured was equivalent to the equilibrium quantity from 60 to 140 milligrams of radium. Usually the amount of emanation pumped off during the experiment corresponded to 20 or 30 milligrams of radium.

Discussion of Experiments

Preliminary experiments showed that the volume of the emanation was certainly much lower than that found by Ramsay and Cameron; the purity of the emanation was examined spectroscopically in the capillary itself without the use of electrodes. Some tin foil was wound round the upper and lower part of the capillary and a discharge produced in the capillary by means of a small induction-coil. The spectrum was examined by a Hilger spectroscope, by means of which the wave-length of the lines could be read off directly. In the preliminary experiments, the spectrum observed was the ordinary band spectrum ascribed to carbon dioxide with some of the mercury lines and occasionally a few other very faint lines. Precautions were then taken to get rid of the carbon dioxide. Phosphorus pentoxide instead of tap grease was used to lubricate the stopcocks. The emanation was always left in the presence of solid caustic potash some time before its introduction into the U-tube. In addition, the emanation after the uncondensed gases were pumped off was left in contact with the tube D which was coated with a layer of caustic potash. Notwithstanding all these precautions it was found at first impossible to remove the last traces of carbon dioxide. Attempts were made to remove the greater part of the residual CO_2 by fractional distillation, using the pentane bath for temperature adjustment, but with only partial success. There appears to be some evidence that the emanation prefers to condense with the CO_2 which is present, and is released with it when the temperature rises. Finally, after a large number of experiments, it was found essential to allow the emanation after purification from other gases to remain 5 or 6 hours, preferably 24 hours, in the presence of the caustic potash tube D. When this was done, it was found that the volume of the gas obtained in the capillary was much reduced and the carbon dioxide spectrum became much fainter. The reason why such a long time of exposure to caustic potash is required is not at all clear. It may be due to the very slow absorption of the last traces of carbon dioxide by caustic potash. It is possible, however, that the spectrum ascribed to CO_2 is in reality due to carbon monoxide in the presence of oxygen. There is considerable difference of opinion among spectroscopists on this point. If the disturbing gas is CO it must first be converted into CO_2 in the presence of oxygen by the action of the emanation before absorption by the caustic potash. This would account for the long time required for complete absorption. As far as my experience has gone, the essential conditions for the purification of the emanation depend upon the pumping off the residual gases at a temperature below the temperature of volatilization of the emanation and considerably above the temperature of liquid air; and the long exposure of the emanation to caustic potash. By the former method practically all the known inactive gases would be pumped off. I think the high value of the volume of the emanation obtained by Ramsay and Cameron must be ascribed to the presence of other gases besides the emanation, which are condensed at the temperature of liquid air.

Changes in Volume

Ramsay and Cameron (*loc. cit.*) have given a number of examples of the changes in volume observed in their experiments to measure the volume of the emanation. The volume of the gas in the capillary usually diminished with time rapidly during the first two hours to about half value and then more slowly. I have observed very similar effects in my experiments, using impure emanation. In some cases the volume diminished in the course of several hours to less than half value, but after this preliminary decrease little change was observed in the volume over the further interval of a week. In other experiments, the volume increased instead of diminishing in the course of a few hours, sometimes increasing to twice the initial volume, followed later by a slow decrease with time. The expansion or contraction of the volume has in many cases no direct connection with the volume changes of the emanation itself, for the true volume of the emanation present was in some experiments certainly not more than 20 per cent of the total.* It is difficult to explain these expansions and contractions except on the supposition

---

* Ramsay and Cameron explained the decrease of volume observed in their experiments on the assumption that the emanation changed from a monatomic to a diatomic gas. Since the volume in these experiments certainly contained less than 20 per cent of emanation, the explanation is inadmissible.

that the gases mixed with the emanation either combine or dissociate under the influence of the powerful radiation from the emanation. Until experiments are made with some known gas or gases added to pure emanation, we can only speculate upon the nature of the gases present and the combinations or dissociations which are effected. There is another possibility which may prove to be an important factor in the volume changes, especially with nearly pure emanation. It is believed that the positive and negative ions produced in a gas at ordinary pressure have a cluster of molecules attached which move with them. Since the emanation itself and the gases associated with it are intensely ionized, it is possible that the effective volume may be decreased due to the production of a large number of these aggregates. On this view, the decrease of volume observed during the first two hours may be partly due to the increase of the number of these aggregates consequent upon the increase of the radiation from radium C.

Experimental Results

We shall now give some typical examples out of a number illustrating the initial changes in volume. The capillary tube used in all the experiments was of Jena borosilicate glass of very uniform bore, 0·58 mm. in diameter. The capillary correction was equal to 14 mm. of mercury. The tube was slowly coloured brown by the emanation. By heating the tube to the temperature of thermo-luminescence, the glass again became quite clear. The capillary was heated at the beginning of each experiment to drive off residual gases. The gases in the capillary in all cases obeyed Boyle's law over the range examined within the limit of experimental error.

Experiment I. This illustrates the increase of volume observed for very impure radium emanation. The amount of emanation in the capillary, determined by direct measurement, corresponded to 67 mgr. of pure radium.

Time after introduction of the emanation into capillary	Volume of gas in capillary at standard pressure and room temperature
2 min.	0·154 cu. mm.
6 ,,	0·169 ,,
17 ,,	0·201 ,,
28 ,,	0·235 ,,
41 ,,	0·260 ,,
50 ,,	0·270 ,,
64 ,,	0·280 ,,
71 ,,	0·291 ,,
81 ,,	0·297 ,,
3·9 hours	0·346 ,,
21 ,,	0·355 ,,

At the end of 24 hours the spectrum of the gas was examined in the capillary. The carbon dioxide spectrum was prominent. In addition to the mercury lines, a few others were observed which were not identified with certainty. The initial volume, 0·154 cu. mm., corresponded to 67 mg. radium, consequently the initial volume corresponding to 1 gr. of radium was 2·3 cu. mm. The final volume after 21 hours corresponds to 5·2 cu. mm. per gram. As the spectrum indicated, the emanation in this case was very impure, containing probably a large proportion of CO or CO_2.

Experiment II. We shall now give an example of the contraction of volume. In this case the emanation was far purer than in experiment I. The emanation was left for four hours in the presence of caustic potash before introduction into the capillary. The amount of emanation in the capillary corresponded to 130 mg. radium.

Time after introduction of the emanation into capillary	Volume of gas in capillary at normal pressure and room temperature
2 min.	0·171 cu. mm.
5 ,,	0·169 ,,
10 ,,	0·165 ,,
18 ,,	0·158 ,,
26 ,,	0·150 ,,
37 ,,	0·135 ,,
56 ,,	0·126 ,,
70 ,,	0·120 ,,
91 ,,	0·106 ,,
5·9 hours	0·097 ,,
17 ,,	0·069 ,,
44 ,,	0·075 ,,
92 ,,	0·079 ,,
161 ,,	0·080 ,,
185 ,,	0·119 ,,
209 ,,	0·125 ,,
257 ,,	0·125 ,,

The initial volume for the emanation from 1 gr. of radium corresponded to 1·32 cu. mm. Correcting for the decay of the emanation, the volume at the minimum after an interval of 17 hours was equal to 0·59 cu. mm. per gram of radium. It will be observed that the volume sank to about 0·4 of its initial value after 17 hours. After passing through a minimum, the volume increased again, though not very regularly. At the conclusion of the experiment, i.e. after the emanation had remained nearly 11 days in the capillary, the spectrum of the gas was examined in the capillary as described above. *A brilliant spectrum was obtained showing all the lines of helium.* The spectrum of CO_2 was also observed, although weak in intensity compared with that of helium. In addition to the mercury lines, a few unidentified bright lines were noted. This result is a confirmation of the well-known experiment of Ramsay and Soddy, who found that the spectrum of helium appeared after some time in a tube containing radium emanation. The cause of the increase of volume after the minimum is now clear. Assuming that the α particles are atoms of helium, the helium would at first be fired into the glass. After a time part of it gradually escaped and added its volume to the emanation and other gases present. It is difficult to be certain how much of the helium was retained in the glass of the capillary. If we take the initial volume of the emanation to be about that observed at the minimum volume, viz. 0·059 cu. mm., the volume of helium to be expected is about three times this amount. This is on the assumption that each α particle expelled is a helium atom. The final volume observed after 11 days was 0·125 cu. mm., and was probably mainly due to the helium.

Experiment III. In this case the emanation was very carefully purified, after standing for 18 hours over caustic potash. The amount of emanation present corresponded to 130 mgs. radium. The initial volume of the emanation was 0·097 cu. mm. This corresponds to a volume of 0·80 cu. mm. per gram of radium. No certain change in volume was observed over an interval of 15 min. The emanation was then recondensed in the U-tube, which was pumped out again using a pentane bath. On introducing the emanation into the capillary again, very nearly the same initial volume as before was observed. In order to test the purity of the emanation, the spectrum of the gas in the capillary was examined. A new spectrum of bright lines, certainly due to the emanation itself, was observed. Some of the bands of the carbon dioxide spectrum were observable. Observations were at once begun to determine the wave-lengths of the new lines with accuracy. Before this was completed, most of the lines due to the emanation suddenly ran out, and the carbon dioxide spectrum became more prominent. The volume of the gas in the capillary was also found to have considerably decreased. It was then observed by the phosphorescence that the emanation was adhering to the walls of the capillary, and only a part of the emanation was free in the gaseous state. The emanation remained fast to the walls for two days, and was only removed finally by a vigorous heating of the tube. It appears that the emanation must have been driven into the walls of the tube or occluded in it under some condition due to the passage of the discharge.

Experiment IV. In this case the emanation, after the initial purification, was left 5 hours over caustic potash. After introduction the initial volume was 0·126 cu. mm. The volume remained nearly stationary for 20 min. and then slowly diminished, reaching a value of 0·076 cu. mm. after 17 hours. The amount of radium emanation initially present was 130 mgs. radium. The initial volume of emanation thus corresponded to 0·97 cu. mm. per gram of radium, and the lowest volume, allowing for the decay of the emanation in the interim, corresponded to 0·66 cu. mm. per gram.

Experiment V. The emanation used in experiment IV was again condensed in the U-tube and then left for 24 hours in the presence of caustic potash. The emanation, after further treatment, was admitted into the capillary. The initial volume was 0·083 cu. mm. As in the case of experiment IV, the volume remained nearly stationary for 20 min. and then slowly decreased. The volume after 4 hours was 0·046 cu. mm. The amount of radium emanation in the tube was equal to 79 mgs. radium. Consequently, the initial volume of the emanation per gram was 1·05 cu. mm. and the volume after 4 hours, 0·58 cu. mm. The spectrum of the gases in the capillary was then examined. As before the carbon dioxide spectrum was seen together with a number of new lines due to the emanation, the wave-lengths of which were measured. No trace of the hydrogen lines was observed in this or in the other experiments. After the discharge had passed at intervals for two hours, most of the lines due to the emanation disappeared. The greater part of the emanation was then found to be sticking to the surface of the capillary, as in the previous experiment. The prominence of the carbon dioxide spectrum cannot, I think, be ascribed to the presence of a considerable amount of this gas mixed with the emanation before the discharge passed, but rather to the production of this gas by the discharge, due probably to the presence of a trace of some organic matter at the surface of the mercury. The correctness of this view was confirmed by the observation that the spectrum of carbon dioxide was unaltered in brightness, after practically all the residual gases and emanation had been removed by the pump.* I hope in a later paper to give a more detailed account of these and other experiments to determine the spectrum of the radium emanation. In these investigations the spectrum has been obtained incidentally in the course of testing the purity of the emanation in the capillary.

Summary of Results

For convenience, the results of experiments II to V on the volume of the emanation are collected below :—

Experiment	Initial volume of emanation per gram of radium	Final volume of emanation per gram of radium
II	1·32 cu.mm.	0·59 cu.mm.
III	0·80 ,, ,,	
IV	0·97 ,, ,,	0·66 ,, ,,
V	1·05 ,, ,,	0·58 ,, ,,

The volumes here given are at normal pressure and room temperature (about 16°C). If corrected to standard temperature, the volumes will be about 5 per cent smaller. A small undetermined correction should also be applied for the heating effect of the emanation. We have already seen

* Later observations have confirmed the correctness of this explanation. The emanation purified after the manner described was introduced in a spectrum-tube with platinum electrodes. No trace of the band spectrum of carbon dioxide has been observed in the spectrum produced by the discharge. (*See* accompanying paper 'On the Spectrum of the Radium Emanation'.) (*This vol.*, p. 84.)

that the calculated volume at normal pressure and temperature is 0·57 cu. mm.

From the above table it is seen that the smallest initial volume of the emanation observed is 0·80 cu. mm. per gram of radium, and the smallest volume after contraction 0·58 cu. mm. The volume before contraction, observed by Ramsay and Cameron, was 7·07 cu. mm. per gram. It was observed that the emanation was not appreciably absorbed in the capillary during the first few hours, provided a discharge was not passed, and was all released on lowering the mercury. For these reasons, it seems probable that the volume after contraction is to be taken as the true volume of the emanation rather than the volume in the beginning. On this view, there is as good an agreement as could be expected from the nature of the experiments between the final volumes, viz. 0·59, 0·66, and 0·58 cu. mm., and the calculated volume, viz. 0·57 cu. mm.

We have already seen that it is difficult to offer a satisfactory explanation of the initial contraction. Before this can be done, a large number of further experiments will be required. The work outlined in this paper is merely preliminary and it is hoped, in a later paper, to give the results of a more complete examination of the volume of the emanation and of the changes it undergoes.

Remarks on the Condensation of the Emanation

When the emanation was obtained in a nearly pure state, it condensed exceedingly rapidly at any point cooled below the temperature of condensation ($-150°C$). If the emanation were contained in the U-tube (Fig. 1), the slow approach of the liquid air to the bottom of the tube caused the condensation in some cases to take place over an extremely small area, probably at a point where the tube was thinnest. A brilliant phosphorescent speck was then observed, and it almost appeared as if one could see the liquid emanation in the form of a flat globule condensed over an area of less than half a square millimetre. This effect was often observed and is very striking. After a few minutes the emanation, even at the temperature of liquid air, gradually diffuses, and the area of distribution becomes much larger. Ramsay has observed that the emanation condensed at the temperature of liquid air can be partly removed by continual pumping, indicating that it has an appreciable vapour-pressure at that temperature. This effect, however, becomes far more noticeable when using a pentane bath whose temperature is not more than 10° or 20°C below the temperature of condensation. Every stroke of the pump then removes a not inconsiderable fraction of the total emanation. There is another effect observed which is very striking. Suppose that the nearly pure emanation contained in the U-tube is condensed over a small area by applying the liquid air only to the bottom of the U-tube. If the U-tube is then fully immersed in liquid air, in the course of about ten minutes it will be observed by the phosphorescence that the emanation is distributed throughout the tube, even though the U-tube is not connected with the pump. In addition, a part of the emanation has condensed above the level of the liquid air. Such experiments bring out clearly that the emanation has a sensible vapour-pressure far below the temperature of condensation. There is continual volatilization of the emanation in one part of the tube and condensation in another part.

I desire to thank Mr T. Royds, M.SC., who very kindly assisted me in many of these experiments.

University, Manchester

Spectrum of the Radium Emanation

by E. RUTHERFORD, F.R.S.,

and T. ROYDS, M.SC., *Beyer Fellow, University of Manchester* *

From the *Philosophical Magazine* for August 1908, ser. 6, XVI, pp. 313–17

THE first determination of the spectrum of the radium emanation was made in 1904 by Ramsay and Collie,† who obtained visual observations of the wave-lengths of eleven lines. They stated that the emanation was a bright line spectrum similar in general character to that observed for other monatomic gases. Since that time, no further information on this important subject has been forthcoming. In a previous paper,‡ one of us has given an account of methods employed in purification of the emanation and determination of its volume. In order to test the purity of the emanation, an electrodeless discharge was passed in the capillary tube in which the volume of the emanation was measured, and visual observations of the wave-lengths of the main ines were made by means of a direct reading Hilger spectroscope. We have observed the spectrum of the radium emanation in this way on four different loccasions during the past two months. It was evident that a number of new lines were present, which were not recorded in the initial observations of Ramsay and Collie. As soon as the measurements of the volume had been completed, arrangements were made to photograph the emanation spectrum in order to determine the wave-lengths of the lines with more accuracy than is possible with visual observations. For this purpose, a quantity of radium emanation was purified as completely as possible by the methods outlined in the last paper. The emanation was first condensed in a U-tube surrounded by liquid air, and the uncondensed gases completely pumped off. The emanation was then left for three hours in contact with a tube coated with caustic potash to remove the last traces of carbon dioxide. Finally, the U-tube was surrounded by a pentane bath cooled down by liquid air, and the uncondensed gases pumped off at a temperature above that of liquid air. In order to obtain the spectrum, a small vacuum tube of capacity about 50 cu. mm., provided with fine platinum electrodes, was used. This was sealed to the connection leading to the pump and completely exhausted, a discharge being passed to free the electrodes of hydrogen. When the emanation had been purified as completely as possible in the manner outlined, it was condensed

---

* A preliminary account of this work was published as a letter in *Nature*, July 8, 1908. (*This vol.*, *p.* 70.)

† *Proc. Roy. Soc.*, lxxiii, p. 470 (1904).

‡ *See* Rutherford, 'Volume of Emanation', *Phil. Mag.*, August 1908. (*This vol.*, *p.* 72.)

in the spectrum-tube by dipping a side tube connected with it in liquid air. When the greater part of the emanation had been condensed, the spectrum-tube was sealed off and removed for observation.

Measurements of the γ ray activity showed that the amount of emanation in the tube corresponded to 130 milligrams of radium. Now the volume of the emanation per gram of radium is 0·57 cu. mm. Consequently, the volume of the pure emanation in the spectrum tube was 0·074 cu. mm. Since the volume of the spectrum-tube was 50 cu. mm., this would give a pressure of emanation in the tube of 1·1 mm. of mercury. In order to photograph the spectrum, a spectrograph with a glass prism of 2 in. base was used. The length of the spectrum on the plate between λ 5,000 and λ 4,000 was 1·5 cm. Arrangements were made so that visual observations of the wave-lengths could be made by the Hilger spectroscope while the plate was being exposed. Two photographs were taken before the emanation spectrum ran out. The first (photograph 1) showed about 30 of the more intense emanation lines. The second (photograph 2), which had a much longer exposure, showed over 100 lines. A helium tube was used for comparison purposes, and its spectrum obtained above and below the emanation spectrum. The plates were measured up with the aid of a Kayser's measuring machine. The wave-lengths were deduced with the aid of the Hartmann dispersion formula.

Remarks on Spectrum

The colour of the discharge through the emanation was bluish and not so intense as the helium tube. The spectrum observed visually was a brilliant one of bright lines. The most noticeable lines were a number of strong lines in the green and another group in the violet. The mercury and hydrogen lines were also observed. In order to be sure that the lines were due to the emanation, the side tube attached to the spectrum-tube was immersed in liquid air. At the moment of condensation, which was readily noticed by the increased brilliancy of the phosphorescence of the glass, practically all the lines except those due to hydrogen vanished. The colour of the discharge then completely changed to a pale rose, and the tube became harder. At the moment of volatilization the emanation lines flashed out again.

The hydrogen lines came out more strongly when the emanation was condensed. In previous experiments with the electrodeless discharge, the hydrogen lines had been absent. Their occurrence in the present experiment was, without doubt, due to the liberation of hydrogen from the platinum electrodes when a strong discharge was passed. This is borne out by the results of another experiment recorded later. The emanation was momentarily condensed at intervals during the experiment in the side tube. From observations of the brilliancy of the phosphorescence at condensation, it was noted that the amount of the free emanation in the tube gradually diminished with increasing time of discharge, while the intensity of the emanation spectrum decreased relatively to that of hydrogen. The emanation lines, however, persisted to the close of the experiment, when practically all the emanation had been driven into the walls of the tube. From observations of the phosphorescence, it was evident that the emanation was approximately uniformly distributed along the line of discharge. As the discharge had been reversed at intervals during the experiment, it was difficult to be certain whether there had been any considerable absorption of the emanation by the electrodes. The occlusion of the emanation had been observed previously on several occasions in the capillary tube using the electrodeless discharge (*see* previous paper). It seems probable that the emanation is in some way driven into the walls of the tube by the discharge. This effect is no doubt similar to that recorded by Campbell Swinton for ordinary gases. It is difficult to remove such occluded emanation even by strongly heating the glass.

After three days, the tube was very much darkened by the emanation, and it was necessary to get rid of the blackening by heat in order to observe the spectrum. The main helium lines were observed, but were faint in comparison with the hydrogen lines.

After a week's interval, the spectrum-tube was again attached to the pump and thoroughly heated above the temperature of thermo-luminescence in order to make the glass as transparent as possible. The spectrum-tube was exhausted, care being taken by heating the tube and by passage of a strong discharge to get rid of most of the hydrogen from the electrodes. About the same quantity of pure emanation as in the first experiment was condensed in the tube. After sealing off the tube, the spectrum was photographed, visual observations being made at the same time.

The same general effect as in the first experiment was observed when the emanation was condensed on the side tube. In this case, however, the hydrogen spectrum was relatively much feebler. On condensing the emanation, the tube became very hard and showed the characteristic green coloration of the cathode-ray vacuum. It was thus clear that the methods employed had been fairly successful in getting rid of the hydrogen from the electrodes. The photograph of the spectrum in this case (photograph 3) showed only the hydrogen line Hβ, although Hα also was observed visually. In the second photograph, already referred to, the stronger lines of the compound line spectrum of hydrogen had been photographed.* The photograph 3 is reproduced in Plates I and II, magnification 3·7 times. Plate I was exposed to bring out clearly the strong lines of the spectrum only. It will be seen that the stronger lines in Plate II are somewhat overexposed in order to bring out some of the less intense lines in the spectrum. This photograph is somewhat better for reproduction purposes than photograph 2, but does not show quite the same number of faint lines.

* No trace of the carbon-dioxide spectrum was observed in either experiment. The occurrence of this spectrum in the electrodeless discharge in nearly pure emanation (*see* paper, *loc. cit.*) was without doubt due to a trace of organic matter on the surface of the mercury.

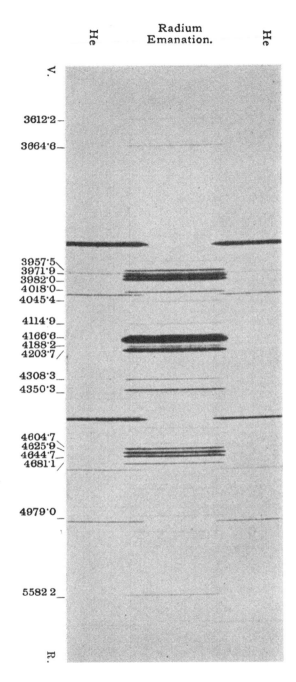

PLATE I

Strongest Lines in the Spectrum of the Radium Emanation

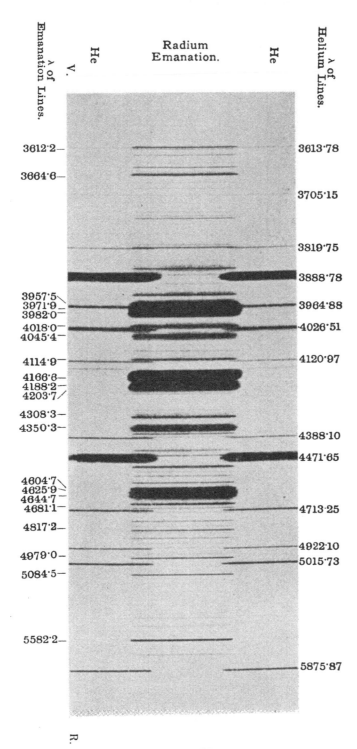

λ of
Emanation Lines.

He

V.

Radium
Emanation.

He

λ of
Helium Lines.

3612·2—

3613·78

3664·6—

3705·15

3819·75

3888·78

3957·5
3971·9
3982·0

3964·88

4018·0
4045·4

4026·51

4114·9—

4120·97

4166·6
4188·2
4203·7

4308·3—
4350·3—

4388·10

4471·65

4604·7
4625·9
4644·7
4681·1—

4713·25

4817·2—

4979·0—

4922·10

5015·73

5084·5—

5582·2—

5875·87

R.

PLATE II

The measurements of the wave-lengths of the lines common to the two plates agreed within the limits of experimental error. We shall consequently only give the measurement of photograph 2, since the hydrogen lines present in this spectrum serve as an indication of the accuracy of the measurements. It will be seen that the error of measurement is certainly not greater than

Wave-lengths of the Emanation Lines

Visual		Photograph		Remarks
Int^y	λ	Int^y	λ	
0	6079			
1	5976			
1	5945			
1	5829			
1	5765			
3	5718	1	5715·0	
5	5582	8	5582·2	
0	5395	0	5392·4	
0	5372			
1	5257			
2	5120			
10	5087	4	5074·5	
2	5060			
10	4985	4	4979·0	
1	4964	0	4965·6	
1	4955	00	4949·4	
1	4917	00	4914·6	
1	4895	0	4889·5	
	4865		4861·3	Hβ 4861·49
0	4831	1	4827·8	
2	4820	4	4817·2	
0	4798	1	4796·7	
1	4772	3	4767·9	
1	4726	5	4721·5	Int^y in photo. 3.
0	4705	2	4701·7	
5	4685	10	4681·1	Does not quite dis-
		1	4671·8	appear when emana-
		1	4659·3	tion condensed.
5	4650	10	4644·7	
6	4631	8	4625·9	
1	4614	7	4609·9	Int^y 4 in photo. 3.
3	4608	4	4604·7	„ 6 „
1	4581	7	4578·7	„ 4 „
0	4550	1	4549·9	
1	4511	9	4509·0	„ 4 „
		2	4504·0	
0	4460	10	4460·0	„ 2 „
		2	4440·6	

Visual		Photograph		Remarks
Int^y	λ	Int^y	λ	
0	4439	8	4435·7	Int^y 2 in photo. 3.
		3	4384·0	
		4	4372·1	
3	4351	15	4350·3	
		7	4340·9	H^y 4340·66. Absent
1	4310	10	4308·3	from 3rd photo.
		2	4225·8	
1	4202	10	4203·7	
		5	4188·2	
1	4169	20	4166·6	
		7	4114·9	
		6	4102·2	Hδ 4101·85. Absent
		2	4088·4	from 3rd photo.
		1	4055·7	
		2	4051·1	
		4	4045·4	
		1	4040·2	
		10	4018·0	
		12	3982·0	
		9	3971·9	Not Hϵ 3970·25.
		7	3957·5	
		3	3952·7	
		3	3933·3	
		1	3927·7	
		2	3905·7	Int^y in photo. 3.
		4	3867·6	
		2	3818·0	
		0	3811·2	
		10	3753·6	Int^y 3 in photo. 3.
		1	3748·6	
		7	3739·9	„ 1 „
		2	3690·4	
		1	3679·2	
		10	3664·6	
		0	3650·0	„ 3 „
		2	3626·6	
		1	3615·4	
		6	3612·2	

None of the emanation lines have been identified in any stellar spectra.

half an Ångström unit. The lines given in the table are common to both photographs. It has not been thought necessary to give the weaker lines observed, for the identity of these with the emanation spectrum requires further confirmation.

In the table the lines observed visually are given in a separate column. In photograph 3, when the hydrogen was far less prominent, the relative intensity of some of the emanation lines differed from that observed in photograph 2.

Manchester University

An Electrical Method of Counting the Number of α-Particles from Radio-active Substances

by E. RUTHERFORD, F.R.S., *Professor of Physics,*

and H. GEIGER, PH.D., *John Harling Fellow, University of Manchester*

from *The Proceedings of the Royal Society, A*, Vol. 81, 1908, pp. 141–61
(Read June 18th; MS. received July 17, 1908)

THE total number of α-particles expelled per second from 1 gramme of radium has been estimated by Rutherford* by measuring the charge carried by the α-particles expelled from a known quantity of radium in the form of a thin film. On the assumption that each α-particle carries the ionic charge $e = 3 \cdot 4 \times 10^{-10}$ electrostatic unit, it was shown that $6 \cdot 2 \times 10^{10}$ α-particles are expelled per second from 1 gramme of radium itself, and four times this number when in radio-active equilibrium with its three α-ray products, viz., the emanation, radium A and C. In order to reconcile the value of e/m found for the α-particle with that to be expected for the helium atom, it was later† pointed out that the α-particle should carry a charge equal to $2e$. On this assumption, the number of α-particles expelled per second per gramme of radium is reduced to one-half the first estimate.

The need of a method of counting the α-particles directly without any assumption of the charge carried by each has long been felt, in order to determine the magnitude of the various radio-active quantities with a minimum amount of assumption. If the number of α-particles expelled from a definite quantity of radio-active matter could be determined by a direct method, the charge carried by each particle could be at once known by measuring the total positive charge carried by the α-particles. In this way, it should be possible to throw some light on the question whether the α-particle carries a charge e or $2e$, and thus settle the most pressing problem in radio-activity, viz., whether the α-particle is an atom of helium.

In considering a possible method of counting the number of α-particles, their well-known property of producing scintillations in a preparation of phosphorescent zinc sulphide at once suggests itself. With the aid of a microscope, it is not very difficult to count the number of scintillations appearing per second on a screen of known area when exposed to a source of α-rays. The doubt, however, at once arises whether every α-particle produces a scintillation, for it is difficult to be certain that the zinc sulphide is homo-

* 'Phil. Mag.,' August, 1905. (*Vol. I, p. 816.*)
† Rutherford. 'Phil. Mag.,' October, 1906. (*Vol. I, p. 880.*)

geneous throughout. No confidence can be placed in such a method of counting the total number of α-particles (except as a minimum estimate) until it can be shown that the number so obtained is in agreement with that determined by some other independent method which does not involve such obvious uncertainties. The results of some observations on the number of scintillations produced by the α-particles from radium will be discussed later.

It has been recognised for several years that it should be possible by refined methods to detect a single α-particle by measuring the ionisation it produces in its path. On the assumption that an α-particle carries an ionic charge e, one of us has shown that the α-particle expelled from radium itself produces 86,000 ions in its path in air before it is stopped. Taking the charge of an α-particle as $2e$, this number is reduced to one-half. Consequently if the α-particle passes through air in a strong electric field, the total quantity of electricity transferred to the electrodes is $43,000e$. Taking $e = 3\cdot4 \times 10^{-10}$ E.S. unit, this corresponds to $1\cdot46 \times 10^{-5}$ E.S. unit. For the purpose of illustration, suppose that a Dolezalek electrometer of capacity 50 E.S. units, which has a sensibility of 10,000 mm. divisions per volt between the quadrants, is used for detection of the ionisation. The quantity $1\cdot46 \times 10^{-5}$ unit transferred to the electrometer system would cause a deflection of the needle of $0\cdot3$ mm. This is small but detectable. In a similar way, if an electroscope of capacity 2 E.S. units be employed instead of an electrometer, the movement of the leaf would correspond to a difference of potential of $2\cdot1 \times 10^{-3}$ volt. While there is no inherent impossibility in detecting such small quantities of electricity by either the electroscope or electrometer, yet the measurement would have to be of a refined character in order to get rid completely of all extraneous sources of disturbance. One difficulty is that the moving system in very sensitive electrometers or electroscopes has a long period of swing and consequently moves very tardily when a small difference of potential is suddenly applied. Some preliminary experiments to detect a single α-particle by its direct ionisation were made by us, using specially constructed sensitive electroscopes. As far as our experience has gone, the development of a certain and satisfactory method of counting the α-particles by their small direct electrical effect is beset with numerous difficulties.

We then had recourse to a method of automatically magnifying the electrical effect due to a single α-particle. For this purpose we employed the principle of production of fresh ions by collision. In a series of papers, Townsend* has worked out the conditions under which ions can be produced by collisions with the neutral gas molecules in a strong electric field. The effect is best shown in gases at a pressure of several millimetres of mercury. Suppose that the current between two parallel plates immersed in a gas at low pressure is observed when the air is ionised by X-rays. The current through the gas for small voltages at first increases with the field and then reaches a saturation value, as is ordinarily observed in ionised gases

* 'Phil. Mag.,' February, 1901; June, 1902; April, 1903; September and November, 1903.

at atmospheric pressure. When the field is increased beyond a certain value, however, the current rises rapidly. Townsend has shown that this effect is due to the production of fresh ions in the gas by the collision of the negative ions with the gas molecules. At a later stage, when the electric field approaches the value required to cause a spark, the positive ions also become effective as ionisers but to a much smaller degree than the negative. Under such conditions, the small current through the gas due to the external ionising agency may be easily increased several hundred times. The magnification of the current depends upon the voltage applied and becomes very large just below the sparking value.

In our experiments to detect a single α-particle, it was arranged that the α-particles could be fired through a gas at low pressure exposed to an electric field somewhat below the sparking value. In this way, the small ionisation produced by one α-particle in passing along the gas could be magnified several thousand times. The sudden current through the gas due to the entrance of an α-particle in the testing vessel was thus increased sufficiently to give an easily measurable movement of the needle of an ordinary electrometer.

Experimental Arrangement.—Before considering the various difficulties that arose in the course of the investigations, a brief description will be given of the method finally adopted. The experimental arrangement is shown in fig. 1. The detecting vessel consisted of a brass cylinder A, from 15 to 25 cm.

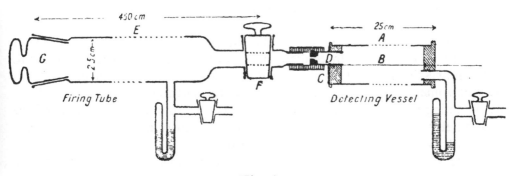

Fig. 1

in length, 1·7 cm. internal diameter, with a central insulated wire B passing through ebonite corks at the ends. The wire B was in most experiments of diameter 0·45 mm. The cylinder, with a pressure gauge attached, was exhausted to a pressure of from 2 to 5 cm. of mercury. The central wire was connected with one pair of quadrants of a Dolezalek electrometer and the outside tube to the *negative** terminal of a large battery of small accumulators, the other pole of which was earthed. In the ebonite cork C

---

* If the tube were connected to the positive pole of the battery, the magnification by collision only became appreciable near the sparking voltage. With the negative pole, the magnification increased more gradually and was far more under control.

was fixed a short glass tube D of internal diameter 5 mm., in the end of which was a circular opening of about 1·5 mm. diameter. This opening, through which the α-particles entered the testing vessel, was covered with a thin sheet of mica tightly waxed over the end of the tube. In most experiments the thickness of mica was equivalent, as regards stopping power of the α-particle, to about 5 mm. of air at atmospheric pressure. Over the tube D was fixed a wide rubber tube, to the other end of which was attached a long glass tube E of 450 cm. in length and 2·5 cm. diameter. A large stop-cock F with an opening 1 cm. in diameter was attached to the end of the glass tube next to the detecting vessel. The other end of the long glass tube was closed by a ground stopper G.

The general procedure of an experiment was as follows. The voltage applied to the testing vessel was adjusted so that the ionisation in the vessel due to an external source of γ-rays was increased by collision several thousand times. The radium tube which served as a source of γ-rays was then removed. Under ordinary conditions, when all external sources of ionisation were absent, there was always a small current passing through the gas. In order to avoid the steady movement of the electrometer needle due to this cause, the current was allowed to leak away through a radio-active resistance attached to the electrometer system. This consisted of two insulated parallel plates, the upper connected with the electrometer and the lower with earth. A layer of radio-active material was placed on the lower plate. As the potential of the electrometer needle rose, equilibrium was soon reached between the current supplied to the electrometer and that which leaked away due to the ionised gas between the plates. This arrangement was of great importance to the success of the experiments, for it practically served to eliminate disturbances due to electrostatic effects or to slow changes in the E.M.F. of the battery. Any sudden rise of potential of the electrometer, for example that due to the entrance of an α-particle in the detecting vessel, then manifested itself as a sudden *ballistic* throw of the electrometer needle. The charge rapidly leaked away and in a few seconds the needle was again at rest in its old position.

The active matter, in the form of a thin film of not more than 1 square cm. in area, was fixed in one end of a hollow soft iron cylinder which could be moved along the glass tube from the outside by means of an electro-magnet. The glass tube was then exhausted by means of a Fleuss pump and, if required, to a still lower pressure by means of a tube of cocoanut charcoal immersed in liquid air.

When the stop-cock was closed, no α-particles could enter the vessel, and the steadiness of the electrometer needle could thus be tested at intervals during an experiment. On opening the stop-cock, a small fraction of the total number of α-particles expelled per second passed through the aperture into the detecting vessel. In practice, it was found convenient to arrange the intensity of the active matter and its distance from the opening so that from three to five α-particles entered the detecting vessel per minute. It became difficult to count a number greater than this with certainty, since the needle had not time to come to rest between successive throws.

The following example serves to illustrate the character of the observations. The source of α-rays in this case was a metal plate about 0·5 square cm. in area made active by exposure for several hours in a large quantity of radium emanation. Fifteen minutes after removal from the emanation, the α-radiation from the plate is due almost entirely to radium C. The active matter is in the form of a thin film, so that all the α-particles are expelled with the same velocity. The intensity of the radiation from radium C decreases with time, falling to half value about one hour after removal and later at a more rapid rate. In this particular case, the detecting tube was filled with carbon dioxide to a pressure of 4·2 cm. The E.M.F. applied was 1320 volts. The active plate was at a distance of 350 cm. from the aperture, which was of diameter 1·23 mm. Observations of the number and magnitude of the throws due to the α-particles were continued over an interval of 10 minutes. The results are shown in the following table:—

	Number of throws	Magnitude of successive throws in scale divisions
1st minute	4	11, 12, 10, 11
2nd ,, ...	3	10, 11, 8
3rd ,, ...	5	10, 9, 13, 8, 12
4th ,, ...	4	18,* 8, 12
5th ,, ...	3	10, 6, 10
6th ,, ...	4	9, 10, 12, 11
7th ,, ...	2	10, 11
8th ,, ...	3	11, 13, 8
9th ,, ...	3	8, 20*
10th ,, ...	4	8, 12, 14, 6
Average per minute = 3·5		Average throw = 10 divisions.

Each scale division was equal to 2·5 mm. The intensity of the α-radiation decreased about 15 per cent. during the time of observation.

When the stop-cock was closed so that no α-particles could enter the detecting vessel, the electrometer needle was very steady, the maximum excursion of the needle from the zero position in the course of 10 minutes being not more than three scale divisions. Only two or three excursions of such amplitude occurred in that interval. We see from the table that the average throw observed with the stop-cock open was 10 divisions.* All small

---

* The magnitude of the throw due to a single α-particle is dependent upon the E.M.F. applied, and can be varied over wide limits.

excursions of magnitude less than three scale divisions are omitted. With the exception of the two numbers marked with asterisks, each of the throws given in the table is due to a single α-particle. The two large throws marked with asterisks are each due to the superposition of the separate effects due to two successive α-particles entering the detecting vessel within a few seconds of each other. This was readily seen from the peculiarity of the motion of the spot of light on the scale. As the electrometer needle was moving slowly near the end of its swing caused by the effect of one α-particle, a second impulse due to the entrance of another was communicated to it, and caused it to move again more rapidly. Such double throws occur occasionally, and are readily recognised, provided the interval between the entrance of successive α-particles is not less than one second.

It will be noted that the number of α-particles entering the opening per minute, and also the interval between successive throws varied within comparatively wide limits. Such a result is to be anticipated on the theory of probability. We may regard a constant source of α-rays as firing off α-particles equally in all directions at a nearly constant rate. The number per minute fired through a small opening some distance away is on the average constant if a large number of throws are counted. When only a small number of throws are observed over a short interval, the number is subject to considerable fluctuations, the probable percentage departure of the observed number from the correct average being greater the smaller the number of α-particles entering within a given time. This phase of the subject is of considerable interest and importance, and will be discussed in more detail later in the paper. It suffices here to say that the variation of the observed number per minute is well within the limits to be anticipated on the general laws of probability.

It is seen that the throws due to an α-particle are somewhat variable in magnitude. Such a result is to be anticipated for several reasons. In the first place, the α-particles do not all pass along the detecting tube at the same distance from the axis. The magnification of the ionisation is less for those that pass closest to the central wire. In addition, as will be shown later, there is always a scattering of the α-rays by the mica screen and by the gas in the detecting vessel. This tends to spread out the pencil of rays in the detecting vessel, and consequently to introduce still greater differences in the effects due to individual α-particles.

Detection of α-Particles from Uranium, Thorium, Radium, and Actinium

The throws of the electrometer observed with the stop-cock open have been ascribed to the α-particles fired into the detecting vessel. This can be readily proved by placing a thin screen between the source of radiation and the detecting vessel. The throws of the electrometer disappear if this screen, together with the mica plate covering the hole, is of the right thickness to stop the α-particles entirely. Under ordinary conditions, the effect due to a β-particle is very small compared to that due to an α-particle, and is not detectable. If a plate coated with the active deposit of radium is used as a source of radiation, it is found that the decay curve obtained by counting the α-particles emitted agrees closely with the ordinary α-ray decay curve.

By this electrical method, we have detected the expulsion of α-particles not only from radium and its products but also from uranium, thorium, and actinium. For example, a plate, made active by exposure to the emanation of a preparation of actinium, gave effects in the testing vessel due to an α-particle of about the same magnitude as that due to an α-particle from radium C. The decay curve obtained by counting the α-particles agreed closely with the known curve. A thin film of radium itself showed a similar effect. As the activity rose, consequent upon the production of fresh emanation and its occlusion in the radium, the number of α-particles entering the detecting vessel increased.

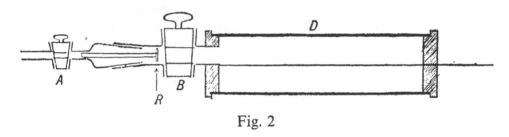

Fig. 2

A special apparatus (see fig. 2) was used to detect the emission of α-particles from weak radio-active substances like uranium and thorium. The active matter spread on a plate R (fig. 2) was placed about 5 cm. from the opening, which in this case was about 1 cm. diameter, and without any mica screen. A stop-cock of wide bore was placed between the active matter and the testing vessel D. With the stop-cock closed, the electrometer needle was very steady. On opening the stop-cock, about two throws per minute of the ordinary magnitude due to an α-particle were observed from the uranium. This was about the number to be expected from known data.

It may be of interest to record an experiment made with a preparation of thorium hydroxide. A small quantity of this (about 3 milligrammes) was wrapped in thin paper, which stopped the α-rays but allowed the emanation to pass through freely. On opening the stop-cock, the emanation diffused into the detecting vessel and immediately a large deflection of the electrometer was observed. After a few minutes an approximately steady radio-active state was reached. The electrometer needle, however, never remained steady, but made wide oscillations on either side of the mean position. Such an effect was to be anticipated, for when occasionally two or three α-particles from the emanation were fired along the cylinder within a second or two of each other, the electrometer needle was widely deflected. When the stop-cock was closed, the mean deflection due to the emanation in the testing vessel decreased with the time at the rate characteristic of the thorium emanation, but the electrometer needle continued to give excursions to and fro until the activity of the emanation had disappeared.

There is no doubt that the principle of automatic increase of the ionisation by collision can be used to extend considerably the range of measurement of minute quantities of radio-active matter.

Experimental Difficulties

The final type of detecting cylinder which was found most satisfactory for counting purposes was of small diameter, viz., 1·7 cm., and of length not more than 25 cm. We shall now discuss the reasons that led us to adopt such a small detecting vessel. In the preliminary experiments, a cylinder of diameter 3·5 cm. and length 1 metre was used. With a pressure of air of 4 cm., the ionisation and stopping power of an α-particle passing the length of the cylinder was equivalent to that due to traversing 5·3 cm. of air at atmospheric pressure. Since the mica screen had a stopping power equal to only 5 mm. of air, an α-particle from radium C (range 7 cm.) produced the major part of its total ionisation in the detecting cylinder. Using such a vessel, it was not difficult to adjust the voltage so that an α-particle entering the vessel produced a throw of at least 100 mm. of the electrometer scale. Under such conditions, however, it was found impossible to avoid natural disturbances of the electrometer needle, when the stop-cock was closed, which were comparable in magnitude and character with those due to the entrance of an α-particle. These sudden movements of the electrometer needle were not numerous, but were sufficient to interfere with an accurate counting of the number of α-particles. These disturbances were inherent in the vessel and could not be got rid of by changing the pressure or nature of the gas or the diameter of the central wire. There is no doubt these irregular movements of the electrometer needle must be ascribed to a slight natural radio-activity of the walls of the brass tube. An α-particle projected near the end of the tube in the direction of the axis of the tube would produce a throw of the electrometer needle of about the same magnitude as that due to an α-particle fired through the opening parallel to the axis of the tube. The great majority of the α-particles emitted by the tube will only travel a short distance before being stopped by the walls, and consequently will only give rise to small individual movements of the needle. The greater part of the current observed by the electrometer with the aperture closed was due to this natural ionisation increased several thousand times by the agency of the strong electric field. The correctness of this conclusion was borne out by the observation that any change of the applied voltage, and consequently of the magnification, altered the magnitude of the natural disturbances, and the throw due to an α-particle in about the same ratio. In addition, it was observed that the number of the natural disturbances fell off rapidly with decrease of the diameter of the detecting tube. For example, the natural movements of the electrometer needle, using a long tube of 5 cm. diameter, were so numerous and so vigorous that it was impossible to use it for counting α-particles at all. With a tube, however, of 1·7 cm. diameter, the natural movements were very occasional, and of magnitude small compared with the effect due to an α-particle. Such a rapid decrease of the disturbances is to be anticipated in the light of the above explanation. If tubes are taken of the same length and of the same natural radio-activity per unit area, but of different diameters, the total number of α-particles shot out is proportional to their radii. Taking corresponding cross sections of the tubes, the fraction of the total number of α-particles emitted, which travel to the end of the tubes without striking the walls, is proportional to the cross sectional area of the tubes. Consequently the number of α-particles which pass along the tube without being stopped by the walls varies directly as the cube of the radius. We thus see that the sudden large movements of the electrometer needle due to the radio-activity of the walls should fall off very rapidly with decrease of the diameter—a result in harmony with the experimental observations.

Since the electrical capacity of the detecting vessel was smaller than the electrometer and its connections, it seemed advisable at first to use long detecting tubes in order to make the ionisation in it due to an α-particle as large as possible. From lack of accumulators at our command, it was not found feasible to work at a higher pressure than about 6 cm. of mercury, for at this pressure about 1500 volts were necessary to obtain the requisite magnification. Experiments were consequently made with a tube 135 cm. long, of diameter 1·7 cm., with a gas pressure varying from 2 to 6 cm. The natural disturbances of the electrometer needle in this vessel were very small, and it was found possible to increase the magnification such that an α-particle produced a throw of several hundred millimetre divisions on the scale. The throws due to successive α-particles were, however, very variable in magnitude. This is illustrated by the following table of observations:—

Air pressure, 3 cm.; radium C. Source of radiation; distance from aperture, 350 cm.

	Number of α-particles	Magnitude of successive throws
1st minute	4	6, 7, 10, 16
2nd „	2	21, 15
3rd „	1	36
4th „	4	6, 25, 17, 11
5th „	4	4, 28, 13, 13
6th „	5	9, 16, 7, 6, 24

D

The great difference in the magnitude of the throws could not be ascribed to several particles entering together, for similar divergencies were noted when on an average only one α-particle entered the detecting vessel per minute. Special experiments were made with sources of radiation of small area at a distance of 4 metres from the aperture and with a small aperture in the detecting vessel. Under such conditions, it was arranged that if the α-particles travelled in straight lines, they should strike the end of the detecting tube within an area of 1 square mm. The use of such a theoretically narrow pencil of rays had no effect, however, in equalising the magnitude of the throws. Finally, after a series of experiments, it was found that this effect was due to the *scattering of the α-particles in their passage through the mica screen and through the gas in the detecting vessel.* In a previous paper by one of us,* attention had been directed to the undoubted scattering of the α-particles in their passage through matter, and the magnitude of this scattering had been determined by the photographic method in special cases. We did not at first realise the importance of this effect in our experimental arrangement. Some of the α-particles, in passing through the thin sheet of mica, are deflected from their rectilinear path, and this deflection is continued in their passage along the gas of the tube. The scattering was sufficiently great to cause a large fraction of the α-particles to impinge on the walls of the tube. The small throws observed were due to α-particles which only traversed a small fraction of the length of the tube before being stopped, while the largest throws were due to those that passed along the tube without striking the walls. A special series of experiments by a new method were made by one of us† to determine the magnitude of this scattering in special cases. An account of these experiments will be published in a separate paper.

As it was not feasible to decrease the scattering by reducing the thickness of the mica screen over the opening, the only way of making the throws more uniform was to diminish the length of the tube. It was for this reason that a tube only 25 cm. long was used. In this short distance, the α-particles were not deflected sufficiently to strike the walls of the tube, and the great majority travelled the whole length of the detecting vessel. Under these conditions, the throws of the electrometer due to the α-particles at once became far more uniform. An example of the throws obtained in the short vessel is given in Table I, p. 93. In the long detecting tube, there was a tendency to overlook the small throws and thus to underestimate the number of α-particles entering into the detecting vessel. The presence of this scattering also makes it necessary to exhaust the long firing tube to a low pressure. The presence of gas in this tube tends to deflect the α-particles from their rectilinear path and, if the tube is narrow near the aperture, to reduce the number entering the detecting vessel. Such a decrease of the number was at

* Rutherford, 'Phil. Mag.,' August, 1906. (*Vol. I, p. 859.*)

† H. Geiger, 'Scattering of the α-Particles by Matter,' Proc. Roy. Soc. A, Vol. 81, 1908, p. 174.

once observed, if the pressure of the gas in the long tube were raised so that its stopping power was equivalent to 2 or 3 cm. of air at atmospheric pressure.

The Number of α-Particles expelled from Radium

A series of experiments was made to determine as accurately as possible by the electrical method the number of α-particles expelled per second from 1 gramme of radium. The arrangement of the apparatus was similar to that shown in fig. 1. A source of homogeneous α-rays was placed at a convenient distance from the detecting vessel in the firing tube, and the average number of α-particles entering the aperture per minute was determined by counting the throws of the electrometer needle.

Let Q be the average number of α-particles expelled per second from the source, consisting of a thin film of active matter. Let A be the area in square cm. of the aperture in the detecting vessel, and r the distance in centimetres of the source of rays from the aperture. It was verified experimentally that the α-particles on an average are projected equally in all directions. Consequently, the fraction of the total number of α-particles expelled from the source which enter the detecting vessel is equal to the area of the aperture divided by the area of the surface of a sphere of radius equal to the distance of the source from the opening. The average number n of α-particles entering the opening per second is thus given by

$$n = \frac{QA}{4\pi r^2}.$$
(1)

This expression holds for all distributions of active matter of dimensions small compared with the distance r, provided that each element of surface of the source can fire directly into the aperture. In practice, the active matter is usually spread on the surface of a body of sufficient thickness to stop the α-particles fired into it, so that only half the total number of α-particles escape from its surface. This in no way interferes with the correctness of the above expression for the number.

After some preliminary experiments with thin films of radium itself, it was decided to employ radium C as a source of α-rays in the counting experiments. If a body is exposed for about three hours in the presence of the radium emanation, the activity imparted to it reaches a maximum value. Fifteen minutes after removal of the body from the emanation, the radiation due to radium A has practically disappeared, and the α-radiation is then due entirely to radium C. Under these conditions all the α-particles escape with the same velocity, and have a range in air of 7 cm. The use of radium C has numerous advantages. The active deposit is in the form of an extremely thin film, and the amount of active matter deposited on a body can readily be varied by altering the amount of emanation or the surface exposed to it. The chief advantage, however, lies in the ease and certainty of measurement

of the quantity of active matter present in terms of the radium standard.*
The penetrating γ-rays from radium in equilibrium arise entirely from its
product, radium C. Consequently, by comparing the γ-ray activity of the
active deposit with the radium standard, the amount of radium C present
may be expressed in terms of the quantity of radium C in equilibrium with
1 gramme of radium. The chief disadvantage lies in the fact that the activity
due to radium C rapidly diminishes, falling to half value in about one hour
and to 14 per cent. of the maximum in two hours.

The shape of the body made active by exposure to the emanation must be
such that each element of the active surface, when in position in the firing
tube, must be in full view of the aperture of the detecting vessel. Examples
of the surfaces employed are shown in fig. 3; *a* and *b* are of glass, and *c* a

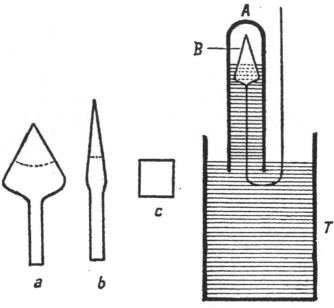

Fig. 3

plane sheet of glass or iron, the dotted lines representing the lower limit of
the active matter. The emanation, mixed with 1 or 2 c.c. of air, was
collected over mercury in the end of the tube A (fig. 3). The body B, to
be made active, was fixed to a glass U-tube, and introduced into the
emanation space by means of the mercury trough T. After remaining in
position for an interval of not less than three hours, the active body was
removed and immediately tested in terms of the radium standard, using a
fixed γ-ray electroscope. The active source was then placed in position in
the firing tube, which was exhausted to a low pressure. In order to follow
the changes of activity of the source, a second travelling γ-ray electroscope

* The radium standard employed in these experiments is one that has been in use for
several years. It is a part of a sample of radium which gave a heating effect of 110 gramme-
calories per hour per gramme.

was employed in which the activity of the source was determined *in situ.* In the counting experiments the active body, as it diminished in activity, was moved nearer the detecting vessel. The electroscope was moved so as to be always directly over the active body and always at the same distance from it. At the end of the counting experiments, the active body was removed from the firing tube and its γ-ray activity again determined on the fixed electroscope. In this way a complete check was obtained on the activity measurements as well as a direct determination of the decay curve of the active body.

The general procedure of an experiment was as follows. After the γ-ray activity had been accurately measured, observations of the number of throws were made continuously for an interval of 10 minutes. The γ-ray activity was then determined again, and then another 10 minutes' count, and so on. When the number of α-particles entering the opening had fallen to between one and two per minute, the active body was brought nearer the opening, and observations continued as before over a total interval of about two hours.

The following table illustrates the results obtained with the same source as it decayed in activity. The detecting vessel contained air at a pressure of 3·75 cm. and about 1200 volts were applied. The diameter of the aperture was 1·23 mm.

Distance of active body from aperture	Mean γ-ray activity of source	Number of throws observed in 10 mins.	Number of α-particles expelled per gramme
350 cm.	0·309 mgr. Ra	45	$3·06 \times 10^{10}$
350 ,,	0·154 ,,	25	$3·33 \times 10^{10}$
350 ,,	0·11 ,,	16	$2·96 \times 10^{10}$
150 ,,	0·055 ,,	49	$3·43 \times 10^{10}$
150 ,,	0·031 ,,	25	$3·11 \times 10^{10}$

Total number of throws = 160.   Average = $3·18 \times 10^{10}$.

The second column gives the mean γ-ray activity of the source in terms of milligrammes of pure radium in equilibrium. The fourth column gives the total number of α-particles from radium C expelled per second in 1 gramme of radium in equilibrium. This number is calculated as follows. We have shown (equation 1) that the total number of α-particles Q emitted per second by the source is given by

$$Q = \frac{4\pi r^2}{A} \cdot n.$$

The total number Q_0 expelled for a γ-ray activity corresponding to 1 gramme of radium is given by

$$Q_0 = \frac{Q}{\rho} = \frac{4\pi r^2}{A} \cdot \frac{n}{\rho},$$

where ρ is the γ-ray activity of the source in terms of 1 gramme of radium. Since n and ρ are determined experimentally, and r and A are known, the value of Q_0 can be at once calculated. The calculated values of Q_0 for each experiment are given in the fourth column, and serve as a comparison of the agreement for the different observations.* The value of $4\pi r^2/A$ for the first three experiments at a distance of 350 cm. is equal to $1\cdot25 \times 10^8$, *i.e.*, on an average, out of 125,000,000 α-particles fired from the source, only one passes through the aperture.

In the course of our experiments, we have verified, as far as possible, the correctness of the assumptions on which the deduction of the number of α-particles expelled from 1 gramme of radium depends. These points are summarised below:—

(1) For a given intensity of radiation at a given distance, the average number of throws observed in the electrometer in a given interval is independent of the pressure or nature of the gas, and also of the magnification of the ionisation.

(2) The number of α-particles entering the aperture is proportional to the activity of the source (measured by the γ-rays) and inversely proportional to the square of the distance of the source from the aperture over the range examined, viz., from 375 to 100 cm.

(3) For a given intensity of radiation at a given distance, the number of α-particles entering the detecting vessel is proportional to the area of the aperture.

(4) Using radium C as a source of rays, the α-particles are, on an average, projected equally in all directions. This has been verified by observing that, within the limit of experimental error, the calculated number of α-particles from radium C in 1 gramme of radium comes out the same whether the α-particles entering the aperture escape nearly tangentially from the active surface, as in fig. 3, *b*, nearly normally, as in fig. 3, *c*, or at an intermediate angle, as in fig. 3, *a*.

The table on page 103 gives the results for a number of separate experiments. The average value of Q_0 for each complete experiment, involving observations for different intensities of the source at different distances, is given in the last column.

Except for the last experiment, in which a tube 21 cm. long and 2·4 cm. diameter was used, the detecting tube was of length 25 cm. and internal diameter 1·7 cm.

In determining the average value of Q_0 given in the last column, which is itself an average of a large series of experiments, the weight to be assigned to each experiment of the series was taken as proportional to the number of α-particles counted. It was found that this differed only slightly from the arithmetic mean. It is seen that the mean value of the collected observations

---

* On account of the probability variation, it is not to be expected that the numbers in the fourth column should agree very closely.

for Q_0 is $3 \cdot 28 \times 10^{10}$. This is subject to a small correction for which it is difficult to assign a definite value. In order to make the throws due to an α-particle as uniform as possible, it was arranged that the α-particle passed obliquely across the detecting tube. A small fraction of the α-particles

Gas	Pressure in detecting vessel	Voltage	Diameter of aperture	Total number of throws counted	Average value of Q_0
Air	$3 \cdot 75$ cm.	1200 volts	$1 \cdot 23$ mm.	161	$3 \cdot 20 \times 10^{10}$
CO_2	$4 \cdot 8$,,	1360 ,,	$1 \cdot 23$,,	59	$3 \cdot 10 \times 10^{10}$
,,	$4 \cdot 8$,,	1360 ,,	$1 \cdot 23$,,	118	$3 \cdot 30 \times 10^{10}$
,,	$4 \cdot 1$,,	1240 ,,	$1 \cdot 23$,,	93	$3 \cdot 13 \times 10^{10}$
,,	$4 \cdot 2$,,	1320 ,,	$1 \cdot 92$,,	194	$3 \cdot 43 \times 10^{10}$
,,	$4 \cdot 2$,,	1320 ,,	$1 \cdot 92$,,	150	$3 \cdot 34 \times 10^{10}$
,,	$3 \cdot 2$,,	1320 ,,	$1 \cdot 92$,,	99	$3 \cdot 43 \times 10^{10}$
				Average =	$3 \cdot 28 \times 10^{10}$

entering the aperture would be stopped by the central wire, diameter $0 \cdot 45$ mm. In counting the number of α-particles, there is a tendency to overlook or put down to natural disturbances all movements which are small compared with the average value. This would be the case if an α-particle were stopped before travelling half the length of the tube. Taking into account the dimensions of the aperture, and of the copper wire and the scattering of the beam in its passage through the mica and the gas, it has been estimated that this correction cannot be more than 3 per cent. Making the correction, the value of Q_0 becomes to the nearest figure $3 \cdot 4 \times 10^{10}$.

We consequently conclude that, on an average, $3 \cdot 4 \times 10^{10}$ α-particles are expelled per second from the radium C present in 1 gramme of radium in equilibrium. From the experiments of Bragg, and the measurements by Boltwood of the ionisation due to the α-particles from each of the products of radium, it appears certain that the same number of α-particles are expelled per second from radium itself and from each of its α-ray products in equilibrium with it.

It follows that 1 gramme of radium itself and each of its ray products in equilibrium with it expels $3 \cdot 4 \times 10^{10}$ α-particles per second. The total number of α-particles emitted per second per gramme of radium in equilibrium with its three α-ray products is $13 \cdot 6 \times 10^{10}$. Taking as the simplest and most probable assumption that one atom of radium in breaking up emits one α-particle, it follows that in 1 gramme of radium $3 \cdot 4 \times 10^{10}$ atoms break up per second.

Counting of Scintillations

It is of importance to compare the number of scintillations produced on a zinc sulphide screen with the number of α-particles counted by the electric method, in order to see whether each scintillation is due to a single α-particle. For this purpose the special zinc sulphide screens provided by Mr. F. H. Glew were used. A thin layer of zinc sulphide is spread over a thin glass plate, and the scintillations produced on the screen are readily seen through the glass by means of a microscope. In order to make the comparison as direct as possible, the same firing tube and aperture, covered with the mica screen, were used. The brass detecting tube was removed, and a small piece of zinc sulphide screen was attached to the end of the glass tube D (fig. 1), with its active surface towards the firing tube. Radium C served as a source of α-rays, as in the electrical method.

Regener* has made a number of observations upon the number of α-particles expelled from an active preparation of polonium by the scintillation method. He has investigated the best conditions for viewing the scintillations and the relation between the focal lengths of the eye-piece and objective to obtain the maximum illumination due to each α-particle. We have found his suggestions very useful in these experiments.

In our experiments a microscope of magnification 50 was used. The small area of screen, struck by the α-particles, covered only about one-half of the field of view. The experiments were made at night in a dark room. As

Diameter of aperture, 1·23 mm. with mica covering. Distance of active source from aperture, 200 cm.

I Calculated number of α-particles per minute	II Observed number of scintillations per minute	III Ratio of observed to calculated number
39	31	0·80
38	49	1·29
34	29	0·85
32	31	0·97
31	32	1·03
28	27	0·96
27	28	1·04
25	21	0·84
23	25	1·09
21	21	1·00
Total number = 294		Average = 0·99

* 'Verh. d. D. Phys. Ges.,' vol. 10, p. 78, 1908.

Regener suggests, it is advisable to illuminate the screen slightly by artificial light, in order to keep the eye focussed on the screen. The distance and intensity of the source were adjusted so that from 20 to 60 scintillations were observed per minute. It is difficult to continue counting for more than two minutes at a time, as the eye becomes fatigued. The zinc sulphide screen usually showed a few scintillations per minute with the stop-cock closed, due to natural radio-activity and other disturbances. These were counted before and after each experiment, and were subtracted from the number counted with the stop-cock open. It was usual to count 100 scintillations and to note the time with a stop-watch. The results of a series of observations for varying intensities of the radiations are given in the table on page 104. The corrected number of scintillations observed per minute is given in Column 2. Taking $3 \cdot 4 \times 10^{10}$ as the number of α-particles expelled per second per gramme, the calculated number of scintillations to be expected from the intensity of the radiation, if each α-particle produces a scintillation, is given in Column 1. The ratio of the observed to the calculated number is given in Column 3.

Another series of observations was made with a fresh piece of zinc sulphide screen with an aperture $3 \cdot 02$ times area of the first and without mica screen.

Calculated number of α-particles per minute	Observed number of scintillations per minute	Ratio of observed to calculated number
36	31	$0 \cdot 86$
34	30	$0 \cdot 88$
31	31	$1 \cdot 00$
30	29	$0 \cdot 97$
27	29	$1 \cdot 07$
Total number $= 150$		Average $= 0 \cdot 96$

Considering the probability error, the agreement between the electrical and optical methods of counting is, no doubt, closer than one would expect. The result, however, brings out clearly that within the limit of experimental error, each α-particle produces a scintillation on a properly prepared screen of zinc sulphide. The agreement of the two methods of counting the α-particles is in itself a strong evidence of the accuracy obtained in counting the α-particles expelled per gramme of radium by the electrical method. It is now clear that we have two distinct methods, one electrical and the other optical, for detecting a single α-particle, and that the employment of either method may be expected to give correct results in counting the number of α-particles.

D*

Since there is every reason to believe that an α-particle is an atom of helium, there are now two distinct methods of detecting the expulsion of a single helium atom, one depending on its electrical effect, the other upon the luminosity produced in crystals of zinc sulphide. It is not necessary here to enter upon a discussion of the mechanism of production of a scintillation. In a previous paper, one of us* has pointed out that there is strong reason to suppose that the molecules of the phosphorescent preparation are dissociated by the α-particle, and that the luminosity observed may accompany either this dissociation, or the consequent recombination of the dissociated parts.

Probability Error

We have previously drawn attention to the fact that the number of α-particles entering a given opening in a given time is conditioned by the laws of probability. E. v. Schweidler† first drew attention to the fact that, according to the theory of probability, the number of α-particles expelled per second from radio-active matter must be subject to fluctuations within certain limits. If z is the average number of atoms of active matter breaking up per second, the average error to be expected in the number is $\sqrt{z}$. The existence of fluctuations in radiations from active matter of the magnitude to be expected on this theory have been shown by the experiments of Kohlrausch,‡ Meyer and Regener,§ and Hans Geiger.‖

In most of the experiments in this paper, an intense source of α-radiation has been used. If, for example, the source had a γ-ray activity equal to 1 milligramme of radium, the average number of α-particles expelled per second is $3 \cdot 4 \times 10^7$. The error to be expected is thus 5830 particles, and the relative error $\sqrt{z}/z$ is 1/5830.

In such a case, we may consider the source as a whole to emit α-particles at a practically constant rate. The probability variation in the number is beyond the limit of detection by ordinary methods. The case, however, is quite different when we consider the number of α-particles entering a small opening at a distance from the source. In the experiments, the number entering the detecting vessel varied between two and six per minute, and the number of α-particles counted in a single experiment varied from 20 to 60. Assuming, for simplicity, that the general theory applies to this case, the probable variation of the observed number from the true mean is equal to $\sqrt{z}$. This amounts to four or five particles for a number 20 and between seven and eight for a number 60. It is not easy to compare accurately theory and experiment in this way, but there is no doubt that the observed variations are of the same order of magnitude as those to be anticipated from the laws of probability.

* Rutherford, 'Phil. Mag.,' July, 1905. (*Vol. I, p. 803.*)
† Schweidler, Congrès International pour l'Étude de la Radiologie, Liège, 1905.
‡ Kohlrausch, 'Wien. Ber.,' p. 673, 1906.
§ Meyer and Regener, 'Ver. d. D. Phys. Ges.,' No. 1, 1908.
‖ Geiger, 'Phil. Mag.,' April, 1908.

Some experiments have been made, both by the electric and scintillation methods, to determine the distribution of the α-particles in time. For this purpose, a thin film of radium was used as a source of rays. A large number of α-particles was counted, the interval between successive entrances of the α-particles in the detecting vessel being noted. A curve is then plotted, the ordinates representing the number of α-particles and the abscissæ the corresponding time intervals between the entrance of successive α-particles. A curve is obtained like that shown in fig. 4, which is similar in general shape

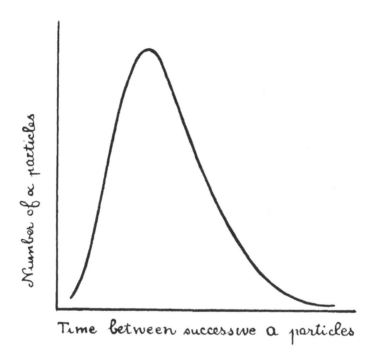

Fig. 4

to the probability-curve of distribution in time. Further experiments are in progress to determine the distribution-curve as accurately as possible, in order to compare theory with experiment.

Summary of Results

(1) By employing the principle of magnification of ionisation by collision, the electrical effect due to a single α-particle may be increased sufficiently to be readily observed by an ordinary electrometer.

(2) The magnitude of the electrical effect due to an α-particle depends upon the voltage employed, and can be varied within wide limits.

(3) This electric method can be employed to count the α-particles expelled from all types of active matter which emit α-rays.

(4) Using radium C as a source of α-rays, the total number of α-particles

expelled per second from 1 gramme of radium have been accurately counted. For radium in equilibrium, this number is $3 \cdot 4 \times 10^{10}$ for radium itself and for each of its three α-ray products.

(5) The number of scintillations observed on a properly-prepared screen of zinc sulphide is, within the limit of experimental error, equal to the number of α-particles falling upon it, as counted by the electric method. It follows from this that each α-particle produces a scintillation.

(6) The distribution of the α-particles in time is governed by the laws of probability.

We have previously pointed out that the principle of magnification of ionisation by collision can be used to extend widely our already delicate methods of detection of radio-active matter. Calculation shows that under good conditions it should be possible by this method to detect a single β-particle, and consequently to count directly the number of β-particles expelled from radio-active substances.

Further work is in progress on this and other problems that have arisen out of these investigations.

The Charge and Nature of the α-Particle

by PROFESSOR E. RUTHERFORD, F.R.S.,

and HANS GEIGER, PH.D., *John Harling Fellow, University of Manchester*

From *The Proceedings of the Royal Society*, A, Vol. 81, 1908, pp. 162–73
(Read June 18; MS. received July 17, 1908)

IN the previous paper, we have determined the number of α-particles expelled per second per gramme of radium by a direct counting method. Knowing this number, the charge carried by each particle can be determined by measuring the total charge carried by the α-particles expelled per second from a known quantity of radium. Since radium C was used as a source of radiation in the counting experiments, it was thought desirable to determine directly the charge carried by the α-particles expelled from this substance. In a paper some years ago*, one of us has investigated the experimental conditions necessary for an accurate determination of the total charge carried by the α-rays, and has measured the charge carried by the α-particles expelled from a thin film of radium itself. In the present experiments the same general method has been used, with certain modifications, rendered necessary by the choice of radium C as a source of α-rays.

The experimental arrangement is clearly seen in fig. 1. A cylindrical glass tube HH of diameter 4 cm. is closed at the ends by ground-glass stoppers D, E. The source of radiation R is attached to the lower stopper E. The radiation from this passes into the testing chamber, which is rigidly attached to the stopper D by means of an ebonite tube F. The testing chamber consists of two parallel plates A and B about 2 mm. apart. A circular opening, 1·92 cm. in diameter, cut in the brass plate B, is covered by a sheet of thin aluminium foil. The upper chamber AC consists of a shallow brass vessel of aperture 2·5 cm., the lower surface of which is covered also with a sheet of aluminium foil.† The plate B is connected through a side glass tube to one terminal of a battery, the other pole of which is earthed. The chamber AC, which is insulated from the plate B, is connected with one pair of quadrants of a Dolezalek electrometer, the other of which is earthed. The whole apparatus is placed between the poles NS of a large electromagnet marked by the dotted lines in the figure, so that the α-rays in their passage from the source R to the testing chamber pass through a strong magnetic field.

When the active matter was placed in position, the apparatus was

---

* Rutherford, 'Phil. Mag.,' August. 1905. (*Vol. I, p. 816.*)
† The stopping power of each aluminium foil corresponded to about 5 mm. of air.

exhausted by means of a Fleuss pump. The evacuation was then completed by means of a tube of cocoanut charcoal immersed in liquid air. A very low vacuum is required in these experiments in order to reduce the ionisation of the residual gas by the α-rays to as low a value as possible. If this is not

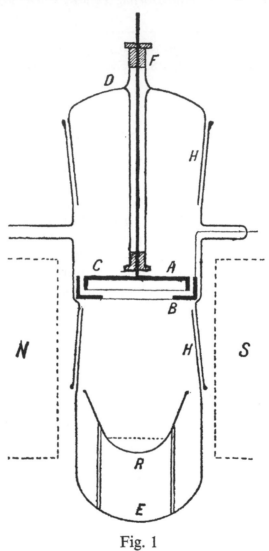

Fig. 1

done, the positive charge communicated to the upper plate by the absorption of the α-particles may very rapidly leak away. In addition to the production of a high vacuum, it is necessary to place the testing chamber in a strong magnetic field. It is well known that the α-particles, in their passage through matter, liberate a large number of slow-velocity electrons, or δ-rays, as they have been termed by J. J. Thomson. The presence of a large number of these negatively-charged particles impinging on the testing chamber completely masks the effect of the positive charge carried by the α-particles. By placing the testing chamber in a strong magnetic field, these slow-moving particles describe very small orbits, and return to the surface from which they were emitted. In this way the disturbing effect due to the δ-rays may be completely eliminated. On account of their very small velocity (about 10^8 cm./sec.) and small mass, a magnetic field of only moderate intensity is required for the purpose. It will be observed that the α-particles are not fired directly into the upper plate AC, but pass first through a thin layer of aluminium foil. This arrangement was adopted in order to diminish as much as possible the number of δ-particles set free in the space between the electrodes. The α-rays pass readily through the thin layer of aluminium at the base of the vessel AC, and are completely stopped by the upper plate. The large number of δ-particles emitted from the plate AC by the impact of the α-rays cannot penetrate back through the aluminium foil, and consequently do not disturb the measurements. It is then only necessary, with the aid of the magnetic field, to get rid of the disturbance due to the δ-rays emitted from the two layers of aluminium foil.

In the present experiments the magnetic field served also for another purpose. Radium C emits β- as well as α-rays, and, in the absence of a magnetic field, these also would be partly absorbed, and give up their negative charges to the upper plate. In the experimental arrangement the magnetic field extended from the source R beyond the testing chamber. The source of radiation was placed about 3·5 cm. below the testing chamber. The strength of the magnetic field was then adjusted, so the β-particles were bent completely away from the lower plate and consequently did not produce any effect in the testing chamber. It was essential for this purpose that the source of radiation was some distance below the plate B, so that the strength of magnetic field obtainable under the experimental conditions and the length of path of the rays were together sufficient to ensure the complete deflection of the β-particles to one side of the glass tube before reaching the plate B.

As the source of radiation was some distance below the testing chamber, it was necessary to use a very active surface of radium C, in order to obtain a reasonably large effect for measurement. For this purpose a small shallow glass cap, represented by the source R in the figure, was attached by a ground-glass joint to a glass tube about 8 cm. long. This was filled with mercury, and the emanation from about 40 milligrammes of radium introduced by the aid of a mercury trough to the top of the cap. The level of the mercury below the top of the cap is represented by the dotted line in the figure. The emanation was left in the cap for more than three hours, when the amount of radium C deposited on the interior walls of the glass cap and on the surface of the mercury reaches its maximum value. By means of the mercury trough, the emanation was then rapidly displaced, the mercury run out, and the cap removed from the glass tube. The inner surface of the cap was washed first with water and then with alcohol, to remove any trace of grease on the inside of the glass. The inner surface of the cap then acted as a source of intense α-radiation. Fifteen minutes after removal the α-radiation is homogeneous, and due entirely to radium C. The active glass cap was then placed in position in the testing vessel, which was then rapidly exhausted by a Fleuss pump. The cocoanut charcoal was then immersed in liquid air and a low vacuum reached in a short time. Usually an interval of 15 to 30 minutes after the removal of the emanation was required for the various operations and to obtain a sufficiently low vacuum for measurements to be started.

In order to determine the amount of radium C deposited in the glass cap, observations of its activity were made by the γ rays *in situ*. For this purpose a γ-ray electroscope was placed some distance on one side of the apparatus, and the rate of discharge observed at intervals during the experiment. The electroscope was standardised in the usual way by means of the standard preparation of radium placed at the same distance as the source R from the electroscope, so that the amount of radium C distributed on the source at any time was determined in terms of the amount in equilibrium with a definite quantity of radium. Such measurements with the γ-rays can be very simply and accurately made, and, with suitable precautions, the error of observation should not be greater than 1 per cent.

Method of Calculation

Using a strong magnetic field, the upper plate received a positive charge, whether the lower plate was charged positively or negatively. The current was first measured with the lower plate charged to a potential $+ V$, and then with the same plate at a potential $- V$. Let i_1 be the current observed in the first case and i_2 in the second case; i_2 is always numerically less than i_1, the ratio depending upon the degree of exhaustion. Let i_0 be the current through the gas due to the ionisation of the residual gas between the plates by the α-rays. Then

$$i_1 = i_0 + nE, \tag{1}$$

where n is the number of α-particles passing into the upper plate per second and E the charge on each. On reversing the voltage, the ionisation current is equal in magnitude but reversed in its direction.

Consequently $$i_2 = nE - i_0 \tag{2}$$

Adding (1) and (2), $$nE = \tfrac{1}{2}(i_1 + i_2).$$

Let Q be the quantity of radium C present at any instant measured in terms of the γ-ray effect due to 1 gramme of radium, and N the number of α-particles of radium C expelled per second and per gramme of radium. The total number of α-particles expelled per second from the source R is QN. Let K be the fraction of the total number of α-particles expelled from the

Rutherford (*on right*) and Geiger with their apparatus for counting α particles; this appears to be the later apparatus of 1912

source which impinge on the upper plate. Then $n = KQN$, where K and Q are measured, and N is known from the counting experiments. Consequently the charge E on each α-particle is given by

$$E = (i_1 + i_2)/2KQN.$$

In preliminary experiments, it was found that the values of i_1 and i_2 were independent of the voltage over the range examined, viz., from 2 to 8 volts. In most of the latter measurements an E.M.F. of ± 2 volts was used. It was found experimentally that the value of $\frac{1}{2}(i_1 + i_2)$ was independent of the strength of the magnetic field beyond a certain limit. For example, an increase of the current in the electromagnet from 10 to 20 amperes made no alteration in the magnitude of i_1 or i_2. A current of 6 amperes gave distinctly smaller values, due to the fact that the strength of the field was not sufficient to bend away all the β-particles completely. In all the final experiments an exciting current of 12 amperes was used. The electromagnet and electrometer connections were well screened and the electrometer readings were remarkably steady. The external γ-ray effect due to the intense source of radiation was screened off as far as possible by plates of thick lead. The apparatus was placed some distance from the electrometer, the insulated connecting wire passing through a long brass tube connected with earth. Notwithstanding these precautions, it was impossible, in consequence of the ionisation due to the γ-rays, to avoid a small back leak of the electrometer system as its potential rose. This was easily corrected for in each observation by observing the rate of movement of the needle over each succeeding 10 divisions of the scale until a deflection of over 150 divisions was reached. The fraction K of the total number of α-particles striking the upper plate was determined on the assumption that the α-particles are emitted equally in all directions. The correctness of this assumption has been verified in other experiments. The distance of the radiant source from the lower plate was determined when in position by a cathetometer. The correction due to the fact that the radiation came from a source of sensible area was determined graphically by dividing up the surface into concentric rings and determining the value of K for each. In the experiments given below the mean value of K was $0 \cdot 0172$. The value of N, as determined by the counting experiments, is $3 \cdot 4 \times 10^{10}$. The following tables illustrate the results obtained in two distinct series of experiments. [See page 114.]

Column I gives a number of successive sets of observations of the values of i_1 and i_2; II, the mean intensity of the γ-ray radiation during the experiment in terms of a milligramme of pure radium; III, the capacity of the electrometer systems in cms.; IV and V, the values of i_1 and i_2 expressed in terms of the number of divisions of the scale moved over by the electrometer needle per second; VI, the mean of i_1 and i_2, also expressed in scale divisions per second; VII, the calculated value of E—the charge on the α-particle—in electrostatic units. The mean value of E in each complete experiment is obtained by giving a weight to each determination of E equal to number of observations of i_1 and i_2. It will be seen that the mean value of E from experiment I is $9\cdot2 \times 10^{-10}$, and from experiment II $9\cdot4\times10^{-10}$. Taking the mean of these, the value of E becomes $9\cdot3\times10^{-10}$. We thus conclude that the positive charge E carried by an α-particle from radium C is $9\cdot3\times10^{-10}$ E.S. units.

Experiment I

I	II	III	IV	V	VI	VII
No. of observations	Intensity of radiation	Capacity	i_1	i_2	$\frac{1}{2}(i_1+i_2)$	E
1	21·0 mg. Ra	495 cms.	2·24 divs./sec.	1·75 divs./sec.	1·99	$8\cdot8\times10^{-10}$
1	18·5 ,,	495 ,,	1·74 ,,	1·55 ,,	1·68	$8\cdot3\times10^{-10}$
2	13·9 ,,	495 ,,	1·61 ,,	1·27 ,,	1·44	$9\cdot2\times10^{-10}$
1	11·4 ,,	495 ,,	1·31 ,,	1·07 ,,	1·19	$9\cdot6\times10^{-10}$
1	10·6 ,,	495 ,,	1·19 ,,	0·92 ,,	1·05	$9\cdot1\times10^{-10}$
2	6·98 ,,	495 ,,	0·856 ,,	0·706 ,,	0·78	$10\cdot0\times10^{-10}$
2	3·08 ,,	146 ,,	1·11 ,,	0·87 ,,	0·99	$8\cdot7\times10^{-10}$
				Mean value		$9\cdot2\times10^{-10}$

Experiment II

I	II	III	IV	V	VI	VII
No. of observations	Intensity of radiation	Capacity	i_1	i_2	$\frac{1}{2}(i_1+i_2)$	E
2	16·1 mg. Ra	495 cms.	1·90 divs./sec.	1·47 divs./sec.	1·68	$9\cdot3\times10^{-10}$
1	14·8 ,,	495 ,,	1·63 ,,	1·28 ,,	1·45	$9\cdot1\times10^{-10}$
2	10·7 ,,	304 ,,	2·06 ,,	1·84 ,,	1·95	$10\cdot0\times10^{-10}$
1	9·8 ,,	304 ,,	1·85 ,,	1·40 ,,	1·62	$9\cdot9\times10^{-10}$
2	6·32 ,,	146·5 ,,	2·22 ,,	1·83 ,,	2·02	$8\cdot7\times10^{-10}$
1	5·16 ,,	146·5 ,,	2·02 ,,	1·46 ,,	1·74	$9\cdot1\times10^{-10}$
				Mean value		$9\cdot4\times10^{-10}$

From other data it is known that the α-particles from all radio-active products which have been examined are identical. Consequently, we may conclude that each α-particle, whatever its source, under normal conditions carries the above charge.

Comparison of the Charge carried by an α-Particle and a Hydrogen Atom

The charge carried by an ion in gases has been determined by a number of observers. Townsend,* from observations on the electrified gas liberated by the electrolysis of oxygen, concluded that each particle carried a charge of about 3×10^{-10} E.S. unit. Measurements of the charge carried by an ion in gases have been made by J. J. Thomson,† H. A. Wilson,‡ Millikan and Begeman,§ using the now well-known method of causing a deposition of water on each ion by a sudden expansion. The final value of e obtained by J. J. Thomson was $3 \cdot 4 \times 10^{-10}$ unit, by Wilson $3 \cdot 1 \times 10^{-10}$, and by Millikan $4 \cdot 06 \times 10^{-10}$.

From the values found by these experimenters, it will be seen that the value E of the charge carried by an α-particle ($9 \cdot 3 \times 10^{-10}$ unit) is between $2e$ and $3e$. On the general view that the charge e carried by an hydrogen atom is the fundamental unit of electricity, we conclude that the charge carried by an α-particle is an integral multiple of e and may be either $2e$ or $3e$.

We shall now consider some evidence based on radio-active data, which indicates that the α-particle carries a charge $2e$ and that the ordinarily accepted values of e are somewhat too small.‖

First Method.—We shall first of all calculate the charge E carried by an α-particle on the assumption that the heating effect of radium is a measure of the kinetic energy of the α-particles expelled from it. There is considerable indirect evidence in support of this assumption, for it is known that the heating effect of the β- and γ-rays together is not more than a few per cent. of that due to the α-rays. If m be the mass of an α-particle and u its initial velocity of projection, the kinetic energy of the α-particle

$$= \tfrac{1}{2}mu^2 = \tfrac{1}{2}\,\frac{mu^2}{\mathrm{E}} \cdot \mathrm{E}.$$

Now, in a previous paper,¶ one of us has accurately determined, from the electrostatic deflection of the α-rays, the values of $\tfrac{1}{2}\dfrac{mu^2}{\mathrm{E}}$. E for each of the four sets of α-particles expelled from radium in equilibrium, and has shown

---

* Townsend, 'Phil. Mag.,' February, 1898; March, 1904.
† J. J. Thomson, 'Phil. Mag.,' March, 1903.
‡ H. A. Wilson, 'Phil. Mag.,' April, 1903.
§ Millikan and Begeman, 'Phys. Rev.,' Feb., 1908, p. 197.
‖ In a recent paper, Regener ('Verh. d. D. Phys. Ges., vol. 10, p. 78, 1908) has deduced from indirect data that an α-particle carries a charge $2e$. The number of scintillations from a preparation of polonium were counted and assumed to be equal to the number of α-particles emitted. A comparison was then made with the number of α-particles deduced from measurements of the ionisation current, and from the data given by Rutherford of the number of ions produced by an α-particle.
¶ Rutherford, 'Phil. Mag.,' October, 1906. (*Vol. I, p. 880.*)

that the kinetic energy of the α-particles from 1 gramme of radium in equilibrium is $4 \cdot 15 \times 10^4 NE$ ergs,* where N is the number of radium atoms breaking up per second.

Now the heating effect of the standard preparation of radium was 110 gramme-calories per gramme per hour. This is mechanically equivalent to $1 \cdot 28 \times 10^6$ ergs per second. Equating the kinetic energy of the α-particles to the observed heating effect,

$$4 \cdot 15 \times 10^4 NE = 1 \cdot 28 \times 10^6.$$

Substituting the known value $N = 3 \cdot 4 \times 10^{10}$,

$$E = 9 \cdot 1 \times 10^{-10} \text{ E.S. unit.}$$

The agreement of the calculated with the observed value is somewhat closer than one would expect, taking into consideration the uncertainty of the data within narrow limits.

Second Method.—We shall now calculate the charge e carried by a hydrogen atom from the known period of transformation of radium. As a result of a series of experiments, Boltwood †has shown that the period of transformation of radium can be very simply measured. He concludes that radium is half transformed in 2000 years. Let P be the number of hydrogen atoms present in 1 gramme of hydrogen. Then the number of atoms of radium present in 1 gramme of radium is P/226, since, according to the latest determinations the atomic weight of radium is about 226. If λ is the transformation constant of radium, the number of atoms breaking up per second per gramme of radium is $\lambda P/226$. On the probable assumption that each atom breaks up with the expulsion of one α-particle, this is equal to the number N of α-particles expelled per second per gramme. The value of N from the counting experiments is $3 \cdot 4 \times 10^{10}$, consequently $\lambda P/226 = 3 \cdot 4 \times 10^{10}$. From data of the electrolysis of water, it is known that

$$Pe = 9 \cdot 6 \times 10^4 \text{ electromagnetic units,}$$
$$= 2 \cdot 88 \times 10^{14} \text{ E.S. units,}$$

where e is the charge carried by the hydrogen atom. Dividing one equation by the other, and substituting the value of $\lambda = 1 \cdot 09 \times 10^{-11}$ deduced from Boltwood's measurements, we have $e = 4 \cdot 1 \times 10^{-10}$ E.S. unit.

This is a novel method of determining e from radio-active data. If two α-particles instead of one are expelled during the breaking up of the radium atom, the value of e is twice the above value, or $8 \cdot 2 \times 10^{-10}$. This is a value more than twice as great as that determined by other methods, and is inadmissible.

* The value of E in the original paper is given in electromagnetic units. For uniformity, it is reduced here to electrostatic units.

† Boltwood, 'Amer. Journ. Sci.,' June, 1908.

Discussion of the Accuracy of the Methods of Determination of e.

We have found, experimentally, that the α-particle carries a positive charge E of $9 \cdot 3 \times 10^{-10}$ unit. If the α-particle has a charge equal to $2e$, the value of e, the charge on a hydrogen atom, becomes $4 \cdot 65 \times 10^{-10}$. This is a somewhat higher value than those found in the measurements of J. J. Thomson, H. A. Wilson and Millikan. It is also somewhat greater than the value deduced above from considerations based on the life of radium. As an accurate knowledge of the value of e is now of fundamental importance, we shall briefly review some considerations which indicate that the values of e found by the old methods are probably all too small. It is far from our intention to criticise in any way the accuracy of the measurements made by such careful experimenters, but we merely wish to draw attention to a source of error which was always present in their experiments, and which is exceedingly difficult to eliminate. In the experiments referred to, the number of ions present in the gas are deduced by observing the rate of fall of the ions when water has been condensed upon them by an adiabatic expansion. It is assumed that there is no sensible evaporation of the drops during the time of observation of the rate of fall. There is no doubt, however, that evaporation does occur, and that the diameter of the drops steadily decreases. A little consideration of the methods of calculation used in the experiment shows that the existence of this effect gives too large a value for the number of ions present, and, consequently, too small a value of e. The correction to be applied for this effect is no doubt a variable, depending upon the dimensions of the expansion vessel and other considerations. If the error due to this effect were about 30 per cent. in the experiments of J. J. Thomson and H. A. Wilson, and 15 per cent. in the experiments of Millikan, the corrected value of e would agree with the value $4 \cdot 65 \times 10^{-10}$ deduced from measurements of the charge carried by an α-particle.

The determination of $e = 4 \cdot 1 \times 10^{-10}$ from the period of transformation of radium is for other reasons probably also too small. The method adopted by Boltwood is very simple, and involves only the comparison of two quantities of radium by the emanation method. Suppose that we take a quantity of an old mineral containing 1 gramme of uranium and determine by the emanation method the quantity R of radium present. Since the uranium is in equilibrium with ionium—the parent of radium—and radium itself, the rate of production q of radium by the disintegration of its parent ionium must be equal to the rate of disintegration λR of radium itself. Now by chemical methods the ionium is separated from the mineral and the rate of growth q of radium from it determined. Consequently, $q = \lambda R$, or $\lambda = q/R$. The ratio q/R can be determined with considerable accuracy by the emanation method and does not involve any consideration of the purity of the radium standard. As Boltwood points out, the accuracy of the method of determination is mainly dependent upon the completeness of the separation of ionium from the mineral. If all the ionium is not separated, the value of λ is too small and the period of transformation consequently too long. For example, if 10 per cent. of the ionium had remained unseparated in the experiments, the period of radium would be 1800 years instead of 2000, and the charge carried by the hydrogen atom calculated from this data would be nearly $4 \cdot 6 \times 10^{-10}$ instead of $4 \cdot 1 \times 10^{-10}$.

Considering the data as a whole, we may conclude with some certainty that the α-particle carries a charge $2e$, and that the value of e is not very different from $4 \cdot 65 \times 10^{-10}$ E.S. unit.*

Atomic Data

We have seen that the method of counting the α-particles and measuring their charge has supplied a new estimate of the charge carried by the α-particle and the charge carried by a hydrogen atom. The atomic data deduced from this are for convenience collected below:—

Charge carried by a hydrogen atom $= 4 \cdot 65 \times 10^{-10}$ E.S. unit.
Charge carried by an α-particle $= 9 \cdot 3 \times 10^{-10}$ E.S. unit.
Number of atoms in 1 gramme of hydrogen $= 6 \cdot 2 \times 10^{23}$.
Mass of the hydrogen atom $= 1 \cdot 61 \times 10^{-24}$ gramme.
Number of molecules per cubic centimentre of any gas at standard pressure and temperature $= 2 \cdot 72 \times 10^{19}$.

Nature of the α-Particle

The value of E/M—the ratio of the charge on the α-particle to its mass— has been measured by observing the deflection of the α-particle in a magnetic and in an electric field, and is equal to $5 \cdot 07 \times 10^3$ on the electromagnetic system.† The corresponding value of e/m for the hydrogen atom set free in the electrolysis of water is $9 \cdot 63 \times 10^3$. We have already seen that the evidence is strongly in favour of the view that E $= 2e$. Consequently M $= 3 \cdot 84m$, *i.e.*, the atomic weight of an α-particle is $3 \cdot 84$. The atomic weight of the helium atom is $3 \cdot 96$. Taking into account probable experimental errors in the estimates of the value of E/M for the α-particle, we may conclude that *an α-particle is a helium atom*, or, to be more precise, *the α-particle, after it has lost its positive charge, is a helium atom*.

Some of the consequences of this conclusion have already been discussed some time ago in some detail by one of us.‡ It suffices to draw attention here to the immediate deduction from it of the atomic weight of the various products of radium. There is direct evidence in the case of radium that each of the α-ray changes is accompanied by the expulsion of one α-particle from each atom. Consequently, since the atomic weight of radium is 226, the

---

* It is of interest to note that Planck deduced a value of $e = 4 \cdot 69 \times 10^{-10}$ E.S. unit from a general optical theory of the natural temperature-radiation.

† Rutherford, 'Phil. Mag.,' October, 1906. (*Vol. I, p. 889.*)

‡ 'Radio-activity,' 2nd Edition, pp. 479—486; 'Radio-active Transformations,' Chapter VIII.

atomic weight of the emanation is 222 and of radium A 218. Our information is at present too scanty to decide with certainty whether a mass equal or comparable with that of an α-particle is expelled in the β-ray or rayless changes.

It is of interest to note that a recent determination by Perkins* of the molecular weight of the emanation from a comparison of its rate of diffusion with that of mercury vapour gives a value 235. The earlier estimates of the molecular weight from diffusion data were much lower, but more weight is to be attached to the recent value since mercury, like the emanation, is monatomic, and has an atomic weight comparable with it.

Calculation of Radio-active Data.

We are now in a position to calculate the magnitude of some important radio-active quantities.

(1) *The Volume of the Emanation.*—One atom of radium, in breaking up, emits one α-particle and gives rise to one atom of emanation of atomic mass 222. Since $3 \cdot 4 \times 10^{10}$ α-particles are expelled per second per gramme of radium, the number of atoms of emanation produced per second is the same. Now we have shown that there are $2 \cdot 72 \times 10^{19}$ molecules in 1 c.c. of any gas at standard pressure and temperature. The volume of the emanation produced per second per gramme is consequently $1 \cdot 25 \times 10^{-9}$ c.c. The maximum volume is equal to the rate of production divided by the value of the radio-active constant λ, which is equal to 1/468000. The maximum volume of the emanation from 1 gramme of radium is consequently $0 \cdot 585$ cubic mm.

(2) *Rate of Production of Helium.*—Since an α-particle is an atom of helium, the number of atoms of helium produced per second per gramme of radium in equilibrium is $4 \times 3 \cdot 4 \times 10^{10}$. The factor 4 is introduced, since there are 4 α-ray products in radium in equilibrium, each of which emits the same number of α-particles per second. Consequently the volume of helium produced per gramme is $5 \cdot 0 \times 10^{-9}$ c.c. per second, which is equal to $0 \cdot 43$ cubic mm. per day, or 158 cubic mms. per year. An accurate experimental determination of the rate of production of helium by radium would be of great interest.

(3) *Heating Effect of Radium.*—If the main fraction of the heat emission of radium is a result of the kinetic energy of the expelled α-particles, its value can at once be calculated. The converse problem has already been discussed earlier in the paper. It will be seen from the numbers there given that the heat emission of radium should be slightly greater than 113 gramme-calories per gramme per hour.

(4) *Life of Radium.*—From the inverse problem discussed earlier, this works out to be 1760 years, supposing the charge on a hydrogen atom equals $4 \cdot 65 \times 10^{-10}$.

* Perkins, 'Amer. Journ. Sci.,' June, 1908.

For convenience, the calculated values of some of the more important radio-active quantities are given below:—

Charge on an α-particle $= 9 \cdot 3 \times 10^{-10}$ E.S. unit.

Number of α-particles expelled per gramme of radium itself $= 3 \cdot 4 \times 10^{10}$.

Number of atoms of radium breaking up per second $= 3 \cdot 4 \times 10^{10}$.

Volume of emanation per gramme of radium $= 0 \cdot 585$ cubic mm.

Production of helium per gramme of radium per year $= 158$ cubic mms.

Heating effect per gramme of radium $= 113$ gramme-calories per hour.

Life of radium $= 1760$ years.

Calculations of the magnitude of a number of other radio-active quantities can be readily made from the experimental data given in this paper. For lack of space we shall not refer to them here.

The Nature and Charge of the α Particles from Radio-active Substances

From *Nature*, **79,** 1908, pp. 12–15

THE development of our knowledge of radio-activity has emphasised the primary importance of the α particles, which are projected in great numbers from most of the active substances. As Rutherford showed in 1903, the α particles are veritable atoms of matter which are ejected from radio-active matter at a speed of about 10,000 miles per second. The great number of α particles which are projected from radium is well illustrated by the multitude of scintillations observed when the α particles from a trace of radium fall on a screen of zinc sulphide. We shall see later that 136 million α particles are expelled every second from one milligram of radium in radio-active equilibrium. From the point of view of modern theory, the appearance of an α particle is the sign of a violent atomic explosion in which a fragment of the atom—an α particle—is ejected at a high speed. In the majority of the known active substances, the expulsion of an α particle accompanies the transformation of one substance into another, and the decrease of atomic mass consequent upon the loss of an α particle at once offers a reasonable explanation of the appearance of an entirely new kind of matter in place of the old.

Space does not allow us here to discuss the very interesting facts that have been brought to light by the work of Bragg and Kleeman and others in regard to the character of the absorption of the α particle by matter. It suffices to say that it has been found that the α particles from one kind of active matter are all projected initially at an identical speed, but that this initial velocity varies within comparatively narrow limits for different kinds of matter. The α particle, in consequence of its great energy of motion, plunges through the molecules of matter in its path, leaving in its train a large number of dissociated or ionised molecules. Some important questions at once arose when it was found that the α particle was an atom of matter of mass comparable with the hydrogen atom, viz., Are the α particles expelled from different kinds of matter identical in constitution, and are the α particles atoms of a known element or some new kind of matter?

These problems were attacked by determining the velocity and the value of E/M—the ratio of the charge carried by an α particle to its mass—of α particles expelled from different kinds of matter. These quantities can be determined by measuring the deflection of a pencil of α rays when passing through strong magnetic and electric fields. Experiments of this kind, which are difficult on account of the small deflection of the α rays under normal experimental conditions, have been made by Rutherford, Des Coudres, Mackenzie, and Huff. The former determined the velocity and value of E/M for each of a number of products of radium and actinium, while Rutherford and Hahn made similar measurements for some of the products of thorium. The results were of great interest, for while it was found that the initial velocity of projection of the α particles from different kinds of matter varied from about 14,000 to 10,000 miles per second, the value of E/M was the same for all. This shows that the α particle. whether expelled from radium, thorium, or actinium, is identical in mass and constitution, and that all the radio-active substances which emit α particles have a common product of disintegration. As the result of a number of experiments, Rutherford found that the value E/M for the α particle was 5070 in electromagnetic units. Now, from experiments on the electrolysis of water, it is known that the corresponding value of *e/m* for the hydrogen atom is 9600, or nearly twice as large. The charge *e* carried by the H atom is believed to be the fundamental unit charge of electricity, so that the charge carried by any body must be an integral multiple of *e*. If we suppose the charge carried by an α particle is equal to the charge carried by an hydrogen atom, the mass of the α particle is, in round numbers, twice that of the hydrogen atom, *i.e.* is equal to the molecule of hydrogen. If, however, we suppose that E $=$ 2*e*, *i.e.* the α particle carries two unit charges, the mass of the α particle is equal to about four. Now, it is known that the atomic mass of helium is 3·96 in terms of hydrogen, so that on this supposition the α particle would appear to be an atom of helium carrying two unit charges. We must now consider some indirect evidence bearing on the question. As the result of the experiments of Ramsay and Soddy and others, it is now well substantiated that helium is produced from radium. Debierne has shown that helium is produced also from actinium. Unless the helium is the result of the accumulated α particles, it is difficult to account for the production of the helium observed. In addition, as we have shown, the α particle is the only known common product of the disintegration of radium and actinium, which both give rise to helium. For these and other reasons, Rutherford suggested in 1905 that it was very probable that the α particle was an atom of helium carrying two unit charges. It has been found exceedingly difficult experimentally either to prove or disprove the correctness of this hypothesis, although the settlement of this question has been for the last few years the most important problem in radio-activity, for, as will be seen, the proof that the α particle is an atom of helium carries numerous consequences of the first importance in its train.

We shall now describe some novel experiments by Rutherford and H. Geiger, which have not only thrown further light on this question, but have led to important conclusions in several directions. An account of this work is contained in two papers published in the Proceedings of the Royal Society, entitled 'An Electrical Method of Counting the α Particles from Radio-active

Matter,' and 'The Charge and Nature of the α Particle' (A. vol. lxxxi., 141–174, 1908).*

In the first paper an account is given of a method for the detection of a single α particle and for counting the number of α particles emitted from one gram of radium.

The current due to the ionisation of the gas produced by a single α particle is too small to detect except by exceedingly refined methods. To overcome this difficulty, recourse was had to a method of automatic magnification of this current, based on the principle of generation of ions by collision—a subject which has been investigated in detail by Townsend and others. Space does not allow us to enter into a description of the methods employed for this purpose or of the various experimental difficulties that arose during the investigation. The general method employed was to allow the α particles to be fired through a small opening into a detecting vessel containing gas at low pressure exposed to an electric field not far from the sparking value. The entrance of an α particle into the detecting vessel was marked by a sudden ballistic throw of the electrometer needle. By adjustment of the electric field, it was found possible to obtain so large a magnification that the entrance of a single α particle was marked by a large excursion of the electrometer needle.

In this way the expulsion of α particles was detected from uranium, thorium, radium, and actinium. In order to count accurately the number of α particles expelled from one gram of radium, not radium itself, but its product radium C was used as a source of radiation. A surface was coated with a thin film of radium C by its exposure for some hours in the presence of the radium emanation. The use of radium C as a source of rays had several advantages, especially as regards the ease and certainty of measurement of the amount of active matter present by means of the γ rays. The number of α particles passing through an opening of known area at a known distance from the active source was counted for a definite interval by noting the excursions of the electrometer needle. From this the total number of α particles expelled per second fiom the source was deduced. In this way it was found that $3 \cdot 4 \times 10^{10}$ α particles were expelled per second from the radium C present in one gram of radium in equilibrium. It is known from other data that radium itself and each of its products, viz. the emanation, radium A and radium C, expel the same number of α particles per second when in equilibrium. Consequently in one gram of radium in equilibrium $3 \cdot 4 \times 10^{10}$ α particles are expelled from each of the products per second, and the total number expelled is $1 \cdot 36 \times 10^{11}$ per second. On the most probable assumption, that one atom of radium in breaking up emits one α particle, $3 \cdot 4 \times 10^{10}$ atoms of radium break up per second per gram.

It was a matter of interest to compare the number of scintillations observed on a properly prepared screen of zinc sulphide with the number of α particles striking it. Within the limit of experimental error, it was found that the number of scintillations was equal to the number of impinging α particles

* *This vol., pp. 89, 109.*

counted by the electric method. Consequently each α particle on striking the screen produces a scintillation. It is thus obvious that, using proper screens, the scintillation method as well as the electric method may be employed to count the number of α particles emitted by a radio-active substance.

Apart from the importance of these results for radio-active data, the experiments are of themselves noteworthy, for it is the first time that it has been found possible to detect a single atom of matter. This, as we have seen, can be done in two ways, one electrical and the other optical. The possibility of detection of a single atom of matter is in this case, of course, due to the great energy of motion of the α particle.

In the second paper, an account is given of experiments to measure the charge carried by the α particles. Since the number of α particles is known from the counting experiments, the charge on each α particle can be determined by measuring the charge carried by the α particles expelled from a known quantity of radium. As in the counting experiments, radium C was used as a source of rays. It was found that each α particle carried a positive charge of $9 \cdot 3 \times 10^{-10}$ electrostatic units. Now the charge carried by an ion in gases has been determined by several observers, using the well-known method of making each ion the nucleus of a visible drop of water by a sudden expansion. J. J. Thomson obtained a value $3 \cdot 4 \times 10^{-10}$, H. A. Wilson $3 \cdot 1 \times 10^{-10}$, and Millikan and Begeman $4 \cdot 06 \times 10^{-10}$.

The mean of these three determinations of e is $3 \cdot 5 \times 10^{-10}$. The charge E on an α particle on this data thus lies between $2e$ and $3e$.

Some calculations of the value of E and e are then made from radio-active data based on simple and very probable assumptions. Taking the half-period of transformation of radium as 2000 years—the value found by direct measurement by Boltwood—it is shown, on the assumption that each atom of radium in breaking up emits one α particle, that the charge e carried by a hydrogen atom comes out to be $4 \cdot 1 \times 10^{-10}$. Similarly, supposing that the heating effect of radium is a measure of the kinetic energy of the α particles, the charge carried by an α particle comes out at $9 \cdot 1 \times 10^{-10}$—a value close to that found experimentally. A discussion is then given of the methods employed in the previous determination of e, and it is shown that in consequence of certain sources of error which are very difficult to eliminate, the values previously obtained tend to be too small. It is concluded that the unit charge e is not very different from E/2 or $4 \cdot 65 \times 10^{-10}$, and that an α particle carries twice the unit charge. From the previous discussion of the interpretation of the value of E/M for the α particle, it follows that an α particle must be an atom of helium carrying a double charge, or, in other words, that an α particle when its charge is neutralised is a helium atom.

It seems at first sight contradictory that an atom of a monatomic gas like helium can carry two unit charges, It must be borne in mind that in this case the α particle plunges at a great speed through the molecules of matter, and must itself be ionised by collision. If two electrons can be removed by this process, the double positive charge is at once explained.

We thus see that by a direct method we have been enabled to count the number of α particles and to determine the charge caused by each, and from other evidence to deduce that the unit charge e is half the charge carried by the α particle.

With the aid of this data we can at once deduce the magnitudes of some important atomic quantities. The value of e/m for the hydrogen atom is $2\cdot88 \times 10^{14}$ electrostatic units. Substituting the value of $e = 4\cdot65 \times 10^{-10}$, it follows that the mass of a hydrogen atom is $1\cdot61 \times 10^{-24}$ gram. From this it follows that there are $6\cdot2 \times 10^{23}$ atoms in one gram of hydrogen, and that there are $2\cdot72 \times 10^{19}$ molecules in a cubic centimetre of any gas at standard pressure and temperature.

From the data already given we can predetermine the magnitude of some important radio-active quantities. Let us first consider the rate of production of helium by radium. One gram of radium in equilibrium contains four α-ray products, each of which expels $3\cdot4 \times 10^{10}$ α particles, *i.e.* atoms of helium, per second. Consequently, since there are $2\cdot72 \times 10^{19}$ atoms of helium in a cubic centimetre, the volume of helium produced per second is

$$\frac{4 \times 3\cdot4 \times 10^{10}}{2\cdot72 \times 10^{19}},$$

or $5\cdot0 \times 10^{-6}$ c.mm. per second. This corresponds to a production of helium of $0\cdot43$ c.mm. per day, or 158 c.mm. per year.

In a similar way, the maximum volume of the emanation in one gram of radium can be calculated. Since one atom of radium in breaking up emits one α particle and gives rise to one atom of emanation, the volume of emanation produced per second is one-quarter the volume of helium, or $1\cdot25 \times 10^{-6}$ c.mm. per second. Since the average life of the emanation is 468,000 seconds, the maximum volume of the emanation comes out to be $0\cdot585$ c.mm. In a recent paper Rutherford (*Phil. Mag.*, August)* has measured the volume of the emanation and obtained a value not very different from the calculated volume. In a similar way, it is not difficult to calculate the period of transformation of radium and the heating effect of radium. The former comes out at 1750 years, which is somewhat shorter than the value 2000 years found experimentally by Boltwood. As Boltwood points out, however, the probable experimental errors are such as to tend to give too high a value for the period. The latter is deduced on the hypothesis that the heating effect is a measure of the kinetic energy of the expelled α particles. The heating effect is calculated to be about 113 gram calories per gram per hour, while the observed heating effect of the sample of radium from which the standard preparation was taken was found to be 110 gram calories per hour. For convenience, the data obtained in this paper are collected below:—

Charge carried by a hydrogen atom	$= 4\cdot65 \times 10^{-10}$ electrostatic units.
Charge carried by α particle	$= 9\cdot3 \times 10^{-10}$ electrostatic units.

* *This. vol., p. 72.*

Mass of H atom	$= 1 \cdot 61 \times 10^{-24}$ gram.
Number of atoms per gram of H	$= 6 \cdot 2 \times 10^{23}$
Number of molecules per c.c. of any gas at standard pressure and temperature	$\left.\right\} = 2 \cdot 72 \times 10^{19}$
Number of α particles expelled per sec. per gram of radium itself	$\left.\right\} = 3 \cdot 4 \times 10^{10}$
Number of atoms breaking up per sec. per gram of radium	$\left.\right\} = 3 \cdot 4 \times 10^{10}$
Calculated volume of emanation per gram of radium	$\left.\right\} = 0 \cdot 585$ c.mm.
Production of helium per gram of radium per year	$\left.\right\} = 158$ c.mm.
Calculated heating effect of radium per gram	$= 113$ gr. cal. per hour.
Calculated period of radium	$= 1750$ years.

We have already seen that there is a substantial agreement between the calculated values of the heating effect, the life of radium and the volume of the emanation, and the experimentally determined values. A still further test would lie in a comparison of the calculated and observed rates of production of helium by radium. Data on this subject will probably soon be forthcoming.*

Some very important consequences follow from the proof that the α particle is a helium atom. It must be concluded that the atoms of the known radio-active elements are in part at least constituted of helium atoms which are liberated at definite stages during the disintegration. It will be seen that in many cases the atomic weights of the various products can be deduced. In the succession of products produced by the disintegration of the uranium-radium series, there occur several rayless products and β-ray products. Assuming, as is not improbable, that the atomic products undergo an internal rearrangement without the expulsion of a mass comparable with the hydrogen atom, we can calculate the atomic weights of the successive products, taking the atomic weight of helium as 4. From the known range of the α particles from uranium and the ionisation it produces compared with the radium associated with it, there is no doubt that uranium expels two α particles to one from radium itself. Whether this is a peculiarity of uranium itself or due to an unseparated product in uranium is not settled.

Taking the atomic weight of uranium as $238 \cdot 5$, the atomic weights of the different products are as follows:—Uranium X $230 \cdot 5$, ionium $230 \cdot 5$, radium $226 \cdot 5$, emanation $222 \cdot 5$, radium A $218 \cdot 5$, radium B $218 \cdot 5$, radium C $214 \cdot 5$, radium D, E, and F (radio-lead) $210 \cdot 5$, radium G (polonium) $210 \cdot 5$. It will be seen that the calculated value of the atomic weight of radium is in good agreement with the most recent experimental values. The end product of

* (Footnote, added September 12, 1908.) In a paper just to hand (Proc. Roy. Soc., A., vol. lxxxi., p. 280) Sir James Dewar has shown experimentally that $0 \cdot 37$ c.mm. of helium is produced per gram of radium per day. This is in excellent agreement with the calculated rate, $0 \cdot 43$ c.mm. per day.

radium after the transformation of polonium has an atomic weight of 206·5 —a value close to that of lead (206·9). Boltwood long ago suggested, from examination of the amount of lead in old radio-active minerals, that lead was the probable final product of the disintegration of the uranium-radium series.

We cannot at the moment apply the same method of calculation to thorium products, for Bronson (*Phil. Mag.*, August, 1908) has recently brought strong evidence that the disintegration of the atoms of some of the products is accompanied by the expulsion of more than one α particle.

In conclusion, it may be of interest to note that the experimental results recorded in this article lead to an experimental proof—if proof be needed—of the correctness of the atomic hypothesis with reference to the discrete structure of matter. The number of α particles expelled from radium can be directly counted, and the corresponding volume of helium determined. In this way it is possible to determine directly the number of atoms in a cubic centimetre of helium quite independently of any measurements of the charge carried by the α particles.

E. RUTHERFORD

The Action of the Radium Emanation upon Water

by PROFESSOR E. RUTHERFORD, F.R.S.,

and T. ROYDS, M.SC., 1851 *Exhibition Science Scholar*

From the *Philosophical Magazine* for November 1908, ser. 6, xvi, pp. 812–8

SINCE the initial experiments of Ramsay and Soddy in 1903, the production of helium from radium and its emanation have been completely substantiated by a number of independent observers. On the view that the α particle is a helium atom, the appearance of helium from radioactive matter in general receives a simple and satisfactory explanation.

Recently Mr. Cameron and Sir William Ramsay have attacked the important question as to whether the radiations from a large quantity of radium emanation are effective in transforming the atoms of ordinary matter. They have published results to prove that under the influence of the radium emanation, copper is transformed into lithium and possibly into sodium and potassium. In addition they have given evidence that in the presence of copper solutions the emanation disintegrates into argon, and in the presence of water into neon. A complete and satisfactory proof of these transformations is attended by great experimental difficulty. In a recent paper,* Cameron and Ramsay sum up the results of their experiments as follows:—'In carrying out such work it is extremely difficult to prevent traces of air leaking into the apparatus during the considerable length of time which must elapse before an experiment is completed. Since $0 \cdot 1$ c.c. of atmospheric nitrogen contains sufficient argon to be detected spectroscopically after the nitrogen has been removed by sparking, the proof that the presence of argon is due to some other cause is rendered extremely difficult. Similarly, it is far from easy to free copper solutions completely from traces of lithium, and to prove convincingly that lithium did not pre-exist in such solution treated with emanation. The detection of neon is open to no such objection. The only possible source, other than transmutation, is the aluminium of the electrodes. The vacuum-tube had been previously run a great number of times at different pressures, washed out with air, and finally showed traces of hydrogen, and hydrogen alone. It is inconceivable that neon can have resulted thus by chance from two experiments with water, where in each case the residues were tested between those from numerous similar experiments with other solutions in which no neon was detected. We must regard the transformation of emanation into neon, in

* Journ. Chem. Soc. June 1908, p. 992.

presence of water, as indisputably proved, and, if a transformation be defined as a transformation brought about at will, by change of conditions, then *this is the first case of transmutation of which conclusive evidence is put forward.*'

Mme. Curie and Mlle. Gleditsch* have recently repeated the experiments of Cameron and Ramsay with reference to the transformation of copper into lithium, and have obtained no certain evidence of the production of lithium.

With the aid of the radium loaned by the Vienna Academy of Sciences, we have made experiments to see whether neon is produced when the radium emanation disintegrates in the presence of water. Thanks to the use of the absorbing properties of charcoal, the detection of a minute quantity of neon is now a comparatively simple matter. Coconut charcoal at the temperature of liquid air absorbs all gases except neon and helium. The spectroscopic test of the presence of neon is unmistakable on account of *its* very characteristic spectrum of a bright line in the yellow and a group of bright lines in the red. Since there could be no possible doubt that neon was present in the experiments of Cameron and Ramsay, the question arose whether the neon observed could have been derived from the air; for Strutt[†] has recently shown that the presence of neon can be spectroscopically detected in $\frac{1}{10}$ c.c. of atmospheric air. Cameron and Ramsay state that in their experiments it was impossible to avoid a small leakage of air into their apparatus during the week or more that an experiment was in progress.

Before beginning the main investigation, a number of experiments were made to determine the amount of neon that could be detected spectroscopically. The apparatus, already described for the purification of the emanation,[‡] was used for this purpose. The arrangement will be clearly seen from fig. 1.

The whole apparatus was first completely exhausted by means of a mercury pump. A known volume of air contained in a small burette over mercury was introduced into the reservoir R by means of the mercury trough. The stopcocks A and B were closed and C opened, and the mercury raised to the level D by raising the reservoir S. The stopcock A was opened to allow the gas to expand into the reservoir E of capacity about 30 c.c. and then closed. To the top of E was attached a small spectrum-tube F, of length about 7 cms., and of diameter 1 mm., provided with thin platinum electrodes. A tube H containing a small quantity of coconut charcoal was attached to the side tube through a stopcock L. Before the experiment the charcoal tube was thoroughly heated and exhausted. On surrounding the tube H by liquid air, all the gases present in the air except neon and helium were absorbed. The stopcock L was then closed, and by raising the mercury reservoir the gases remaining in E were compressed into the vacuum-tube. On passing a discharge, the hydrogen and mercury lines were usually prominent, the former probably due to a trace of water vapour present. A pad of cotton-wool was then wrapped round the spectrum-tube just above the level of the mercury

* Acad. Sciences, Aug. 10, 1908.   † Proc. Roy. Soc. A. lxxx. p. 572 (1908).
‡ Rutherford, Phil. Mag. Aug. 1908. (*This vol., p. 72.*)

E

and was soaked with liquid air. The spectrum of hydrogen and mercury disappeared, and the tube became very hard and showed only the neon spectrum.

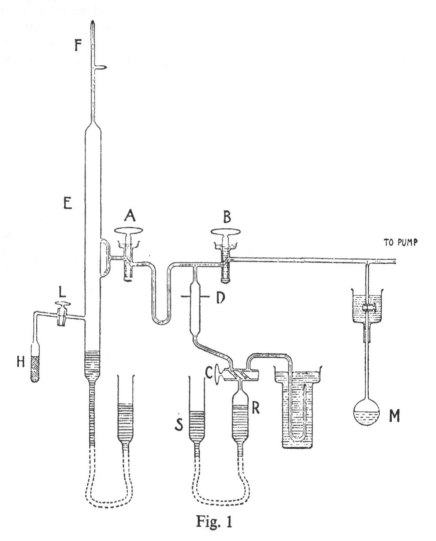

Fig. 1

Proceeding in this way, we were able to detect easily the neon yellow line when $\frac{1}{15}$ c.c. was introduced into the reservoir E. The conditions to bring out the spectrum most clearly could readily be tested by passing a discharge at short intervals through the tube as it slowly warmed up from evaporation of the liquid air. At a certain stage, the neon yellow line and the stronger reds flashed out clearly. With $\frac{2}{15}$ c.c. of air the red lines of neon were clearly visible, and with $\frac{1}{5}$ c.c. a brilliant spectrum of neon was obtained, showing most of the lines. These experiments were repeated on several occasions, the vacuum-tube being changed between each experiment, and in every case the same results were obtained.

According to the measurements of Ramsay neon is present in air in about one part in 100,000 by volume. The experiments thus show that one can readily detect the presence of neon in amount less than one millionth of a cubic centimetre. It is quite probable that with a still better choice of conditions, a still smaller quantity could be spectroscopically detected.

A Watts binocular grating spectroscope was found very convenient for a rapid examination of the spectra; measurements of wave-lengths were carried out by means of a Hilger direct-reading spectroscope.

Experiments with the Emanation

In order to collect the gases formed by the action of radium emanation on water, we employed a method very similar to that used by Cameron and Ramsay. For clearness, we shall describe with some detail the progress of a complete experiment. A glass bulb of about 4 c.c. capacity, provided with a stopcock, was half filled with distilled water and then placed in an evaporation-bath to expel the air from the bulb and the water. The stopcock was then closed, and the bulb sealed on to a side-tube connected with the mercury-pump, as shown at M. The water in the bulb was frozen by surrounding it with liquid air and the last trace of air then pumped out. In the meantime the emanation, corresponding to the equilibrium amount from 150 mg. of radium, was carefully purified after the manner described in a previous paper and stored in the U-tube. By opening the stopcock B the emanation was then all condensed in the frozen bulb. The stopcock was thereupon closed and surrounded by a mercury seal, to prevent any possible leakage of air into the bulb. At the end of three days, when it was judged that the pressure of hydrogen and oxygen formed by the emanation from the water was about an atmosphere, the bulb was again frozen in liquid air and the non-condensable gases pumped out and collected. About 4 c.c. of mixed gases* were obtained, consisting mainly of hydrogen and oxygen. The gases were then introduced into a separate burette and exploded. A small quantity of pure electrolytic oxygen was then added, and the gases again exploded to remove any excess of hydrogen. The residual gases (about 0·2 c.c.) were not further treated, but passed directly into the receiver E in the manner already described, and the gases not absorbed by the charcoal were compressed into the vacuum-tube and spectroscopically examined. A complete and brilliant spectrum of helium was observed, but not a trace of the neon spectrum.

The greater part of the emanation had remained condensed in the bulb when the mixed gases were pumped out. The stopcock was closed and the seal again placed in position and the liquid air removed. The bulb was allowed to stand for six days, when about 4 c.c. of the mixed gases were again pumped

* This amount of mixed gases is one third of that experimented with by Cameron and Ramsay, but the amount of neon to be expected from their results could have easily been detected with the smaller quantity. In experiments recorded later in the paper 30 c.c. of mixed gases were used.

out. As it might be possible that any neon formed would be dissolved in the water, the bulb was allowed to warm up to atmospheric temperature and then connected with the pump. The rapid evaporation of the water then gave a chance for removal of any absorbed gases into the pump. The small quantity of gas thus obtained was pumped out and added to the first quantity. The mixed gases were treated as before. A complete spectrum of helium was obtained but again no sign of the neon lines.

It might be thought that the presence of the helium in the discharge-tube would interfere with the detection of a small quantity of neon. A special experiment was made to test this point. At the end of the last experiment, the mercury was lowered below the level of the carbon tube and $\frac{1}{10}$ c.c. of air passed into the reservoir. After absorption by the charcoal, the residual gases were tested spectroscopically as before. The neon spectrum was now clearly seen in addition to that of helium. When the lower part of the vacuum-tube was cooled to the temperature of liquid air by the pad of cotton-wool, the yellow line of helium was distinctly brighter than the yellow line of neon, but at a certain stage of the warming up, the neon and helium spectra were about equal in brilliancy.

In all five experiments were made by this method, but in only one case was the presence of neon observed. This was in the first experiment of the series. In order to see whether the presence of neon in this experiment could be ascribed to a leak of air into the apparatus, the charcoal was warmed up and the gases again pumped out and collected in a burette. The oxygen present was removed by melted phosphorus. About $\frac{1}{15}$ c.c. of gas remained, consisting mainly of nitrogen. Now $\frac{1}{15}$ c.c. of nitrogen corresponds to a leakage of about $\frac{1}{10}$ c.c. of air. The brightness of the neon yellow line compared with that of helium in the experiment was about that to be expected if this quantity of air had been added. This was confirmed by the observation that when an additional $\frac{1}{10}$ c.c. of air was added, the neon spectrum was increased notably in brightness.

Experiments with the Gases produced from the Radium Solution

Cameron and Ramsay point out that if the emanation produces neon in the presence of water, neon should be found in the water solution of the radium salt from which the emanation is pumped off, and they state that they have observed the spectrum of neon from the gases thus obtained from the radium solution. We have made experiments to test this point. The radium salt containing over 200 mgrs. of radium was dissolved in water to which a small quantity of hydrochloric acid had been added. The emanation was allowed to collect for about five days, and was then pumped off with about 30 c.cs. of hydrogen and oxygen. These gases were treated as before, and the emanation removed by condensation. The residual gas, mostly oxygen, was tested, by the method described, for neon. We have made four experiments in all, but in no case have we found a trace of neon, although a brilliant spectrum of helium was always observed.

Conclusions

From the rate of production of helium by radium recently measured by Sir James Dewar,* the quantity of helium produced by a known quantity of emanation can be readily estimated. In most of the experiments with water a quantity of emanation corresponding to the equilibrium amount from about 150 mgr. of radium was introduced and left for three days. This should lead to the production of about $\frac{1}{10}$ c.mm. of helium. We have seen that in these experiments we could detect with certainty a quantity of neon less than $\frac{1}{1000}$ c.mm. Consequently, even supposing that the emanation does change into neon, the amount so produced cannot be more than one per cent. of the helium which is also formed. Cameron and Ramsay were apparently not aware of the delicacy of the spectroscopic detection of neon in small quantities of air. In the experiment described in their last paper, they state that they were unable to avoid a leakage of air into their apparatus, and working with 12·6 c.c. of mixed gases, they found after the removal of hydrogen and oxygen that the residue consisted of 0·292 c.c. of nitrogen and carbon dioxide. Assuming that this residue consisted mainly of nitrogen, it would show that there was a leakage of air of about 0·36 c.c. In our experience, the admission of such a quantity of air into the apparatus gives a brilliant spectrum of neon comparable in brightness with the companion helium spectrum. Consequently, the experiment described by Cameron and Ramsay is quite inadequate as a proof of the production of neon from the emanation.

University of Manchester
Oct. 6, 1908

* Proc. Roy. Soc. A. lxxxi. p. 280 (1908).

The Nature of the α Particle

by PROFESSOR E. RUTHERFORD, F.R.S., *and* T. ROYDS, M.SC.

From the *Memoirs* of the Manchester Literary and Philosophical Society, 1908,
IV, vol. liii, No. 1, pp. 1–3
(Received and read November 3rd, 1908)

THE nature of the α particle from radioactive substances has, for several years, been one of the most important questions in Radioactivity. The evidence as a whole indicates that the α particle is an atom of helium carrying a positive charge. Recent experiments of Rutherford and Geiger (*Proc. Roy. Soc.*, 1908)* have substantiated this conclusion. An additional proof of the correctness of this point of view is afforded by the good agreement between the rate of production of helium calculated by Rutherford and Geiger, and the rate of production recently measured by Sir James Dewar (*Proc. Roy. Soc.*, 1908).

This evidence is, however, of too indirect a character to prove decisively that the α particle is an atom of helium. It might be possible, for example, that the expulsion of an α particle led to the liberation of helium from the active matter, but that the α particle itself was not an atom of helium. In order to give a definite proof of the identity of the α particle with a helium atom, it is necessary to show that helium can be obtained from accumulated α particles, quite independently of the active matter from which they are expelled. This has been done in the following way:—

Purified emanation, corresponding to the equilibrium amount from 150 mgs. of radium, was compressed by raising a column of mercury into a fine glass tube about 1·5 cms. long. The walls of this glass tube were sufficiently strong to withstand atmospheric pressure but thin enough to allow the greater part of the expelled α particles to be fired through them. After a number of trials, Mr. Baumbach succeeded in blowing a number of such fine tubes for us. The emanation tube was surrounded by a larger cylindrical glass tube about 8 cms. long and 1·5 cms. diameter. This was first exhausted by a pump and the exhaustion completed by means of a charcoal tube immersed in liquid air. By means of another side tube connected with a mercury reservoir, the gases formed within the outside tube could be compressed into a small vacuum tube attached to the top and their spectra examined.

The tube containing emanation was about 1/100 mm. thick. The stopping power of the glass for the α particle corresponded to less than 2 cms. of air,

* *This vol., p. 109.*

so that the α particles expelled from the emanation itself, radium A and radium C escaped through the emanation tube, and were fired into the walls of the outer glass tube. Twenty-four hours after the introduction of the emanation, no trace of helium was detected on compression of the gases into the vacuum tube; at the end of two days the helium yellow line was seen faintly; after four days, the yellow and green lines came out brightly, and after six days practically the whole helium spectrum was observed.

An experiment was then made to test whether the helium observed could have *diffused* from the emanation through the thin glass walls. For this purpose, the emanation was replaced by about ten times its volume of helium and a new outer tube and vacuum tube placed in position. No trace of helium was observed in the outer tube over a period of eight days. Emanation was again introduced, and after four days, the helium spectrum was again observed.

In these experiments, every precaution was taken to prevent possible contamination of the apparatus with helium. Freshly distilled mercury and fresh glass apparatus was used. No trace of helium was observed unless the emanation was introduced into the fine capillary.

This experiment affords a conclusive proof that the α particle after losing its charge is an atom of helium. Other evidence indicates that the positive charge on the α particle is twice that carried by the hydrogen atom.

Some Properties of the Radium Emanation

by E. RUTHERFORD, F.R.S.

From the *Memoirs* of the Manchester Literary and Philosophical Society, 1908,
IV, vol. liii, No. 2, pp. 1–2
(Received and read November 3rd, 1908)

IN 1906 (*Nature*, Oct. 25),* I drew attention to the fact that the emanations of radium, thorium, and actinium were completely absorbed by cocoanut charcoal at ordinary temperatures. I have had occasion recently to repeat these experiments with much larger quantities of radium emanation and have found that the actual volume of emanation capable of absorption by charcoal at room temperature is very small. For example, several grams of cocoanut charcoal are required to absorb completely the emanation from 200 mgs. of radium at ordinary temperature although the volume of the gas is only one-tenth of a cubic millimetre. As was to be expected, the absorptive power of charcoal for the emanation increases rapidly with lowering of the temperature. This was investigated as follows:—A quantity $0 \cdot 8$ gram of cocoanut charcoal, which absorbed about 4 c.cms. of air at the temperature of liquid air, was connected with a pump and the air completely removed by heating the charcoal. The charcoal was then surrounded by a pentane bath at $-150°$ C., and the purified emanation from 83 mg. of radium (about $0 \cdot 05$ cubic mm.) absorbed in it. As the temperature of the bath slowly rose, the unabsorbed emanation was allowed to expand into an exhausted receiver of about 50 c.cms. capacity. This was pumped out at different temperatures of the charcoal, and the emanation collected and afterwards measured by the γ ray method. At $-50°$ C. the amount of unabsorbed emanation was less than $\frac{1}{10}$ per cent of the total. Above $-40°$ C., the emanation commenced to escape rapidly, and half had been pumped off at a temperature of $10°$ C. About 19 per cent. remained in the charcoal at $100°$ C., but practically all was released at a temperature of the softening of glass. It is seen from these results, that at $10°$ C. the charcoal absorbs about $0 \cdot 03$ cubic mm. of emanation per gram. and at $-40°$ C. about $0 \cdot 06$ cubic mm. per gram.

An experiment was shown to illustrate the rapidity of condensation of pure emanation contained in an exhausted vessel when one point was cooled to the temperature of liquid air.

* *Vol. I, p. 876.*

The Chemical Nature of the α-Particles from Radioactive Substances

NOBEL LECTURE

Delivered by E. RUTHERFORD

Before The Royal Academy of Science at Stockholm Dec. 11, 1908

THE study of the properties of the α-rays has played a notable part in the development of Radioactivity and has been instrumental in bringing to light a number of facts and relationships of the first importance. With increase of experimental knowledge there has been a growing recognition that a large part of radioactive phenomena is intimately connected with the expulsion of the α-particles. In this lecture an attempt will be made to give a brief historical account of the development of our knowledge of the α-rays and to trace the long and arduous path trodden by the experimenter in the attempts to solve the difficult question of the chemical nature of the α-particles. The α-rays were first observed in 1899 as a special type of radiation and during the last six years there has been a persistent attack on this great problem, which has finally yielded to the assault when the resources of the attack seemed almost exhausted.

Shortly after his discovery of the radiating power of uranium by the photographic method, BECQUEREL showed that the radiations from uranium like the Röntgen-rays possessed the property of discharging an electrified body. In a detailed investigation of this property, I examined the effect on the rate of discharge by placing successive layers of thin aluminium foil over the surface of a layer of uranium oxide and was led to the conclusion that two types of radiation of very different penetrating power were present. The conclusions at that period were summed up as follows:

'These experiments show that the uranium radiation is complex and that there are present at least two distinct types of radiation—one that is very readily absorbed, which will be termed for convenience the α-radiation, and the other of a more penetrative character, which will be termed the β-radiation.' (RUTHERFORD 'Uranium Radiation and the Electrical Conduction produced by it.' Phil. Mag. Jan. 1899. p. 116.)* When other radioactive substances were discovered, it was seen that the types of radiation present were analogous to the α- and β-rays of uranium and when a still more penetrating type of radiation from radium was discovered by VILLARD, the term

* Vol. I, p. 169.

E*

γ-rays was applied to them. The names thus given soon came into general use as a convenient nomenclature for the three distinct types of radiation emitted from uranium, radium, thorium and actinium. On account of their insignificant penetrating power, the α-rays were at first considered of little importance and attention was mainly directed to the more penetrating β-rays. With the advent of active preparations of radium, GIESEL in 1899 showed that the β-rays from this substance were easily deflected by a magnetic field in the same direction as a stream of cathode rays and consequently appeared to be a stream of projected particles carrying a negative charge. The proof of the identity of the β-particles with the electrons constituting the cathode rays was completed in 1900 by BECQUEREL, who showed that the β-particles from radium had about the same small mass as the electrons and were projected at a speed comparable with the velocity of light. Time does not allow me to enter into the later work of KAUFMANN and others on this subject, which has greatly extended our knowledge of the constitution and mass of electrons.

In the meantime, further investigation had disclosed that the α-particles produced most of the ionization observed in the neighbourhood of an unscreened radioactive substance, and that most of the energy radiated was in the form of α-rays. It was calculated by RUTHERFORD and MCCLUNG in 1901 that one gram of radium radiated a large amount of energy in the form of α-rays.

The increasing recognition of the importance of the α-rays in radioactive phenomena led to attempts to determine the nature of this easily absorbed type of radiation. STRUTT in 1901 and Sir WILLIAM CROOKES in 1902 suggested that they might possibly prove to be projected particles carrying a positive charge. I independently arrived at the same conclusion from consideration of a variety of evidence. If this were the case, the α-rays should be deflected by a magnetic field. Preliminary work showed that the deflection was very slight if it occurred at all. Experiments were continued at intervals over a period of two years and it was not until 1902, when a preparation of radium of activity 19,000 was available, that I was able to show conclusively that the particles were deflected by a magnetic field, though in a very minute degree compared with the β-rays. This showed that the α-rays consisted of projected charged particles while the direction of deflection indicated that each particle carried a positive charge. The α-particles were shown to be deflected also by an electric field and from the magnitude of the deflection, it was deduced that the velocity of the swiftest particles was about $2 \cdot 5 \times 10^9$ cms. per second, or one-twelfth the velocity of light, while the value of e/m—the ratio of the charge carried by the particle to its mass—was found to be 6,000 electromagnetic units. Now it is known from the data of the electrolysis of water that the value of e/m for the hydrogen atom is 9650. If the α-particle carried the same positive charge as the unit fundamental charge of the hydrogen atom, it was seen that the mass of the α-particle was about twice that of the hydrogen atom. On account of the complexity of the rays it was recognised that the results were only approximate, but the experiments indicated clearly that the α-particle was atomic in mass and might prove ultimately to be either a hydrogen or a helium atom or the atom of some unknown element of light atomic weight. These experiments were repeated by DES COUDRES in 1903 with similar results, while BECQUEREL showed the deflection of the α-rays in a magnetic field by the photographic method.

This proof that the α-particles consisted of actual charged atoms of matter projected with an enormous velocity at once threw a flood of light on radioactive processes, in particular upon another important series of investigations which were being contemporaneously carried on in the Laboratory at Montreal in conjunction with MR. F. SODDY. Had time permitted, it would have been of interest to consider in some detail the nature of these researches which placed on a firm foundation the now generally accepted 'transformation theory' of radioactivity. From a close examination of the substances thorium, radium and uranium, RUTHERFORD and SODDY had reached the conclusion that radioactive bodies were in a state of transformation, as a result of which a number of new substances were produced entirely distinct in chemical and physical character from the parent element. From the independence of the rate of transformation of chemical and physical agencies, it was recognised that the transformation was atomic and not molecular in character. Each of these new bodies was shown to lose its radioactive properties according to a definite law. Even before the discovery of the material nature of the α-rays, it had been considered probable that the radiation from any particular substance accompanied the breaking up of its atoms. The proof that the α-particle was an ejected atom of matter at once strengthened this conclusion and at the same time gave a more concrete and definite representation of the processes occurring in radioactive matter. The point of view reached by us at that time is clearly seen from the following quotation, which with little alteration holds good to-day. 'The results obtained so far point to the conclusion that the beginning of the succession of chemical changes taking place in radioactive bodies is due to the emission of the α-rays, i.e. the projection of a heavy charged mass from the atom. The portion left behind is unstable, undergoing further chemical changes which are again accompanied by the emission of α-rays, and in some cases also of β-rays.

The power possessed by the radioactive bodies of apparently spontaneously projecting large masses with enormous velocities supports the view that the atoms of these substances are made up, in part at least, of rapidly rotating or oscillating systems of heavy charged bodies large compared with the electron. The sudden escape of these masses from their orbit may be due either to the action of internal forces or external forces of which we have at present no knowledge.' (RUTHERFORD. Phil. Mag. Feb. 1903, p. 106.)*

Consider for a moment the explanation of the changes in radium. A minute fraction of the radium atoms is supposed each second to become unstable, breaking up with explosive violence. A fragment of the atom—an α-particle —is ejected at a high speed, and the residue of the atom, which has a lighter

† *Vol. I, p. 557.*

weight than before, becomes an atom of a new substance, the radium emanation. The atoms of this substance are far more unstable than those of radium and explode again with the expulsion of an α-particle. As a result the atom of radium A makes its appearance and the process of disintegration thus started continues through a long series of stages.

I can only refer in passing here to the large amount of work done by various experimenters in analysing the long series of transformations of radium and thorium and actinium; the linking up of radium with uranium and the discovery by BOLTWOOD of the long looked-for and elusive parent of radium, viz. ionium. This phase of the subject is of unusual interest and importance but has only an indirect bearing on the subject of my lecture. It has been shown that the great majority of the transition elements produced by the transformation of uranium and thorium break up with the expulsion of α-particles. A few, however, throw off only β-particles while some are 'rayless', i.e. undergo transformation without the expulsion of high-speed α- and β-particles. It is necessary to suppose that in these latter cases the atoms break up with the expulsion of α-particles at a speed too low to be detected, or, as is more probable, undergo a process of atomic rearrangement without the expulsion of material particles of atomic dimensions.

Another striking property of radium was soon seen to be connected with the expulsion of α-particles. In 1903 P. CURIE and LABORDE showed that radium was a self-heating substance and was always above the temperature of the surrounding air. It seemed probable from the beginning that the effect must be the result of the heating effect due to the impact of the α-particles on the radium. Consider for a moment a pellet of radium enclosed in a tube. The α-particles are shot out in great numbers equally from all parts of the radium and in consequence of their slight penetrating power are all stopped in the radium itself or by the walls of the tube. The energy of motion of the α-particles is converted into heat. On this view the radium is subject to a fierce and unceasing bombardment by its own particles and is heated by its own radiation. This was confirmed by the work of RUTHERFORD and BARNES in 1903, who showed that three quarters of the heating effect of radium was not directly due to the radium but to its product, the emanation, and that each of the different substances produced in radium gave out heat in proportion to the energy of the α-particles expelled from it. These experiments brought clearly to light the enormous energy, compared with the weight of matter involved, which was emitted during the transformation of the emanation. It can readily be calculated that one kilogram of the radium-emanation and its products would initially emit energy at the rate of 14,000 horse-power, and during its life would give off energy corresponding to about 80,000 horse-power for one day.

It was thus clear that the heating effect of radium was mainly a secondary phenomenon resulting from the bombardment by its own α-particles. It was evident also that all the radioactive substances must emit heat in proportion to the number and energy of the α-particles expelled per second.

We must now consider another discovery of the first importance. In discussing the consequences of the disintegration theory, RUTHERFORD and SODDY drew attention to the fact that any stable substances produced during the transformation of the radio-elements should be present in quantity in the radioactive minerals, where the processes of transformation have been taking place for ages. This suggestion was first put forward in 1902 (Phil. Mag. p. 582). 'In the light of these results and the view that has already been put forward of the nature of radioactivity, the speculation naturally arises whether the presence of helium in minerals and its invariable association with uranium and thorium, may not be connected with their radioactivity'* and again (Phil. Mag. p. 453, April) 'It is therefore to be expected that if any of the unknown ultimate products of the changes of a radioactive element are gaseous, they would be found occluded, possibly in considerable quantities, in the natural minerals containing that element. This lends support to the suggestion already put forward (Phil.Mag. 1902, p. 582), that possibly helium is an ultimate product of the disintegration of one of the radioactive elements, since it is only found in radioactive minerals'.†

It was at the same time recognised that it was quite possible that the α-particle itself might prove to be a helium atom. As only weak preparations were then available, it did not seem feasible at that time to test whether helium was produced from radium. About a year later, thanks to Dr. GIESEL of Braunschweig, preparations of pure radium bromide were made available to experimenters. Using 30 milligrams of GIESEL's preparation, Sir WILLIAM RAMSAY and SODDY in 1903 were able to show conclusively that helium was present in radium some months old and that the emanation produced helium. This discovery was of the greatest interest and importance, for it brought to light that in addition to a series of transition elements, radium also gave rise in its transformation to a stable form of matter.

A fundamental question immediately arose as to the position of helium in the scheme of transformations of radium. Was the helium the end or final product of transformation of radium or did it arise at some other stage or stages? In a letter to Nature(Aug. 20, 1903)‡ I pointed out that probably helium was derived from the α-particles fired out by the α-ray products of radium and made an approximate estimate of the rate of production of helium by radium. It was calculated that the amount of helium produced per gram of radium should lie between 20 and 200 cubic millimetres per year and probably nearer the latter estimate. The data available for calculation at that time were imperfect, but it is of interest to note that the rate of production of helium recently found by Sir JAMES DEWAR, in 1908, viz. 134 cubic mm. per year, is not far from the value calculated as most probable at that time.

These estimates of the rate of production of helium were later modified as new and more accurate data became available. In 1905, I measured the charge carried by the α-particles from a thin film of radium. Assuming that each

* *Vol. I, p. 506.*   † *Vol. I, p. 572.*   ‡ *Vol. I, p. 609.*

α-particle carried the ionic charge measured by J. J. Thomson, I showed that $6 \cdot 2 \times 10^{10}$ α-particles were expelled per second per gram of radium itself and four times this number when radium was in equilibrium with its three α-ray-products. The rate of production of helium calculated on these data was 240 cubic mms. per gram per year.

In the meantime, by the admirable researches of Bragg and of Bragg and Kleeman in 1904, our knowledge of the character of the absorption of the α-particles by matter had been much extended. It had long been known that the absorption of α-particles by matter was different in many respects from that of the β-rays. Bragg showed that these differences arose from the fact that the α-particle, on account of its great energy of motion, was not deflected from its path like the β-particle, but travelled in nearly a straight line, ionizing the molecules in its path. From a thin film of matter of one kind, the α-particles were all projected at the same speed and lost their power of producing ionization suddenly, after traversing a certain definite distance of air. The velocity of the α-particles in this view were reduced by their passage through matter by equal amounts. These conclusions of Bragg were confirmed by experiments I made by the photographic method. As a source of rays, a thin film of radium C, deposited from the radium-emanation on a thin wire, was used. By examining the deflection of the rays in a magnetic field, it was found that the rays were homogeneous and were expelled from the surface of the wire at an identical speed. By passing the rays through a screen of mica or aluminium, it was found that the velocity of all the α-particles were reduced by the same amount and the issuing beam was still homogeneous.

A remarkable result was noted. All α-particles apparently lost their characteristic properties of ionization, phosphorescence and photographic action, at exactly the same point while they were still moving at a speed of about 9,000 kilometres per second. At this critical speed, the α-particle suddenly vanishes from our ken and can no longer be followed by the methods of observation at our command.

The use of a homogeneous source of α-rays like radium C at once suggested itself as affording a basis for a more accurate determination of the value of e/m for the α-particle and for seeing whether the value was consistent with the view that the α-particle was a charged atom of helium. In the course of a long series of experiments, I proved that the α-particles, whether expelled from radium, thorium or actinium, were identical in mass and must consist of the same kind of matter.

The velocity of expulsion of the α-particles from different kinds of active matter varied over comparatively narrow limits but the value of e/m was constant and equal to 5,070. This value was not very different from the one originally found. A difficulty at once arose in interpreting this result. We have seen that the value of e/m for the hydrogen atom is 9,650. If the α-particle carried the same positive charge as the hydrogen atom, the value of e/m for the α-particle would indicate that its mass was twice that of the hydrogen atom, i.e. equal to the mass of a hydrogen molecule. It seemed very improbable that hydrogen should be ejected in a molecular and not an atomic state as a result of the atomic explosion. If, however, the α-particle carried a charge equal to twice that of the hydrogen atom, the mass of the α-particle would work out at nearly four, i.e. a mass nearly equal to that of the atom of helium.

I suggested that, in all probability, the α particle was a helium atom which carried two unit charges. On this view, every radioactive substance which emitted α-particles must give rise to helium. This at once offered an explanation of the fact observed by DEBIERNE that actinium as well as radium produced helium. It was pointed out that the presence of a double charge of helium-atom was not altogether improbable for reasons to be given later.

While the evidence as a whole strongly supported the view that the α-particle was a helium atom, it was found exceedingly difficult to obtain a decisive experimental proof of the relation. If it could be shown experimentally that the α-particle did in reality carry two unit charges, the proof of the relation would be greatly strengthened. For this purpose an electrical method was devised by RUTHERFORD and GEIGER for counting directly the α-particles expelled from a radioactive substance. The ionization produced in a gas by a single α-particle is exceedingly small and would be difficult to detect electrically except by a very refined method. Recourse was had to an automatic method of magnifying the ionization produced by an α-particle. For this purpose it was arranged that the α-particles should be fired through a small opening into a vessel containing air or other gas at a low pressure, exposed to an electric field near the sparking value. Under these conditions the ions produced by the passage of the α-particle through the gas generate a large number of fresh ions by collision. In this way it was found possible to magnify the electrical effect due to an α-particle several thousand times. The entrance of an α-particle into the testing vessel was then indicated by a sudden deflection of the electrometer needle. This method was developed into an accurate method of counting the number of α-particles fired in a known time through the small aperture of the testing vessel. From this was deduced the total number of α-particles expelled per second from any thin film of radioactive matter. In this way it was shown that $3\cdot4 \times 10^{10}$ α-particles are expelled per second from one gram of radium itself and from each of its α-ray products in equilibrium with it.

The correctness of this method was indicated by another, quite distinct method of counting. Sir WILLIAM CROOKES and ELSTER and GEITEL had shown that the α-particles falling on a screen of phosphorescent zinc sulphide produced a number of scintillations. Using specially prepared screens, RUTHERFORD and GEIGER counted the number of these scintillations per second with the aid of a microscope. It was found that, within the limit of experimental error, the number of scintillations per second on a screen agreed with the number of α-particles impinging on it, counted by the electrical method. It was thus clear that each α-particle produced a visible scintillation on the screen, and that either the electrical or the optical method could be used for counting the α-particles. Apart from the purpose for which these experiments were made, the results are of great interest and importance, for it is the first time that it has been found possible to detect a single atom of matter by its electrical and optical effect. This is of course only possible because of the great velocity of the α-particle.

Knowing the number of α-particles expelled from radium from the counting experiment, the charge carried by each α-particle was determined by measuring the total positive charge carried by all the α-particles expelled. It was found that each α-particle carried a positive charge of $9 \cdot 3 \times 10^{-10}$ electrostatic units. From a consideration of the experimental evidence of the charge carried by the ions in gases, it was concluded that the α-particle did carry two unit charges, and that the unit charge carried by the hydrogen atom was equal to $4 \cdot 65 \times 10^{-10}$ units. From a comparison of the known value of e/m for the α-particle with that of the hydrogen atom, it follows that an α-particle is a projected atom of helium carrying two charges, or, to express it in another way, the α-particle, after its charge is neutralized, is a helium atom.

The data obtained from the counting experiments allow us to calculate simply the magnitude of a number of important radioactive quantities. It was found that the calculated values of the life of radium, of the volume of the emanation, and of the heating effect of radium were in excellent agreement with the values found experimentally. A test of the correctness of these methods of calculation was forthcoming shortly after the publication of these results. RUTHERFORD and GEIGER calculated, on the assumption that the α-particle was a helium atom, that one gram of radium in equilibrium should produce a volume of 158 cub. mms. of helium per year. Sir JAMES DEWAR in 1908 carried out a long experimental investigation on the rate of production of helium by radium, and showed that one gram of radium in equilibrium produced about 134 cub. mms. per year. Considering the difficulty of the investigation, the agreement between the experimental and calculated values is very good and is strong evidence in support of the identity of the α-particle with a helium atom.

While the whole train of evidence we have considered indicates with little room for doubt that the α-particle is a projected helium atom, there was still wanting a decisive and incontrovertible proof of the relationship. It might be argued, for example, that the α-particle appeared as a result of the disintegration of the radium atom in the same way as the atom of the emanation and had no direct connection with the α-particle. If one helium atom were liberated at the same time that an α-particle was expelled, experiment and calculation might still agree and yet the α-particle might be an atom of hydrogen or of some unknown substance.

In order to remove this possible objection, it is necessary to show that the α-particles, collected quite independently of the active matter from which they are expelled, give rise to helium. With this purpose in view some experiments were recently (1908) made by RUTHERFORD and ROYDS.* A large quantity of emanation was forced into a glass tube which had walls so thin that the

* *This vol., p. 134, and p. 163.*

α-particles were fired right through them, though the walls were impervious to the emanation itself. The α-particles were projected into the glass walls of an outer sealed vessel and were gradually released into the exhausted space between the emanation tube and the outer vessel. After some days a bright spectrum of helium was observed in the outer vessel. There is, however, one objection to this experiment. It might be possible that the helium observed had diffused through the thin glass walls from the emanation. This objection was removed by showing that no trace of helium appeared, when the emanation was replaced by a larger volume of helium itself. We may thus confidently conclude that the α-particles themselves give rise to helium, and are atoms of helium. Further experiments showed that when the α-particles were fired through the glass walls into a thin sheet of lead or tin, helium could always be obtained from the metals after a few hours' bombardment.

Considering the evidence together, we conclude that the α-particle is a projected atom of helium, which has, or in some way during its flight acquires, two unit charges of positive electricity. It is somewhat unexpected that the atom of a monatomic gas like helium should carry a double charge. It must not however be forgotten that the α-particle is released at a high speed as a result of an intense atomic explosion, and plunges through the molecules of matter in its path. Such conditions are exceptionally favourable to the release of loosely attached electrons from the atomic system. If the α-particle can lose two electrons in this way, the double positive charge is explained.

We have seen that there is every reason to believe that the α-particles, so freely expelled from the great majority of radioactive substances, are identical in mass and constitution and must consist of atoms of helium. We are consequently driven to the conclusion that the atoms of the primary radioactive elements like uranium and thorium must be built up in part at least of atoms of helium. These atoms are released at definite stages of the transformations at a rate independent of control by laboratory forces. There is good reason to believe that in the majority of cases, a single helium atom is expelled during the atomic explosion. This is certainly the case for radium itself and its series of products. On the other hand, BRONSON has drawn attention to certain cases, viz. the emanations of actinium and of thorium, where apparently two and three atoms of helium respectively are expelled at one time. No doubt these exceptions will receive careful investigation in the future. It is of interest to note that uranium itself appears to expel two α-particles for one from each of its products. Knowing the number of atoms of helium expelled from the atom of each product, we can at once calculate the atomic weights of the products. For example, in the uranium-ionium-radium series, uranium expels two α-particles and each of the six following α-ray-products one, i.e. eight in all. Taking the atomic weight of uranium as 238,5, the atomic weight of ionium should be 230,5, of radium 226,5 of the emanation 222,5 and so on. It is of interest to note that the atomic weight of radium deduced in this way is in close agreement with the latest experimental values. The atomic weight of the end product of radium, resulting from the transformation of radium F (polonium) should be $238{,}5 - 8 \times 4 = 206{,}5$ or a value close to that for lead. Long ago, BOLTWOOD suggested from examination of analyses of old uranium minerals, that lead was in all probability a transformation product of the uranium-radium-series. The coincidence of numbers is certainly striking. but a direct proof of the production of lead from radium will be required before this conclusion can be considered as definitely established.

It is very remarkable that a chemically inert element like helium should play such a prominent part in the constitution of the atomic systems of uranium and thorium and radium. It may well be that this property of helium of forming complex atoms is in some way connected with its inability to enter into ordinary chemical combinations. It must not be forgotten that uranium and thorium and each of their transformation products must be regarded as distinct chemical elements in the ordinary sense. They differ from ordinary elements in the comparative instability of their atomic systems. The atoms break up spontaneously with great violence, expelling in many cases an atom of helium at a high speed. All the evidence is against the view that uranium or thorium or radium can be regarded as an ordinary molecular compound of helium with some known or unknown element, which breaks up into helium. The character of the radioactive transformations and their independence of temperature and other agencies have no analogy in ordinary chemical changes.

Apart from their radioactivity and high atomic weight, uranium, thorium, and radium show no specially distinctive chemical behaviour. Radium for example is closely allied in general chemical properties to barium. It is consequently not unreasonable to suppose that other elements may be built up in part of helium, although the absence of radioactivity may prevent us from obtaining any definite proof. On this view, it may prove significant that the atomic weights of many elements differ by four—the atomic weight of helium —or a multiple of four. Time is too limited to discuss in greater detail these and other interesting questions which have been raised by the proof of the chemical nature of the α-particle.

The Discharge of Electricity from Glowing Bodies

by PROFESSOR E. RUTHERFORD, F.R.S.

From *The Electrician*, December 11, 1908, pp. 343–4
(Abstract of a lecture delivered before the Manchester Section of the
Institution of Electrical Engineers)

THE study of the discharge of electricity from glowing bodies is so closely associated with the ionisation of gases that it will be advisable to refer briefly to some of the effects of ionisation.

Ionised gas may be produced in a variety of ways—*e.g.*, by the application of X-rays, cathode rays, high temperatures or ultra-violet light. If two metal plates, separated by air, be connected to a battery of electrical accumulators no current will flow under normal conditions, but if the dielectric be subjected to the influence of X-rays ionisation takes place, and a current of electricity actually passes through the gas. This effect is explained by the fact that the X-rays cause a dissociation of the air molecules, giving rise to positive and negative particles, and immediately these particles are liberated they tend to move in two streams, the positive particles going to the negative plate and vice versa. Ionised gas, therefore, becomes a conductor, the ions themselves being the carriers of the electricity. If a curve be plotted showing how the current through the gas varies with the P.D. between the plates, it will be found that up to a certain point the current is directly proportional to the voltage. This is explained by stating that at low voltages the speed of the ions from plate to plate is relatively slow, and many re-combinations take place before the ions reach their destination. As the voltage is increased, however, a saturation point is reached when there is no further increase in current, and at this stage all the ions present succeed in reaching the electrodes, hence they are all actively engaged as current carriers, and no further increase in current can take place. Ohm's law fails in application when dealing with ionised gas; for example, in the case just mentioned, if the distance between the plates be reduced by half, but the P.D. kept constant, only half the number of ions will be formed, hence the current will be halved instead of doubled.

Similar effects are obtained for low pressures of the gas, the only difference being that the saturation point is reached much more quickly owing to the more rapid motion of the ions. Townsend investigated the cause of breakdown of the dielectric by an electric spark passing from plate to plate, and this he attributed to ionisation by collision. A negative ion on making its way to the positive plate collides with a molecule, and if its velocity is greater than, say, 10^8 cm. per second, it is able to generate new ions by collision with the neutral gas molecules.

The rate of production of ions by movement of the negative ions increases rapidly with the strength of the electric field. When the field approaches the value required for the passage of a spark, the positive ions also become effective as ionisers, though to a much smaller degree than the negative. Under these conditions, the current through the gas rises rapidly with the voltage until the gas breaks down with the passage of an electric spark. The theory and experiments of Townsend afford a very complete explanation of the phenomena until the passage of the spark. After the spark discharge has started to pass, the temperature and other conditions of the gas are much altered and the processes occurring in the gas can no longer be followed with certainty.

A good illustration of ionisation is formed by placing two platinum wires forming the terminals of a battery in different parts of a Bunsen flame. The flame itself, being an ionised gas, permits an appreciable current to flow, and this current can be further increased by the addition of potassium or sodium salts to the flame. It is interesting to note, also, that one of the platinum wires may be withdrawn from actual contact with the flame and still permit the current to flow, owing to a number of ions which are drawn out of the flame by the electric field.

The effect of high temperatures in producing ionisation can be readily explored by surrounding a platinum wire by a metal cylinder, so as to form a condenser in *vacuo*. Now, if the wire and cylinder form the electrodes of a battery (the wire being at negative potential relatively to the cylinder) and a separate source of current be employed to heat the wire, negative ions escape from the wire. The current increases rapidly with rises of temperature, thereby permitting a current to flow between the wire and cylinder. The value of this current at any temperature can be derived from the expression

$$C = a\theta^{-\frac{1}{2}}e^{-b/\theta},$$

where a and b are constants, θ is the absolute temperature, e is the base of Naperian logarithms.

J. J. Thomson has given a general explanation of the escape of negative electricity from hot bodies by supposing that the body contains a large number of mobile negative particles or electrons which are kept from escaping from the body by the action of a 'double layer' of electricity formed on its surface. As the temperature rises, some of the electrons acquire sufficient energy of movement to escape from the body and are conveyed to the positive electrode by the electric field. Obviously, when the wire is positive it tends to retain the particles. These effects were demonstrated some years ago by Fleming in his well-known carbon filament experiments.

Wehnelt found that a platinum plate coated with calcium or barium salts has a greatly increased power of emitting electrons, and by this method currents at the rate of 1 ampere per square centimetre of surface are easily obtained. A special vacuum tube designed to demonstrate these effects consists of a long glass cylinder, having at one end a simple platinum wire electrode and a similar starting electrode about midway in the tube. The negative electrode situated at the remaining end of the tube consists of a strip of platinum coated with lime, which can be heated electrically by an external battery. It is found that in order to start the flow of current through the tube the two platinum wire electrodes must be short-circuited, thereby affording a shorter path for the ions emitted from the heated lime. After starting, however, ionisation by collision takes place, and permits the end electrode only to be used. This apparatus permits eight electric lamps, taking 5 amperes, to be lighted through the tube from a 200 volt supply circuit, the area of the lime surface being less than one square cm. By this method very heavy currents can be passed through a gas at very low pressure. The striations produced in the tube are very brilliant and sharply defined. Such a tube acts as a rectifier for alternating currents.

Der Ursprung des Radiums (Bericht)

von E. RUTHERFORD

From: *Jahrbuch der Radioaktivität und Elektronik*. V. Band, 1908, Heft 2, pp. 153–166
Verlag von S. Hirzel in Leipzig

IN der vorliegenden Arbeit will ich das allmähliche Anwachsen unserer Kenntnis vom Ursprung und von der Umwandlung des Radiums, soweit es angängig ist, in historischer Entwicklung verfolgen; indessen werde ich in einem so kurzen Überblick das Augenmerk nur auf die hervorstechendsten Tatsachen und Hypothesen lenken können. Das Suchen nach der hypothetischen Substanz, welche unmittelbar in Radium umgewandelt wird, ist von geradezu dramatischem Interesse gewesen und illustriert schlagend den Wert der Umwandlungstheorie als eines Führers für den experimentellen Forscher. Der gegenwärtige Standpunkt, das Radium als ein in langsamer Umwandlung befindliches Element anzusehen, wurde erstmalig von Rutherford und Soddy in einer Arbeit mit dem Titel „Radioactive change" („Radioaktive Umwandlung")* mit folgenden Worten bestimmt ausgesprochen:

„Im Falle des Radiums muß sich indessen derselbe Betrag (nämlich ungefähr 1 Milligramm auf ein Gramm jährlich) umwandeln. Die „Lebensdauer" des Radiums kann infolgedessen nicht mehr als ein paar tausend Jahre nach dieser niedrigsten Schätzung betragen, die sich auf die Annahme gründet, daß jedes Teilchen bei jeder Umwandlung einen Strahl hervorbringt. . . . Es erscheint darnach sicher, daß das in einem Mineral vorhandene Radium nicht so lange existiert hat als wie das Mineral selbst, vielmehr fortgesetzt durch radioaktive Umwandlung hervorgebracht wird."

Nach dieser Theorie wird das Radium als ein typisches radioaktives Produkt betrachtet, welches nach demselben Gesetze umgewandelt wird wie alle übrigen Produkte. Da die Mineralien, in denen sich Radium findet, in vielen Fällen geologisch ein hohes Alter haben, so ist die Annahme vernünftig, daß die gegenwärtige Existenz von Radium eine Folge seiner fortwährenden Erzeugung seitens einer anderen in dem Mineral vorhandenen Substanz mit langer Umwandlungsperiode ist.

Anfangs schien es wahrscheinlich, daß das Uranium diese Muttersubstanz sei,† denn es erfüllte die notwendigen Bedingungen, immer mit Radium zugleich vorzukommen und eine im Vergleich zu diesem lange radioaktive Lebensdauer zu haben. Diese Hypothese läßt sich experimentell auf mehrere

---

* Phil. Mag., Mai 1903. (*Vol. I, p. 596.*)

† Vergl. Rutherford, „Radioactivity", 1. Aufl., S. 332—335, 1904.

verschiedene Weisen prüfen. Wenn das Uranium der Erzeuger des Radiums wäre, so müßte eine Lösung von Uranium, welche anfänglich durch chemische Mittel von Radium frei gemacht worden ist, im Laufe der Zeit eine Entstehung frischen Radiums zeigen. Die in einer Uraniumlösung zu irgendeiner Zeit vorhandene Radiummenge kann man in sehr einfacher und direkter Weise bestimmen, indem man die Menge der erzeugten Radiumemanation mißt.

Ehe wir die Versuchseinzelheiten betrachten, wollen wir einige Berechnungen über die in einem Gramm Uranium zu erwartende Zunahme an Radium für den Fall anstellen, daß das Uranium unmittelbar in Radium umgewandelt wird. Bekanntlich ist die in einem Mineral vorkommende Radiummenge immer dem Uraniumgehalt proportional. Ich habe im Verein mit Boltwood* gezeigt, daß $3,84 \cdot 10^{-7}$ Gramm Radium auf ein Gramm Uranium vorkommen. Nun muß in den Mineralien ein Gleichgewichtszustand zwischen Uranium und Radium bestehen, und infolgedessen muß das Radium in dem gleichen Maße erzeugt werden wie sich Uranium umwandelt. Als Rechnungsergebnis aus verschiedenen Zahlenwerten habe ich im Jahre 1904 gefunden†, daß die halbe Umwandlungsperiode des Radiums vermutlich etwa 1500 Jahre sei. Ich habe dann bei einer neueren Rechnung‡, die sich auf genauere Angaben stützt, eine Periode von 2600 Jahren erhalten. Wie wir weiter unten sehen werden, hat Boltwood§ neuerdings die Periode des Radiums experimentell bestimmt und zu ungefähr 1900 Jahren gefunden. Da diese Periode vermutlich dem wahren Werte nahe kommt, so wollen wir sie als Grundlage unserer Berechnungen anwenden. Da das Radium nach demselben Gesetz umgewandelt wird wie alle radioaktiven Produkte, so wird die Menge des jährlich umgewandelten Radiums gegeben durch λQ, wo λ die Umwandlungskonstante bedeutet, welche gleich $3,7 \cdot 10^{-4}$ ist, und wo Q die vorhandene Radiummenge darstellt. Nun muß die aus einem Gramm Uranium in einem Mineral jährlich gebildete Radiummenge gleich dem Bruchteil des mit dem Uranium vergesellschafteten Radiums sein, der in derselben Zeit umgewandelt wird. Demzufolge müßten wir, falls sich Uranium unmittelbar in Radium verwandelt, erwarten, daß aus einem Gramm Uranium jährlich

$$3,7 \cdot 10^{-4} \times 3,8 \cdot 10^{-7} = 1,37 \cdot 10^{-10}$$

Gramm Radium werden.

Nun sind wir, dank geeigneter Methoden, in der Lage, nicht nur das Vorhandensein einer außerordentlich geringen Radiummenge in einer Substanz nachzuweisen, sondern diese Menge auch mit beträchtlicher Genauigkeit zu messen. Mit Hilfe von Emanationselektroskopen können wir eine Menge von 10^{-12} Gramm Radium noch nachweisen, während

---

* Sill. Journ., Juli 1906. (*Vol. I, p. 856.*)
† Rutherford, „Radioactivity", 1. Aufl., S. 333, 1904; 2. Aufl., S. 458, 1905
‡ Phil. Mag., Okt. 1906. (*Vol. I, p. 880.*)
§ Nature, 15. Nov. 1906; 25. Juli 1907.

wir 10^{-10} Gramm Radium bis auf wenige Prozent genau messen können. Folglich müßte, falls Uranium unmittelbar in Radium überginge, das entstehende Radium in einer anfänglich von Radium vollkommen freien Lösung eines Grammes Uranium im Laufe eines Jahres leicht gemessen werden können, und wenn man ein Kilogramm Uranium verwenden würde, müßte das Entstehen von Radium in einem Tage leicht beobachtet werden können.

Die ersten Versuche, welche zu entscheiden bezweckten, ob in einer Uraniumlösung Radium auftritt, sind von Soddy* angestellt worden. Er verwandte ein Kilogramm Uranylnitrat. Die in käuflichen Uraniumsalzen immer vorhandene geringe Radiummenge wurde zum großen Teil durch Ausfällen einer kleinen Spur Bariumsulfat aus der Lösung beseitigt. Darnach wurde die in der Lösung zurückgebliebene Menge Radium nach der Emanationsmethode in der Weise gemessen, daß Luft durch die Lösung in ein Emanationselektroskop hineinaspiriert wurde. Anfangs ergaben die Beobachtungen keine merkliche Radiumzunahme, späterhin aber wurde ein Anwachsen des Radiumgehaltes bemerkt. Eine ähnliche Beobachtung hat Whetham† gemacht. Zu jener Zeit waren die heutigen genauen Methoden zur Messung kleiner Radiummengen nach der Emanationsmethode noch nicht ausgebildet. Jetzt wissen wir, daß die Aspirationsmethode keine befriedigenden Ergebnisse liefert, weil es sehr schwer ist, mit dieser Methode alle Emanation herauszubefördern. Boltwood hat später gezeigt‡, daß die beste Methode zur Bestimmung der Emanationsmenge in einer Lösung darin besteht, die Lösung heftig zu kochen und dann die gesammelten Gase in ein evakuiertes Emanationselektroskop zu leiten. Die Ergebnisse von Soddy haben gezeigt, daß die beobachtete Radiumzunahme sehr klein war gegenüber der, welche theoretisch zu erwarten wäre, falls Uranium unmittelbar in Radium überginge. Inzwischen hat nun Boltwood§ gleichfalls das Problem angegriffen: Eine Menge von 100 Gramm Uraniumnitrat wurde sorgfältig gereinigt und praktisch alles Radium aus ihr entfernt. Sie wurde dann beiseite gestellt, und in langen Zwischenräumen wurde die vorhandene Radiummenge nach der Emanationsmethode bestimmt. Während des Zeitraumes von einem Jahre wurde keine merkliche Radiumzunahme beobachtet, während spätere Beobachtungen, die sich über $2^{1}/_{2}$ Jahre erstreckten, zeigten, daß die Radiumzunahme, falls eine solche überhaupt vorhanden war, sicherlich weniger als 10^{-11} Gramm Radium betrug. Falls sich Uranium unmittelbar in Radium umwandeln würde, so hätte die Zunahme in diesem Zeitraum mehr als tausendmal größer sein müssen als dieser Betrag.

Später haben dann Soddy und Mackenzie‖ eine Reihe von Beobach-

---

* F. Soddy, Nature, 12. Mai 1904; 19. Jan. 1905; Juni 1905. — Phil. Mag., Juni 1905.
† Whetham, Nature, 5. Mai 1904; 26. Jan. 1905.
‡ Bertram B. Boltwood, Sill. Journ **18**, 379, 1904.
§ Bertram B. Boltwood, Sill. Journ. **20**, 239, 1905.
‖ Soddy and Mackenzie, Phil. Mag., August 1907.

tungen über die Radiummenge in solchen Uraniumlösungen gemacht, die mit großer Sorgfalt gereinigt worden waren. Während eines Zeitraumes von 600 Tagen haben sie kein sicheres Anzeichen für eine Zunahme beobachtet. Wenn eine solche überhaupt vorliegt, ist sie sicherlich kleiner als $1/_{1000}$ des theoretischen Wertes.

Die Versuche von Boltwood und von Soddy zeigen also, daß in sorgfältig gereinigten Uraniumlösungen keine sicher nachweisbare Radiumzunahme im Verlaufe eines Jahres oder länger auftritt, und daß die Radiumzunahme, sofern eine solche überhaupt stattfindet, sicherlich nicht größer ist als $1/_{1000}$ des theoretischen Wertes. Wie wir sahen, schloß Soddy aus seinen ersten Versuchen, daß eine langsame Radiumzunahme auftritt. Diese Beobachtung steht nicht notwendig im Widerspruch mit den später von ihm und von Boltwood gemachten Versuchen, denn zwischen den Versuchsbedingungen in beiden Fällen bestanden wesentliche Verschiedenheiten. Bei den ersten Beobachtungen war das Uranium nur teilweise von Radium befreit worden und nicht chemisch gereinigt. Im Lichte neuerer Ergebnisse, die weiter unten erörtert werden sollen, darf eine langsame Radiumzunahme in alten ungereinigten Uraniumlösungen vernunftgemäß erwartet werden.

Der Umstand, daß ein Enstehen von Radium aus Uranium in einem begrenzten Zeitraume nicht beobachtet worden ist, zeigt überzeugend, daß Uranium, oder vielmehr dessen erstes Produkt, Uranium X, sich nicht unmittelbar in Radium verwandelt. Damit ist jedoch die Möglichkeit nicht ausgeschlossen, daß ein anderes unbekanntes Produkt mit langsam verlaufender Umwandlung zwischen Uranium X und Radium vermittelt. Ist beispielsweise ein Produkt mit einer Periode von 1000 Jahren vorhanden, so wird eine Uraniumlösung, die vollständig von diesem Produkt befreit ist, nur eine sehr geringe Menge Radium im Verlaufe eines Jahres entstehen lassen. Der Grund hierfür ist klar, denn am Ende des Jahres würde die von dem Uranium hervorgebrachte Menge dieser Zwischensubstanz nur $1/_{1400}$ ihres Endbetrages im Gleichgewichtszustande ausmachen. Das Verhältnis, in welchem Radium enstände, würde folglich nur einen kleinen Bruchteil des Wertes ausmachen, der zu erwarten wäre, falls Uranium unmittelbar in Radium überginge.

Die Richtigkeit dieses Gesichtspunktes hat durch eine Reihe wichtiger Versuche, die nach einer anderen Richtung hin angestellt worden waren, eine kräftige Stütze erfahren. Wenn Radium genetisch mit Uranium verbunden ist, so muß die in irgendeinem Material vorhandene Radiummenge immer der Uraniummenge proportional sein. Das muß offenbar der Fall sein, denn die Mineralien sind sehr alt im Vergleich zu der Umwandlungsperiode des Radiums, und folglich muß die in irgendeinem Mineral vorhandene Radiummenge schon längst ihren Gleichgewichtswert erreicht haben. Die Konstanz dieses Verhältnisses ist nur in kompakten Mineralien zu erwarten, wo keine Möglichkeit dafür vorhanden ist, daß Radium oder Uranium infolge der selektiven Wirkung einer Lösung oder anderer in der

Natur vorkommender Agenzien entweichen könnte. Die Frage nach dem Zusammenhang zwischen der Radiummenge und der Uraniummenge haben unabhängig voneinander McCoy*, Strutt† und Boltwood‡ angefaßt. Bei den Versuchen von McCoy wurde die Aktivität dünner Schichten radioaktiver uraniumhaltiger Mineralien verglichen. Ein großer Teil der an Uraniummineralien beobachteten Aktivität rührt von dem in ihnen enthaltenen Radium her. Folglich muß, wenn Radium in genetischem Zusammenhange mit Uranium steht, die beobachtete Aktivität dem vorhandenen Uraniumgehalt proportional sein. Es ergab sich, daß dies innerhalb der Versuchsfehlergrenzen der Fall ist. Die Konstanz des beobachteten Verhältnisses würde also darauf hindeuten, daß auch das in Uraniummineralien gefundene Aktinium von Uranium erzeugt ist. Direktere Versuche, welche später von Boltwood angestellt worden sind, haben ergeben, daß dies der Fall ist.

Bei den Versuchen von Strutt und von Boltwood gelangte eine direktere Methode zur Anwendung. Die in Mineralien vorhandene Radiummenge wurde nach der Emanationsmethode bestimmt, während der Prozentgehalt an Uranium durch die chemische Analyse bestimmt wurde. Beide Forscher folgerten, daß der Radiumgehalt in Mineralien dem Uraniumgehalt proportional ist. Die Versuche von Boltwood zeigen eine besonders enge Übereinstimmung des Verhältnisses für verschiedene Mineralien mit stark verschiedenem Radiumgehalt und von verschiedener Herkunft und verschiedenem geologischen Alter. Die Schwankungen im Werte dieses Verhältnisses betrugen in den meisten Fällen nur wenige Prozent.

Durch die Versuche ist somit bündig dargetan, daß der in irgendeinem Mineral auf ein Gramm Uranium vorhandene Radiumgehalt eine bestimmte Konstante ist. Der Wert dieser Konstanten ist von mir und Boltwood§ nach der Emanationsmethode und von Eve‖ nach der γ-Strahlenmethode bestimmt worden. Beide Methoden ergeben übereinstimmend, daß sich mit einem Gramm Uranium $3,8 \cdot 10^{-7}$ Gramm Radium vergesellschaftet finden. Die Radiummenge ist hier auf eine Radiumprobe bezogen, welche eine Wärmewirkung von 110 Grammkalorien per Gramm und Stunde lieferte und vermutlich nahezu rein ist.

Diese Verwandtschaft zwischen den beiden Elementen Uranium und Radium steht in der Chemie einzig da, denn es ist das erste Mal, daß man die vorhandene Menge des einen Elementes voraussagen kann, wenn man die des andern kennt. Die Konstanz dieses Verhältnisses in allen Mineralien ist ein vollgültiger Beweis dafür, daß Radium aus Uranium erzeugt wird. Sie wirft jedoch kein Licht auf die Frage nach der Existenz von Zwischenprodukten zwischen Uranium X und Radium. Wir haben gesehen, daß

---

* McCoy, Chem. Ber. 1904, S. 2641.
† Strutt, Nature, 17. März 1904; 7. Juli 1904. — Proc. Roy. Soc. 1905.
‡ Boltwood, Nature, 25. Mai 1904. — Phil. Mag., April 1905.
§ Sill. Journ. Juli 1906. (*Vol. I v. 856*)   ‖ Eve Sill. Journ., Juli 1906.

reines Uranium selbst anfänglich nicht in meßbaren Mengen zu Radium wird. Es war daher notwendig, das Zwischenprodukt zwischen Uranium und Radium zu suchen, welches unmittelbar in Radium umgewandelt wird. Diese unbekannte Substanz mußte in Uraniummineralien vorhanden sein, doch fehlten alle Anzeichen dafür, ob sie unter der Emission von Strahlen umgewandelt wurde oder nicht. Nach dieser Muttersubstanz hat Boltwood* gesucht. Er stellte einige Versuche an Aktiniumpräparaten an, die aus einem Kilogramm des Minerals Carnotit nach einer besonderen Methode abgeschieden worden waren. Nach einer entsprechenden Behandlung wurde eine Spur Thorium zusammen mit etwas Oxalsäure zugesetzt. Letztere fällte das Aktinium mit dem Thorium zusammen aus. Die Anwesenheit von Aktinium ist leicht durch die charakteristische kurzlebige Emanation festzustellen, welche es aussendet. Eine Lösung dieses Aktiniumpräparates wurde zunächst so weit wie möglich von Radium befreit, und der zurückbleibende Radiumgehalt bestimmt. Nach Verlauf von 193 Tagen wurde die Lösung wieder untersucht; sie zeigte jetzt einen Zuwachs von $8{,}5 \cdot 10^{-9}$ Gramm Radium. Berücksichtigt man den Uraniumgehalt in dem Mineral, aus welchem das Aktinium abgeschieden worden war, so erfolgte die Erzeugung von Radium ungefähr in demselben Maße, wie theoretisch zu erwarten wäre, wenn der unmittelbare Vorfahre des Radiums in dem Aktiniumpräparat vorhanden wäre. Wie Boltwood erklärt hat, erschien es wahrscheinlich, daß Aktinium unmittelbar in Radium umgewandelt würde, und daß Aktinium und seine Familie von Zerfallsprodukten zwischen Uranium X und Radium vermittelten.

Die schnelle Erzeugung von Radium in Aktiniumpräparaten wurde durch meine eigenen Versuche† bestätigt. Im Jahre 1904 hatte ich einen Versuch eingeleitet, um zu untersuchen, ob aus Aktinium Radium erzeugt würde. Ich löste ein Gieselsches Aktiniumpräparat in Säure und entfernte aus der Lösung den größten Teil des Radiums durch Ausfällen des Baryums als Sulfat. Der vorhandene Radiumgehalt wurde nach der Emanationsmethode beobachtet; dazu wurde Luft durch die Lösung hindurch aspiriert. Während einer Periode von drei Monaten war keine Zunahme mit Sicherheit zu bemerken. Das Präparat wurde dann beiseite gestellt und aus verschiedenen Gründen erst wieder untersucht, nachdem Boltwood seine Entdeckung veröffentlicht hatte. Eine Prüfung des Präparates ergab, daß in dem inzwischen verflossenen Zeitraum von 2,7 Jahren eine große Zunahme des Radiumgehalts eingetreten war.‡ Ich begann dann Versuche, um genau zu bestimmen, in welchem Maße die Radiumzunahme erfolgte, und um endgültig festzustellen, ob Aktinium der unmittelbare Vorfahre des Radiums sei. Es lag zunächst ein Einwand gegen die Auffassung vor, daß Aktinium der Erzeuger des Radiums sei. Wenn jedes Atom radioaktiver Materie beim Zerfall ein α-Teilchen aussendet, so müßte das in der

---

* Boltwood, Nature, 15. Nov. 1906.

† Rutherford, Bakerian Lecture, Phil. Trans. (A), 1904, S. 218. (*Vol. I, p. 671.*)

‡ Rutherford, Nature, 17. Jan. 1907. (*Vol. I, p. 907.*)

Pechblende vorhandene Aktinium nebst seinen Produkten nahezu denselben Anteil zu der gesamten α-Strahlenaktivität des Uraniumminerals beitragen wie das Radium und seine Familie von Produkten. Boltwood hatte jedoch gezeigt, daß dies nicht der Fall wäre, da Aktinium mit seinen Produkten nur etwa ein Viertel der von der Radiumfamilie herrührenden Aktivität besäße. Um dieser Schwierigkeit zu begegnen, mußte man nach dem Vorschlage Boltwoods* notwendig annehmen, daß jedes Atom der verschiedenen Radiumprodukte vier α-Teilchen gegenüber dem einen des Aktiniums aussende. Ich habe zunächst einen Versuch angestellt, um festzustellen, ob das Anwachsen des Radiums in Aktiniumlösungen konstant sei. Das in den Aktiniumlösungen ursprünglich vorhandene Radium wurde durch Ausfällen mit Ammoniumsulfid fast vollständig entfernt. Dann wurde die Lösung beiseite gestellt und die vorhandene Radiummenge in kurzen Zwischenräumen untersucht. Es zeigte sich, daß die Schnelligkeit der Radiumerzeugung konstant war, denn der Radiumgehalt der Lösung nahm proportional mit der Zeit zu. Im Verlaufe weiterer Beobachtungen an anderen Lösungen† bemerkte ich, daß die Geschwindigkeit der Erzeugung von Radium nur etwa ein Drittel des Wertes hatte, der ursprünglich für dieselbe Aktiniummenge beobachtet worden war. In einem andern Falle erhielt ich ein Aktiniumpräparat, welches während eines Zeitraumes von 20 Tagen keine merkliche Zunahme an Radium zeigte. Die Entstehung von Radium, falls eine solche überhaupt eintrat, blieb sicherlich unter $1/_{500}$ des an der ersten Lösung beobachteten Betrages. Diese Ergebnisse führten zu dem Schlusse, daß in dem Aktiniumpräparate eine neue Substanz vorhanden sei, welche sich unmittelbar in Radium verwandelte, und daß Aktinium selbst dabei weiter keine besondere Rolle spielte, als daß die neue Substanz gewöhnlich mit ihm zusammen abgeschieden wurde. Das Fehlen einer Entstehung von Radium in der einen Aktiniumlösung zeigt, daß die neue Substanz sich vom Aktinium hinsichtlich ihrer chemischen Eigenschaften unterscheidet und durch geeignete chemische Verfahren vollständig beseitigt werden kann.

Die Richtigkeit dieser Schlußfolgerung wurde weiterhin noch durch den Nachweis dafür bekräftigt, daß das Maß, in welchem Radium für gewöhnlich in Aktiniumpräparaten ensteht, durch Entfernung der Produkte des Aktinium X nicht beeinflußt wird. Außerdem wurde nachgewiesen, daß der aktive Radiumniederschlag, welcher das Endprodukt des Aktiniums enthält, eine merkliche Radiumzunahme aufwies. Die Versuche wurden von Boltwood‡ mit dem gleichen Ergebnisse wiederholt und erweitert, und zwar an den Aktiniumpräparaten, die er aus Carnotit gewonnen hatte. Boltwood fand, daß in einem Präparate das Radium während eines Zeitraumes von 600 Tagen in konstantem Maße zunahm. Dieses Tempo der Entstehung blieb unbeeinflußt durch die Beseitigung des Radioaktiniums

---

* Boltwood, Nature, 3. Jan. 1907.

† Rutherford, Nature, 6. Juni 1907. — Phil. Mag., Dez. 1907. (*This vol., p. 34 and p. 42.*)   ‡ Boltwood, Nature, 26. Sept. 1907; 10. Okt. 1907.

und des Aktinium *X* aus den Lösungen. Boltwood war nicht imstande, eine Abscheidung des radiumerzeugenden Produktes mit dem von ihm verwandten Reagens, nämlich Ammoniumsulfid, zu bewerkstelligen; dagegen fand er Natriumthiosulfat wirksam. Er konnte damit den Erzeuger des Radiums vollständig vom Aktinium trennen und eine genügende Menge dieser Substanz erhalten, um ihre radioaktiven Eigenschaften zu untersuchen. Diesem neuen Körper, welcher unmittelbar in Radium umgewandelt wird, hat Boltwood den Namen „Ionium" beigelegt. Er hat die chemischen Eigenschaften des Thoriums und ist von diesem nur schwer vollständig zu trennen. Er sendet α-Strahlen aus, die unter allen bekannten das geringste Durchdringungsvermögen besitzen. Nach der Szintillationsmethode und der elektrischen Methode ergab sich, daß diese Stahlen in einer Luftschicht von etwa 3 cm Dicke absorbiert werden. Die Reichweiten der α-Strahlen aus Uranium und Radium betragen ungefähr 3,5 cm. Dieses Ergebnis steht in Einklang mit der Beobachtung, daß das neue Element, mit Uranium im Gleichgewicht, eine Aktivität von 0,8 der des eigentlichen mit ihm vergesellschafteten Radiums hat. W. Marckwald und Keetman* haben die Abscheidung des Ioniums aus Uraniummineralien nach einer Methode bewerkstelligt, die von der Boltwoodschen etwas abweicht; dabei haben sie die Ergebnisse Boltwoods bestätigt gefunden. Es ist noch nicht genügend Zeit verflossen, um Beobachtungen über die Änderung der Aktivität des Ioniums mit der Zeit zu machen. Wir werden weiter unten sehen, daß, wenn es eine einfache Substanz ist, Gründe zu der Annahme vorhanden sind, daß es eine Umwandlungsperiode hat, die mit der des eigentlichen Radiums vergleichbar ist.

Die Voraussagungen der Umwandlungstheorie sind in schlagender Weise bestätigt worden. Radium ist ein Element, das sich verwandelt und fortwährend durch die Umwandlung eines anderen Elements, des Ioniums, erzeugt wird. Dieses wiederum entsteht aus der Umwandlung des Uraniums. Uranium hat eine so lange Umwandlungsperiode, vermutlich ungefähr 5000 Millionen Jahre, daß es für den Augenblick nicht Erfolg verspricht, einen noch früheren Vorfahren des Radiums zu suchen. Die Stellung des Radiums in der Umwandlungsreihe des Uraniums ist in nachfolgender Zeichnung

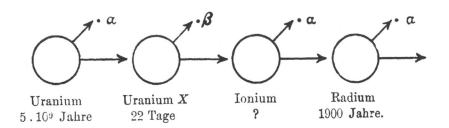

Uranium	Uranium *X*	Ionium	Radium
5 . 10⁹ Jahre	22 Tage	?	1900 Jahre.

---

* W. Marckwald und Keetman, Chem. Ber. **51**, 49, 1908.

dargestellt, in welcher die Perioden der einzelnen Produkte unter ihren Namen angegeben sind.

Alle einschlägigen Versuche stimmen somit gut überein.

Reines Uranium ergibt keine merkliche Radiumzunahme in begrenzter Zeit, da eine andere Substanz von langer Umwandlungsperiode, „Ionium", zwischen ihnen steht. Die Radiummenge und ebenso die Ioniummenge sind beide der Uraniummenge in einem Mineral proportional, da Uranium mit beiden genetisch verbunden ist.*

Ein sehr interessantes Schlaglicht werfen auf dieses Problem einige neuere Beobachtungen von Hahn.† Wir haben allen Grund zu der Annahme, daß Thorium und Uranium vom radioaktiven Gesichtspunkt aus vollkommen unabhängige Elemente sind; Hahn hat indessen gefunden, daß eine beständige Radiumzunahme in Thoriumpräparaten erfolgt, die anfangs von Radium frei waren. Bei Untersuchung käuflicher Thoriumpräparate fand er ferner, daß der vorhandene Radiumgehalt dem Alter des Präparates proportional war. Hahn hat diese Beobachtungen in sehr einfacher Weise erklärt. Ein großer Teil des käuflichen Thoriums ist aus Monazitsand hergestellt, welcher eine kleine Spur Uranium, etwa 0,3 v. H., enthält und, wie Boltwood gezeigt hat, auch die entsprechende Menge Radium. Nun haben wir gesehen, daß Boltwood am Ionium chemische Eigenschaften gefunden hat, die denen des Thoriums sehr nahe verwandt sind. Bei den Methoden zur fabrikmäßigen Abtrennung des Thoriums aus dem Monazitsand wird das Ionium mit ihm zusammen ausgeschieden. Das Ionium in den Thoriumpräparaten, und nicht das Thorium selbst gibt die Veranlassung zu dem beobachteten Enstehen von Radium. Die Richtigkeit dieser Schlußfolgerung wurde durch den Nachweis bestätigt, daß das Produkt des Thoriums, das Radiothorium, keine Enstehung von Radium aufwies. Dieses Produkt war chemisch frei von Ionium und erzeugte folglich kein Radium. Eine weitere Bestätigung für diese Erklärung lieferte der Nachweis, daß die auf Grund der verfügbaren rohen Zahlen berechnete Periode des Radiums ungefähr von der richtigen Größe war.

Es ist sehr lehrreich zu bemerken, wie gelegen diese Entdeckung der Enstehung von Radium in Thoriumpräparaten erfolgte. Wenn die Entdeckung, wie es leicht der Fall hätte sein können, einige Jahre früher gemacht worden wäre, so würde sie vermutlich das Problem arg verwirrt und die richtige Erklärung der Abstammung des Radiums verzögert haben.

Um die Kette der experimentellen Beweise zu schließen, ist noch der Nachweis erforderlich, daß Uranium Ionium erzeugt, und daß anfänglich radiumfreies Uranium im Laufe der Jahre ein Anwachsen von Radium

---

* Einstweilen deutet nichts darauf hin, daß Aktinium in irgendwelcher direkten radioaktiven Verbindung mit Radium steht. Boltwood hat gefunden, daß der Aktiniumgehalt in verschiedenen Mineralien dem Uraniumgehalt proportional ist. Dieser Zusammenhang deutet darauf hin, daß das Aktinium in genetischer Verbindung mit dem Uranium steht. Seine richtige Stellung im Schema der Umwandlungsformen des Uraniums ist noch eine offene Frage.

† Hahn, Nature **77**, 30—31 1907 (14. Nov.) — Chem. Ber. **40**, 4415, 1907.

aufweisen wird. Es dürfte lehrreich sein, auf Grund der uns zur Verfügung stehenden Daten die wahrscheinliche Umwandlungsperiode des Ioniums zu berechnen, da von dieser die Möglichkeit abhängt, innerhalb einer begrenzten Zeit die Erzeugung von Ionium und das Auftreten von Radium in Uraniumlösungen nachzuweisen.

Es seien P_0, Q_0, R_0 die Anzahl der Atome Uranium, Ionium und Radium, wenn sie in dem Mineral im Gleichgewicht sind, λ_1, λ_2, λ_3 ihre einzelnen Umwandlungskonstanten. Es folgt, daß

$$\lambda_1 P_0 = \lambda_2 Q_0 = \lambda_3 R_0$$

sein muß. Für die Rechnung können wir ohne merklichen Fehler das kurzlebige Produkt Uranium X unberücksichtigt lassen und annehmen, daß sich Uranium unmittelbar in Ionium und Ionium unmittelbar in Radium verwandle. Der Einfachheit halber wollen wir annehmen, daß die Periode des Uraniums im Vergleich zu der des Ioniums oder des Radiums sehr lang sei. Die Anzahl Q der Atome Ionium, die aus einer Menge P_0 reinen Uraniums in einer gegenüber der Periode des Ioniums oder des Radiums sehr kurzen Zeit t gebildet werden, ist $\lambda_1 P_0 t$. Die in der kurzen Zeit dt erzeugte Radiummenge ist nun $\lambda_2 Q dt$, folglich ist die in dem Zeitraum gebildete Radiummenge R gegeben durch die Gleichung:

$$R = \int_0^t \lambda_1 \lambda_2 P_0\, t\, dt = \frac{1}{2}\lambda_1 \lambda_2 P_0 t^2 = \frac{1}{2}\lambda_2 \lambda_3 R_0 t^2.$$

Folglich wird die gebildete Radiummenge anfänglich proportional dem Quadrate der Zeit zunehmen.

Wir können leicht einen unteren Grenzwert für die Periode des Ioniums aus den Versuchsdaten berechnen. Boltwood fand, daß in 100 Gramm Uraniumnitrat, die ungefähr 50 Gramm reinen Uraniums entsprachen, die Bildung von Radium jedenfalls nicht mehr betrug als 10^{-11} Gramm in 2,5 Jahren. Nun ist die Gleichgewichtsmenge R_0 des Radiums, die 50 Gramm Uranium entspricht, $50 \cdot 3,84 \cdot 10^{-7}$ Gramm. Nehmen wir als Zeiteinheit ein Jahr, so ist λ_3, die Konstante des Radiums, gleich $3,7 \cdot 10^{-4}$. Setzen wir diese Werte

$$R = 10^{-11}$$
$$\lambda^3 = 3,7 \cdot 10^{-4},$$
$$R_0 = 1,92 \cdot 10^{-5},$$
$$t = 2,5$$

in die Gleichung ein, so sehen wir, daß λ_2 nicht kleiner ist als $4,6 \cdot 10^{-4}$, das heißt, die Periode des Ioniums beträgt nicht weniger als 1500 Jahre. Dies ist die niedrigste Schätzung des Wertes. Die tatsächliche Periode mag sich schließlich als viel länger erweisen.

Da der Radiumgehalt im Uranium proportional dem Quadrat der Zeit zunimmt, so kann eine Bildung von Radium, die nach einem Jahre unnachweisbar ist, nach Verlauf von 10 Jahren sehr wohl im Bereich der Meßbarkeit liegen.

Die Periode des Radiums

Es dürfte von Interesse sein, die Einzelheiten der Methode zu betrachten, welche B o l t w o o d angewandt hat, um die Umwandlungsperiode des Radiums durch direkte Versuche zu bestimmen. Auf den ersten Blick erscheint es einigermaßen überraschend, daß die Periode einer sich so langsam umwandelnden Substanz wie Radium aus Beobachtungen abgeleitet werden kann, die sich nur über wenige Monate erstrecken. In dem Mineral, wo Radium mit seinem Vater Ionium im Gleichgewicht ist, muß die Geschwindigkeit der Bildung frischen Radiums aus Ionium sehr nahezu dem Betrage der Umwandlung des vorhandenen Radiums gleich sein. Es sei Q die in dem Mineral vorhandene Radiummenge, gemessen nach der Emanationsmethode an der Bewegung des Goldblättchens eines Elektroskops. Wenn nun alles Ionium durch chemische Verfahren isoliert ist, wird es sofort anfangen, zu Radium zu werden, und zwar in dem Maße wie vor der Trennung in dem Mineral. Diese Bildungsgeschwindigkeit ist, wie wir gesehen haben, während einer Periode von mehreren Jahren praktisch konstant und wird an der Bewegung des Goldblättchens desselben Elektroskops wie zuvor gemessen. Da die Bildungsgeschwindigkeit q gleich der Umwandlungsgeschwindigkeit λR des Radiums ist, so ist

$$\lambda = \frac{q}{R}.$$

Da q und R in Angaben desselben Instruments ausgedrückt sind, so ist λ bekannt. Auf diese Weise fand B o l t w o o d für λ einen Wert von ungefähr $3,7 \cdot 10^{-4}$/Jahr und demgemäß für die Periode des Radiums ungefähr 1900 Jahre. Das Ergebnis ist unabhängig von allen Fragen nach der Reinheit der Radiumpräparate, wie sie bei meinen früheren Berechnungen der vermutlichen Periode des Radiums mitspielten. Die von B o l t w o o d gefundene Periode ist nicht sehr verschieden von der Periode von 2600 Jahren die ich auf Grund der Annahme berechnet habe, daß das α-Teilchen ein Heliumatom ist, das eine Ionenladung trägt.

R a m s a y und C a m e r o n* haben kürzlich, gestützt auf Messungen des Volumens der Emanation, die Periode des Radiums zu 236 Jahren berechnet. Diese Periode ist von einer ganz anderen Größenordnung als die von B o l t w o o d bestimmte und läßt sich unmöglich mit den sonstigen radioaktiven Daten in Einklang bringen. Es bedarf weiterer Messungen über das Volumen der reinen Emanation, um zu entscheiden, ob die große Verschiedenheit zwischen den nach den beiden Versuchsmethoden bestimmten Werten der Periode nur scheinbar oder reell ist.

(Aus dem Englischen übersetzt von M a x   I k l é.)

(Eingegangen 11. April 1908.)

* Ramsay and Cameron, Trans. Chem. Soc. **91**, 1266, 1907.

The Boiling Point of the Radium Emanation

From *Nature*, **79**, 1909, pp. 457–8

IT was shown by Rutherford and Soddy in 1903 that the radium emanation was condensed from the gases with which it was mixed at a temperature of about $-150°$ C. From observations of the range of temperature of condensation and volatilisation it was concluded that the condensed emanation exerted a sensible vapour pressure. This has been confirmed by later experiments, using much larger quantities of emanation. Sir William Ramsay and Cameron have pointed out that the emanation, condensed in a glass tube kept at the temperature of liquid air, can be removed by continuous pumping, thus indicating appreciable vapour pressure even at that low temperature. I have found that the rate of removal of the emanation in this way increases rapidly as the temperature of complete volatilisation is approached.

In the initial experiments of Rutherford and Soddy only very small quantities of radium were available, and the partial pressure of the emanation in the experiments was exceedingly small. If the emanation behaves like an ordinary gas, it is to be expected that the boiling point of pure emanation at atmospheric pressure should be much higher. I have recently made experiments to test this point. As the volume of pure emanation available in the present experiments was only about 1/20 cubic millimetre it was necessary to employ special methods to investigate the boiling point of the emanation at various pressures. Purified emanation corresponding to the equilibrium amount from about 100 milligrams of radium was compressed into a fine glass capillary of about 1/20 millimetre diameter. The end of the capillary dipped into a pentane bath, which was cooled down to any desired temperature, measured by means of a thermocouple. The point of initial condensation was marked by the appearance of a brilliant point of phosphorescent light, due to condensed emanation, at the coldest part of the capillary. In this way I have found that the temperature of initial condensation of the emanation rises from about $-150°$ C. at a very low pressure to about $-65°$ C. at atmospheric pressure. This fixes the boiling point of the emanation at atmospheric pressure at about $-65°$ C., or $208°$ absolute.

As it is a difficult matter to purify completely the small volume of emanation and to keep it pure, the observed pressure of the emanation and mixed gases at the temperature of condensation was corrected for by taking the true volume of the emanation from 1 gram of radium in equilibrium as $0 \cdot 585$ cubic millimetre. This calculated volume is in excellent agreement with the minimum value which I have found experimentally. As the emanation is apparently an inert gas of atomic weight 222, it is of interest to compare its

F

boiling point with those of the heavier inert gases found in the atmosphere. The boiling points of argon, krypton, xenon, and emanation are, respectively 86·9, 121·3, 163·9, and 208 degrees absolute. It will be noted that as the boiling point of krypton is about intermediate between that of argon and xenon, so the boiling point of xenon is nearly the mean between that of krypton and emanation.

If the capillary tube containing pure emanation is quickly placed in the pentane bath, cooled well below the temperature of initial condensation, under a microscope small drops of liquid emanation are seen on the walls of the capillary. The position of each globule is marked by a brilliant local phosphorescence of the glass of the capillary.

E. RUTHERFORD

University, Manchester
February 13

The Nature of the α Particle from Radioactive Substances

by PROFESSOR E. RUTHERFORD, F.R.S.,
and T. ROYDS, M.SC., 1851 *Exhibition Science Scholar*

From the *Philosophical Magazine* for February 1909, ser. 6, xvii, pp. 281–6

THE experimental evidence collected during the last few years has strongly supported the view that the α particle is a charged helium atom, but it has been found exceedingly difficult to give a decisive proof of the relation. In recent papers, Rutherford and Geiger* have supplied still further evidence of the correctness of this point of view. The number of α particles from one gram of radium have been counted, and the charge carried by each determined. The values of several radioactive quantities, calculated on the assumption that the α particle is a helium atom carrying two unit charges, have been shown to be in good agreement with the experimental numbers. In particular, the good agreement between the calculated rate of production of helium by radium and the rate experimentally determined by Sir James Dewart, is strong evidence in favour of the identity of the α particle with the helium atom.

The methods of attack on this problem have been largely indirect, involving considerations of the charge carried by the helium atom and the value of e/m of the α particle. The proof of the identity of the α particle with the helium atom is incomplete until it can be shown that the α particles, accumulated quite independently of the matter from which they are expelled, consist of helium. For example, it might be argued that the appearance of helium in the radium emanation was a result of the expulsion of the α particle, in the same way that the appearance of radium A is a consequence of the expulsion of an α particle from the emanation. If one atom of helium appeared for each α particle expelled, calculation and experiment might still agree, and yet the α particle itself might be an atom of hydrogen or of some other substance.

We have recently made experiments to test whether helium appears in a vessel into which the α particles have been fired, the active matter itself being enclosed in a vessel sufficiently thin to allow the α particles to escape, but impervious to the passage of helium or other radioactive products.

The experimental arrangement is clearly seen in the figure. The equilibrium quantity of emanation from about 140 milligrams of radium was purified and compressed by means of a mercury-column into a fine glass tube A about 1·5 cms. long. This fine tube, which was sealed on a larger capillary tube B,

---

* Proc. Roy. Soc. A. lxxxi. pp. 141–173 (1908). (*This vol., pp. 89–120.*)
† Proc. Roy. Soc. A. lxxxi. p. 280 (1908).

was sufficiently thin to allow the α particles from the emanation and its products to escape, but sufficiently strong to withstand atmospheric pressure. After some trials, Mr. Baumbach succeeded in blowing such fine tubes very

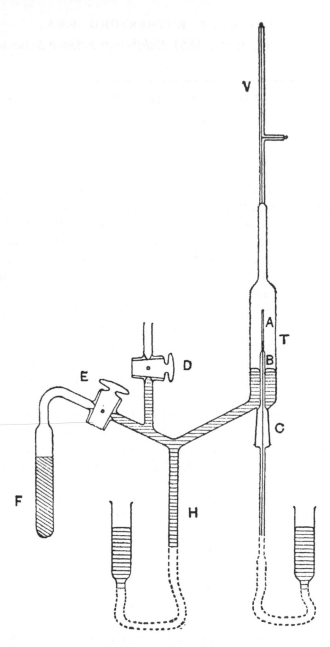

uniform in thickness. The thickness of the wall of the tube employed in most of the experiments was less than $\frac{1}{100}$ mm., and was equivalent in stopping power of the α particle to about 2 cms. of air. Since the ranges of the α particles from the emanation and its products radium A and radium C are 4·3, 4·8, and 7 cms. respectively, it is seen that the great majority* of the α particles expelled by the active matter escape through the walls of the tube. The ranges of the α particles after passing through the glass were determined with the aid of a zinc-sulphide screen. Immediately after the introduction of the emanation the phosphorescence showed brilliantly when the screen was close to the tube, but practically disappeared at a distance of 3 cms. After an hour, bright phosphorescence was observable at a distance of 5 cms. Such a result is to be expected. The phosphorescence initially observed was due mainly to the α particles of the emanation and its product radium A (period 3 mins.). In the course of time the amount of radium C, initially zero, gradually increased, and the α radiations from it of range 7 cms. were able to cause phosphorescence at a greater distance.

The glass tube A was surrounded by a cylindrical glass tube T, 7·5 cms. long and 1·5 cms. diameter, by means of a ground-glass joint C. A small vacuum-tube V was attached to the upper end of T. The outer glass tube T was exhausted by a pump through the stopcock D, and the exhaustion completed with the aid of the charcoal tube F cooled by liquid air. By means of a mercury column H attached to a reservoir, mercury was forced into the tube T until it reached the bottom of the tube A.

Part of the α particles which escaped through the walls of the fine tube were stopped by the outer glass tube and part by the mercury surface. If the α particle is a helium atom, helium should gradually diffuse from the glass and mercury into the exhausted space, and its presence could then be detected spectroscopically by raising the mercury and compressing the gases into the vacuum-tube.

In order to avoid any possible contamination of the apparatus with helium, freshly distilled mercury and entirely new glass apparatus were used. Before introducing the emanation into A, the absence of helium was confirmed experimentally. At intervals after the introduction of the emanation the mercury was raised, and the gases in the outer tube spectroscopically examined. After 24 hours no trace of the helium yellow line was seen; after 2 days the helium yellow was faintly visible; after 4 days the helium yellow and green lines were bright; and after 6 days all the stronger lines of the helium spectrum were observed. The absence of the neon spectrum shows that the helium present was not due to a leakage of air into the apparatus.

There is, however, one possible source of error in this experiment. The helium may not be due to the α particles themselves, but may have *diffused* from the emanation through the thin walls of the glass tube. In order to test this point the emanation was completely pumped out of A, and after some hours a quantity of helium, about 10 times the previous volume of the emanation, was compressed into the same tube A.

The outer tube T and the vacuum-tube were removed and a fresh apparatus substituted. Observations to detect helium in the tube T were made at

---

* The α particles fired at a very oblique angle to the tube would be stopped in the glass. The fraction stopped in this way would be small under the experimental conditions.

intervals, in the same way as before, but no trace of the helium spectrum was observed over a period of eight days.

The helium in the tube A was then pumped out and a fresh supply of emanation substituted. Results similar to the first experiment were observed. The helium yellow and green lines showed brightly after four days.

These experiments thus show conclusively that the helium could not have diffused through the glass walls, but must have been derived from the α particles which were fired through them. In other words, the experiments give a decisive proof that the α particle after losing its charge is an atom of helium.

Other Experiments

We have seen that in the experiments above described helium was not observed in the outer tube in sufficient quantity to show the characteristic yellow line until two days had elapsed. Now the equilibrium amount of emanation from 100 milligrams of radium should produce helium at the rate of about $0 \cdot 03$ c.mm. per day. The amount produced in one day, if present in the outer tube, should produce a bright spectrum of helium under the experimental conditions. It thus appeared probable that the helium fired into the glass must escape very slowly into the exhausted space, for if the helium escaped at once, the presence of helium should have been detected a few hours after the introduction of the emanation.

In order to examine this point more closely the experiments were repeated, with the addition that a cylinder of thin sheet lead of sufficient thickness to stop the α particles was placed over the fine emanation tube. Preliminary experiments, in the manner described later,showed that the lead-foil did not initially contain a detectable amount of helium. Twenty-four hours after the introduction into the tube A of about the same amount of emanation as before, the yellow and green lines of helium showed brightly in the vacuum-tube, and after two days the whole helium spectrum was observed. The spectrum of helium in this case after one day was of about the same intensity as that after the fourth day in the experiments without the lead screen. It was thus clear that the lead-foil gave up the helium fired into it far more readily than the glass.

In order to form an idea of the rapidity of escape of the helium from the lead some further experiments were made. The outer cylinder T was removed and a small cylinder of lead-foil placed round the thin emanation-tube surrounded the air at atmospheric pressure. After exposure for a definite time to the emanation, the lead screen was removed and tested for helium as follows. The lead-foil was placed in a glass tube between two stopcocks. In order to avoid a possible release of the helium present in the lead by pumping out the air, the air was displaced by a current of pure electrolytic oxygen.* The stopcocks were closed and the tube attached to a subsidiary apparatus

* That the air was completely displaced was shown by the absence of neon in the final spectrum.

similar to that employed for testing for the presence of neon and helium in the gases produced by the action of the radium emanation on water (Phil. Mag. Nov. 1908*). The oxygen was absorbed by charcoal and the tube then heated beyond the melting-point of lead to allow the helium to escape. The presence of helium was then spectroscopically looked for in the usual way. Using this method, it was found possible to detect the presence of helium in the lead which had been exposed for only four hours to the α rays from the emanation. After an exposure of 24 hours the helium yellow and green lines came out brightly. These experiments were repeated several times with similar results.

A number of blank experiments were made, using samples of the lead-foil which had not been exposed to the α rays, but in no case was any helium detected. In a similar way, the presence of helium was detected in a cylinder of tinfoil exposed for a few hours over the emanation-tube.

These experiments show that the helium does not escape at once from the lead, but there is on the average a period of retardation of several hours and possibly longer.

The detection of helium in the lead and tin foil, as well as in the glass, removes a possible objection that the helium might have been in some way present in the glass initially, and was liberated as a consequence of its bombardment by the α particles.

The use of such thin glass tubes containing emanation affords a simple and convenient method of examining the effect on substances of an intense α radiation quite independently of the radioactive material contained in the tube.

We can conclude with certainty from these experiments that the α particle after losing its charge is a helium atom. Other evidence indicates that the charge is twice the unit charge carried by the hydrogen atom set free in the electrolysis of water.

University of Manchester
Nov. 13, 1908

* *This vol., p. 128.*

Differences in the Decay of the Radium Emanation

by PROFESSOR E. RUTHERFORD, F.R.S., *and* Y. TUOMIKOSKI

From the *Memoirs* of the Manchester Literary and Philosophical Society, 1909,
IV, vol. liii, No. 12, pp. 1–2
(Received and read March 23rd, 1909)

DETERMINATIONS of the rate of decay of activity of the radium emanations have been made by various methods by a number of observers. The activity has been found to decrease exponentially with the time with a period (*i.e.*, time taken to fall to half value), varying in different cases between 3·75 and 3·99 days.

Experiments were begun a few months ago by Mr. Tuomikoski to determine the decay of the radium emanation over a wide range of activity. For this purpose a large amount of emanation was enclosed in sealed tubes, and the decay of activity measured by the γ rays, using for the purpose an electroscope surrounded by lead. The rate of decay of the emanation has been found to be irregular, depending upon the treatment to which the emanation had been subjected. For example, a sample of emanation, purified by condensation with liquid air, commenced to decay for the first five days with an average period of 3·58 days. Between 5 and 20 days the average period was 3·75 days, while between 20 and 40 days the decay has been nearly exponential with a period of 3·85 days. Another preparation of emanation was found to decay exponentially from the beginning with a period of 4·4 days. Similar differences have been observed in a number of experiments. These variations in period of decay must be ascribed to differences of quality of the emanation in the various cases.

As a result of a series of experiments, it has been found that samples of emanation which decay most rapidly are more easily absorbed in water and more easily condensed by liquid air than the more slowly decaying fractions. For example, if the radium emanation is allowed to stand over water for some hours, the part absorbed in the water decays more rapidly than the part unabsorbed. In a similar way on condensing the emanation, the part removed by pumping has a longer period than the part condensed.

We have been unable to find any evidence that radium produces two emanations, or that the products of transformation of the emanation of slow period are in any way different from those of the emanation of quick period.

The results indicate that the emanation is a non-homogeneous chemical substance. So far as our observations have gone, it appears probable that the physical and chemical properties of the emanation atoms vary to some extent with their life, *i.e.*, on the length of time after production before disintegration. It seems probable that the atoms of emanation undergo a progressive change in properties before disintegration. Further experiments are in progress to test the correctness of this point of view.

F*

Condensation of the Radium Emanation

by E. RUTHERFORD, F.R.S.,

*Professor of Physics, University of Manchester**

From the *Philosophical Magazine* for May 1909, ser. 6, xvii, pp. 723-9

RUTHERFORD and SODDY first showed in 1903 that the radium emanation condensed from the gases with which it was mixed at a temperature of about −150° C. At that time only small quantities of radium preparations were available, so that the partial pressure of the emanation with the gases with which it was conveyed was exceedingly small. Notwithstanding the very minute quantity of emanation present, the temperatures of complete condensation and of complete volatilization were found to be sharply marked, and did not differ from each other by more than a few degrees. Some evidence was obtained that the emanation had a vapour-pressure like an ordinary gas.

This property of condensation of the emanation in liquid air has proved invaluable in all later researches as a means of separating the emanation from the inactive gases with which it is mixed.

Using large quantities of radium, Sir William Ramsay and Cameron observed that the emanation, condensed in a glass tube surrounded by liquid air, could be gradually removed by continuous pumping, indicating that the emanation exerted a sensible vapour-pressure even at that low temperature. The writer has found that the rate of removal of the emanation by pumping increases rapidly as the temperature of the emanation-tube approaches the temperature of complete volatilization of the emanation.

The temperature of condensation, viz. −150° C., found by Rutherford and Soddy corresponded to the liquefaction-point of the emanation under a very low pressure. If the emanation behaves like an ordinary gas, the temperature of initial condensation should rise with increase of pressure of the emanation. It was consequently of interest to examine how the condensation-point of the emanation varied with pressure, and to fix its boiling-point under atmospheric pressure.

Special experimental methods are necessary in order to determine the vapour-pressure of the very small volume of emanation available. It has been shown in the experiments of the writer† that the volume of pure emanation from one gram of radium in equilibrium is about 0·6 cubic mm. at normal

---

* A preliminary account of the results was communicated as a letter to 'Nature,' Feb. 18, 1909. (*This vol., p. 161.*)   † Phil. Mag. Aug. 1908. (*This vol., p. 72.*)

pressure and temperature. This is in good agreement with the calculated value, viz. 0·585 cubic mm., which has been deduced by Rutherford and Geiger.* In the present experiments, the amount of emanation, available after the process of purification, was equivalent to the equilibrium amount from 140 mgs. of radium. Taking the calculated volume of the emanation, this corresponds to a volume of pure emanation of 0·082 cubic mm.

In order to obtain a column of gas of several centimetres length at atmospheric pressure, it was consequently necessary to employ capillary tubes of fine bore. In the experiments recorded later, glass capillary tubes were employed of diameter varying between about 0·05 mm. and 0·15 mm.

After purification of the emanation in the manner described in a previous paper,† the emanation was allowed to expand into a vertical glass reservoir and then compressed by raising the mercury into the capillary tube fixed at the top. This tube of length nearly 20 cms. was bent twice at right angles, so that the free end of length about 8 cms. was vertical and dipped downwards. The end of the capillary, of external diameter about 1 mm., was immersed in liquid pentane contained in a small unsilvered Dewar cylinder. The temperature of the bath could be varied by circulating liquid air through a glass U-tube placed in the liquid. The temperature of the bath, which was kept well stirred, was determined by means of a nickel-iron thermo-junction in series with a D'Arsonval galvanometer. The deflexions on the scale were calibrated by keeping one junction in a freezing-mixture and immersing the other successively in (1) a paste of solid carbon-dioxide and ether ($-78°·2$), (2) boiling ethylene ($-103°·5$), and (3) liquid air whose percentage of oxygen was determined.

As the complete purification of the emanation is a long and tedious process, and it is difficult to keep it pure over a wide range of observations, many of the experiments were made with emanation of about 50 to 60 per cent. purity. The true volume of the emanation present was determined by comparing its γ ray activity by means of an electroscope with that of a standard radium preparation, assuming that the true volume of the emanation from one gram of radium is 0·585 cubic mm. The actual volume occupied by the emanation and impurities was measured in the capillary at atmospheric pressure. By comparison of the calculated with the observed volume, the percentage of impurity was determined and also the correction to be applied to the observed pressure to give the true partial pressure of the emanation.

In the experiments on the condensation-point of the emanation for pressures above 5 cms., a capillary tube of mean diameter 0·05 mm. was used. The cross-section of the capillary was found to be elliptical in shape, the axes being 0·048 mm. and 0·052 mm. respectively. In this capillary the emanation from 100 mgrs. of radium would occupy a length of 3·0 cms. It was found necessary to use such a fine bore in order to be able to cool down

---

* Proc. Roy. Soc. A. lxxxi. p. 162 (1908). (*This vol., p. 119.*)
† Rutherford, Phil. Mag. Aug. 1908. (*This vol., p. 72.*)

the end of the capillary to the temperature of the pentane bath and yet to keep the mercury column from freezing.

The point of condensation was found in most cases to be well defined. At the moment of condensation a brilliant phosphorescent point of light due to condensed emanation appeared at the extreme end of the capillary. The temperature of the pentane bath was noted at the moment the phosphorescent point disappeared. This corresponded to the temperature of initial condensation at that particular pressure of the emanation.

Preliminary experiments made in this way showed that the condensation temperature of the emanation like that of all gases rose with increase of pressure of the emanation. If the emanation started condensing at a particular temperature, lowering of the pressure at once caused a rapid volatilization of the condensed emanation.

A number of experiments were made of the initial condensation-point of the emanation at atmospheric pressure. This was found to be about $-65°$ C., the temperature of the pentane bath being measured, both by the thermocouple and a pentane thermometer, and the pressure being kept constant. This fixes the true boiling-point of the emanation as $-65°$ C. or $208°$ absolute. Since the emanation gives out heat, the temperature of the inside of the capillary was no doubt slightly higher than that of the pentane bath. The experiments are not, however, of sufficient precision to introduce small corrections of this kind.

A number of experiments were made on the vapour-pressure of the emanation, using a paste of solid carbon dioxide and ether to give a constant temperature ($-78°·2$ C.). As it was difficult to view the end of the capillary through the opaque paste, it was found convenient to remove rapidly the refrigerant, and observe at the moment of removal whether the emanation was condensed in the tube. It was found in this way possible to fix the point of initial condensation with considerable accuracy. An increase of one per cent. in the pressure was sufficient to cause a transition from no condensation to well-marked condensation.

The following table gives the vapour-pressure of the emanation at various temperatures.

Vapour-Pressure	Temperature
76 cms.	$-65°$ C.
25 ,,	$-78°$
5 ,,	$-101°$
0·9 ,,	$-127°$

The mean of experiments showed that the emanation commenced to condense at $-78°·2$ at a pressure of 23 cms. It was found that the amount of impurity with the emanation had no influence on its condensation-point when the true partial pressure of the emanation was deduced in the manner previously discussed. An interesting effect was noticed in these experiments.

On removal of the refrigerant, occasionally some of the paste adhered to the capillary. The rapid evaporation of this caused a local lowering of the temperature sufficient to cause a marked condensation of the emanation at points on the tube, even though no condensation was observed in the tube when in the bath itself.

A number of experiments have been made at still lower pressures, but the results of these will be reserved for a later paper. The vapour-pressure curve of the emanation is similar in general respects to that of carbon dioxide although, as we have seen, the boiling-point of the emanation is somewhat higher.

An unexpected effect was observed in these experiments which is still under examination. It was found that when the pressure was kept constant, the emanation did not all condense when once the condensation had started, but a considerable raising of the pressure was necessary to produce complete condensation. I do not think this effect could be ascribed to slowness of diffusion of the emanation in the capillary tube. It appeared as if the emanation were not homogeneous and that some of the emanation condensed at a lower temperature than the remainder. An investigation* in conjunction with Mr. Tuomikoski has lent support to this point of view. It has been found that, on condensation of the emanation, the uncondensed part which is pumped off has on the average a slower rate of decay than the part condensed. A more complete determination of the vapour-pressure curve of the emanation is withheld until this point has been more completely examined.

Experiments with Liquefied Emanation

If the emanation in the capillary tube at about atmospheric pressure is plunged suddenly in a refrigerant well below the temperature of initial condensation, the emanation is condensed locally at several parts of the tube, probably at points where the glass is thinnest. The liquid emanation causes an intense greenish coloured phosphorescence on the walls of the tube. If the liquid emanation is concentrated in the bottom of the capillary, apart from the local phosphorescence of the glass, it appears practically colourless when viewed by a microscope by transmitted light. A very different effect is observed when the emanation is condensed in liquid air at the bottom of the capillary. A few seconds after the beginning of condensation, the emanation shows a reddish tinge and rapidly becomes orange-coloured. It is difficult to be certain whether the emanation under these conditions is in the liquid or solid state. The term 'liquid emanation' will be used for convenience in describing the effects observed. If kept in liquid air, the condensed emanation viewed under its own light in a microscope retains its colour unchanged with time. Viewed with the naked eye, the end of the capillary containing the liquid at first appears rose-coloured, but in the course of an hour or two becomes greenish.

* A preliminary account of this work was communicated to Manch. Lit. and Phil. Soc. March 23, 1909. (*This vol., p. 168.*)

This is no doubt due to the increased local phosphorescence of the glass, due to the radiation from the active deposit formed by the condensed emanation. If the emanation is volatilized its colour instantly disappears, showing that the colour is a property of the liquid or solid emanation at low temperature when bombarded by its own α particles.

The extreme end of the capillary where it had been drawn off was conical in shape. With care, the whole of the liquid could be concentrated in this glass cone. This was most simply done by applying a pad of cotton-wool soaked in liquid air to the end of the capillary. The moment the emanation was all condensed, the capillary was plunged into liquid air. Under these conditions, the volume of the capillary occupied by the liquid could be examined at leisure by a microscope. The colour of the condensed emanation made it comparatively easy to locate its distribution. After several trials, the whole of the emanation was liquefied in the extreme tip of the capillary and did not occupy a length of more than a fifth of a millimetre. Knowing the diameter of the capillary the volume occupied by the liquid could be estimated approximately. This volume was certainly not greater than $1 \cdot 2 \times 10^{-4}$ cubic millimetre.

The amount of emanation in the capillary corresponded to 100 milligrams of radium, and was consequently equal to about $0 \cdot 06$ c.mm. We thus see that the volume of the liquid emanation in liquid air was certainly not greater than 1/500 of the volume of the gas at normal pressure and temperature. Taking the emanation as a monatomic gas of atomic weight 222, it can readily be calculated from the above data that the density of liquid emanation is not less than 5. No doubt if a much finer capillary were used in which to condense the emanation, a more accurate estimate could be made.

It has so far not been found possible to determine directly the density of the emanation on account of the very small quantity available for experiment. Its atomic weight can, however, be deduced with considerable confidence from radioactive considerations. It has recently been shown by Rutherford and Royds* in a decisive experiment that the α particle is an atom of helium. Since the atom of the emanation is derived from the radium atom (atomic weight 226) by the expulsion of an α particle, its atomic weight should be 222. The absence of combining properties of the emanation indicates that it is an inert gas, and similar in that respect to the group of monatomic gases. Taking the view that it is monatomic, the emanation is the heaviest gas known with a density 111 times that of hydrogen.

For purposes of comparison, the atomic weight, boiling-point, and density of liquid of the heavier monatomic gases are given below.

	Argon	Krypton	Xenon	Radium Emanation
Atomic Weights	39·9	82	128	222
Absolute Boiling-point	86°·9	121°·3	163°·9	208°
Density of liquid at Boiling-point	1·212	2·155	3·52	5?

* Phil. Mag. Feb. 1909. (*This vol.*, p. 163.)

It is seen from the above table that the boiling-point of xenon is about a mean between that of krypton and the emanation. From the increase of density of the liquid with atomic weight, it might reasonably be expected that the density of liquid emanation should be about 6—a result, as we have seen, not inconsistent with experiment. In a similar way, it is possible to form some idea of the probable critical pressure and temperature of the emanation.

I desire to express my thanks to the Radium Commission of the Vienna Academy of Sciences for the loan of the radium preparation which has made this and other work on the emanation possible.

The Action of the α Rays on Glass

by PROFESSOR E. RUTHERFORD, F.R.S.

From the *Memoirs* of the Manchester Literary and Philosophical Society, 1909,
IV, vol. liv, No. 5, p. 1
(Received and read November 30th, 1909)

IT has been shown by Joly that the pleochroic halos observed in mica are to be ascribed to the action on the mica of the α rays emitted by small inclusions of radio-active material. The radius of the halo (about 0·04 mm.) indicates that the effect is due in large part to the rays from radium. I have recently reproduced the conditions under which such halos would be formed by enclosing a large quantity of radium emanation in a fine capillary tube of soda glass. After the emanation had decayed, the inner wall of the capillary was seen under the microscope to be surrounded by an area of distinct colouration extending about 0·04 millimetre from the walls. The outer boundary of the coloured region was sharply defined, and the depth of the colouration was equivalent to the maximum distance of the most penetrating α particle from the active matter. This result strongly confirms the correctness of the explanation of halos given by Joly.

Production of Helium by Radium

by PROFESSOR E. RUTHERFORD, F.R.S., and DR. B. B. BOLTWOOD

From the *Memoirs* of the Manchester Literary and Philosophical Society, 1909;
IV, vol. liv, No. 6, pp. 1–2
(Received and read November 30th, 1909)

SINCE the demonstration by Ramsay and Soddy of the production of helium by radium, it has been of great importance to determine accurately the amount of helium produced by a known quantity of radium. It has been shown by Rutherford, Geiger, and Royds that the α particle emitted by radium and its products is an atom of helium. In addition Rutherford and Geiger have calculated by counting the α particles and determining their charge that one gram of radium in equilibrium should produce 158 cubic mms. of helium per year. The first systematic measurements of the production of helium by radium were made last year by Sir James Dewar.* His experiments indicate that radium in equilibrium produces helium at a constant rate equivalent to 135 cubic mms. per gram per year—a result in fair agreement with the calculated quantity.

Another determination has recently been made by the writers using a barium-radium salt containing about 200 milligrams of radium, loaned by the Vienna Academy of Sciences. The salt, chemically treated to remove polonium and radium D, was placed in a platinum capsule which was in turn sealed in an exhausted tube of hard glass. At the end of 83 days, the gases were completely removed by heating and exposed to charcoal cooled in liquid air. The unabsorbed gases were pumped out and repeatedly exposed to fresh cooled charcoal. The residual gas consisting of essentially pure helium was found to have a volume corresponding to the production of helium at the rate of 163 cubic mms. per gram of radium per year.

With the experience gained in these preliminary experiments, it is hoped ultimately to determine the value of this constant with considerable precision.

Other experiments which have been made to test whether helium is produced by polonium have shown conclusively that such is the case.

* *Proc. Roy. Soc.*, A, 81, 280 (1908).

Action of the α Rays on Glass

by PROFESSOR E. RUTHERFORD, F.R.S., *University of Manchester*

From the *Philosophical Magazine* for January 1910, ser. 6, xix, pp. 192–4

IN 1907 Professor Joly* showed that the pleochroic halos observed in brown mica were in all probability to be ascribed to the action of the α rays from radium. These coloured areas, the origin of which had long puzzled geologists, are always found to contain at their centre a minute crystal of zirconite, or more rarely of apatite. Strutt has shown that both of these minerals are rich in radium. According to the explanation of Joly, the α rays emitted for long periods of time from the crystal have gradually darkened the mica and have given rise to the halo observed. The evidence indicated that the halos are spherical in shape with an average radius of about 0·04 millimetre. It is well known that the α rays from each radioactive product have a definite range of action depending upon the density of the material. The most penetrating α rays from the uranium-radium series arise from radium C, and have a range in air of about 7·06 cm. In a material like mica the corresponding range is about 0·04 millimetre. The range of the α particles in mica thus agreed closely with the observed radius of the halo. The evidence seems conclusive that these halos are of radioactive origin.

Recently, in the course of some experiments with the radium emanation, I had occasion to allow a large quantity of purified emanation to decay in a fine capillary tube of soda glass. The emanation was introduced into the capillary under pressure, and the tube then sealed off. The greater part of the emanation was then condensed near the end of the capillary and left surrounded by liquid air for four days. The tube was then removed from the liquid air and allowed to stand for a month, in which time the greater part of the emanation was transformed. The base of the capillary was not uniform, but gradually tapered over a distance of 3 centimetres to nearly a point. The external diameter of the glass tube near the end of the capillary was about 0·6 mm.

Notwithstanding that a large quantity of emanation, corresponding to the equilibrium amount from 150 milligrams of radium, had been kept in the tube, the thin-walled part of the capillary was only slightly coloured when viewed by the eye. The capillary was placed under a microscope to examine whether any indication of the effect of the α rays from the emanation and its products in the glass could be observed. A reddish coloured area was seen to surround the whole length of the capillary. The line of demarcation of the coloured area from the almost colourless outside glass was quite sharp. The

* Phil. Mag. March 1907.

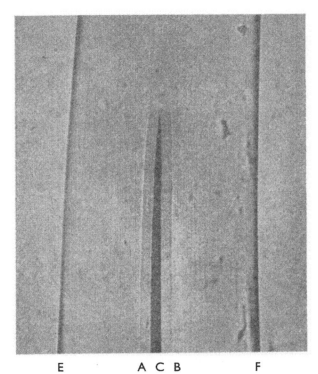

E   A C B   F

C, bore of capillary;   A, B edges of coloured area;
E, F external walls of capillary

distance from the edge of the capillary bore to the edge of the altered area was equal in all parts of the tube, notwithstanding wide variations in the bore of the capillary along its length. The altered area was most marked near the end of the capillary where the emanation had been most concentrated, but was appreciable along the whole tube. The effect observed will be clearly understood from the accompanying photograph of the end part of the capillary tube. The magnification may be judged from the fact that the external diameter of the capillary tube shown in the photograph was at its middle point about 0·6 mm. This micro-photograph was kindly taken for me by Dr. Stansfield by immersing the end of the capillary tube in glycerine, and using a point source of light.

It will be seen that the depth of the darkening measured from the walls of the capillary is the same at all points along the tube, and even follows a slight bulge in the capillary tube.

The width of the coloured area was measured from the photograph and also from longitudinal sections of a part of the tube. The mean depth of the coloured area was 0·039 millimetre. From the law of absorption of the α rays found by Bragg, viz., that the stopping power of an atom for an α particle is proportional to the square root of its atomic weight, it can be calculated from a knowledge of the chemical composition of the glass, that the range of the α particle from radium C in the glass should be about 0·041 millimetre. The agreement of theory and experiment is sufficiently close to show clearly that the coloured area is due to the α rays, and that the edge of the colouration marks the extreme points that the swiftest α particles were able to penetrate.

The sharpness of the edge of the coloured area is a result of a difference of refractive index between the altered and unaltered glass. Dr. Stansfield kindly made an examination of this point for me, and found that the altered glass had a refractive index several parts in ten thousand higher than the unaltered part. Whether this change in refractive index results from chemical changes in the composition of the glass due to the action of the α rays, or is due to the helium fired into the glass, is difficult to fix with certainty. It can readily be calculated from the magnitudes involved in the experiments that if the helium fired into the glass in the form of α particles remained in the glass, its presence would increase the refractive index of the glass several parts in ten thousand.

The darkening observed in the glass tube is not to be ascribed to the result that one end was kept for several days in liquid air. A similar effect was noted in the part not exposed in the liquid air, and also in other glass tubes which had contained sufficient radium emanation to strongly colour the glass. There can be no doubt that the effect artificially produced in glass corresponds to the halos observed in mica, and confirms the correctness of the interpretation of Professor Joly. It is intended to continue experiments of a similar character, using other materials beside glass to contain the emanation.

I desire to thank Dr. Stansfield for his assistance in the photographs, and Mr. Hickling for his kindness in cutting sections of the glass tubes.

Properties of Polonium

From *Nature*, **82**, 1910, pp. 491–2

THE statements regarding polonium which appeared in the report from Paris reprinted from the *Times* in NATURE of February 17, must have surprised many readers to whom polonium has been a familiar substance for the last ten years. It may be of interest to review briefly our present knowledge of polonium and the bearing of the recent work of Mme. Curie and Debierne upon it.

Polonium was the first of the active substances separated from pitchblende residues by Mme. Curie. Various methods of concentration were devised by her, with the result that preparations of polonium mixed with bismuth were early obtained many thousand times more active than uranium. Marckwald later separated from 15 tons of pitchblende about 3 milligrams of intensely active material which he called radio-tellurium, since it was separated initially with tellurium as an impurity. By dipping a copper plate into a solution of this substance, he obtained a deposit of weight not more than 1/100 milligram, which was far more active than an equal weight of radium. It was soon recognised that this preparation was identical with polonium, for it gave off the typical α radiation, and had the characteristic rate of decay of that substance. Unfortunately, Marckwald was not aware at the time of separation of the great importance of testing whether lead appeared as a product of transformation of polonium. Before such an experiment could be made, the polonium had to a large extent been transformed.

Polonium is one of the numerous transition elements produced during the transformation of the uranium-radium series. It is half-transformed in about 140 days, emitting α particles during the process. Rutherford showed in 1904 that polonium was in reality a transformation product of radium itself. Radium at first changes into the emanation, and then successively into radium A, B, C, D, E, F, radium F being identical in all respects with the polonium directly separated from a radio-active mineral. When the radium emanation is allowed to decay in a sealed glass tube, the walls of the tube are coated with an invisible deposit of pure radium D, radium E, and radium F, but the amount of the latter to be obtained in this way is far too small to be weighable.

The amount of polonium present in any radio-active mineral can easily be calculated. Since the radium and polonium (radium F) in a mineral are in radio-active equilibrium, the same number of α particles are expelled from each per second. Since polonium is half-transformed in 140 days and radium in 2000 years, the former breaks up 5000 times faster than the latter. The maximum amount of polonium to be obtained from a mineral is in consequence only 1/5000 of the amount of radium. In 1000 kilos. of pitchblende containing 50 per cent. of uranium, there are present 170 milligrams of radium.

The weight of polonium is about 1/5000 of this, or about 1/30 milligram. It is thus obvious that to obtain 1/10 of a milligram of pure polonium, several tons of high-grade pitchblende must be worked up. The most natural source of polonium is radium D (radio-lead), which grows polonium and has a period of half-transformation of about twenty years. Since polonium breaks up about 5000 times faster than radium, its activity, weight for weight, should be about 5000 times greater than that of radium. There is nothing surprising in this, for the radium emanation has an activity about 200,000 times that of radium, while radium A (period three minutes) must have an activity 400 million times that of radium itself. Since the radiation from polonium is entirely in the form of α rays, it is to be expected that the radiation from it would show chemical and physical effects identical with those observed for pure emanation, the only difference being that the products of the latter emit β and γ rays as well.

Apart from the interest of obtaining a weighable quantity of polonium in a pure state, the real importance of the present investigations of Mme. Curie lies in the probable solution of the question of the nature of the substance into which the polonium is transformed. This problem has been much discussed in recent years. Since polonium emits α particles, one of its products of decomposition, as for all the other α-ray products, should be helium. The production of helium from a preparation of polonium has been observed by Rutherford and Boltwood (Manchester Lit. and Phil. Society, November 30, 1909),* and also by Mme. Curie and Debierne in their present experiments. Boltwood several years ago suggested that the end product of the radium series was lead, and has collected strong evidence in support of this view by comparing the amount of helium and lead in old radio-active minerals. Since polonium is the last of the active products observed in the radium series, it is to be expected that polonium should be transformed into helium and lead, one atom of polonium producing one atom of helium and one atom of lead. This point of view receives additional weight from consideration of the atomic weight to be expected for the end product of radium. Since in the uranium-radium series, seven α particles, each of which is an atom of helium of atomic weight four, are successively expelled before radium F is reached, the atomic weight of polonium should be $7 \times 4 = 28$ units less than uranium (atomic weight $238 \cdot 5$). This gives an atomic weight of polonium of $210 \cdot 5$, and after the loss of an α particle, a final product of atomic weight $206 \cdot 5$—a value very close to the atomic weight of lead.

It is a matter of very great interest and importance to settle definitely whether polonium changes into lead. The evidence as a whole has long been in favour of that supposition. The outlook is very promising that the experiments of Mme. Curie and Debierne will settle this question conclusively. No doubt, an interval must elapse to allow the polonium to decay before the final examination of the residual substance can be made.

E. RUTHERFORD

* *This vol., p. 177.*

Theory of the Luminosity produced in certain Substances by α Rays

by E. RUTHERFORD, F.R.S., *Professor of Physics, University of Manchester*

From the *Proceedings of the Royal Society*, 1910, ser. A, lxxxiii, pp. 561–72
(Received and read February 17, 1910.)

IN the preceding paper, Mr. Marsden has examined the decay of the luminosity excited by α-rays in zinc sulphide, willemite, and barium platinocyanide, when subjected to an intense bombardment by α-particles. He has shown that the luminosity decreases with continued bombardment to a very small fraction of its initial value. For a given bombardment, the rate of decay of luminosity is about the same for zinc sulphide and willemite, but is especially rapid in barium platinocyanide. The action of the α-particles on phosphorescent zinc sulphide is of special interest and importance on account of the marked scintillations observed, and the fact that each α-particle under suitable conditions produces a visible scintillation. Mr. Marsden has brought out the essential fact that the actual number of scintillations observed for a constant source of α-rays changes very little with continued bombardment, but the brightness of the scintillations rapidly diminishes.

It is well known that the α-particles exert a marked dissociation effect in complex molecules on which they fall. For example, the α-rays from radium or its emanation, dissolved in water, dissociate the water molecules, producing hydrogen and oxygen at a rapid rate. I have shown elsewhere ('Radio-active Transformations,' p. 253), that the magnitude of this effect is in agreement with the view that each α-particle dissociates as many molecules of water as it produces ions in its path in air. The loss of energy of the α-particle in passing through a gas is mainly used up in producing ions in the gas. The laws of absorption of α-particles, which have been so carefully worked out by Bragg, show that no definite distinction as regards absorption can be drawn between a solid and a gas. It is reasonable to suppose that the α-particle produces ions in a solid as well as in a gas, and that the absorption of the α-particle is due mainly to the energy used up in this process. If the solid matter is composed of complex molecules, the latter will be dissociated by the α-particles. As Marsden has pointed out, pure zinc sulphide does not exhibit the scintillation effect, but this only appears in zinc sulphide to which certain impurities have been added. Since the amount of impurity present is of the order of 1 per cent., it is probable that only a small fraction of the total number of molecules give rise to the scintillation effect. These 'active centres', as they will be called, will on the average be uniformly distributed among the inactive molecules.

In the theory of the decay of luminosity outlined in this paper, it is supposed that a scintillation results from the dissociation of a number of these active centres which lie in the path of the α-particle. Each active centre, when dissociated by an α-particle, emits light and is no longer effective in producing light when struck again by an α-particle. The effect of continued bombardment by α-particles is thus to destroy gradually the active centres. Since, probably, initially many thousands of these centres lie in the path of an α-particle, the effect of a continued bombardment will not cause a diminution in the number of scintillations, but a decrease in the intensity of each scintillation, for the intensity of a scintillation on an average will be proportional to the number of active centres remaining. This point of view is in agreement with the experimental results observed by Mr. Marsden.

Theory of Decay of Luminosity with Time

Since the destruction of active centres is connected with their ionisation and consequent dissociation by the α-particles, it is necessary to consider how the ionisation varies within the layer of phosphorescent material. In the experiments of Marsden, a glass tube, coated inside with a layer of zinc sulphide or willemite, was exposed to the action of the α-rays from the emanation and its products. About three hours after the introduction of the emanation into the tube, the emanation and its products, Radium A, B, and C, are in equilibrium, and the same number of α-particles is emitted per second from the emanation, Radium A, and Radium C. The ranges of penetration of the α-particles from these three products are $4 \cdot 33, 4 \cdot 85$, and $7 \cdot 06$ cm. respectively in air. Since the emanation was usually introduced at low pressure, the α-particles were not appreciably stopped by the gas in the tube. Each part of the surface of the phosphorescent material was exposed to a bombardment of α-particles falling equally from all directions. The thickness of phosphorescent matter was more than sufficient to stop the α-particles completely.

It was early observed that the total ionisation produced by a thin layer of active material from which α-particles were fired equally in all directions decreased approximately according to an exponential law when uniform screens of absorbing matter were placed over the active matter.

The variation of ionisation with distance under these conditions has been calculated by Bragg.* If R is the range of the α-particles in the material and D the thickness, then the value i/i_0 is given in the following table for different ratios of D/R. In this case i_0 represents the total ionisation due to the α-particles when no screen is interposed, and i the value of the total ionisation produced *after* traversing a distance D of the material.

D/R .	0	0·061	0·124	0·250	0·357	0·500	0·690	0·833	1·00
i/i_0 ..	1·00	0·807	0·672	0·467	0·335	0·193	0·077	0·023	0

In this calculation, a correction is made for the variation of the ionisation

* Bragg, 'Phil. Mag.,' vol. 11, p. 754, 1906.

due to an α-particle along its path. In the experimental case under consideration, it is required to find the variation of ionisation with distance for a thin layer of radio-active matter emitting three sets of α-particles, equal in number. Taking R as the range in the material of the α-particles from Radium C, the range of the α-particles from the emanation and Radium A are 0·613 R and 0·684 R respectively. Geiger* has shown that an α-particle from the emanation, Radium A, and Radium C produces $1·74 \times 10^5$, $1·87 \times 10^5$, $2·37 \times 10^5$ ions respectively. By drawing a curve giving the variation of ionisation with distance traversed for each of the three sets of rays and adding them together, the variation of ionisation with distance for the three sets together can be deduced.

The results are collected in the following table:—

D/R	0	0·05	0·1	0·2	0·3	0·4	0·5	0·6	0·7	0·8	0·9	1·0
i/i_0	1·00	0·80	0·65	0·45	0·295	0·18	0·10	0·056	0·027	0·012	0·005	0
e^{-KD}	1·00	0·80	0·64	0·41	0·26	0·165	0·105	0·067	0·042	0·027	0·017	0·011

It will be seen that the value i/i_0 falls off roughly according to an exponential law. This is brought out in the third column, where the values of e^{-KD} are plotted for a value $K = 4·5/R$. With this value of K, the ionisation falls off to half value after the α-particles have traversed a distance 0·154 R. We can consequently assume, as a first approximation, that the ionisation decreases with distance traversed according to an exponential law. This assumption makes the final equations much more easily integrable.

Consider a layer of phosphorescent material of area one square centimetre and thickness dx at a depth x from the interior surface. Suppose P α-particles from the emanation and also from each of the α-ray products strike each square centimetre of surface per second. The number of ions produced in thickness dx is easily seen to be

$$Ki_0e^{-Kx}dx, \text{ where } i_0 = P . q,$$

where q is the sum of the number of ions produced by an α-particle from the three α-ray products.

The number of molecules of material (supposed mainly zinc sulphide) in thickness dx is Ndx, if N is the number of molecules of zinc sulphide per cubic centimetre. Consequently, the fraction of the total number of molecules ionised in the time dt is

$$\frac{Ki_0e^{-Kx} . dx}{N . dx} \cdot dt = \frac{Ki_0e^{-Kx}}{N} \cdot dt. \tag{1}$$

If the active centres be supposed to be of the same diameter as the inactive, so that the chance of being struck by an α-particle is the same, this must

* Geiger, 'Roy. Soc. Proc.,' A, vol. 82, p. 486, 1909.

represent the fraction $-dn/n$ of the active molecules which are destroyed in the time dt, n being the number of active molecules present per unit volume.

Thus
$$\frac{dn}{n} = -\frac{Ki_0}{N} e^{-Kx} \cdot dt. \tag{2}$$

This equation will obviously hold if the active centres constitute only a small part of the total number present. If the active centres consist of molecular aggregates or small crystals of dimensions large compared with that of the zinc sulphide molecule, each of which is destroyed as a scintillation centre by the impact from an α-particle, it is obvious that their chance of being struck by an α-particle is much greater than if they were of molecular dimensions. In such a case it would be necessary to introduce a constant factor on the left-hand side of equation (2). We shall, however, first work out the theory on the simple hypothesis that the active and inactive molecules have the same chance of being struck by an α-particle. A comparison of theory with experiment will at once disclose whether such a correction is necessary.

Since, in the experiments of Mr. Marsden, the emanation in a sealed tube was used as a source of radiation, the radiation from it decays with time. If P_0 be the initial number of α-particles expelled per second from the emanation itself, the number P after a time t is given by $P = P^{-\lambda t}$, where λ is the constant of decay of the emanation. Under these conditions equation (2) becomes

$$\frac{dn}{n} = -\frac{i_0 K}{N} e^{-Kx} \cdot e^{-\lambda t} \cdot dt. \tag{3}$$

Putting $i_0 K/N = A$, and integrating between the limits t and 0,

$$\log_e \frac{n}{n_0} = -\frac{A}{\lambda}(1 - e^{-\lambda t})e^{-Kx},$$

where n_0 is the value of n when $t = 0$. Thus

$$\frac{n}{n_0} = e^{-\frac{A}{\lambda}(1-e^{-\lambda t}) \cdot e^{-Kx}}. \tag{4}$$

Now the luminosity dI supplied by the layer dx is proportional to $-dn/dt$, and may be taken for simplicity as equal to $-dn/dt$. Then the total illumination I due to the whole layer of material acted on by α-particles is given by

$$I = \int_0^\infty -\frac{dn}{dt} \cdot dx.$$

Substituting the value of dn/dt from (3),

$$I = Ae^{-\lambda t} \int_0^\infty e^{-Kx} \cdot e^{-\frac{A}{\lambda}(1-e^{-\lambda t}) \cdot e^{-Kx}} \cdot dx$$

Putting $e^{-Kx} = y$, this reduces to

$$I = \frac{A}{K} e^{-\lambda t} \int_1^0 e^{-\frac{A}{\lambda}(1-e^{-\lambda t})y} \, . \, dy$$

$$= \frac{\lambda e^{-\lambda t}}{K(1 - e^{-\lambda t})} \{1 - e^{-\frac{A}{\lambda}(1-e^{-\lambda t})}\}.$$

When $t = 0$, the value I_0 of the intensity of luminosity is given by

$$I_0 = A/K,$$

and
$$\frac{I}{I_0} = \frac{e^{-\lambda t}}{\frac{A}{\lambda}(1 - e^{-\lambda t})} \{1 - e^{-\frac{A}{\lambda}(1-e^{-\lambda t})}\}. \tag{5}$$

This gives the equation of decay of luminosity, and is seen to involve only one unknown constant A, since the decay constant λ of the emanation is known.

Simple Cases

We can at once deduce from this equation the decay of luminosity for certain special cases of importance.

Case 1.—Suppose the bombardment is constant. In this case $\lambda = 0$. It is at once seen that

$$\frac{I}{I_0} = \frac{1}{At}(1 - e^{-At}). \tag{6}$$

Case 2.—Suppose the layer of material is very thin compared with the range of the α-particles in it. In this case, the ionisation may be taken as sensibly constant and equal to Ki_0 through the thin layer.

Equation (3) in this case reduces to

$$\frac{dn}{n} = \frac{-Ki_0}{N} e^{-\lambda t} \, . \, dt,$$

and equation (4) becomes
$$\frac{n}{n_0} = e^{-\frac{A}{\lambda}(1-e^{-\lambda t})}.$$

From this it follows that I, which is equal to $-dn/dt$, is given by

$$\frac{I}{I_0} = e^{-\lambda t} \, . \, e^{-\frac{A}{\lambda}(1-e^{-\lambda t})}.$$

For a constant source of bombardment $I/I_0 = e^{-At}$, *i.e.* the luminosity decays according to an exponential law with the time. Such a result is to be anticipated from the general theory without calculation.

Case 3.—Suppose a screen is exposed to a narrow pencil of α-rays falling normally. In this case all the α-particles penetrate a distance equal to their range. Taking the ionisation as uniform along the path of the α-particles, it is clear that this case is similar to Case 2. With a decaying source of radiation

$$\frac{I}{I_0} = e^{-\lambda t} \cdot e^{-\frac{A}{\lambda}(1 - e^{-\lambda t})},$$

and for a constant source $I/I_0 = e^{-At}$.

It is, however, well known that the ionisation in a gas due to a narrow pencil of α-rays increases at first, passes through a maximum, and then falls off rapidly to zero at the end of the range. The equations for I/I_0 deduced above consequently serve only as a first approximation. If required it would not be difficult to make a correction for the variation of ionisation along the path.

For the purpose of comparing theory with experiment it is convenient to express equation (5) in the form

$$\frac{I}{I_0} = \frac{e^{-\lambda t}}{B(1 - e^{-\lambda t})} \{1 - e^{-B(1 - e^{-\lambda t})}\}. \tag{7}$$

The value of $$B = \frac{A}{\lambda} = \frac{i_0 K}{N\lambda} = \frac{P_0 q \cdot K}{N\lambda}. \tag{8}$$

The value of the constant B in different experiments is consequently proportional to the initial number P_0 of α-particles falling per second per square centimetre on the phosphorescent material.

It is of interest to observe that

$$B(1 - e^{-\lambda t}) = \frac{q K}{N} \cdot P_0 \left(\frac{1 - e^{-\lambda t}}{\lambda}\right),$$

where the variable factor $\frac{P_0}{\lambda}(1 - e^{-\lambda t})$ is equal to the total number of

α-particles which have traversed the material.

It is obvious on this theory that the fraction of the active molecules destroyed by a bombardment depends only on the total number of α-particles which have struck the screen and not on their distribution with regard to time.

Comparison of Experiment with Theory

In the experiment of Mr. Marsden, the bombardment of the screen by the α-particles increased for about three hours after the introduction of the emanation, due to the growth of the α-ray products Radium A and C. Systematic measurements were usually begun after the three hours' interval. For a comparison of experiment with theory it has been necessary to correct for the bombardment during this interval. This has been done by continuing the observed curve backwards, corresponding to an interval of two hours. This estimated correction is small for small bombardments, but becomes of more importance in the case of intense bombardments where the luminosity falls considerably in the course of a few hours.

The agreement between the experimental values and those deduced by taking a suitable value of the constant B in the equation

$$\frac{I}{I_0} = \frac{1}{B(1 - e^{-\lambda t})} \{1 - e^{-B(1 - e^{-\lambda t})}\}$$

is shown in the following tables. The first column gives the time in hours calculated from one hour after the introduction of the emanation; the second column the value of $e^{-\lambda t}$ for the emanation, taking the half period as $3 \cdot 85$ days.

It will be seen that there is a substantial agreement between theory and experiment for both zinc sulphide and willemite over a considerable range of intensity of bombardment. Taking into account the approximate nature of some of the assumptions and the experimental difficulty of accurate measurement of the weak luminosity, the agreement between the calculated and observed numbers is as close as could be expected. In the simple theory no account has been taken of the effect of the β-rays on the material, within

Zinc Sulphide

Time in hours	$e^{-\lambda t}$	A Bombardment, 4×10^7 α-particles per sec. per sq. cm.		B Bombardment, 2×10^8 α-particles per sec. per sq. cm.	
		Observed value, I/I_0	Calculated value, I/I_0 B = 3	Observed value, I/I_0	Calculated value, I/I_0 B = 12
0	1·00	1·00	1·00	1·00	1·00
5	0·963	0·90	9·86	0·78	0·77
10	0·928	0·82	0·83	0·63	0·61
20	0·861	0·69	0·70	0·44	0·42
30	0·798	0·59	0·60	0·315	0·30
40	0·741	0·51	0·52	0·22	0·23
50	0·687	0·45	0·445	0·185	0·18
60	0·638	0·395	0·39	0·15	0·15
80	0·549	0·31	0·30	0·10	0·10
100	0·472	0·25	0·24	0·06	0·08
120	0·407	0·22	0·19	0·045	0·06

or outside the range of penetration of the α-particles, though it is probable that the luminosity due to this cause becomes important when most of the active centres in the surface layer have been destroyed. In one experiment of

Mr. Marsden where the bombardment was very intense (10^9 α-particles per square centimetre per second), the agreement between theory and experiment is not so good. This is not unexpected, for the bombardment was so intense

Willemite

Time in hours	$e^{-\lambda t}$	A Bombardment, 6×10^7 α-particles per sec. per sq. cm.		B Bombardment, $1\cdot3 \times 10^8$ α-particles per sec. per sq. cm.	
		Observed value, I/I_0	Calculated value, I/I_0 B $= 1\cdot4$	Observed value, I/I_0	Calculated value, I/I_0 B $= 3\cdot1$
0	$1\cdot00$	$1\cdot00$	$1\cdot00$	$1\cdot00$	$1\cdot00$
5	$0\cdot963$	$0\cdot92$	$0\cdot93$	$0\cdot905$	$0\cdot91$
10	$0\cdot928$	$0\cdot86$	$0\cdot88$	$0\cdot825$	$0\cdot83$
20	$0\cdot861$	$0\cdot76$	$0\cdot78$	$0\cdot69$	$0\cdot695$
30	$0\cdot798$	$0\cdot68$	$0\cdot69$	$0\cdot59$	$0\cdot585$
40	$0\cdot741$	$0\cdot615$	$0\cdot62$	$0\cdot51$	$0\cdot50$
50	$0\cdot687$	$0\cdot555$	$0\cdot56$	$0\cdot44$	$0\cdot44$
60	$0\cdot638$	$0\cdot505$	$0\cdot50$	$0\cdot39$	$0\cdot38$
80	$0\cdot549$	$0\cdot41$	$0\cdot405$	$0\cdot31$	$0\cdot29$
100	$0\cdot472$	$0\cdot34$	$0\cdot335$	$0\cdot255$	$0\cdot23$
120	$0\cdot407$	$0\cdot28$	$0\cdot27$	$0\cdot215$	$0\cdot19$
140	$0\cdot349$	—	—	$0\cdot18$	$0\cdot15$

that the luminosity increased little, if any, after the introduction of the emanation. This shows that a considerable fraction of the active centres in the surface layers had been destroyed before the emanation was in equilibrium with its products and regular measurements begun.

It will be noted that on this theory the active centres in the surface layer where the ionisation is most intense are destroyed far more rapidly than those in the deeper layers. In consequence of this, the luminosity with continued bombardment is supplied more and more from the deeper layers of the material.

Value of the Constant B.—The value of the constant B in the formula can be calculated on certain assumptions. Its value is given by

$$B = \frac{P_0 q K}{\lambda N} = \frac{4\cdot5 P_0 q}{\lambda N R},$$

since we have already shown earlier that $K = 4\cdot5/R$, where R is the range of the α-particle from Radium C in the material under investigation.

It is known that the ionisation due to an α-particle is approximately the same for all simple gases, but increases somewhat for complex gases and vapours. In the absence of definite information, it will be assumed for the purposes of calculation that the number of ions produced in the solid material is equal to the number of ions produced in a simple gas, for example, in air.

Taking the values of the ionisation found by Geiger (*loc. cit.*) for an α-particle from the emanation, Radium A and C, the value of

$$q = (1 \cdot 74 + 1 \cdot 87 + 2 \cdot 37)10^5 = 6 \times 10^5.$$

Taking the half period of the emanation as $3 \cdot 85$ days, $\lambda = 1/481000 \, (\text{sec.})^{-1}$. The values of R and N vary for each substance, but can be simply determined. Taking the density of zinc sulphide (ZnS), willemite (Zn_2SiO_4), barium-platinocyanide ($BaPt(CN)_4$) as equal to four, it can easily be shown that the range of the α-particle from Radium C is

For zinc sulphide . . . . $R = 3 \cdot 7 \times 10^{-3}$ cm.
,, willemite . . . . $R = 3 \cdot 0 \times 10^{-3}$,,
,, barium platinocyanide. . $R = 4 \cdot 2 \times 10^{-3}$,,

The range for each substance is calculated from Bragg's rule that the stopping power of a complex molecule is very approximately proportional to the sum of the square roots of the atomic weights of the constituent atoms. The range of the α-particle from Radium C in aluminium ($0 \cdot 004$ cm.) is taken as a basis of comparison. Taking the charge on a hydrogen atom as $4 \cdot 65 \times 10^{-10}$ electrostatic unit, it can readily be calculated that the value of N, the number of molecules per unit volume, is

For zinc sulphide . . . . $2 \cdot 57 \times 10^{22}$
,, willemite . . . . $1 \cdot 12 \times 10^{22}$
,, barium platinocyanide. . $5 \cdot 7 \times 10^{21}$

The value of the constant B is independent of the density of the material, since the value of N varies directly, and the value of R varies inversely as the density. In a similar way, the value of the charge for a hydrogen atom is not involved, for q and N vary inversely as the value of this quantity. Collecting results, it is seen that

For zinc sulphide . . . . $B = 1 \cdot 37 \times 10^{-8} . P_0$
,, willemite . . . . $B = 3 \cdot 9 \times 10^{-8} . P_0$
,, barium platinocyanide. . $B = 5 \cdot 4 \times 10^{-8} . P_0$

P_0 here represents the number of α-particles per second per square centimetre from the emanation alone, or one-third the actual bombardment when in equilibrium with its products. The calculated and observed values of B are shown in the following table:—

	$3P_0$	B (calc.)	B (obs.)	$\dfrac{\text{B (obs.)}}{\text{B (calc.)}}$
Willemite	6×10^7	$0 \cdot 78$	$1 \cdot 4$	$1 \cdot 8$
	$1 \cdot 3 \times 10^8$	$1 \cdot 7$	$3 \cdot 1$	$1 \cdot 8$
Zinc sulphide	4×10^7	$0 \cdot 18$	$3 \cdot 0$	$17 \big\}15$
	2×10^8	$0 \cdot 9$	$12 \cdot 0$	13
	About			
Barium platinocyanide.	$1 \cdot 5 \times 10^7$	$0 \cdot 3$	1400	4700

The decay of the luminosity of barium platinocyanide takes place much more rapidly than for willemite and zinc sulphide. The numbers for this substance given in the above table were deduced from data obtained by Marsden. A rod coated with the material was inserted through mercury in a space containing emanation in equilibrium. With a bombardment of $1 \cdot 5 \times 10^7$ α-particles per square centimetre per second, the luminosity of the barium platinocyanide decayed to half value in 8 minutes.

The calculated values of B given in the above table are deduced on the assumption that the same number of ions is produced in the material as in air, and that the active centre has the same chance of being struck by an α-particle as a molecule of the material. The latter condition implies that the active centre is of the same dimensions as the molecule.

In interpreting the relation between the observed and calculated values of B, there are two points of view that may be considered. In the first place, it may be supposed that the active centre is a molecular aggregate of greater dimensions than the molecule, and has, consequently, a greater chance than a molecule of being struck by an α-particle. If we suppose that the chance of being struck is proportional to the cross-section of the aggregate, it is seen from the above table that for willemite, zinc sulphide, and barium platino-cyanide, the active centres have cross-sections about 2, 15, and 5000 times that of the corresponding molecules. On such a point of view, the number of active centres must be small compared with the number of inactive molecules.

In the second place, it may be supposed that in substances like zinc sulphide, where the molecular arrangements are very easily disturbed by the action of light or other weak stimuli, the region of dissociation of an α-particle is not confined to the molecules and active centres directly in its path, but extends some distance from it. In such a case, the dissociation of the neighbouring active centres is to be ascribed to the effect of the secondary radiations set up by the α-particle in the molecules in its direct path. On this theory, the region of disturbance is small for willemite, but in zinc sulphide is sufficiently large to cause about fifteen active centres to be dissociated for one directly struck by an α-particle. In barium platinocyanide the active centres are so numerous or so unstable that the ratio is nearly 5000.

This point of view of the action of the α-particle may be expressed in another way without taking into consideration the ionisation produced by the α-particle. Suppose that an α-particle, in its passage through the material, destroys all the active centres in a cylinder of average cross-section A and of length equal to its range in the material. For simplicity, take the case of a very thin layer of the material on which P α-particles fall per second per square centimetre. The fraction of the volume destroyed in the first second is PA. It is readily seen from equation (2) that, under the conditions specified, $PA = -dn/n = A = B\lambda$, where B is the constant deduced experimentally for an initial bombardment P, and $\lambda = 1/481000$ is the constant of the radium emanation. If d be the diameter of the cylinder of action of the α-particle, $d = \sqrt{(4B\lambda/\pi P)}$.

We shall now calculate the value of d for special examples—

Zinc sulphide	$P = 2 \times 10^8$,	$B = 12$,	$d = 1 \cdot 3 \times 10^{-7}$ cm.
Willemite	$P = 1 \cdot 3 \times 10^8$,	$B = 3 \cdot 1$,	$d = 2 \cdot 5 \times 10^{-7}$,,
Barium platinocyanide..	$P = 1 \cdot 5 \times 10^7$,	$B = 1400$,	$d = 1 \cdot 6 \times 10^{-5}$,,

It is seen that the diameter of the cylinder of action of each α-particle is, for willemite and zinc sulphide, somewhat greater than the diameter of the sphere of action of the molecule. The diameter of the latter for ordinary gaseous molecules is calculated by Meyer ('Kinetic Theory of Gases') to be between 2×10^{-8} and 10^{-7} cm. The value of d for barium platinocyanide is undoubtedly very large compared with the diameter of a molecule.

On both points of view, it must be supposed that the luminosity due to α-rays accompanies the dissociation of the active centres and cannot be ascribed to the recombination of the dissociated parts, for there is an undoubted chemical alteration of the material due to continued bombardment.

Whatever point of view is taken, it is clear that for willemite and zinc sulphide the active centres have dimensions comparable with a molecule. This excludes the suggestion put forward at one time that the scintillations of zinc sulphide due to α-particles were a consequence of the mechanical cleavage of small crystals.

I am indebted to Mr. Marsden for the use of the data obtained by him, and for his assistance in calculation of the decay curves.

Radium Standards and Nomenclature

From *Nature*, **84**, 1910, pp. 430–1

THE International Congress of Radiology and Electricity, held at Brussels, September 12th to 15th, afforded an excellent opportunity of discussing several important questions of general interest to workers in radio-activity. The need of a definite radium standard, in which all results should be expressed, has been growing more acute with the increase of accuracy of radio-active measurements. At the present time, scientific results are expressed in many cases in terms of arbitrary radium standards kept in each laboratory, and it has been difficult to be certain of the accuracy or relative value of such standards. Mr. C. E. S. Phillips several years ago pointed out to the Röntgen Society the desirability of adopting a fixed radium standard, and arranged for the preparation of several small radium standards which were compared with the working standard adopted by Rutherford and Boltwood. Duplicates of the latter standard have been used for several years by a number of English, American, and Continental workers.

At the opening meeting, Prof. Rutherford read a report on the desirability of establishing an international radium standard. He pointed out that he had compared by the γ ray method the radium standards used by several important European laboratories, and had found that there was a considerable difference amongst them, amounting in some cases to 20 per cent. It is now possible to measure with considerable precision a number of magnitudes connected with radium; for example, the volume of the emanation, the heating effect, the rate of production of helium, and the rate of emission of α and β particles. The value of each of these quantities is dependent on the accuracy of the radium standard in which the results are expressed. For the comparison of results obtained by workers in different laboratories, it is necessary that they should all be expressed in terms of the same standard. For example, at the present time it is not possible to compare the results obtained on the heat emission of radium by various observers until the radium standards employed have been accurately compared. When once a standard has been adopted, it is relatively a simple matter to determine the radium contents of substandards by the γ ray method or modification of it, without opening the tube containing the radium.

A special international committee was appointed to report to the congress on the best means to be adopted to fix an international radium standard. This committee comprises the following workers in radio-activity representative of a number of countries: Mme. Curie, Debierne, Rutherford, Soddy, Hahn, Geitel, Meyer, Schweidler, Eve, and Boltwood. No doubt representatives of

G

other countries who are prepared to assist in the work will be added later. This committee reported to the congress at its final meeting and their suggestions were adopted by the congress. As a member of the committee, Mme. Curie agreed to prepare a radium standard containing about 20 milligrammes of radium enclosed in a suitable sealed tube. This standard is somewhat large, but the amount was considered necessary on account of the difficulty of weighing small quantities of radium salt with the requisite accuracy. The thanks of all workers in this subject are due to Mme. Curie in undertaking the full responsibility of preparation of a standard, and for the large expenditure of time and labour its preparation will involve. The committee agreed to reimburse Mme. Curie for the cost of the radium and its preparation, after which the standard becomes the property, and is under the control, of the international committee. It was suggested that the standard should be suitably preserved in Paris. The initial cost of preparation of this standard will be somewhat heavy (about 500*l.*), but it is hoped that scientific societies and Governments of various countries will assist in defraying the expenses.

As soon as the primary standard has been prepared, it is proposed to approach through the committee the various national laboratories to ask them to acquire a radium standard accurately determined in terms of the primary standard. In this way it was thought that any Government interested in the question could acquire an accurate radium standard to be used as a basis for standardisation of quantities of radium in use in scientific laboratories, or to be sold commercially. As the primary standard is somewhat large for use in ordinary laboratories, the committee propose to investigate the question of the best method of comparing accurately in terms of the primary standard smaller substandards containing one or two milligrams of radium.

The committee also has under consideration the question of the preparation of very small substandards to be used for the determination of minute quantities of radium and of radium emanation. It is proposed that special investigations be made by the committee to determine the most suitable method of preparation and preservation of such standards. There is at present some uncertainty of how far radium solutions are affected by time in consequence of the tendency of radium to be precipitated out of the solution. No doubt before long it should be possible to secure accurate standard solutions to distribute amongst scientific workers.

In the course of the congress it was suggested that the name Curie, in honour of the late Prof. Curie, should, if possible, be employed for a quantity of radium or of the emanation. This matter was left for the consideration of the standards committee; the latter suggested that the name Curie should be used as a new unit to express the quantity or mass of radium emanation in equilibrium with one gram of radium (element). For example, the amount of emanation in equilibrium with one milligram of radium would be called 1/1000 Curie or one millicurie. The adoption of this unit will avoid much circumlocution, and will prove useful since the radium emanation is now so widely used in all kinds of experiments.

The committee has under consideration the question whether special names should be given to a very small quantity of radium, and also to the emanation in equilibrium with it. For example, the quantity 10^{-12} gram radium seems a natural unit for expression of the radium content of rocks and soils. At the same time, the large amount of investigation on the emanation content of springs and waters may make it desirable to adopt a convenient unit for expression of such quantities.

The committee pointed out that its recommendations were tentative, as all the members of the standards committee were not present at the congress, and had no opportunity of expressing their opinions. It is intended that the preparation of the radium standard should be proceeded with as soon as possible, and it is hoped that the standardisation of substandards will be possible before a year has elapsed. Prof. Stefan Meyer, of the University of Vienna, was appointed secretary of the international committee, and all communications relative to standards should be addressed to him.

The question of the nomenclature of radio-active products was informally discussed at the congress. There was a general consensus of opinion that it was not desirable to alter materially the present system of nomenclature, although it was recognised that it is far from perfect. It was felt that the gain to be obtained by a possibly more systematic nomenclature was more than counterbalanced by the confusion that would arise in consequence of a change of names. It was pointed out that the present system of nomenclature was capable of extension to include possible new products. For example, if future investigation should disclose that the product radium C consists of several products these could be named radium C_1, radium C_2, radium C_3, &c., but the term radium C would be used generally to represent the group of products as they normally always occur together. Reference was made to the undesirability of individual workers assuming the right to give new and fancy names to well-known substances.

A number of suggestions in regard to general nomenclature in radio-activity and ionisation were also made to the congress. For example, it is proposed that the term 'half-value period' should be used in all cases to represent the term required for a substance to be transformed to half its original value. It is suggested that the terms 'induced' and 'excited' activity should be abandoned and the term 'active deposit' employed in its stead, as reference is usually made to the radio-active matter itself and not to its radiations. There was a good deal of informal discussion amongst members as to the exact use of a number of scientific terms arising in radio-activity and allied subjects. Such discussions are of great importance in preventing unnecessary confusion in nomenclature due to the development of a rapidly growing subject.

A more general account of the meetings and deliberations of the congress, prepared by Dr. Makower, will appear in another issue of NATURE.

E. RUTHERFORD

The Number of α Particles emitted by Uranium and Thorium and by Uranium Minerals

by HANS GEIGER, PH.D., *and* PROFESSOR E. RUTHERFORD, F.R.S.

From the *Philosophical Magazine* for October 1910, ser. 6, xx, pp. 691–8

IN previous papers we have shown that the number of α particles emitted per second from radioactive materials can be counted either by the electrical or scintillation method. It has been shown that one gram of radium itself, and each of the three α ray products in equilibrium with it, emits $3 \cdot 4 \times 10^{10}$ α particles per second. Since Rutherford and Boltwood* have shown that in an old unaltered mineral there is $3 \cdot 4 \times 10^{-7}$ gram of radium per gram of uranium, it is possible to deduce the number of α particles emitted per second from one gram of uranium and also from a mineral containing one gram of uranium. In this calculation it is supposed that uranium is the ultimate parent of radium, and that the mineral is in radioactive equilibrium. If a uranium atom, like a radium atom, emits one α particle in its transformation, the number of α particles emitted per second per gram of uranium should be $3 \cdot 4 \times 10^{10} \times 3 \cdot 4 \times 10^{-7}$, or 11,600. We shall for convenience call this number N.

As a result of a very careful analysis of the radioactive constituents of uranium minerals, Boltwood† has shown that the total activity of uranium, measured by the electric method, is about twice as great as would be expected if uranium emits one α particle for one from the radium itself in equilibrium with it. This suggests that the uranium atom in its transformation emits at least two α particles. In the present state of our knowledge it is not certain whether this can be ascribed to the existence of an additional α ray product which is always separated with the uranium, or to the expulsion of two or more α particles in the transformation of the uranium atom.

Supposing, for the purpose of calculation, that the uranium in a mineral emits two α particles for one from each of the subsequent six α ray products, viz. ionium, radium, emanation, radium A, radium C, radium F (polonium), the number of α particles emitted per second per gram of uranium in a mineral is 8 N, or four times the number emitted by ordinary purified uranium. In this calculation no account has been taken of the actinium which occurs in all uranium minerals, and which Boltwood has shown stands in a

---

* Amer. Journ. Sci. vol. xxii. p. 2 (1906); (*Vol. I, p. 856*); also Boltwood, Amer. Journ. Sci. vol. xxv. p. 296 (1908).

† Boltwood, Amer. Journ. Sci. vol. xxv. p. 270 (1908).

genetic relation with uranium. However, Boltwood (*loc. cit.*) has found that the actinium and its four α ray products contributes an activity to the mineral equal to only $0 \cdot 21$ of that of the uranium. The relative number of α particles is still smaller, for the α particles from actinium have an average range of about $5 \cdot 7$ cms. of air, while the α rays of uranium, according to Bragg, have a range of $3 \cdot 5$ cms. Taking as a first approximation that the ionization due to an α particle is proportional to its range, the number of α particles emitted by the actinium in a mineral should be about $0 \cdot 17$ of that from uranium. The total number of α particles emitted by a mineral containing one gram of uranium should consequently be $2 \cdot 34 \, N + 6 \, N = 8 \cdot 34 \, N$. Since N by calculation is 11,600, the total number of α particles emitted per second from a mineral containing one gram of uranium should be $9 \cdot 67 \times 10^4$, and the number per second from one gram of ordinary purified uranium should be $2 \cdot 32 \times 10^4$.

It was the object of the present experiments* to determine the number of α particles experimentally, and to test the agreement with the calculated number.

Arrangement of Experiment

The scintillation method was adopted in order to count the number of α particles from a known weight of active material. A small quantity of the material under examination was finely powdered in an agate mortar, and then mixed with alcohol or ether and deposited as a thin uniform film on a thin sheet of aluminium or glass. The method adopted was similar to that first used by McCoy. Care was taken that the powder suspended in the liquid was well stirred in order to avoid a separation of the lighter from the denser portions. The weight of the active film was determined by weighing the plate before and after the active material had been removed. It was desirable to use very thin films in order that all the α particles might emerge without much loss of their range. In the case of uranium, however, the number of α particles emitted was so small that they were difficult to count with accuracy. For this reason thicker films were in some cases purposely employed. The efficiency of the zinc sulphide screen was tested by counting the number of α particles emitted from a definite quantity of radium C. The number of scintillations observed was found to be 8 per cent. less than the actual number of α particles incident on the screen. The latter value was calculated from the known result that one gram of radium and each of its products emits $3 \cdot 4 \times 10^{10}$ α particles per second. In the initial experiments the number of scintillations was counted by placing the screen close to the active material. In this case, the number of α particles striking the screen is equal to one half the total number emitted from an area of the active film equal to the area of screen seen in the micro-

* The experiments described later were, for the most part, completed more than a year ago. Recently, J. N. Brown (Proc. Roy. Soc. vol. A. lxxxiv. p. 151, 1910) has counted the scintillations from a uranium mineral and found a value per gram of uranium of $7 \cdot 36 \times 10^4$, which is somewhat smaller than our experimental value given later, viz. $9 \cdot 6 \times 10^4$.

scope. This method is open to some objections, for it requires that the film should be very uniformly spread and, in addition, very thin, for otherwise the particles emitted at an oblique angle suffer a considerable loss of range in the active material itself. The lack of uniformity of the film can be corrected for by counting at different points parts of the film, but this involves much labour.

In most of the experiments the active matter was spread in a circular area, and the small zinc sulphide screen was placed parallel to the film and opposite to its centre.

If $a =$ radius of circular film,
   $d =$ distance of screen from centre of film,
   $A =$ area of screen observed in field of microscope,
   $\sigma =$ total number of particles emitted per second per square centimetre of surface of film,

then, by a simple integration, it can be shown that the number n of α particles incident per second on the area A is given by

$$n = \frac{\sigma A}{2}\left(1 - \frac{d}{\sqrt{a^2 + d^2}}\right).$$

A simple example will serve to illustrate the method of calculation. The uranium film No. 1 (see table later) contained $10 \cdot 43$ milligrams of uranium oxide (U_3O_8) spread on an area of $5 \cdot 9$ square cms. 515 scintillations were counted, and the average number of scintillations observed corresponded to $5 \cdot 16$ per minute, and per second $0 \cdot 086$. Making the 8 per cent. correction for the imperfection of the screen, the corrected value becomes $0 \cdot 093$. This is the value of n to be substituted in the formula.

   $A = 3 \cdot 16$ sq. mms. $d = 2 \cdot 06$ cms. $a = 1 \cdot 37$ cms.

Substituting these values in the formula,

$$\sigma = 35 \cdot 0.$$

Now the weight of film per square centimetre was $1 \cdot 77$ mg. U_3O_8, or $1 \cdot 50$ mg. uranium. Consequently, from this experiment, the total number of α particles emitted per second per gram of uranium is $2 \cdot 33 \times 10^4$.

The chief difficulty of the experiments lay in counting accurately a sufficiently large number of scintillations. The number of scintillations observed in the microscope varied from one to five per minute in the case of uranium or thorium. While different observers agreed closely in counting scintillations due to radium or polonium when 30 to 50 scintillations were seen per minute, the agreement was not so good for uranium films. This difference is in part due to the fact that the eye becomes quickly fatigued when only a few scintillations appear on the screen per minute. This was especially marked in counting the scintillations from uranium, which are relatively much fainter than those from radium C. In the case of uranium and thorium minerals, where the scintillations are on the average much brighter than those from uranium, the counting was relatively easy. The brightness of scintillations of course depends on the range of the α particle striking the screen. We shall see later that the range of the α particle, and consequently the intensity of the scintillations from uranium, is less than from any other radioactive substance.

The active materials used in these investigations were kindly presented to us by Professor Boltwood, and were fractions of larger quantities analysed by him. We desire to express our indebtedness to Professor Boltwood for the use of these materials.

(1) Uranic-uranose oxide (U_3O_8) prepared from uranium nitrate which had been crystallized fifteen times. The least soluble fraction was taken and ignited at a high heat in a current of oxygen.

(2) Uraninite—a selected sample from Joachimsthal. This contained $61 \cdot 7$ per cent. of uranium. The mineral, when finely powdered, lost $6 \cdot 2$ per cent. of its emanation. The sample employed had been finely ground for several years, and during this time the emanation had steadily escaped. Under these conditions it can be simply deduced that the emission of α particles from the mineral is about three per cent. less than if the mineral had retained all its emanation. A correction of this amount has consequently been made to the counted number of α particles.

(3) Thorium oxide prepared from thorite. This was tested five weeks after its chemical separation. Since, in the chemical process of purification, the mesothorium is removed from the thorium, the α-ray activity of the purified thorium decays with time due to the decay of its product radiothorium. Since the half period of decay of the latter is about 737 days, a positive correction of about two per cent. is necessary to give the correct number of α particles emitted from thorium oxide in radioactive equilibrium. The activity of the thorium oxide in the form of a thin film was compared with that of a film of the mineral thorite of known composition, and gave nearly the ratio to be expected from their relative content of thorium.

The results of the observations are included in the following Table (p. 200).

Since only about 900 scintillations were counted altogether, the agreement between the three uranium films is closer than could be expected, considering the possible errors in the experiment. In the case of the mineral films 2000 scintillations were counted in all, and about an equal number for the thorium films. Before and after each set of observations the screen was carefully tested to determine the number of scintillations observed when the active material was removed. The correction for the screen employed was small, and usually corresponded to one scintillation in three or four minutes. All the counting experiments were checked among themselves by measuring the activity of the films in an α-ray electroscope. The activity measured in this way was found to be proportional to the weight of the film for thin films, but for the thicker films the activity was relatively smaller on account of absorption.

Radioactive Substances	Number of α particles emitted per second per gram of Uranium or Thorium	
Uranium film No. 1 10·43 mgrs. U_3O_8 on area 5·9 cm.^{2}	$2 \cdot 33 \times 10^4$	Average $2 \cdot 37 \times 10^4$
Uranium film No. 2 2·85 mgrs. U_3O_8 on area 12·8 cm.^{2}	$2 \cdot 36 \times 10^4$	
Uranium film No. 3 3·04 mgrs. U_3O_8 on area 14·9 cm.^{2}	$2 \cdot 43 \times 10^4$	
Mineral film No. 1 10·95 mgrs. Uraninite (Joachimsthal) on area 5·9 cm.^{2}	$9 \cdot 5 \times 10^4$	Average $9 \cdot 6 \times 10^4$
Mineral film No. 2 12·73 mgrs. Uraninite (Joachimsthal) on area 5·9 cm.^{2}	$9 \cdot 7 \times 10^4$	
Thorium film No. 1 4·43 mgrs. ThO_2 on 6·1 cm.^{2}	$2 \cdot 55 \times 10^4$	Average $2 \cdot 7 \times 10^4$
Thorium film No. 2 1·21 mgrs. ThO_2 on 6·4 cm.^{2}	$2 \cdot 84 \times 10^4$	
Thorium film No. 3 3·58 mgrs. ThO_2 on 6·15 cm.^{2}	$2 \cdot 65 \times 10^4$	

It will be seen that there is a good agreement between the experiments and the numbers calculated on the assumption considered in the beginning of this paper. This is brought out by the Table below.

	Number of α particles per gram of Uranium per second	
	Calculated	Observed
Uranium	$2 \cdot 32 \times 10^4$	$2 \cdot 37 \times 10^4$
Uranium mineral	$9 \cdot 67 \times 10^4$	$9 \cdot 6 \ \times 10^4$
Thorium, number of α particles per gram:		$2 \cdot 7 \ \times 10^4$

No doubt the agreement is closer than would be expected under the conditions of the experiments.

The agreement between theory and experiment confirms in another way the correctness of Boltwood's conclusion that uranium emits two α particles for one from each of its later products. The experiments are not of sufficient accuracy to confirm the data on the relative activity of actinium and radium. There is no doubt, however, that the number of α particles to be ascribed to actinium is very small compared with that to be expected if actinium and its series of products emitted one α particle for one from radium. The connexion of actinium with the uranium-radium series is difficult to determine, and remains one of the chief outstanding problems in the analysis of radioactive changes.

Production of Helium by Uranium, Uranium Minerals, and Thorium

Since the α particle is a charged atom of helium, it is a simple matter to deduce the rate of production of helium from the active materials considered. Calculation and experiment show that one gram of radium in equilibrium with its three α-ray products produces 158 cubic mm. of helium per year. Since radium and each of its products emits $3\cdot4 \times 10^{10}$ α particles per gram per second, uranium, which emits $2\cdot37 \times 10^4$ α particles per gram per second, produces $2\cdot75 \times 10^{-5}$ cubic mm. per year. The rate of production of helium for the different materials is given below.

	Production of Helium per gram per year
Uranium	$2\cdot75 \times 10^{-5}$ cubic mm.
Thorium	$3\cdot1 \ \times 10^{-5}$,,
Uranium mineral in equilibrium	$11\cdot0 \ \times 10^{-5}$,,
Radium in equilibrium..............	158 ,,

A simple calculation allows us to estimate the production of helium for a mineral like thorianite containing both uranium and thorium.

Range of the α particles from Uranium

The range of the α particles from uranium has been difficult to determine directly on account of the smallness of the activity of the thin films of the substance. By observations of the decrease of the ionization due to a layer of uranium when sheets of thin aluminium were placed over it, Bragg* deduced that the range in air of the α particle from uranium was about $3\cdot5$ cms. In the course of counting the scintillations from a thin film of ionium, it was observed that the scintillations were as bright if not brighter than those from a thin film of uranium. Boltwood has found that the range of the α particle from ionium is $2\cdot8$ cms., so that it appeared probable that the range of the α particles from uranium had been overestimated. This conclusion was confirmed by finding that the α rays from a thin film of uranium were more readily absorbed by aluminium than those from ionium. By a special method,

* Bragg, Phil. Mag. 1906, xi. p. 754.

G*

the range of the α particle from uranium has been measured and found to be about 2·7 cms., while the range of the α particle from ionium is a millimetre or two longer. Further experiments are in progress to determine the range of the α particle from uranium accurately, and to examine carefully whether two sets of α particles of different range can be detected.

University of Manchester
July 1910

The Probability Variations in the Distribution of α Particles

by PROFESSOR E. RUTHERFORD, F.R.S., *and* H. GEIGER, PH.D.

With a Note by H. BATEMAN

From the *Philosophical Magazine* for October 1910, ser. 6, xx, pp. 698–707

IN counting the α particles emitted from radioactive substances either by the scintillation or electric method, it is observed that, while the average number of particles from a steady source is nearly constant, when a large number is counted, the number appearing in a given short interval is subject to wide fluctuations. These variations are especially noticeable when only a few scintillations appear per minute. For example, during a considerable interval it may happen that no α particle appears; then follows a group of α particles in rapid succession; then an occasional α particle, and so on. It is of importance to settle whether these variations in distribution are in agreement with the laws of probability, *i.e.* whether the distribution of α particles on an average is that to be anticipated if the α particles are expelled at random both in regard to space and time. It might be conceived, for example, that the emission of an α particle might precipitate the disintegration of neighbouring atoms, and so lead to a distribution of α particles at variance with the simple probability law.

The magnitude of the probability variations in the number of α particles was first drawn attention to by E. v. Schweidler.* He showed that the average error from the mean number of α particles was $\sqrt{N \cdot t}$, where N was the number of particles emitted per second and t the interval under consideration. This conclusion has been experimentally verified by several observers, including Kohlrausch[†], Meyer and Regener,[‡] and H. Geiger,[§] by noticing the fluctuations when the ionization currents due to two sources of α rays were balanced against each other. The results obtained have been shown to be in good agreement with the theoretical predictions of von Schweidler.

The development of the scintillation method of counting α particles by Regener, and of the electric method by Rutherford and Geiger, has afforded a more direct method of testing the probability variations. Examples of the

* v. Schweidler, Congrès International de Radiologie, Liège, 1905.
† Kohlrausch, *Wiener Akad.* cxv. p. 673 (1906).
‡ Meyer and Regener, *Ann. d. Phys.* xxv. p. 757 (1907).
§ Geiger, Phil. Mag. xv. p. 539 (1908).

distribution of α particles in time have been given by Regener[*] and also by Rutherford and Geiger.[†] It was the intention of the authors initially to determine the distribution of α particles in time by the electric method, using a string electrometer of quick period as the detecting instrument. Experiments were made in this direction, and photographs of the throws of the instrument were readily obtained on a revolving film; but it was found to be a long and tedious matter to obtain records of the large number of α particles required. It was considered simpler, if not quite so accurate, to count the α particles by the scintillation method.

Experimental Arrangement

The source of radiation was a small disk coated with polonium, which was placed inside an exhausted tube, closed at one end by a zinc sulphide screen. The scintillations were counted in the usual way by means of a microscope on an area of about one sq. mm. of screen. During the time of counting (5 days), in order to correct for the decay, the polonium was moved daily closer to the screen in order that the average number of α particles impinging on the screen should be nearly constant. The scintillations were recorded on a chronograph tape by closing an electric circuit by hand at the instant of each scintillation. Time-marks at intervals of one half-minute were also automatically recorded on the same tape.

After the eye was rested, scintillations were counted from 3 to 5 minutes. The motor running the tape was then stopped and the eye rested for several minutes; then another interval of counting, and so on. It was found possible to count 2000 scintillations a day, and in all 10,000 were recorded. The records on the tape were then systematically examined. The length of tape corresponding to half-minute marks was subdivided into four equal parts by means of a celluloid film marked with five parallel lines at equal distances. By slanting the film at different angles, the outside lines were made to pass through the time-marks, and the number of scintillations between the lines corresponding to 1/8 minute intervals were counted through the film. By this method correction was made for slow variations in the speed of the motor during the long interval required by the observations.

In an experiment of this kind the probability variations are independent of the imperfections of the zinc sulphide screen. The main source of error is the possibility of missing some of the scintillations. The following example is an illustration of the result obtained. The numbers, given in the horizontal lines, correspond to the number of scintillations for successive intervals of 7·5 seconds.[‡]

* Regener, *Verh. d. D. Phys. Ges.* x. p. 78 (1908); *Sitz. Ber. d. K. Preuss. Akad. Wiss.* xxxviii. p. 948 (1909).

† Rutherford and Geiger, Proc. Roy. Soc. A. lxxxi. p. 141 (1908).   (*This vol., p. 89*)

‡ *Editor's footnote:* Some minor mistakes in this and the following two tables in the original are here corrected.

The length of tape was about 14 cms. for one minute interval. The average number of particles deduced from counting 10,000 scintillations was 31·0 per minute. It will be seen that for the 1/8 minute intervals the number of scintillations varied between 0 and 10; for one minute intervals between 25 and 42.

									Total per minute
1st minute: 3	7	4	4	2	3	2	0		25
2nd ,, 5	2	5	4	3	5	4	2		30
3rd ,, 5	4	1	3	3	1	5	2		24
4th ,, 8	2	2	2	3	4	2	6		29
5th ,, 7	4	2	6	4	5	10	4		42

Average for 5 minutes	30·0
True average	31·0

The distribution of α particles according to the law of probability was kindly worked out for us by Mr. Bateman. The mathematical theory is appended as a note to this paper. Mr. Bateman has shown that if x be the true average number of particles for any given interval falling on the screen from a constant source, the probability that n α particles are observed in the same interval is given by $\frac{x^n}{n!} e^{-x}$. n is here a whole number, which may have all positive values from 0 to ∞. The value of x is determined by counting a large number of scintillations and dividing by the number of intervals involved. The probability for n α particles in the given interval can then at once be calculated from the theory. The following table contains the results of an examination of the groups of α particles occurring in 1/8 minute interval. For convenience the tape was measured up in four parts, the results of which are given separately in horizontal columns I to IV.

For example (see column I), out of 792 intervals of 1/8 minute, in which 3179 α particles were counted, the number of intervals giving 3 α particles was 152. Combining the four columns, it is seen that out of 2608 intervals containing 10,097 particles, the number of times that 3 α particles were observed was 525. The number calculated from the equation was the same, viz. 525. It will be seen that, on the whole, theory and experiment are in excellent accord. The difference is most marked for four α particles, where the observed number is nearly 5 per cent. larger than the theoretical. The number of α particles counted was far too small to fix with certainty the number of groups to be expected for a large value of n, where the probability of the occurrence is very small. It will be observed that the agreement between theory and experiment is good even for 10 and 11 particles, where the probability of the occurrence of the latter number in an interval is less than 1 part in 600. The closeness of the agreement is no doubt accidental. The relation between theory and experiment is shown in fig. 1 for the results given in Table I,

Number of α particles	0	1	2	3	4	5	6	7	8	9	10	11	12	13	14	Number of α particles	Number of intervals	Average number
I	15	56	106	152	170	122	88	50	17	12	3	0	0	1	0	3179	792	4·01
II	17	39	88	116	120	98	63	37	4	9	4	1	0	0	0	2334	596	3·92
III	15	56	97	139	118	96	60	26	18	3	3	1	0	0	0	2373	632	3·75
IV	10	52	92	118	124	92	62	26	6	3	0	2	0	0	1	2211	588	3·76
Sum	57	203	383	525	532	408	273	139	45	27	10	4	0	1	1	10097	2608	3·87
Theoretical values	54	210	407	525	508	394	254	140	68	29	11	4	1	1	1			

where the ◌ represent observed points and the broken line the theoretical curve.

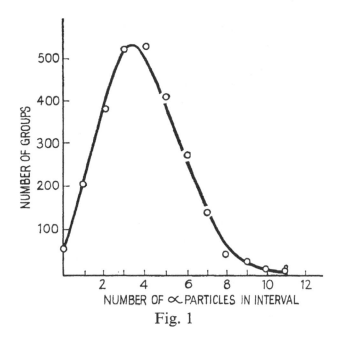

Fig. 1

The results have also been analysed for 1/4 minute intervals. This has been done in two ways, which give two different sets of numbers. For example, let A, B, C, D, E represent the number of α particles observed in successive 1/8 minute intervals. One set of results, given in Table A, is obtained by adding A + B, C + D, &c.; the other set, given in Table B, by starting 1/8 minute later and adding B + C, D + E, &c. The results are given in the appended Tables. In the final horizontal columns are given the sum of the occurrences in Tables A and B and the corresponding theoretical values.

In the cases for 1/4 minute intervals, the agreement between theory and experiment is not so good as in the first experiment with 1/8 minute interval. It is clear that the number of intervals during which particles were counted was not nearly large enough to give the correct average even for the maximum parts of the probability curve, and much less for the initial and final parts of the curve, where the probability of an occurrence is small. However, taking the results as a whole for the 1/8 minute and the 1/4 minute intervals, there is a substantial agreement between theory and experiment, and the errors are not greater than would be anticipated, considering the comparatively small number of intervals over which the α particles were counted. We may consequently conclude that the distribution of α particles in time is in agreement with the laws of probability and that the α particles are emitted at random. As far as the experiments have gone, there is no evidence that the variation in number of α particles from interval to interval is greater than would be expected in a random distribution.

TABLE A

Number of particles	0	1	2	3	4	5	6	7	8	9	10	11	12	13	14	15	16	17	18	19	20	21	Whole number of scintillations	Whole number of intervals	Average number in one interval
I	0	3	4	7	17	35	42	60	71	49	46	22	19	11	2	5	1	2	0	0	0	0	3182	396	8·04
II	0	2	3	6	19	33	34	56	32	34	27	24	11	7	5	3	1	1	0	0	0	0	2330	298	7·82
III	0	0	6	8	25	39	39	51	42	38	26	17	11	6	5	2	1	0	0	0	0	0	2373	316	7·51
IV	0	1	7	11	14	30	40	47	48	36	27	18	8	4	1	0	1	1	0	0	1	0	2214	295	7·53
Sum	0	6	20	32	75	137	155	214	193	157	126	81	49	28	13	10	4	4	0	1	1	0	10099	1305	7·74

TABLE B

Number of particles	0	1	2	3	4	5	6	7	8	9	10	11	12	13	14	15	16	17	18	19	20	21	Whole number of scintillations	Whole number of intervals	Average number in one interval
I	0	2	4	9	21	35	46	56	68	48	35	30	15	12	6	8	0	0	0	1	0	0	3180	396	8·03
II	0	1	5	7	21	27	38	46	40	38	28	16	12	9	3	5	0	1	1	0	0	0	2333	298	7·83
III	0	0	5	12	32	38	32	41	36	50	32	14	11	6	5	1	1	0	0	0	1	1	2371	316	7·50
IV	0	0	3	18	25	26	35	44	36	37	36	15	6	8	2	0	0	2	2	0	0	0	2210	294	7·52
Sum	0	3	17	46	99	126	151	187	180	173	131	75	44	35	16	14	1	2	2	1	1	1	10094	1304	7·74

Number of particles	0	1	2	3	4	5	6	7	8	9	10	11	12	13	14	15	16	17	18	19	20	21	Whole number of scintillations	Whole number of intervals	Average number in one interval
Sum of Tables A and B	0	9	37	78	174	263	306	401	373	330	257	156	93	63	29	24	5	5	2	1	2	1	20193	2609	7·74
Theoretical values	1·1	9	34	88	170	263	339	372	363	312	242	170	110	65	36	19	9	4	1·8	·72	·28	·10			

Apart from their bearing on radioactive problems, these results are of interest as an example of a method of testing the laws of probability by observing the variations in quantities involved in a spontaneous material process.

University of Manchester
July 22nd, 1910

NOTE

On the Probability Distribution of α Particles

by H. BATEMAN

LET λdt be the chance that an α particle hits the screen in a small interval of time dt. If the intervals of time under consideration are small compared with the time period of the radioactive substance, we may assume that λ is independent of t. Now let $W_n(t)$ denote the chance that n α particles hit the screen in an interval of time t, then the chance that $(n + 1)$ particles strike the screen in an interval $t + dt$ is the sum of two chances. In the first place, $n + 1$ α particles may strike the screen in the interval t and none in the interval dt. The chance that this may occur is $(1 - \lambda dt)W_{n+1}(t)$. Secondly, n α particles may strike the screen in the interval t and one in the interval dt; the chance that this may occur is $\lambda dt W_n(t)$. Hence

$$W_{n+1}(t + dt) = (1 - \lambda dt)W_{n+1}(t) + \lambda dt W_n(t).$$

Proceeding to the limit, we have

$$\frac{dW_{n+1}}{dt} = \lambda(W_n - W_{n+1}).$$

Putting $n = 0, 1, 2 \ldots$ in succession we have the system of equations:

$$\frac{dW_0}{dt} = -\lambda W_0,$$

$$\frac{dW_1}{dt} = \lambda(W_0 - W_1),$$

$$\frac{dW_2}{dt} = \lambda(W_1 - W_2),$$

$$\cdot \quad \cdot \quad \cdot \quad \cdot \quad \cdot$$

which are of exactly the same form as those occurring in the theory of radioactive transformations,* except that the time-periods of the transformations would have to be assumed to be all equal.

* Rutherford, 'Radioactivity,' 2nd edition, p. 330. The chance that an atom suffers n disintegrations in an interval of time t is equal to the ratio of the amount of the nth product present at the end of the interval to the amount of the primary substance present at the commencement.

The equations may be solved by multiplying each of them by $e^{\lambda t}$ and integrating. Since $W_0(0) = 1$, $W_n(0) = 0$, we have in succession:

$$W_0 = e^{-\lambda t},$$

$$\frac{d}{dt}(W_1 e^{\lambda t}) = \lambda, \quad \therefore W_1 = \lambda t e^{-\lambda t},$$

$$\frac{d}{dt}(W_2 e^{\lambda t}) = \lambda^2 t, \quad \therefore W_2 = \frac{(\lambda t)^2}{2!} e^{-\lambda t},$$

and so on. Finally, we get

$$W_n = \frac{(\lambda t)}{n!} e^{-\lambda t}.$$

The *average* number of α particles which strike the screen in the interval t is λt. Putting this equal to x, we see that the chance that n α particles strike the screen in this interval is

$$W_n = \frac{x^n}{n!} e^{-x}.$$

The particular case in which $n = 0$ has been known for some time (Whitworth's 'Choice and Chance,' 4th ed. Prop. 51).

If we use the above analogy with radioactive transformation, the theorem simply tells us that the amount of primary substance remaining after an interval of time t is $e^{-\lambda t}$ if a unit quantity was present at the commencement.

The *probable* number of α particles striking the screen in the given interval is

$$p = \sum_{n=1}^{\infty} nW_n = xe^{-x} \sum_{n=1}^{\infty} \frac{x^{n-1}}{(n-1)!} = x.$$

The *most probable* number is obtained by finding the maximum value of W_n.

Since $\dfrac{W_n}{W_{n-1}} = \dfrac{x}{n}$, this ratio will be greater than 1 so long as $n < x$. Hence if $n \lessgtr x$,

$$W_n \lessgtr W_{n-1};$$

if $n = x$, $W_n = W_{n-1}$. The most probable value of n is therefore the integer next greater than x; if, however, x is an integer, the numbers $x - 1$ and x are equally probable, and more probable than all the others.

The value of λ which is calculated by counting the total number of α particles which strike the screen in a large interval of time T, will not generally be the true value of λ. The mean deviation from the true value of λ is calculated by finding the mean deviation of the total number N of α particles observed in time T from the true average number λT. This mean deviation D (mittlerer

Fehler) is, according to the definition of Bessel and Gauss, the square root of the probable value of the square of the difference $N - \lambda T$, and so is given by the series

$$D^2 = \sum_{n=0}^{\infty} (N - \lambda T)^2 \frac{(\lambda T)^N}{N!} e^{-\lambda T}$$

$$= e^{-\lambda T} \sum_{N=0}^{\infty} \left[\frac{(\lambda T)^N}{(N-2)!} + \frac{(\lambda T)^N}{(N-1)!} - 2\frac{(\lambda T)^{N+1}}{(N-1)!} + \frac{(\lambda T)^{N+2}}{(N)!} \right] = \lambda T.$$

Hence $D = \sqrt{\lambda T}$, and the mean deviation from the value of λ is accordingly

$$\frac{D}{T} = \sqrt{\frac{\lambda}{T}};$$

it thus varies inversely as the square root of the length of the interval of time. This result is of the same form as the classical one used by E. v. Schweidler in the paper referred to earlier.

The probable value of $|N - \lambda T|$ (der durchschnittliche Fehler) is much more difficult to calculate.

The Scattering of the α and β Rays and the Structure of the Atom

by PROFESSOR E. RUTHERFORD, F.R.S.

From the *Proceedings* of the Manchester Literary and Philosophical Society, IV, 55, pp. 18–20
Abstract of a paper read before the Society on March 7, 1911

IT is well known that the α and β particles are deflected from their rectilinear path by encounters with the atoms of matter. On account of its smaller momentum and energy, the scattering of the β particles is in general far more pronounced than for the α particles. There seems to be no doubt that these swiftly moving particles actually pass through the atomic system, and a close study of the deflexions produced should throw light on the electrical structure of the atom. It has been usually assumed that the scattering observed is the result of a multitude of small scatterings. Sir J. J. Thomson (*Proc. Cam. Phil. Soc.*, 15, Pt. 5, 1910) has recently put forward a theory of small scattering, and the main conclusions of the theory have been experimentally examined by Crowther for β rays (*Proc. Roy. Soc.*, 84, p. 226, 1910). On this theory, the atom is supposed to consist of a positive sphere of electrification containing an equal quantity of negative electricity in the form of corpuscles. By comparison of theory with experiment, Crowther concluded that the number of corpuscles in an atom is equal to about three times its atomic weight in terms of hydrogen. There are, however, a number of experiments on scattering, which indicate that an α or β particle occasionally suffers a deflexion of more than 90° in a single encounter. For example, Geiger and Marsden (*Proc. Roy. Soc.*, 82, p. 495, 1909) found that a small fraction of the α particles incident on a thin foil of gold suffers a deflexion of more than a right angle. Such large deflexions cannot be explained on the theory of probability, taking into account the magnitude of the small scattering experimentally observed. It seems certain that these large deviations of the α particle are produced by a single atomic encounter.

In order to explain these and other results, it is necessary to assume that the electrified particle passes through an intense electric field within the atom. The scattering of the electrified particles is considered for a type of atom which consists of a central electric charge concentrated at a point and surrounded by a uniform spherical distribution of opposite electricity equal in amount. With this atomic arrangement, an α or β particle, when it passes close to the centre of the atom, suffers a large deflexion, although the probability of such large deflexions is small. On this theory, the fraction of the number of electrified particles which are deflected between an angle ϕ and $\phi + d\phi$ is given by $\dfrac{\pi}{4} ntb^2 \cot \phi/2 \operatorname{cosec}^2 \phi/2 d\phi$ where n is the number of atoms per unit volume of the scattering material, t the thickness of material supposed small, and $b = \dfrac{2NeE}{mu^2}$ where Ne is the charge at the centre of the atom, E the charge on the electrified particle, m its mass, and u its velocity.

It follows that the number of scattered particles per unit area for a constant distance from the point of incidence of the pencil of rays varies as $\operatorname{cosec}^4 \phi/2$. This law of distribution has been experimentally tested by Geiger for α particles, and found to hold within the limit of experimental error.

From a consideration of general results on scattering by different materials, the central charge of the atom is found to be very nearly proportional to its atomic weight. The exact value of the central charge has not been determined, but for an atom of gold it corresponds to about 100 unit charges. From a comparison of the theories of large and small scattering, it is concluded that the effects are mainly controlled by the large scattering, especially when the fraction of the number of particles scattered through considerable angles is small. The results obtained by Crowther are for the most part explained by this theory of large scattering although no doubt they are to a certain extent influenced by small scattering. It is concluded that for different materials the fraction of particles scattered through a large angle is proportional to NA^2 where N is the number of atoms per unit volume, and A the atomic weight of the material.

The main results of large scattering are independent of whether the central charge is positive or negative. It has not yet been found possible to settle this question of sign with certainty.

This theory has been found useful in explaining a number of results connected with the scattering and absorption of α and β particles by matter. The main deductions from the theory are at present under examination in the case of the α rays by Dr. Geiger using the scintillation method.

Untersuchungen über die Radiumemanation:
II. Die Umwandlungsgeschwindigkeit

von E. RUTHERFORD

Aus den Sitzungsberichten der kaiserl. Akademie der Wissenschaften in Wien.
Mathem.-naturw. Klasse; Bd. CXX. Abt. IIa. März 1911, pp. 303–12
(Vorgelegt in der Sitzung am 16. März 1911.)

In Anbetracht der Wichtigkeit, welche der Radiumemanation bei Experimentaluntersuchungen zukommt, ist es wünschenswert, ihre Umwandlungsgeschwindigkeit mit größtmöglichster Genauigkeit zu bestimmen. Seit dem Jahre 1902 sind bereits eine Reihe von Bestimmungen der Halbwertszeit der Emanation, d. h. der Zeit, in welcher sie zur Hälfte sich umwandelt, ausgeführt worden. Die Resultate, welche hierbei von verschiedenen Beobachtern erhalten wurden, sind aus folgender Tabelle ersichtlich:

		Halbwertszeit der Emanation
P. Curie*	1902	3·99 Tage
Rutherford und Soddy[†]	1903	3·77 »
Bumstead und Wheeler[‡]	1904	3·88 »
Sackur[§]	1905	3·86 »
Rümelin[‖]	1907	3·75 »
Madame Curie[¶]	1910	3·85 »

Einige von diesen Beobachtungen wurden in der Weise ausgeführt, daß die α-Strahlenaktivität der Emanation selbst gemessen wurde, andere wieder in der Weise, daß man die Emanation in einem Gefäß abdichtete und die β- und γ-Strahlenaktivität der Zerfallsprodukte bestimmte, welche mit der Emanation im radioaktiven Gleichgewichte standen.

Im Jahre 1909 begann Herr Tuomikoski in meinem Laboratorium Versuche über die Zerfallsgeschwindigkeit der Radiumemanation nach der γ-Strahlenmethode. Hierzu wurde ein Aluminiumblattelektroskop benutzt, welches mit einer 3 *mm* dicken Bleischicht umgeben war. Um die aktive

---

* C. r., 135, 857 (1902).
† Phil. Mag. [6], 5, 445 (1903). (*Vol. I, p. 565.*)
‡ Amer. Journ. Sc., Februar 1904.
§ Ber. der Deutschen chem. Ges., *38*, 1754 (1905).
‖ Phil. Mag. [6], *14*, 550 (1907).
¶ Radioactivité, p. 219 bis 226 (Gauthier-Villars, Paris).

Substanz in einer Röhre von möglichst kleinen Dimensionen zu erhalten, wurde die Emanation zuerst gereinigt, indem man sie durch flüssige Luft kondensieren ließ und die nicht verflüssigten Gase abpumpte. Die so gewöhnlich erhaltene Emanationsmenge entsprach der Gleichgewichtsmenge von zirka 50 *mg* Radium. Das mit Emanation gefüllte Röhrchen wurde in passender Entfernung (1 *m* oder mehr) vom Elektroskop aufgestellt und die Änderungen der Abfallsgeschwindigkeit der Blättchen in regelmäßigen Zwischenräumen untersucht. Die kleinen Schwankungen der Empfindlichkeit des Elektroskops von Tag zu Tag wurden auskorrigiert, indem jedesmal die Abfallsgeschwindigkeit bestimmt wurde, die ein Radium-Standardpräparat von konstanter Aktivität in bestimmter Entfernung vom Elektroskop hervorbrachte. Die ersten Beobachtungen schienen darauf hinzuweisen, daß die Unwandlungsgeschwindigkeit der Emanation durch physikalische und chemische Behandlung beeinflußt werde. So schien z. B. gereinigte Emanation anfangs rascher zu zerfallen als der Bruchteil, welcher während der Reinigung weggepumpt worden war. Ähnlicherweise schien die Emanation, welche nicht durch Wasser absorbiert worden war, langsamer abzuklingen als der absorbierte Teil. Diese Resultate würden — wenn sie korrekt wären — von allergrößter Wichtigkeit sein, da sie darauf hindeuten würden, daß nicht alle Atome der Emanation in ihren physikalischen und chemischen Eigenschaften gleich wären.

Die Versuche wurden von Tuomikoski fortgeführt und so eine große Anzahl von Abfallskurven von Emanationspräparaten erhalten, die in der verschiedensten Weise behandelt worden waren. In den späteren Versuchen waren die Verschiedenheiten der Abfallsgeschwindigkeit verschiedener Emanationsproben nicht mehr so ausgeprägt und viel unregelmäßiger als bei den zuerst beobachteten. In allen Fällen ergab sich, daß mehrere Tage nach der Herstellung der Zerfall der Emanation exponentiell mit einer Halbwertszeit zwischen $3 \cdot 8$ und $3 \cdot 9$ Tagen vor sich ging.

Beim Vergleich der Zerfallsperioden von Bruchteilen einer und derselben Emanationsprobe, wo sich doch identische Zerfallsperioden ergeben sollten, wurden gleichfalls anscheinende Abweichungen in der Periode bemerkt, was darauf hindeutet, daß die Meßmethode doch nicht so genau war, als man ursprünglich angenommen hatte. Wenn sich derartige Beobachtungen über mehrere Wochen erstrecken, wobei die meteorologischen Bedingungen stark variieren können, ist es schwierig, die Zerfallsgeschwindigkeit mit der geforderten Genauigkeit zu ermitteln. Kleine Änderungen in der Empfindlichkeit des Elektroskops oder Schwierigkeiten bei Erreichung des Sättigungsstromes können unerwartete Fehler einschleppen.

Die Kompensationsmethode

Zur Vermeidung dieser Schwierigkeiten wurden zunächst einige Experimente mit einer Kompensationsmethode gemacht. Die Ionisationsströme, welche von den γ-Strahlen zweier verschiedener Emanationsproben herrühren.

wurden gegeneinander ausgeglichen. Es wurde ein Elektrometer benutzt. Die zwei Ionisationsgefäße, mit Zentralelektroden versehen, waren aus 2 *mm* dickem Bleiblech verfertigt und waren voneinander geschieden durch einen 5 *cm* dicken Bleischirm. Die zwei Gefäße wurden zu entegegengesetzt gleichen Sättigungspotentialen (zirka 100 Volt) aufgeladen und die zentralen Elektroden miteinander und mit dem Elektrometer verbunden. Die mit Emanation gefüllten Röhrchen (in welchen bereits radioaktives Gleichgewicht herrschte) wurden in solcher Entfernung von den Ionisationsgefäßen angebracht, daß zu Beginn des Versuches die Ionisationseffekte sich genau kompensierten. Wenn die beiden Emanationsproben mit genau derselben Geschwindigkeit abklingen, so sollte das Gleichgewicht ungestört bleiben. Wenn jedoch eine kleine Verschiedenheit der Abklingungsgeschwindigkeiten besteht, müßte sie sich sofort durch Störung des Elektrometergleichgewichtes verraten. Wenn die Verschiedenheit der Perioden klein ist, so müßte unter diesen Umständen, wie sich leicht berechnen läßt, der durch die Wanderungsgeschwindigkeit der Elektrometernadel gemessene Effekt zuerst mit der Zeit ansteigen, nach 5·5 Tagen ein Maximum erreichen und dann allmählich abnehmen. Einige vorläufige Versuche zeigten, daß das Gleichgewicht sehr empfindlich gestaltet werden konnte und daß eine Änderung in der Intensität von $^1/_{400}$ noch mit Sicherheit konstatiert werden konnte. Wenn zwei Emanationsproben in dieser Weise miteinander verglichen wurden, ergab sich in einigen Fällen kaum irgendeine kleine Störung des Gleichgewichtes, in anderen Fällen ergab sich eine allmähliche Änderung des Gleichgeswichtes um etwa 1%. Ich konnte nicht die Überzeugung gewinnen, daß diese Störungen wirklichen Verschiedenheiten in der Periode der Emanation zuzuschreiben sind. Wahrscheinlicher erscheint es vielmehr, daß sie durch kleine Temperaturänderungen oder andere Störungen in den Versuchsgefäßen bewirkt wurden. Jedenfalls geben bis jetzt diese Experimente keinen definitiven Anhaltspunkt, daß verschiedene Proben von Emanation mit verschiedenen Perioden zerfallen. Wenn überhaupt reelle Verschiedenheiten der Perioden bestehen, sind sie sicherlich sehr klein.

Der Zerfall der Emanation in längeren Zeiträumen

Bei den meisten früheren Versuchen wurde der Zerfall der Emanation über einen verhältnismäßig nur kurzen Zeitraum verfolgt, wobei die anfängliche Intensität schließlich auf den hundertsten Teil gesunken war.* Bei den gengenwärtigen Versuchen wurde eine Methode ausgearbeitet, um die mittlere Zerfallsgeschwindigkeit über einen Zeitraum von mehr als 100 Tagen messen zu können, wobei die Intensität endlich bis auf den hundertmillionsten Teil abgenommen hatte. Es ist sehr wichtig festzustellen, ob der Zerfall in solch langen Zeiten im Mittel gleich rasch erfolgt als in kürzeren Intervallen. Eine

---

* Mad. Curie hat bei ihren neuesten Versuchen den Abfall der Emanation durch 43 Tage verfolgt; während dieser Zeit fällt die Aktivität auf $\dfrac{1}{2500}$ ihres ursprünglichen Wertes.

große Menge Emanation wurde in einem kleinen Glasröhrchen abgeschlossen und ihr γ-Strahleneffekt mittels eines Elektroskops aufs genaueste verglichen mit einem im Gleichgewicht befindlichen Radiumstandard. Dann ließ man den Zerfall der Emanation durch mehr als 100 Tage vor sich gehen, ohne daß an der ursprünglichen Lage des Röhrchens etwas geändert wurde. Dann wurde das Röhrchen geöffnet und der darin noch befindliche Emanationsrest durch einem Luftstrom in ein passendes Emanationselektroskop übergeführt. Dieses Elektroskop war nach bekannten Emanationsquanten geeicht, indem man die Emanation von einer Radiumstandardlösung durch Kochen ausgetrieben hatte (diese Lösung enthielt einen ganz bestimmten kleinen Bruchteil des Radiumstandards, mit dem zuerst der γ-Strahleneffekt verglichen worden war). Die Standardlösungen waren für mich vor mehreren Jahren von Dr. Boltwood und Prof. Eve angefertigt worden. Es wurden auch eigene Versuche angestellt, um die Genauigkeit dieser Standardlösungen zu prüfen; die Methode, nach welcher dies ausgeführt wurde, wird später dargelegt.

Um die Genauigkeit obiger Methode zu prüfen, wollen wir annehmen, daß in einer Zeit von t Tagen die Emanation auf den Bruchteil 10^{-8} ihres Anfangswertes gesunken sei; dann ist also

$$e^{-\lambda t} = 10^{-8} \quad \lambda t = 18\cdot 42.$$

Angenommen, daß der Fehler der Beobachtung des Betrages von $10^{-8} \pm 10\%$ sei, so ist

$$\lambda t = 18\cdot 42 \pm \log_e 1\cdot 10 = 18\cdot 42 \pm 0\cdot 0953.$$

Folglich verursacht ein Fehler von 10% beim Vergleich der Emanationsmengen einen Fehler von nur $\frac{1}{2}\%$ beim Werte von λ. Demnach würde, da unter den besten Versuchsbedingungen der Beobachtungsfehler beim Vergleich der anfänglichen und der zum Schlusse vorhandenen Emanationsmenge nicht 1 oder 2% überschreitet, der Fehler bei λ kaum $^1/_{10}\%$ übersteigen.

Die folgende Tabelle gibt die Resultate bei einer Anzahl von Beobachtungen mit verschiedenen Emanationsproben.

Anfängliche Menge von Emanation	Am Schlusse des Versuches vorhandene Emanationsmenge	Gesamtes Intervall	λ (Tage)$^{-1}$
$53\cdot 1$ mg Ra	$3\cdot 24 . 10^{-7}$ mg Ra	$105\cdot 2$ Tage	$0\cdot 1800$
$56\cdot 3$	$2\cdot 57 . 10^{-7}$	$106\cdot 3$	$0\cdot 1807$
$46\cdot 2$	$5\cdot 52 . 10^{-7}$	$101\cdot 2$	$0\cdot 1802$
$47\cdot 0$	$2\cdot 83 . 10^{-7}$	$105\cdot 15$	$0\cdot 1800$
$48\cdot 8$	$2\cdot 44 . 10^{-7}$	$106\cdot 2$	$0\cdot 1800$

Mittelwert $\lambda = 0\cdot 1802$ (Tage)$^{-1}$

Dies ergibt eine mittlere Halbwertszeit von $3\cdot 84_6$ Tagen oder rund $3\cdot 85$ Tage. Eine Anzahl von ähnlichen Beobachtungen der Periode der Emanation wurde mit Proben ausgeführt, die in verschiedenster Weise gereinigt worden waren; ich habe jedoch keine Verschiedenheiten beobachtet, welche die Beobachtungsfehler überstiegen. Die gefundene Zahl für die Halbwertszeit steht in ausgezeichneter Übereinstimmung mit der letzten Bestimmung von Madame Curie (3·85 Tage).

Die Gehaltsbestimmung an den Radiumlösungen

Die Genauigkeit der Bestimmung des Endbetrages der Emanation durch das Emanationselektroskop hängt nur von der Genauigkeit der Standardlösung ab, ausgedrückt in Bruchteilen des Radiumstandards, das bei der anfänglichen Vergleichung nach der γ-Strahlenmethode benutzt wurde. Da seit der Bereitung dieser Standardlösungen mehrere Jahre verstrichen waren, war es notwendig, diesen Punkt zu untersuchen. Dies geschah in folgender Weise:

Eine Menge Emanation, deren γ-Strahleneffekt nahezu gleich war dem des Radiumstandards (3·69 *mg* RaBr$_2$), wurde in einem Glasröhrchen verschlossen. Sobald das Gleichgewicht erreicht war, wurde der γ-Strahleneffekt genau verglichen mit dem des Standards, und zwar nach der elektroskopischen Methode. Das Emanationsröhrchen wurde dann geöffnet und sein Inhalt in ein vorher evakuiertes Gefäß von zirka 1 *l* Inhalt getrieben. Dann wurde Luft hinzugefügt bis zur Erreichung des atmosphärischen Druckes und der Inhalt gut durchmischt. Ein bestimmter Bruchteil dieser Quantität von Emanation (ungefähr ein Tausendstel) wurde weggenommen, indem die Emanation in eine kleine evakuierte Röhre, die mit zwei Hähnen versehen war, abgelassen wurde. Die Emanation in dieser Röhre wurde in der gleichen Weise in ein anderes Gefäß von ebenfalls zirka 1 *l* Fassungsraum übertragen; abermals wurde ein Tausendstel davon entfernt und diese Menge wurde in das Emanationselektroskop eingeführt. So wurde ungefähr ein Milliontel der ganzen Emanationsmenge ins Elektroskop gebracht. Es wurde besondere Sorgfalt darauf verwendet, die Volumina der verschiedenen Gefäße genau zu bestimmen und für eine vollkommen gleichförmige Mischung der Emanation mit Luft sowie für Gleichheit der Temperatur während der Wegnahme der Bruchteile garantieren zu können.

Die Radiumstandardlösungen, welche zum Vergleich genommen wurden, enthielten 0·785, 1·57 und 3·14 . 10^{-6} *mg* Ra. Die Emanation wurde durch Kochen vertrieben und in einem Apparate gesammelt, der ähnlich dem von Boltwood verwendeten war. Die Resultate für die drei Standards waren sehr gut übereinstimmend und gaben am Elektroskop 3 Stunden nach der Einbringung der Emanation in dasselbe eine Abfallsgeschwindigkeit von 6·1 Skalenteilen. Dieses Resultat stimmte innerhalb 2% mit der Abfallsgeschwindigkeit überein, die mit den Unterteilungen der ursprünglichen Emanationsmenge erhalten wurde, und beweist, daß innerhalb der Grenzen der Beobachtungsfehler die Radiumlösungen richtig waren und sich seit ihrer Herstellung nicht verschlechtert hatten.

Auch bei dieser Methode der Unterteilungen ist es möglich, die Periode der

Emanation über einen Zeitraum von 30 bis 40 Tagen zu bestimmen. Eine Menge Radiumemanation, entsprechend zirka 2 *mg* Ra, wurde in einem Röhrchen abgeschlossen. Etwa 30 Tage später wurde sie in ein Gefäß von zirka 1 *l* Inhalt geleitet, ein bestimmter Bruchteil (zirka ein Tausendstel) weggenommen und in das Elektroskop eingeleitet. Die Abfallsgeschwindigkeit innerhalb dieser Zeit ergab sich genau übereinstimmend mit der, welche für längere Zeiträume gefunden worden war.

Einfluß der Konzentration auf die Periode der Emanation

Die Frage, ob die Gesamtaktivität oder die Zerfallsperiode eines radioaktiven Elementes durch den Grad seiner Konzentration beeinflußt wird, ist von großer Wichtigkeit in der Theorie der Radioakivität. Die allgemeine Übereinstimmung der Zerfallsperioden der verschiedenen Produkte, welche über einen sehr weiten Bereich der Aktivität verfolgt worden sind, scheinen zu beweisen, daß der Effekt — sofern er überhaupt existiert — sehr klein sein muß. In dem Werke »Radioactivity« (second edition, p. 466) wurde ein Experiment vom Verfasser beschrieben, bei welchem er fand, daß der Konzentrationsgrad keinen Einfluß auf die γ-Strahlenaktivität des Radiums hatte. Ein ähnliches Experiment wurde mit Radiumemanation gemacht: Eine Emanationsmenge, welche ungefähr 100 *mg* Radium entsprach, wurde gereinigt und an dem Ende eines kleinen Kapillarröhrchens zusammengedrängt. Der γ-Strahleneffekt im Gleichgewichte wurde in einem kleinen, zirka 2 *m* entfernt aufgestellten, geschlossenen Elektroskop mit Bleiwänden gemessen. Durch Erniedrigung des Quecksilberstandes wurde sodann bewirkt, daß die Emanation auf ein 2000 mal größeres Volumen verteilt wurde. 3 Stunden später, wenn die Emanation wieder im Gleichgewichte mit ihrem aktiven Beschlage war, wurde der γ-Strahleneffekt wiederum gemessen. Wenn man den kleinen Abfall der Emanation während der Zeit berücksichtigte, ergab sich derselbe γ-Strahleneffekt wie vorher.

Durch Betrachtung der Dimensionen des Raumes, welcher die Emanation enthielt, fand man, daß die Verschiedenheit in der Verteilung des aktiven Niederschlages in den beiden Experimenten keinen nennenswerten Einfluß bei der elektroskopischen Messung hervorbringen konnte. Aus dem Experimente können wir folglich erschließen, daß die Aktivität der Emanation und daher auch ihre Zerfallsperiode durch den Grad der Konzentration nicht beeinflußt wird. Dies beweist, daß sogar, wenn die Emanation in sehr konzentrierter Form vorliegt, ihre Strahlung keine merkliche Änderung der Umwandlungsgeschwindigkeit bewirkt.

Einfluß der Temperatur

Es wurden auch einige Experimente nach der im vorhergehenden beschriebenen Kompensationsmethode ausgeführt, um zu untersuchen, ob der Zerfall der Emanation durch Erniedrigung der Temperatur nicht beeinflußt werde.

Zu diesem Behufe wurde die γ-Strahlung von zwei ähnlichen Emanations-proben, welche in Glasröhrchen eingeschlossen waren, gegeneinander kompensiert. Das eine dieser Röhrchen war innerhalb eines zylindrischen Dewar'schen Gefäßes so aufgestellt, daß es die gegen das Versuchsgefäß liegende Seitenwand berührte. Nachdem die Kompensation erreicht war, wurde flüssige Luft eingeleitet. Es wurde keine augenblickliche Störung des Kompensationszustandes beobachtet. Bei einigen Experimenten mehr provisorischer Natur waren die die Emanation enthaltenden Röhrchen etwa 4 *cm* lang und senkrecht aufgestellt in einer Entfernung von zirka 20 *cm* von den Untersuchungsgefäßen. Mit so großen Röhrchen wurde beobachtet, daß die Änderung in der Verteilung des aktiven Niederschlages bei der Temperatur der flüssigen Luft den Kompensationszustand störte. Der Betrag der Störung wurde nach Verlauf von 3 Stunden aber nicht, wie erwartet werden sollte, konstant, sondern änderte sich noch weiter einen ganzen Tag lang, sogar noch länger. Erst in späteren Experimenten wurde es klar, daß der Effekt doch einer allmählichen Änderung in der Verteilung der Emanation und des aktiven Niederschlages im Verlaufe der Zeit zuzuschreiben war. Um diese Störung zu vermeiden, wurde gereinigte Emanation in gleichen Teilen auf zwei kleine Glaskugeln von zirka 5 *mm* Durchmesser verteilt. Es ergab sich, daß unter diesen Umständen der Kompensationszustand selbst eine ganze Woche hindurch ungeändert blieb. Gelegentlich wurden zwar kleine Störungen beobachtet, doch waren diese zweifellos äußeren Ursachen zuzuschreiben. Wir können folglich schließen, daß die Emission der γ-Strahlen und die Zerfallsperiode der Emanation ungeändert bleibt, wenn man die Temperatur von der gewöhnlichen Zimmertemperatur auf −186° erniedrigt.

Zusammenfassung

1. Die Zerfallsperiode der Radiumemanation wird durch chemische oder physikalische Prozesse nicht beeinflußt.

2. Die Zerfallsperiode von je einer und derselben Emanationsprobe wurde gemessen über einen Zeitraum, innerhalb dessen sie bereits auf den hundert-millionsten Teil gesunken war, und wurde mit großer Annäherung zu 3·85 Tagen (Halbwertszeit) festgestellt. Der Wert von λ ist 0·1802 (Tage)$^{-1}$.

3. Die Zerfallsperiode der Emanation wird durch den Grad ihrer Konzentration nicht beeinflußt.

4. Die Zerfallsperiode der Emanation ist bei Zimmertemperatur dieselbe wie bei der Temperatur der flüssigen Luft.

Zum Schlusse spreche ich meinen Dank aus Herrn Y. Tuomikoski für seine Unterstützung bei dem ersten Teile dieser Arbeit und Herrn W. C Lantsberry für seine Hilfe bei der Vergleichung der Emanationsmengen.

Universität Manchester
1910

Die Erzeugung von Helium durch Radium

von B. B. BOLTWOOD *und* E. RUTHERFORD*

Aus den Sitzungsberichten der kaiserl. Akademie der Wissenschaften in Wien.
Mathem.-naturw. Klasse; Bd. CXX. Abt. IIa. März 1911, pp. 313–36
(Vorgelegt in der Sitzung am 16. März 1911.)

Einleitung

Der Zusammenhang des Heliums mit der Umwandlung der radioaktiven Stoffe ist in den letzten acht Jahren ein Problem von größtem Interesse und hoher Bedeutung gewesen und wurde Gegenstand einer ganzen Anzahl von Untersuchungen.

Im Jahre 1902 veröffentlichten Rutherford und Soddy† die Theorie des Zerfalles der radioaktiven Materie und sprachen die Vermutung aus, daß das Helium, welches in relativ großen Mengen in den radioaktiven Mineralien vorkommt, eines der Umwandlungsprodukte der in diesen Mineralien vorkommenden Radioelemente sei.

Im Jahre 1903 zeigten Ramsay und Soddy‡ auf experimentellem Wege, daß bei Radiumsalzen stets Helium nachweisbar ist und daß Helium von der Radiumemanation erzeugt wird. Bei diesen ersten Experimenten war Helium nur in minimalen Mengen vorhanden und wurde spektroskopisch nachgewiesen. Diese Beobachtungen wurden kurz darauf von einer Anzahl voneinander unabhängigen Beobachtern bestätigt. Etwas später fand Debierne, daß Helium auch von Actiniumpräparaten erzeugt wird.

Zur Zeit der Entdeckung der Erzeugung von Helium durch Radium war es von fundamentaler Wichtigkeit, die Stellung des Heliums in der allgemeinen Zerfallsreihe der Radioelemente festzustellen. Im Jahre 1903 zeigte Rutherford,§ daß die vom Radium ausgeschleuderten α-Strahlen aus positiv geladenen Partikeln von hoher Geschwindigkeit bestehen. Die Bestimmung von $\frac{e}{m}$, des Verhältnisses der Ladung zur Masse der α-Partikel, ergab, daß das α-Partikel etwa die Dimension der gewöhnlichen Atome besitzen müsse mit einer scheinbaren Masse, die ungefähr zweimal so groß war wie die des Wasserstoffatoms. Von allem Anfang an schien es wahr-

* Eine vorläufige Mitteilung über diesen Gegenstand wurde bereits im November 1909 der Philosophical Society in Manchester vorgelegt. (*This vol., p. 177.*)
† Phil. Mag. [6], *4*, 582 (1902). (*Vol. I, p. 506.*)
‡ Proc. Roy. Soc., *72*, A., 204 (1903).
§ Phil. Mag. [6], *5*, 177 (1903). (*Vol. I, p. 549.*)

scheinlich, daß, wenn die α-Partikel aus irgendeinem bekannten Stoffen bestehen, sie entweder geladene Teilchen von Wasserstoff oder von Helium sein müßten. Und Rutherford* hatte, unmittelbar nachdem die Heliumbildung durch Radium experimentell gezeigt worden war, dargelegt, daß die Quelle des Heliums aller Wahrscheinlichkeit nach die angehäuften α-Partikel waren, die vom Radium und dessen Zerfallsprodukten ausgesendet wurden. Auf Grund dieser Hypothese wurde unter Benutzung der damals unvollkommenen erreichbaren Daten eine schätzungsweise Berechnung der Heliummenge durchgeführt, die von 1 g Radium im Gleichgewicht produziert wird; darnach sollte die von 1 g Radium pro Jahr erzeugte Heliummenge zwischen 20 und 200 mm^3 liegen.

Die Wichtigkeit der Frage nach der wahren Natur der α-Partikel führte zu einer genaueren Messung der Geschwindigkeiten und der Verhältnisse $\frac{e}{m}$ bei den α-Partikeln von den verschiedenen radioaktiven Elementen. Der Wert von $\frac{e}{m}$ ergab sich nach Rutherford† zu

$$\frac{e}{m} = 5 \cdot 07 \cdot 10^3 \text{ elektromagnetischen Einheiten.}$$

Die Resultate dieser Untersuchung wiesen darauf hin, daß die α-Partikel entweder Wasserstoffmoleküle mit einfacher Ionenladung oder Heliumatome mit doppelter Ionenladung wären. Die letztere Annahme schien die wahrscheinlichere zu sein und ließ vermuten, daß Helium ein Zerfallsprodukt aller Arten von radioaktiven Stoffen, welche α-Strahlen aussenden, sei.

Die Frage wurde weiter durchforscht durch Rutherford und Geiger,‡ welche eine Methode zur direkten Zählung der von einer radioaktiven Substanz ausgesendeten α-Teilchen entdeckten. Es ergab sich,§ daß 1 g Radium (ohne Zerfallsprodukte) pro Sekunde $3 \cdot 4 \cdot 10^{10}$ α-Partikel aussendet und daß Radium samt seinen drei α-strahlenden Zerfallsprodukten im Gleichgewichte viermal so viel α-Teilchen abschleudert. Gleichzeitig wurde die von jedem α-Partikel mitgeführte Ladung zu $9 \cdot 3 \cdot 10^{-10}$ elektrostatischen Einheiten bestimmt. Aus verschiedenen Gründen wurde geschlossen, daß die α-Partikel die doppelte Einheitsladung mitführen und daß der Wert der Einheitsladung, d. h. der Ladung, welche das Wasserstoffatom besitzt, gleich $4 \cdot 65 \cdot 10^{-10}$ elektrostatischen Einheiten ist.

Der endgültige Beweis der Identität von α-Partikel und Heliumatom wurde von Rutherford und Royds‖ erbracht, welche zeigten, daß, wenn α-Partikel in ein Vakuum oder in einen festen Körper eindringen, dort stets

* Nature, *68*, 366 (1903). (*Vol. I, p. 609.*)
† Phil. Mag. [6], *12*, 358 (1906). (*Vol. I, p. 889.*)
‡ Proc. Roy. Soc., *81*, A., 141 (1908). (*This vol., p. 89.*)
§ Proc. Roy. Soc., *81*, A., 161 (1908). (*This vol., p. 108.*)
‖ Phil. Mag. [6], *17*, 281 (1909). (*This vol., p. 163.*)

sich Helium ansammelt. So war definitiv bewiesen, daß das α-Partikel ein fortgeschleudertes Heliumatom ist, welches zwei Einheitsladungen mitführt.

Rutherford und Geiger* berechneten auf Grund der oben angeführten Daten die zu erwartende Heliumproduktion durch Radium. 1 *g* Radium im Gleichgewicht mit seinen Zerfallsprodukten emittiert $13 \cdot 6 \cdot 10^{10}$ α-Partikel, i. e. Heliumatome pro Sekunde. Nimmt man die Ladung des Wasserstoffatoms zu $4 \cdot 65 \cdot 10^{-10}$, so kann man sofort aus dem elektrochemischen Äquivalent für Wasserstoff ausrechnen, daß 1 *cm³* jedes Gases bei normalem Druck und Temperatur $2 \cdot 72 \cdot 10^{19}$ Moleküle enthält. Da Helium einatomig ist, so ergibt sich die von 1 *g* Radium erzeugte Heliummenge zu $\dfrac{13 \cdot 6 \cdot 10^{10}}{2 \cdot 72 \cdot 10^{19}}$ *cm³* pro Sekunde oder 158 *mm³* pro Jahr.

Die erste direkte Bestimmung der Heliumproduktion durch Radium wurde ausgeführt von Sir James Dewar,† welcher 70 *mg* wasserfreien Radiumchlorids benutzte, einen Teil des Materials, welches Dr. Thorpe bei seiner Bestimmung des Atomgewichtes des Radiums bereitet hatte. Das krystallisierte Radiumsalz wurde in ein evakuiertes Gefäß gegeben, das mit einem Mac-Leod-Manometer in Verbindung stand. Das Salz wurde von Zeit zu Zeit erhitzt, um das angesammelte Helium auszutreiben. Andere Gase, welche etwa anwesend waren, wurden durch Absorption mittels eines Stückes in flüssiger Luft gekühlter Kokosnußkohle entfernt. Es wurde auch eine schätzungsweise Bestimmung der in dem Gasgemenge anwesenden Warsser stoffmenge erzielt, indem man die Kokosnußkohle weiter zur Temperatu des flüssigen Wasserstoffes abkühlte.

In seiner ersten Abhandlung schloß Dewar, daß 1 *g* Radium im Gleichgewicht pro Tag $0 \cdot 37$ *mm³* Helium erzeugte. Bei einem späteren Versuche‡ war das Radium durch neun Monate verschlossen und die mittlere Heliumproduktion wurde zu $0 \cdot 463$ *mm³* pro Gramm Radium und pro Tag bestimmt. In der zweiten Abhandlung wurde ausgeführt, daß die ursprüngliche Berechnung bei dem ersten Experiment einen Fehler enthielt, nach dessen Beseitigung die Heliumproduktion anstatt zu $0 \cdot 37$ *mm³* zu $0 \cdot 499$ *mm³* pro Gramm Radium und pro Tag sich ergab. Die von 1 *g* Radium pro Jahr erzeugte Heliummenge ist daher nach diesen beiden Experimenten 182 *mm³*, beziehungsweise 169 *mm³*. Diese Werte sind beide etwas höher als der berechnete Wert (158 *mm³*).

Da es sehr wichtig ist, den Wert der Heliumproduktion so genau als möglich zu kennen, wurde von den Verfassern eine unabhängige Bestimmung dieser Größe ausgeführt. Es wurden zwei getrennte Bestimmungen der vom Radium erzeugten Heliummenge gemacht, bei welchen ein beträchtlicher Teil des Radiumpräparates in Verwendung kam, das die kaiserl. Akademie der Wissenschaften in Wien dem einen Verfasser in liebenswürdigster Weise geliehen hat.

---

* Proc. Roy. Soc., *81*, A., 162 (1908). (*This vol., p. 109.*)
† Proc. Roy. Soc., *81*, A., 280 (1908).   ‡ Proc. Roy. Soc., *83*, A., 404 (1910).

Die Reinigung des Präparates

Es war ganz wesentlich, daß das bei diesen Versuchen gebrauchte Radiumsalz keine anderen radioaktiven Substanzen als Radium selbst enthielt. Das Salz bestand aus einer Quantität von Barium-Radiumchlorid, welches etwa 7% des letzteren Elementes enthielt und schon bevor es in unsere Hände kam, einer teilweisen fraktionierten Umkrystallisation unterworfen worden war. Es war daher in hohem Grade unwahrscheinlich, daß andere radioaktive Substanzen als Radium und seine Zerfallsprodukte anwesend waren. Das Salz stand über ein Jahr lang in wässeriger, schwach saurer Lösung, bevor diese Experimente begannen, und früher war es durch einen etwas undefinierten Zeitraum in Form eines krystallinischen Salzes gewesen. Um nun das Radioblei und das Polonium zu entfernen, das sich in dieser Zeit angesammelt hatte, wurde die Lösung folgender Behandlung unterworfen: Etwa 50 *mg* Antimontrichlorid, 10 *mg* Bleinitrat und 5 *mg* Wismutnitrat wurden in einer kleinen Menge verdünnter Salzsäure gelöst und diese Lösung wurde zur Lösung des Radiumsalzes hinzugefügt. Das Ganze wurde dann mit Wasser verdünnt auf ein Volumen von zirka 100 *cm*³ und ein Überschuß von Schwefelwasserstoff eingeleitet. Infolge der Hinzufügung des Antimonsalzes koagulierte die Fällung der Sulfide und konnte leicht durch Filtration abgetrennt werden; nachher wurde das Präzipitat gut mit destilliertem Wasser gewaschen.

Die Fällung auf dem Filtrierpapier wurde aufgelöst durch Erwärmen mit verdünnter Salzsäure, zu welcher einige kleine Krystalle von Kaliumchlorat hinzugefügt wurden, und nach Verdünnung mit Wasser die Behandlung mit Schwefelwasserstoff wiederholt. Die zweite Fällung der Sulfide wurde abfiltriert und das Filtrat zum Filtrate der ersten Fällung hinzugefügt. Die Hinzufügung der Blei- und Wismutsalze zur ursprünglichen Lösung sollte die Vollständigkeit der Fällung des Radiobleies und Poloniums sichern.*

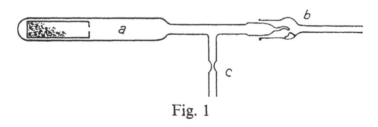

Fig. 1

Die vereinigten Filtrate von den Sulfidniederschlägen wurden in einer Quarzschale im Wasserbade zur Trockene eingedampft und der Rückstand von Radium-Bariumchlorid sanft erhitzt, um den Überschuß von Salzsäure und Wasser auszutreiben. Das trockene Salz wurde sodann in eine zylindrische Platinkapsel gegeben, welche mit einem durchlochten Deckel verschlossen war. Die Kapsel wurde darauf eingeschlossen in eine Röhre von Jenaer Hartglas (*a*, Fig. 1). Das eine Ende dieser Röhre paßte in eine Weichglasröhre

* Boltwood, Amer. Journ. Science, *25*, 228 (1908).

b und das Endstück der Röhre *a*, das sich bis zum Verbindungsstück *b* erstreckte, mündete in eine enge gebogene Kapillare. Ein innen vorstehendes Stück in der Röhre *b* in der Nähe des Verbindungsstückes ermöglichte es, durch Drehung der Röhre *a* im Schliffstück das Kapillarende der Röhre *a* abzubrechen und so gegebenenfalls die Verbindung zwischen *a* und *b* herzustellen. Diese Anordnung wurde getroffen, um die Entfernung des Heliums am Ende der Ansammlungszeit zu erleichtern. Die Jenaglasröhre wurde vollkommen luftleer gepumpt durch ein Seitenrohr *c*, welches dann zugeschmolzen wurde.

Bestimmung des Radiumgehaltes im Salze

Die in dem Salze vorhandene Radiummenge wurde bestimmt durch Messung der γ-Strahlung nach einem Zeitraum von über zwei Monaten nach dem Einschmelzen des Präparates.

Mit Hilfe eines dickwandigen Bleielektroskops wurde die γ-Strahlenaktivität des Salzes in der Röhre verglichen mit der γ-Strahlenaktivität eines Standardpräparates unseres Laboratoriums mit $3 \cdot 69 \, mg$ $RaBr_2$, sowohl direkt als auch indirekt durch Vergleichung mit einem dritten Radiumpräparat, dessen Gehalt mit Hilfe des kleinen Standards zu $32 \, mg$ $RaBr_2$ bestimmt worden war. Die Vergleichung der Radiummengen wurde in verschiedenen Entfernungen und unter verschiedenen Versuchsbedingungen ausgeführt und ergab, daß die in dem bei diesen Experimenten gebrauchten Röhrchen enthaltene Radiummenge $191 \, mg$ (Metall) betrug (gleichwertig $326 \, mg$ Radiumbromid).

Als die Experimente zur Bestimmung der Heliumproduktion beendet waren, wurde noch eine weitere Messung der in dem Salz anwesenden Radiummenge in folgender Weise ausgeführt: Das Salz wurde in destilliertem Wasser gelöst, das eine kleine Quantität Salzsäure enthielt, und die Lösung zu einem Endvolumen von $50 \, cm^3$ verdünnt. Ein kleiner, bestimmter Bruchteil ($1 \cdot 22 \%$) dieser Lösung wurde entfernt, zur Trockene eingedampft in einem kleinen Glasröhrchen und hermetisch verschlossen durch Zuschmelzen des Glases. Nach einem Monat wurde der kleine Bruchteil direkt mit Hilfe der γ-Strahlenmethode mit dem $3 \cdot 69 \, mg$ $RaBr_2$-Standardpräparat unseres Laboratoriums verglichen und ergab sich als äquivalent $4 \cdot 02 \, mg$ $RaBr_2$. Aus diesem Vergleich ergäbe sich für die in dem gebrauchten Präparat ursprünglich anwesende Radiummenge $193 \, mg$.

Ansammlung und Messung des Heliums

Die vom Radiumsalz erzeugte Heliummenge wurde gesammelt, gereinigt und gemessen in dem Apparat, der schematisch in der Fig. 2 dargestellt ist. Dieser enthielt eine Geißlerpumpe *P*, um die anderen Teile des Apparates evakuieren zu können. Die durch diese Pumpe weggeschafften Gase konnten, wenn erforderlich, bei *D* aufgefangen werden. Das Gefäß *M* enthielt Phosphor-

H

pentoxyd als Trockenmittel. Das Gefäß A mit angeschmolzenen Kapillar-
röhren E und e war im wesentlichen ein Mac-Leod-Manometer, wurde aber
bei diesen Experimenten zur Messung der kleinen Volumina, nicht der Drucke
benutzt, wie es sonst üblich ist.

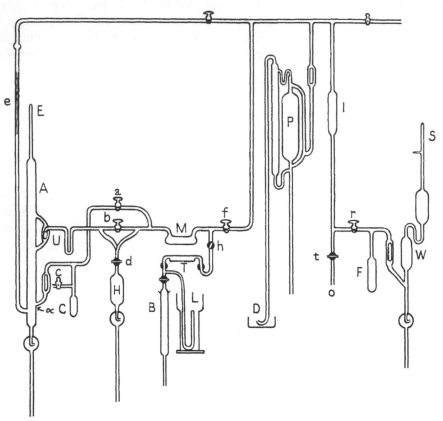

Fig. 2

Das Volumen der Kapillarröhre E war genau bestimmt in jedem Teile
seiner ganzen Länge, und zwar durch direkte Kalibration mit einem Queck-
silberfaden. Die Vergleichsröhre e war von demselben Rohrstück gefertigt
wie E, hatte also denselben Durchmesser und denselben Querschnitt, so
daß, wenn der Druck des in E befindlichen Gases gemessen wurde, keine
Korrektion wegen der Kapillardepression notwendig war.

In die Röhre A mündeten von rechts her zwei Seitenröhren; diese konnten
durch Glasschwimmerventile gegen A hin gesperrt werden; das in der unteren
Röhre war von gewöhnlicher Type, das obere mit Spezialkonstruktion, die
verhinderte, daß zwischen dem Glasschwimmer und der Außenwand Gas sich
verfangen konnte. Die untere Röhre erstreckte sich bis zur Phosphor-
pentoxydvorlage M, war mit einem Hahn bei a versehen und mit der Kohlenröhre C verbunden, welche einige Gramme Kokosnußkohle enthielt. Ein
Seitenrohr c ermöglichte die Einführung kleiner Volumina von reinem,

trockenem, elektrolytischem Sauerstoff an dieser Stelle des Apparates. Die obere der zwei von *A* nach rechts führenden Röhren hatte ein dünnwandiges Biegungsstück *U* und führte durch den Hahn *b* zur Röhre *M* und durch den Hahn *d* zur Übertragungspumpe *H*. Diese diente dazu, um das Gas von der Vorlage *M* zu entfernen und in *A* und seine Verbindungsstücke einzuleiten.

An diesen Teil des Apparates grenzte weiter an die Bürette *B* mit eingeschmolzenen Platindrähten, zwischen welchen Funken eingeleitet werden konnten. Die Bürette diente dazu, um Gase in den Apparat einzuleiten mit Hilfe der Röhre *L*, welche unter Quecksilber endigte. Nach Passierung der Funkenstrecke konnten die Gase durch die Röhre *T* streichen, welche aus Hartglas bestand und zur Rotglut erhitztes Kupferoxyd enthielt.

Der übrige Teil des Apparates (siehe rechte Seite der Fig. 2) diente zur Untersuchung des Spektrums des Gases, welches vorher in *A* volumetrisch gemessen worden war. Die Einzelheiten der Konstruktion sollen hier nicht beschrieben werden; es genügt zu sagen, daß dieser Teil des Apparates gestattete, die Gasreste aus *A* zu entfernen und in die kleine Röhre *S* einzuführen, wo das Spektrum bequem untersucht werden konnte.

Alle die verschiedenen Gefäße *A*, *H*, *B* etc. waren mit Bunsen'schen Verschlüssen versehen, um den Eintritt von Luft entlang den Quecksilbersäulen zu verhindern und die damit verbundenen Glasröhren erstreckten sich nach unten über 76 *cm*, wie es üblich ist. Die Ausdehnung dieser Rohre, die Kautschukschläuche und Quecksilberreservoirs, welche am unteren Ende angefügt waren, sind in der Fig. 2 nicht ersichtlich gemacht.

Kalibration der zur Volummessung bestimmten Teile des Apparates

Wie schon erklärt, ähnelte das Gefäß *A* im allgemeinen einem gewöhnlichen Mac-Leod-Manometer. Zum Zwecke vorliegender Versuche wurde es in der folgenden Weise kalibriert. Der ganze Apparat war vollkommen luftleer, das Gefäß wurde dann noch zu sanfter Rotglut erhitzt, um die in der Kokosnußkohle absorbierten Gase zu entfernen, und der Druck im Apparat wurde durch fortgesetztes Pumpen mit der großen Pumpe *P* aufs äußerste erniedrigt. Das Quecksilberniveau in *A* wurde dann gehoben bis zum Punkte *α*, die Hähne *f*, *d*, *a* und *c* geschlossen und eine kleine Menge von Sauerstoff, gemengt mit reinem Helium (aus krystallinischem Uraninit), in das Gefäß *A* durch die Röhre *L* und ihre Verbindungen eingelassen. Der Hahn *b* wurde nun geschlossen und die Röhre *C* mit flüssiger Luft umgeben. Nach zirka 20 Minuten wurde das Quecksilber in *A* gehoben und Volumen und Druck des Gases in der Kapillare *E* gemessen. Nach abermaligem Heben des Quecksilbers bis zum obersten Rande von *A* wurden die Hähne *a*, *b*, *f* geöffnet und die flüssige Luft rings um *C* entfernt. Die Röhre *C* erwärmte sich, wurde schließlich zur schwachen Rotglut erhitzt und die Röhren *C*, *U*, *M* etc. gründlich ausgepumpt. Die Hähne *a* und *b* wurden dann wieder geschlossen, die Röhren *C* und *U* mit flüssiger Luft umgeben und das Quecksilberniveau in *A* bis zum Punkte α gesenkt. Infolgedessen konnte das Helium, welches vorher in *E* eingeschlossen war, sich auch auf die Röhren *C* und *U* ausdehnen. Das Quecksilber wurde abermals gehoben in *A* und das Volumen des Heliums in *E* gemessen. Das Verhältnis des in *E* befindlichen Volumens Helium bei der zweiten Messung zu dem im ersten Versuche war gleich dem Verhältnisse der durch Heben des Quecksilbers in *A* in *E* eingeschlossenen Heliummenge zum Gesamtvolumen des in *A* und seinen Verbindungen enthaltenen Heliums.

Nach abermaliger Hebung des Quecksilbers in *A* wurden die anderen Röhren wieder ausgepumpt und die Expansion des Heliums von *E* in die Verbindungsröhren *U*, *C* abermals vorgenommen. In dieser Weise wurden viele Versuche hintereinander ausgeführt, welche miteinander in ausgezeichneter Weise übereinstimmten und das Resultat lieferten, daß 71·5% des Gesamtvolumens von Helium im Apparat in der Kapillare *E* abgeschlossen wurden, wenn das Quecksilber in *A* bei sonst normalen Verhältnissen gehoben wurde.

Durch Auspumpen der Röhre *A* und ihrer Verbindungen bis zum äußersten Vakuum (mittels der Pumpe *P*), Schließen der Hähne *a*, *b*, *c* und Eintauchen der Röhre *C* in flüssige Luft durch mindestens 20 Minuten war es möglich, in der Röhre *A* ein so niedriges Vakuum herzustellen, daß nach Heben des Quecksilbers in *A* keine meßbare Druckzunahme in *E* erfolgte, wenn das Quecksilber in *E* bis zu 1 *mm* vom geschlossenen Ende der Kapillare gehoben wurde. Unter diesen Verhältnissen war es auch möglich, durch eine Serie von Vergleichsexperimenten die Ablesungen in *E* und *e* hinsichtlich der kleinen Kaliberdifferenzen von *E* und *e* vollkommen auszukorrigieren. In dieser Weise wurde bestimmt, daß, wenn die Röhre *E* Gas enthielt, sein Druck, der durch die betreffende Höhe der Quecksilbersäulen in *E* und *e* angezeigt wurde, auf weniger als 0·5 *mm* Quecksilber genau geschätzt werden konnte.

Die vom Radiumsalz entwickelte Heliummenge

(Erste Bestimmung.)

Das Radium-Bariumsalz (siehe p. 224), welches noch einen Teil seines Krystallwassers enthielt, wurde in der auf p. 224 und 225 (Fig. 1) beschriebenen Röhre 83 Tage lang verschlossen gehalten. Eine kleine Kugel mit Phosphorpentoxyd und eine andere mit fester Kalilauge wurden an dem Apparate (Fig. 2) im Punkte *O* angeschmolzen. Die Röhre *b* (Fig. 1) wurde an die Kalilaugevorlage angeschlossen (innerhalb der Röhre *b* befand sich das Radiumröhrchen *a*) und der ganze Apparat wurde bis zum Verbindungsstücke von *a* und *b* (Fig. 1) evakuiert. Die Hähne *r* und *f* waren geschlossen. Hierauf wurde das Kapillarende des Radiumröhrchens im Schliffe zerbrochen, die im Radiumröhrchen angesammelten Gase weggepumpt und in *D* gesammelt (Fig. 2). Während des Pumpens wurde das untere Ende des Radiumröhrchens und die Platinkapsel mittels eines Gasgebläses zur Rotglut erhitzt. Das im Salz enthaltene Krystallwasser wurde so ausgetrieben und kondensierte in den KOH- und P_2O_5-Vorlagen. Das Radiumsalz wurde durch etwa 30 Minuten erhitzt. Eine kleine

Menge elektrolytischen Sauerstoffes wurde zum Gasgemenge hinzugefügt und die Gase in *B* eingeleitet (Fig. 2) durch die Röhre *L*. Hierauf wurde die Funkenstrecke in *B* in Tätigkeit versetzt. Die übrigbleibenden Gase wurden dann langsam durch die Röhre *U*, welche in flüssiger Luft eingebettet war, in die Röhre *A* hinübergepumpt.

Die im Gasgemenge enthaltene Radiumemanation wurde in der Röhre *U* natürlich verflüssigt. Wenn das Gas vollkommen in die Röhre *A* übertragen worden war, wurde die Kokosnußkohlenröhre *C* in flüssige Luft getaucht, während das Quecksilberniveau in *A* beim Punkte α stand. Nach etwa 20 Minuten wurde das Quecksilber in *A* hegoben und Volumen und Druck des in *E* gefangenen Gases gemessen. Hierauf wurde der Quecksilberstand in *A* erniedrigt bis unter die Öffnung von *U* nach *A* und mittels der Übertragungspumpe *H* die in *C* und dessen Verbindungsstücken enthaltenen unkondensierten Gase ausgepumpt und durch *U* nach *A* eingeführt. Nach Schließen der Hähne *b* und *d* und Öffnen des Hahnes *a* wurde die flüssige Luft von der Kohlenröhre entfernt, diese erhitzt und die früher in *C* kondensierten Gase durch die große Pumpe abgepumpt.

Der Hahn *a* wurde geschlossen, das Rohr *C* wiederum mit flüssiger Luft umgeben und das Quecksilberniveau in *A* bis zum Punkte α erniedrigt. Die Gase in *A* wurden in dieser Weise wiederum der gekühlten Kokosnußkohle ausgesetzt. Nach zirka 30 Minuten wurde das Quecksilber in *A* gehoben und Volumen und Druck des in *E* befindlichen Gases wiederum gemessen. Darauf wurden die in der Röhre *C* befindlichen unkondensierten Gase nach *A* hinübergeschafft, die flüssige Luft entfernt und die Röhre *C* erhitzt. Die in *C* kondensierten Gase wurden dann wieder durch die große Pumpe abgepumpt. Dieser Kreisprozeß von Operationen wurde im ganzen fünfmal ausgeführt, ergab somit fünf getrennte Messungen von Volumen und Druck. Die Resultate der zweiten und dritten Messung zeigten eine kleine Verminderung der Werte. Doch die letzten drei stimmten befriedigend überein (innerhalb 1 %) und zeigten, daß die Reinigung des Heliums so weit durchgeführt war, als es überhaupt durch Behandlung mit tief abgekühlter Kohle möglich ist.

Das im Apparate befindliche Helium wurde dann in die Spektralröhre *S* eingeführt und das Spektrum untersucht. Es ergab sich im wesentlichen ein reines Heliumspektrum. Bei diesen Experimenten wurden zur Entfernung des Wasserstoffes aus dem Gemenge der vom Radiumsalz abgegebenen Gase keine anderen Vorsichtmaßregeln getroffen als das Einleiten der Funken in dem Gemenge bei Gegenwart reinen Sauerstoffes und die Exposition des übrigen Gases in gekühlter Kokosnußkohle.

Die Kupferoxydröhre *T* wurde erst nach der ersten Bestimmung im Apparat eingesetzt. Aus den wertvollen Untersuchungen von Sir James Dewar und anderen ist wohl bekannt, daß die Behandlung mit Kohle allein nicht ausreicht, um unter diesen Bedingungen die letzten Spuren von Wasserstoff zu entfernen.

Es wurde ein Versuch gemacht, um zu bestimmen, inwieweit die Gegenwart von freiem Wasserstoff die Resultate beeinflußt habe. Zu diesem Behufe wurde ein kleines Volumen Wasserstoff in den Apparat eingeführt und dieselbe Reihe von Operationen nun damit durchgeführt wie früher mit den Gasen aus dem Radiumsalze. Es ergab sich, daß das Gas, welches nach der ersten Behandlung mit gekühlter Kohle übrig blieb, rasch weniger wurde und nach dreimaliger Ausführung des Kreiprozesses zu einem Betrage von weniger als 1% des Gesamtvolumens des vom Radiumsalz entwickelten Heliums herabsank. Es schien daher unwahrscheinlich, daß irgendein ernsterer Fehler dadurch bewirkt werden konnte.

Andrerseits erheischte auch noch die Frage einige Aufmerksamkeit, ob durch das Erhitzen des Salzes das vom Salz entwickelte Helium vollkommen ausgetrieben worden sei.

Bevor das Salz erhitzt wurde, untersuchten wir sorgfältig, welcher Prozentsatz von Radiumemanation aus dem festen Salz entweicht und in den oberen Partien des Radiumröhrchens sich ansammelt. Man kann erwarten, daß eine beträchtliche Menge von α-Partikeln der Zerfallsprodukte solcher freier Emanation in den Wänden des Röhrchens feststeckt und daraus ein schwer vermeidlicher Fehler resultiert.

Es wurden Messungen der γ-Strahlung des oberen Teiles der Röhre ausgeführt, während der untere Teil mit dem Radiumsalz durch dicke Bleiblöcke abgeschirmt war. Diese Messungen zeigten, daß der Betrag von Emanation im oberen Teile der Röhre äußerst gering sein mußte: obwohl 1% der gesamten Emanationsmenge noch leicht meßbar war, ergab der obere Teil keinen meßbaren Effekt.

Es schien also unnötig, das Entwickeln der Emanation aus dem festen Salz in Betracht zu ziehen. Nach dem Erhitzen des festen Salzes wurde eine weitere Reihe von γ-Strahlenmessungen ausgeführt, welche anzeigte, daß die Emanation durch das Erhitzen vollständig vertrieben war. Es konnte daher auch angenommen werden, daß das Helium ebenfalls vollständig ausgetrieben worden sei.

Die vom Radiumsalz entwickelte Heliummenge

(Zweite Bestimmung.)

Um keine Zweifel über etwaigen Einfluß des Wasserstoffes im ersten Experiment zu hinterlassen und jede Möglichkeit einer unvollständigen Austreibung des Heliums aus dem Salze zu vermeiden, wurde eine zweite Bestimmung unter ganz anderen Versuchsbedingungen ausgeführt. Die Röhre T mit Kupferoxyd wurde in den Apparat eingesetzt und es wurde experimentell festgestellt, daß die letzten Spuren von Wasserstoff durch Hindurchstreichenlassen eines Gemenges des Wasserstoffes mit einem Überschuß von Sauerstoff durch die Kupferoxydschicht entfernt werden konnten, während diese zur mäßigen Rotglut erhitzt wurde. Das Radiumsalz wurde auf eine weitere Periode von 132 Tagen abgeschlossen. Nach Verlauf dieser Frist wurden die Gase (bestehend hauptsächlich aus Sauerstoff, welcher vor dem Einschmelzen des Salzes eingeleitet worden war) aus der Röhre ausgepumpt und gesammelt. Das Radiumsalz wurde dann von der Glasröhre entfernt, in der es eingeschmolzen gewesen war. Der durchlochte Deckel wurde von der Platinkapsel weggenommen und diese in einem anderen Glasrohr (Fig. 3) eingeschlossen.

Die Radiumkapsel wurde auf den Boden der zylindrischen Röhre (Fig. 3) gestellt. Die untere Partie dieser Röhre wurde mit nassem Papier überkleidet, um zu verhüten, daß irgendeine Erhitzung das Radiumsalz erreiche, und darauf der obere Teil der Röhre mit der kugelförmigen Ausbauchung zugeschlossen. Diese Kugel war im obersten Teile angebracht, um zu verhindern, daß später beim heftigen Sieden unter vermindertem Druck Radiumlösung hinausspritze.

Diese Röhre wurde an einem Ende durch ein kurzes Stück dickwandigen Kautschukschlauches mittels einer Schraubklemme verschlossen. Die Luft wurde aus der Röhre vollständig entfernt und zirka 30 cm^3 verdünnter Salzsäure durch den Schlauch in solcher Weise eingeführt, daß keine Spur von Luft miteintrat. Das Radiumsalz wurde sodann vollständig gelöst durch sanftes Erwärmen der Röhre und die Gase von der Röhre abgepumpt durch kleine Vorlagen von KOH und P_2O_5 und schließlich durch eine dünnwandige, U-förmige Röhre bei O (Fig. 2), die in flüssiger Luft gekühlt war, um die Emanation an diesem Punkte auszusondern. Diese Gase wurden mit den früher abgepumpten Gasen vereint entfernt und in die Gasbürette B eingeführt. Nach Übergang von Funken zwischen den Elektroden in B wurden die Gase durch das rotglühende Kupferoxyd in T durchgeleitet und in den Meßapparat eingeleitet, wo das Heliumvolumen in derselben Weise wie bei der ersten Bestimmung gemessen wurde.

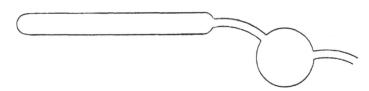

Fig. 3

Der Kreisprozeß der bei der zweiten Beobachtungsreihe ausgeführten Operationen enthielt dreimalige getrennte Expositionen in flüssiger Luft mit jedesmaliger Entfernung aller kondensierten Gase aus der Kohle nach der Bestimmung.

Die erhaltenen Resultate stimmen ausgezeichnet überein und zeigten auch am Beginne keine Veränderung des Volumens, wie das bei der ersten Serie von Experimenten der Fall war, wo der Wasserstoff bei der ersten Einführung der Gase in A nicht vollständig entfernt war.

Die in den beiden Messungsreihen erhaltenen Resultate sind in den folgenden Tabellen zusammengefaßt.

Erste Bestimmung (Ansammlungsdauer 83 Tage)

1. Messung	$8 \cdot 08 \; mm^3$		Helium (bei 0°
2. »	$6 \cdot 79$		und 760 mm
3. »	$6 \cdot 59$		Druck)
4. »	$6 \cdot 54$		
5. »	$6 \cdot 60$		

Mittel der 3., 4. und 5. Messung: $\overline{6 \cdot 58 \; mm^3}$

Zweite Bestimmung (Ansammlungsdauer 132 Tage)

1. Messung	$10 \cdot 32 \; mm^3$		Helium (bei 0°
2. »	$10 \cdot 37$		und 760 mm
3. »	$10 \cdot 45$		Druck)

Mittel aller drei Messungen: $\overline{10 \cdot 38 \; mm^3}$

Berechnung der Heliumproduktion

Sei x das Volumen Helium, welches von dem im Salze vorhandenen Radium (Element) pro Tag erzeugt wird, und y das Volumen Helium, das pro Tag von der Emanation und den zwei α-strahlenden Zerfallsprodukten Radium A und Radium C im Gleichgewichte mit Radium erzeugt wird, dann ist, im Falle die drei Produkte im Gleichgewichte sind,

$$y = 3x,$$

d. h. von den drei α-strahlenden Produkten RaEm, RaA, RaC wird dreimal so viel Helium erzeugt als vom Ra allein.

Zu Beginn jeder Ansammlungsperiode jedoch war vom Radiumsalz die ganze Emanation entfernt worden. Die in einem darauffolgenden Zeitraum von τ Tagen erzeugte Heliummenge würde daher gleich

$$\tau x + y \int_0^\tau (1 - e^{-\lambda t})dt = \tau x + \left(\tau - \frac{e^{-\lambda \tau}}{\lambda}\right)y = \tau x + \left(\tau - \frac{1}{\lambda}\right)y,$$

wobei λ die Zerfallskonstante der Radiumemanation bedeutet und der Tag als Zeiteinheit gewählt ist. Für Werte von τ größer als 40 Tage ist $e^{-\lambda t}$ mit großer Annäherung gleich 1.

Wenn Q die ganze Menge von Helium bedeutet, die in der Zeit τ entwickelt wurde, und der Wert $3x$ für y im obigen Ausdruck eingesetzt wird, erhalten wir

$$Q = \left[\tau + 3\left(\tau - \frac{1}{\lambda}\right)\right]x.$$

Setzt man in diese Gleichung die Werte von Q und τ der beiden Bestimmungen ein, so erhält man

1. Bestimmung $x = 0 \cdot 0209 \ mm^3$,
2. » $x = 0 \cdot 0203 \ mm^3$,

d. h. eine mittlere Heliumproduktion von $0 \cdot 0206 \ mm^3$ pro Tag von dem im Salz anwesenden Radium.

Die im Salze vorhandene Radiummenge war 192 *mg* (Mittel aus 191 und 193). Die Produktion von Helium pro Gramm Radium war daher

$$\frac{0 \cdot 0206}{0 \cdot 192} = 0 \cdot 107 \ mm^3 \ \text{pro Tag}$$

und pro Gramm Radium im Gleichgewichte mit seinen ersten Zerfallsprodukten (Emanation, Radium A und Radium C) pro Jahr

$$0 \cdot 107 \ . \ 365 \ . \ 4 = 156 \ mm^3 \ \text{Helium pro Jahr}.$$

Dieser beobachtete Wert stimmt mit der berechneten Heliumproduktion (158 mm^3 pro Jahr), über welche p. 223 referiert wurde, sehr nahe überein und bestätigt in schlagender Weise die Schlüsse, auf denen die Berechnung basiert war.

Es besteht kein Zweifel mehr, daß das α-Partikel aus einem Heliumatom mit doppelter Elementarladung besteht und daß Helium selbst einatomig ist. Die Übereinstimmung zwischen Berechnung und Theorie verifiziert in bemerkenswerter Weise die wesentliche Richtigkeit der atomistischen Auffassung der Materie. Die Zahl der von 1 *g* Radium pro Sekunde ausgeschleuderten α-Partikel ist direkt gezählt worden und das entsprechende Volumen Helium wurde experimentell bestimmt. Aus diesen zwei Experimenten ist es möglich, mit einem Minimum an Annahmen die Zahl der Heliumatome in 1 cm^3 bei 0° C. und normalem Drucke zu berechnen. Diese Zahl (Loschmidt'sche Zahl) ist danach $2 \cdot 69 \ . \ 10^{19}$. Nach Avogadro's Hypothese gibt dies auch die Zahl der Atome irgendeines anderen Gases unter normalen Bedingungen.

Man ersieht, daß diese Methode die Bestimmung der fundamentalen Einheitsladung e nicht involviert, sondern daß vielmehr der Wert von e daraus direkt abgeleitet werden kann.

Es ist von Wichtigkeit zu bemerken, daß die Übereinstimmung zwischen dem beobachteten und berechneten Werte der Heliumproduktion unabhängig von der Richtigkeit des angewendeten Radiumstandardpräparates ist, denn in beiden Fällen wurde derselbe Standard angewendet. Sobald ein internationaler Radiumstandard hergestellt ist, wird es nicht schwierig sein, die Heliumproduktion auf Grund des neuen Standards umzurechnen.

Durch das Entgegenkommen von Sir James Dewar waren wir in der Lage, die bei seinen Experimenten verwendete Radiummenge mit unserem

H*

Arbeitsstandardpräparat zu vergleichen. Es ergab sich, daß die Radiummenge, welche Sir James Dewar angewendet und zu 70 mg reinen Radiumchlorids angegeben hatte, nach unserem Standard äquivalent 72 mg Radiumchlorids war. Nach unserem Standard würde die von Sir James Dewar ermittelte Heliumproduktion statt 169 mm^3 pro Jahr sich auf 164 mm^3 reduzieren. Dieser Wert ist zwar etwas höher als der unsere, doch stimmt er im wesentlichen überein.

Die Heliumproduktion der Radiumemanation

Es wurde auch eine Bestimmung der Heliummenge gemacht, welche durch den Zerfall einer bekannten Quantität Radiumemanation entsteht. Die Emanation wurde aus einer Radiumlösung gewonnen und durch Kondensation bei der Temperatur der flüssigen Luft abgesondert und gereinigt. Eine Weichglasröhre von zirka 10 mm innerer Weite und 8 cm Länge wurde vorbereitet. Aus einem Ende dieser Röhre war ein eingeschliffenes Verbindungsstück mit Kapillarende, ähnlich wie bei der Röhre (Fig. 1). Dieser Schliff paßte in eine andere Röhre, ähnlich wie b (Fig. 1).

Die Glasröhre wurde etwa halbvoll mit geschmolzenem Schwefel gefüllt und vollständig ausgepumpt, während der Schwefel sich abkühlte. Die gereinigte Radiumemanation wurde in diese Röhre eingeleitet und diese dann zugeschmolzen. Hierauf wurde die Röhre erwärmt, bis der Schwefel schmolz. Während der Abkühlung wurde durch entsprechendes Drehen des Röhrchens erreicht, daß die innere Oberfläche des Glases mit einer gleichmäßigen Schichte von etwa 3 mm Dicke sich überzog. Dieser Schwefelüberzug sollte eine Innenwand bilden, in welcher die von der Emanation und dem aktiven Niederschlag (RaA — RaC) ausgeschleuderten α-Partikel stecken blieben und von welcher das Helium durch Schmelzen des Schwefels leicht wieder ausgetrieben werden konnte.

Etwa 19½ Stunden nach Einführung der Emanation in die Röhre wurde die γ-Strahlung derselben gemessen und mit der des Radiumstandardpräparates verglichen. Die Resultate zeigten, daß die anfänglich im Röhrchen vorhandene Emanationsmenge der Gleichgewichtsmenge von 126 mg Radium äquivalent war.

21 Tage nach Einführung der Emanation wurde die Röhre behufs Schmelzen des Schwefels so lange erwärmt, bis das Kapillarende der Röhre vollkommen durchsichtig wurde. Dann wurde die Röhre vermittels des Schliffstückes an den Meßapparat angeschlossen, der Apparat vollständig bis zu diesem Schliffstück evakuiert, das Kapillarende aufgebrochen, die Gase abgepumpt und gesammelt. Während des Pumpens wurde die Röhre erhitzt, so daß der Schwefel schmolz. Dann ließ man die Röhre sich abkühlen und etwas Sauerstoff einströmen, der darauf wieder abgepumpt und zu den anderen Gasen hinzugefügt wurde. Die Gase wurden in die Bürette B (Fig. 2) eingeführt, über erhitztes Kupferoxyd geleitet, in den Meßapparat gedrängt und dort in der gewöhnlichen Weise das Volumen des Heliums bestimmt. Das

Gas wurde dann in die Spektralröhre S (Fig. 2) übergeführt, wo das Spektrum untersucht und als reines Heliumspektrum erkannt wurde. Das gefundene Volumen des Heliums war $0 \cdot 202 \ mm^3$.

Die anfänglich in der Röhre befindliche Emanationsmenge war die Gleichgewichtsmenge von $0 \cdot 126 \ g$ Radium. Die anfängliche Produktion von Helium würde daher anzusehen sein als

$$0 \cdot 126 \times 3 \times 0 \cdot 107 \ mm^3 \text{ pro Tag}$$

(der Faktor 3 ist wegen der drei α-strahlenden Umwandlungsprodukte Emanation, RaA und RaC eingesetzt). Wegen des Zerfalles der Emanation wird jedoch die Heliumproduktion abnehmen und die Gesamtmenge des gebildeten Heliums wird proportional sein dem Ausdruck

$$\frac{1}{\lambda}(1 - e^{-\lambda t}),$$

wo λ die Zerfallskonstante und t die Zeit bedeutet. Nehmen wir den Wert von λ entsprechend einer Halbwertszeit von $3 \cdot 85$ Tagen und als Zeiteinheit den Tag, so wird der obige Ausdruck gleich $5 \cdot 42$, d. h. die in 21 Tagen produzierte Heliummenge wird sein

$$0 \cdot 126 \times 3 \times 0 \cdot 107 \times 5 \cdot 42 = 0 \cdot 220 \ mm^3.$$

Der experimentell gefundene Wert stimmt mit diesem zu erwartenden Wert in Anbetracht der experimentellen Schwierigkeiten gut überein.

Die Heliumproduktion des Poloniums

Es wurde auch eine Bestimmung der Heliumentwicklung bei Polonium gemacht. Die Sulfide von Blei, Wismut und Antimon, welche bei der Reinigung des Radiumsalzes erhalten wurden (p. 224), wurden durch Digerieren mit konzentrierter Salzsäure und wenig Kaliumchlorat zersetzt. Die Lösung wurde erhitzt, bis der Überschuß an Salzsäure größtenteils verdampft war, und wurde mit schwacher Salzsäure verdünnt. Diese Lösung wurde in eine starke Schwefelammoniumlösung geschüttet, die gefällten Sulfide von Blei und Wismut entfernt und wiederum mit Salzsäure und Kaliumchlorat behandelt. Ein Überschuß von Schwefelsäure wurde hinzugefügt, die Mischung durch Eindampfen konzentriert und so lange erhitzt, bis sich Schwefelsäuredämpfe entwickelten. Nach dem Abkühlen wurde zu dem Rückstand etwas verdünnte Schwefelsäure hinzugefügt und das unlösliche Bleisulfat auf einem kleinen Asbestfilter abgetrennt. Das Bleisulfat wurde dann durch Erwärmen mit starker Salzsäure und Kaliumchlorat aufgeschlossen, zu dieser Lösung verdünnte Schwefelsäure hinzugefügt und die eben beschriebenen Operationen wiederholt. Die zwei Filtrate vom Bleisulfat wurden zusammengegeben, bis zum Kochen erhitzt und ein kleiner Überschuß von Ammoniak zugefügt. Der geringe Wismutniederschlag wurde abfiltriert und in verdünnter Salzäure gelöst. Das Volumen dieser Lösung wurde genau bestimmt und ein kleiner bestimmter Bruchteil davon entnommen. Zur Hauptlösung wurden 5 g metallisches Kupfer in Form von kleinen Stückchen hinzugefügt, die Mischung sanft erwärmt und mehrere Stunden bei öfterem Umrühren belassen. Hierbei setzt sich das in der salzsauren Lösung anwesende Polonium vollständig in einer gut haftenden Schicht am Kupfer ab. Das Kupfer wurde dann gewaschen und getrocknet.

Das Kupfer mit dem Poloniumüberzug wurde in einem kleinen Glasröhrchen (ähnlich, doch kleiner als das beim Radiumsalz verwendete) (a, Fig. 1) eingeführt. Die Luft wurde abgepumpt und wenig reiner Sauerstoff eingeleitet. Hierauf wurde die Röhre verschlossen.

128 Tage später wurde diese Poloniumröhre an den Meßapparat angeschlossen und die innen befindlichen Gase abgepumpt. Während des Pumpens wurde das untere Ende des Röhrchens samt dem Kupfer zu heller Rotglut erhitzt, um das im Metall okkludierte Helium auszutreiben. Das im Gasgemenge enthaltene Helium wurde in der üblichen Weise bestimmt. Sein Volumen ergab sich zu $0 \cdot 009 \ mm^3$. Es wurde leicht spektralanalytisch identifiziert.

Der kleine bestimmte Bruchteil der Poloniumlösung, welcher vor der Behandlung der Hauptlösung mit Kupfer (siehe diese Seite oben) abgesondert worden war, wurde durch schwach salzsauer gemachtes Wasser auf zirka $10 \ cm^3$ verdünnt. Diese Lösung wurde in ein Glasgefäß gegeben, dessen Boden aus einer metallischen Kupferplatte bestand. Das Polonium setzte sich an dieser Platte ab und die von ihm ausgesendete Zahl der α-Partikel wurde von Dr. Geiger für uns liebenswürdigerweise bestimmt. Die pro Sekunde abgegebene Zahl von α-Partikeln war $10 \cdot 7 . 10^5$. Die Lösung war $1/_{131}$ der Hauptlösung. Die Gesamtzahl der vom Polonium in der ganzen Lösung pro Sekunde abgegebenen α-Partikel wäre somit $14 \cdot 0 . 10^7$.

Die von 1 g Radium pro Sekunde ausgesendete Zahl von α-Partikeln ist nach Rutherford und Geiger $3 \cdot 4 . 10^{10}$. Die anwesende Menge Polonium war daher einer Gleichgewichtsmenge von $4 \cdot 1 \ mg$ Radium äquivalent.

Die vom Polonium produzierte Heliummenge sollte proportional sein dem Ausdrucke

$$\frac{1}{\lambda}(1 - e^{-\lambda t}),$$

wo λ die Zerfallskonstante des Poloniums und t die Zeit in Tagen bedeutet. Dieser Ausdruck hat den Zahlenwert 96, d. h. die in 128 Tagen produzierte Heliummenge ist 96 mal so groß als die vom Polonium anfänglich pro Tag entwickelte Heliummenge.

Die Gesamtmenge Helium würde daher sein

$$0 \cdot 0041 \times 0 \cdot 107 \times 96 = 0 \cdot 042 \ mm^3.$$

Die Heliummenge, welche wirklich gefunden wurde, war daher nur etwa ein Viertel von dem nach der Theorie zu erwartenden Betrage. Doch ist diese Diskrepanz zweifellos der Tatsache zuzuschreiben, daß ein sehr großer Teil der vom Polonium ausgeschleuderten α-Partikel im metallischen Kupfer okkludiert geblieben war. Das Experiment zeigte somit nur qualitativ die Heliumproduktion des Poloniums.

Die Produktion von Helium aus Radium *D*

Obgleich Radium *D* selbst keine α-Partikel aussendet, so wird doch, wenn ein Präparat von RaD längere Zeit sich selbst überlassen wird, das Radium *E* und Radium *F* (Polonium) nachgebildet, und von letzterem muß man erwarten, daß es Helium produziert. Das Bleisulfat, welches von der Poloniumlösung abgetrennt worden war (p. 235) und das Radioblei (RaD) enthielt, wurde in einer evakuierten Hartglasröhre zirka 110 Tage abgeschlossen gehalten. Dann wurde das Bleisulfat erhitzt und die Gase aus der Röhre abgepumpt. Das durch Kohle nicht absorbierte, übriggebliebene Gas ergab ein gutes Heliumspektrum, doch wurde sein Volumen nicht bestimmt.

Zusammenfassung der Resultate

Die Heliumproduktion durch ein Radiumsalz wurde aufs genaueste gemessen; sie ergab sich zu etwa $0 \cdot 107 \ mm^3$ Helium pro Tag und pro Gramm Radium (Element); pro Jahr würde dies für 1 *g* Radium im Gleichgewicht mit seinen ersten Zerfallsprodukten (Emanation, Radium *A* und Radium *C*) $156 \ mm^3$ Helium ergeben.

Die von Rutherford und Geiger aus ihren Experimenten über die Zählung der vom Radium emittierten α-Partikel berechnete Heliumproduktion ($158 \ mm^3$) steht mit der obigen experimentell gefundenen Zahl in bester Übereinstimmung.

Es wurde auch die durch den Zerfall einer bestimmten Menge Radiumemanation gebildete Heliummenge gemessen. Sie entsprach vollständig der theoretisch in der betreffenden Zeit zu erwartenden Menge.

Ferner wurde die Heliumproduktion durch Polonium und Radiobleipräparate beobachtet.

The Scattering of α and β Particles by Matter and the Structure of the Atom

by PROFESSOR E. RUTHERFORD, F.R.S., *University of Manchester**

From the *Philosophical Magazine* for May 1911, ser. 6, xxi, pp. 669–88

§ 1. It is well known that the α and β particles suffer deflexions from their rectilinear paths by encounters with atoms of matter. This scattering is far more marked for the β than for the α particle on account of the much smaller momentum and energy of the former particle. There seems to be no doubt that such swiftly moving particles pass through the atoms in their path, and that the deflexions observed are due to the strong electric field traversed within the atomic system. It has generally been supposed that the scattering of a pencil of α or β rays in passing through a thin plate of matter is the result of a multitude of small scatterings by the atoms of matter traversed. The observations, however, of Geiger and Marsden† on the scattering of α rays indicate that some of the α particles must suffer a deflexion of more than a right angle at a single encounter. They found, for example, that a small fraction of the incident α particles, about 1 in 20,000, were turned through an average angle of 90° in passing through a layer of gold-foil about 0·00004 cm. thick, which was equivalent in stopping-power of the α particle to 1·6 millimetres of air. Geiger‡ showed later that the most probable angle of deflexion for a pencil of α particles traversing a gold-foil of this thickness was about 0°·87. A simple calculation based on the theory of probability shows that the chance of an α particle being deflected through 90° is vanishingly small. In addition, it will be seen later that the distribution of the α particles for various angles of large deflexion does not follow the probability law to be expected if such large deflexions are made up of a large number of small deviations. It seems reasonable to suppose that the deflexion through a large angle is due to a single atomic encounter, for the chance of a second encounter of a kind to produce a large deflexion must in most cases be exceedingly small. A simple calculation shows that the atom must be a seat of an intense electric field in order to produce such a large deflexion at a single encounter.

Recently Sir J. J. Thomson§ has put forward a theory to explain the scattering of electrified particles in passing through small thicknesses of

* A brief account of this paper was communicated to the Manchester Literary and Philosophical Society in February, 1911. (*This vol., p. 212.*)

† Proc. Roy. Soc. lxxxii. p. 495 (1909).   ‡ Proc. Roy. Soc. lxxxiii. p. 492 (1910).
§ Camb. Lit. & Phil. Soc. xv. pt. 5 (1910).

matter. The atom is supposed to consist of a number N of negatively charged corpuscles, accompanied by an equal quantity of positive electricity uniformly distributed throughout a sphere. The deflexion of a negatively electrified particle in passing through the atom is ascribed to two causes—(1) the repulsion of the corpuscles distributed through the atom, and (2) the attraction of the positive electricity in the atom. The deflexion of the particle in passing through the atom is supposed to be small, while the average deflexion after a large number m of encounters was taken as $\sqrt{m} \cdot \theta$, where θ is the average deflexion due to a single atom. It was shown that the number N of the electrons within the atom could be deduced from observations of the scattering of electrified particles. The accuracy of this theory of compound scattering was examined experimentally by Crowther* in a later paper. His results apparently confirmed the main conclusions of the theory, and he deduced, on the assumption that the positive electricity was continuous, that the number of electrons in an atom was about three times its atomic weight.

The theory of Sir J. J. Thomson is based on the assumption that the scattering due to a single atomic encounter is small, and the particular structure assumed for the atom does not admit of a very large deflexion of an α particle in traversing a single atom, unless it be supposed that the diameter of the sphere of positive electricity is minute compared with the diameter of the sphere of influence of the atom.

Since the α and β particles traverse the atom, it should be possible from a close study of the nature of the deflexion to form some idea of the constitution of the atom to produce the effects observed. In fact, the scattering of high-speed charged particles by the atoms of matter is one of the most promising methods of attack of this problem. The development of the scintillation method of counting single α particles affords unusual advantages of investigation, and the researches of H. Geiger by this method have already added much to our knowledge of the scattering of α rays by matter.

§ 2. We shall first examine theoretically the single encounters† with an atom of simple structure, which is able to produce large deflexions of an α particle, and then compare the deductions from the theory with the experimental data available.

Consider an atom which contains a charge $\pm Ne$ at its centre surrounded by a sphere of electrification containing a charge $\mp Ne$ supposed uniformly distributed throughout a sphere of radius R. e is the fundamental unit of charge, which in this paper is taken as $4 \cdot 65 \times 10^{-10}$ E.S. unit. We shall suppose that for distances less than 10^{-12} cm. the central charge and also the charge on the α particle may be supposed to be concentrated at a point.

---

* Crowther, Proc. Roy. Soc. lxxxiv. p. 226 (1910).

† The deviation of a particle throughout a considerable angle from an encounter with a single atom will in this paper be called 'single' scattering. The deviation of a particle resulting from a multitude of small deviations will be termed 'compound' scattering.

It will be shown that the main deductions from the theory are independent of whether the central charge is supposed to be positive or negative. For convenience, the sign will be assumed to be positive. The question of the stability of the atom proposed need not be considered at this stage, for this will obviously depend upon the minute structure of the atom, and on the motion of the constituent charged parts.

In order to form some idea of the forces required to deflect an α particle through a large angle, consider an atom containing a positive charge Ne at its centre, and surrounded by a distribution of negative electricity Ne uniformly distributed within a sphere of radius R. The electric force X and the potential V at a distance r from the centre of an atom for a point inside the atom, are given by

$$X = Ne\left(\frac{1}{r^2} - \frac{r}{R^3}\right)$$

$$V = Ne\left(\frac{1}{r} - \frac{3}{2R} + \frac{r^2}{2R^3}\right).$$

Suppose an α particle of mass m and velocity u and charge E shot directly towards the centre of the atom. It will be brought to rest at a distance b from the centre given by

$$\tfrac{1}{2}mu^2 = NeE\left(\frac{1}{b} - \frac{3}{2R} + \frac{b^2}{2R^3}\right).$$

It will be seen that b is an important quantity in later calculations. Assuming that the central charge is $100\ e$, it can be calculated that the value of b for an α particle of velocity $2 \cdot 09 \times 10^9$ cms. per second is about $3 \cdot 4 \times 10^{-12}$ cm. In this calculation b is supposed to be very small compared with R. Since R is supposed to be of the order of the radius of the atom, viz. 10^{-8} cm., it is obvious that the α particle before being turned back penetrates so close to the central charge, that the field due to the uniform distribution of negative electricity may be neglected. In general, a simple calculation shows that for all deflexions greater than a degree, we may without sensible error suppose the deflexion due to the field of the central charge alone. Possible single deviations due to the negative electricity, if distributed in the form of corpuscles, are not taken into account at this stage of the theory. It will be shown later that its effect is in general small compared with that due to the central field.

Consider the passage of a positive electrified particle close to the centre of an atom. Supposing that the velocity of the particle is not appreciably changed by its passage through the atom, the path of the particle under the influence of a repulsive force varying inversely as the square of the distance will be an hyperbola with the centre of the atom S as the external focus. Suppose the particle to enter the atom in the direction PO (fig. 1), and that

Rutherford's first rough note on the nuclear theory of atomic structure; written, probably, in the winter of 1910–11

Chance

Chance of deflection through the electrons?

$$= \pi b^2 . N . t \quad \text{where } N = \text{number of ...}$$

Chance for collision for a big deflection

$$= \frac{a^2}{b^2}$$

∴ Chance of big deflection in distance t

$$= \sigma a^2 . N . t$$

This chance ... for a deflection a such that $\frac{N_0 e}{a}$ in ...

$$\frac{N_0 e}{a} = \frac{4\pi}{3} . \tfrac{1}{2} m v^2 . K \quad \text{where } K \text{ is } < 1$$

$$\frac{N_0 e}{a} = \frac{4\times 10^4}{3} . K$$

$$a = \frac{3 N_0 e}{4\times 10^4 . K} = \frac{3\times 200 \times 4.65}{4\times 10^4 . \times 10^{10} . K} = \frac{7}{10^{12} K}$$

Chance of big deflection $= \sigma \times \frac{4\pi}{10^{24}} . N . t$

$$= \frac{1.5}{10^{22}} . \frac{N.t}{K}$$

N for gold

$$= 6.3 \times 10^{22} = \frac{9.4}{10} . 10^{22} . t$$

$$= \frac{9.4 \, t}{K}$$

Another page of these rough notes, in which Rutherford estimated the chance of a large deflection by gold

the direction of motion on escaping the atom is OP′. OP and OP′ make equal angles with the line SA, where A is the apse of the hyperbola. $p = $ SN $=$

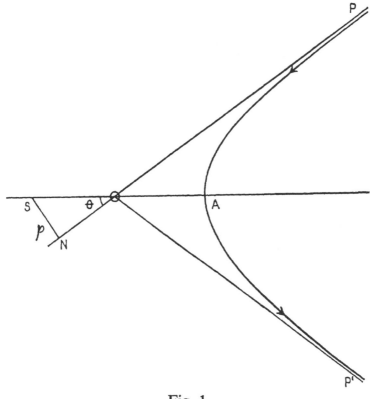

Fig. 1

perpendicular distance from centre on direction of initial motion of particle.

Let angle POA $= \theta$.

Let V $=$ velocity of particle on entering the atom, v its velocity at A, then from consideration of angular momentum

$$p\text{V} = \text{SA} . v.$$

From conservation of energy

$$\tfrac{1}{2}m\text{V}^2 = \tfrac{1}{2}mv^2 - \frac{Ne\text{E}}{\text{SA}},$$

$$v^2 = \text{V}^2\Big(1 - \frac{b}{\text{SA}}\Big).$$

Since the eccentricity is sec θ,

$$\text{SA} = \text{SO} + \text{OA} = p \ \text{cosec} \ \theta(1 + \cos \theta)$$
$$= p \ \cot \ \theta/2,$$
$$p^2 = \text{SA} \ (\text{SA} - b) = p \ \cot \ \theta/2(p \ \cot \ \theta/2 - b),$$
$$\therefore \quad b = 2p \ \cot \ \theta.$$

The angle of deviation ϕ of the particles is $\pi - 2\theta$ and

$$\cot \phi/2 = \frac{2p^*}{b} \qquad \qquad \text{. . . . . . . (1)}$$

This gives the angle of deviation of the particle in terms of b, and the perpendicular distance of the direction of projection from the centre of the atom.

For illustration, the angle of deviation ϕ for different values of p/b are shown in the following table:—

p/b	10	5	2	1	0·5	0·25	0·125
ϕ	5°·7	11°·4	28°	53°	90°	127°	152°

§ 3. *Probability of single deflexion through any angle*

Suppose a pencil of electrified particles to fall normally on a thin screen of matter of thickness t. With the exception of the few particles which are scattered through a large angle, the particles are supposed to pass nearly normally through the plate with only a small change of velocity. Let $n =$ number of atoms in unit volume of material. Then the number of collisions of the particle with the atom of radius R is $\pi R^2 nt$ in the thickness t.

The probability m of entering an atom within a distance p of its centre is given by

$$m = \pi p^2 nt.$$

Chance dm of striking within radii p and $p + dp$ is given by

$$dm = 2\pi pnt \,.\, dp = \frac{\pi}{4} ntb^2 \cot \phi/2 \,\operatorname{cosec}^2 \phi/2 \, d\phi, \qquad \text{. . . (2)}$$

since $\qquad\qquad\qquad\qquad \cot \phi/2 = 2p/b.$

The value of dm gives the *fraction* of the total number of particles which are deviated between the angles ϕ and $\phi + d\phi$.

The fraction ρ of the total number of particles which are deflected through an angle greater than ϕ is given by

$$\rho = \frac{\pi}{4} ntb^2 \cot^2 \phi/2. \qquad\qquad \text{. . . . . . . (3)}$$

The fraction ρ which is deflected between the angles ϕ_1 and ϕ_2 is given by

$$\rho = \frac{\pi}{4} ntb^2 \left(\cot^2 \frac{\phi_1}{2} - \cot^2 \frac{\phi_2}{2} \right). \qquad \text{. . . . . (4)}$$

---

* A simple consideration shows that the deflexion is unaltered if the forces are attractive instead of repulsive.

It is convenient to express the equation (2) in another form for comparison with experiment. In the case of the α rays, the number of scintillations appearing on a *constant* area of a zinc sulphide screen are counted for different angles with the direction of incidence of the particles. Let $r =$ distance from point of incidence of α rays on scattering material, then if Q be the total number of particles falling on the scattering material, the number y of α particles falling on unit area which are deflected through an angle ϕ is given by

$$y = \frac{Q dm}{2\pi r^2 \sin \phi . d\phi} = \frac{ntb^2 . Q . \cosec^4 \phi/2}{16 r^2} \quad . \quad . \quad . \quad (5)$$

Since $b = \dfrac{2NeE}{mu^2}$, we see from this equation that the number of α particles (scintillations) per unit area of zinc sulphide screen at a given distance r from the point of incidence of the rays is proportional to

(1) $\cosec^4 \phi/2$ or $1/\phi^4$ if ϕ be small;
(2) thickness of scattering material t provided this is small;
(3) magnitude of central charge Ne;
(4) and is inversely proportional to $(mu^2)^2$, or to the fourth power of the velocity if m be constant.

In these calculations, it is assumed that the α particles scattered through a large angle suffer only one large deflexion. For this to hold, it is essential that the thickness of the scattering material should be so small that the chance of a second encounter involving another large deflexion is very small. If, for example, the probability of a single deflexion ϕ in passing through a thickness t is 1/1000, the probability of two successive deflexions each of value ϕ is $1/10^6$, and is negligibly small.

The angular distribution of the α particles scattered from a thin metal sheet affords one of the simplest methods of testing the general correctness of this theory of single scattering. This has been done recently for α rays by Dr. Geiger,* who found that the distribution for particles deflected between 30° and 150° from a thin gold-foil was in substantial agreement with the theory. A more detailed account of these and other experiments to test the validity of the theory will be published later.

§ 4. *Alteration of velocity in an atomic encounter*

It has so far been assumed that an α or β particle does not suffer an appreciable change of velocity as the result of a single atomic encounter resulting in a large deflexion of the particle. The effect of such an encounter in altering the velocity of the particle can be calculated on certain assumptions. It is supposed that only two systems are involved, viz., the swiftly moving particle and the

---

* Manch. Lit. & Phil. Soc. 1910.

atom which it traverses supposed initially at rest. It is supposed that the principle of conservation of momentum and of energy applies, and that there is no appreciable loss of energy or momentum by radiation.

Let m be mass of the particle,

$v_1 =$ velocity of approach,
$v_2 =$ velocity of recession,
$M =$ mass of atom,
$V =$ velocity communicated to atom as result of encounter.

Let OA (fig. 2) represent in magnitude and direction the momentum mv_1 of the entering particle, and OB the momentum of the receding particle which has been turned through an angle $AOB = \phi$. Then BA represents in magnitude and direction the momentum MV of the recoiling atom.

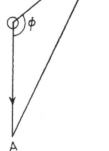

$$(MV)^2 = (mv_1)^2 + (mv_2)^2 - 2m^2v_1v_2 \cos\phi. \quad . \quad (1)$$

By the conservation of energy

$$MV^2 = mv_1{}^2 - mv_2{}^2. \quad . \quad . \quad . \quad . \quad (2)$$

Suppose $M/m = K$ and $v_2 = \rho v_1$, where ρ is <1.
From (1) and (2),

$$(K + 1)\rho^2 - 2\rho \cos\phi = K - 1,$$

or $\qquad \rho = \dfrac{\cos\phi}{K+1} + \dfrac{1}{K+1}\sqrt{K^2 - \sin^2\phi}.$ $\qquad$ Fig. 2

Consider the case of an α particle of atomic weight 4, deflected through an angle of 90° by an encounter with an atom of gold of atomic weight 197.
Since $K = 49$ nearly,

$$\rho = \sqrt{\frac{K-1}{K+1}} = 0\cdot979,$$

or the velocity of the particle is reduced only about 2 per cent. by the encounter.

In the case of aluminium $K = 27/4$ and for $\phi = 90°$ $\rho = 0\cdot86$.

It is seen that the reduction of velocity of the α particle becomes marked on this theory for encounters with the lighter atoms. Since the range of an α particle in air or other matter is approximately proportional to the cube of the velocity, it follows that an α particle of range 7 cms. has its range reduced to $4\cdot5$ cms. after incurring a single deviation of 90° in traversing an aluminium atom. This is of a magnitude to be easily detected experimentally. Since the value of K is very large for an encounter of a β particle with an atom, the reduction of velocity on this formula is very small.

Some very interesting cases of the theory arise in considering the changes of velocity and the distribution of scattered particles when the α particle encounters a light atom, for example a hydrogen or helium atom. A discussion of these and similar cases is reserved until the question has been examined experimentally.

§ 5. *Comparison of single and compound scattering*

Before comparing the results of theory with experiment, it is desirable to consider the relative importance of single and compound scattering in determining the distribution of the scattered particles. Since the atom is supposed to consist of a central charge surrounded by a uniform distribution of the opposite sign through a sphere of radius R, the chance of encounters with the atom involving small deflexions is very great compared with the chance of a single large deflexion.

This question of compound scattering has been examined by Sir J. J. Thomson in the paper previously discussed (§ 1). In the notation of this paper, the average deflexion ϕ_1 due to the field of the sphere of positive electricity of radius R and quantity Ne was found by him to be

$$\phi_1 = \frac{\pi}{4} \cdot \frac{NeE}{mu^2} \cdot \frac{1}{R}.$$

The average deflexion ϕ_2 due to the N negative corpuscles supposed distributed uniformly throughout the sphere was found to be

$$\phi_2 = \frac{16}{5} \frac{eE}{mu^2} \cdot \frac{1}{R} \sqrt{\frac{3N}{2}}.$$

The mean deflexion due to both positive and negative electricity was taken as

$$(\phi_1^2 + \phi_2^2)^{1/2}.$$

In a similar way, it is not difficult to calculate the average deflexion due to the atom with a central charge discussed in this paper.

Since the radial electric field X at any distance r from the centre is given by

$$X = Ne\left(\frac{1}{r^2} - \frac{r}{R^3}\right),$$

it is not difficult to show that the deflexion (supposed small) of an electrified particle due to this field is given by

$$\theta = \frac{b}{p}\left(1 - \frac{p^2}{R^2}\right)^{3/2},$$

where p is the perpendicular from the centre on the path of the particle and b has the same value as before. It is seen that the value of θ increases with diminution of p and becomes great for small values of ϕ.

Since we have already seen that the deflexions become very large for a particle passing near the centre of the atom, it is obviously not correct to find the average value by assuming θ is small.

Taking R of the order 10^{-8} cm., the value of p for a large deflexion is for α and β particles of the order 10^{-11} cm. Since the chance of an encounter involving a large deflexion is small compared with the chance of small deflexions, a simple consideration shows that the average small deflexion is practically unaltered if the large deflexions are omitted. This is equivalent to integrating over that part of the cross section of the atom where the deflexions are small and neglecting the small central area. It can in this way be simply shown that the average small deflexion is given by

$$\phi_1 = \frac{3\pi}{8}\frac{b}{R}.$$

This value of ϕ_1 for the atom with a concentrated central charge is three times the magnitude of the average deflexion for the same value of Ne in the type of atom examined by Sir J. J. Thomson. Combining the deflexions due to the electric field and to the corpuscles, the average deflexion is

$$(\phi_1^2 + \phi_2^2)^{1/2} \quad \text{or} \quad \frac{b}{2R}\left(5 \cdot 54 + \frac{15 \cdot 4}{N}\right)^{1/2}.$$

It will be seen later that the value of N is nearly proportional to the atomic weight, and is about 100 for gold. The effect due to scattering of the individual corpuscles expressed by the second term of the equation is consequently small for heavy atoms compared with that due to the distributed electric field.

Neglecting the second term, the average deflexion per atom is $\frac{3\pi b}{8R}$.

We are now in a position to consider the relative effects on the distribution of particles due to single and to compound scattering. Following J. J. Thomson's argument, the average deflexion θ_t after passing through a thickness t of matter is proportional to the square root of the number of encounters and is given by

$$\theta_t = \frac{3\pi b}{8R}\sqrt{\pi R^2 \cdot n \cdot t} = \frac{3\pi b}{8}\sqrt{\pi n t},$$

where n as before is equal to the number of atoms per unit volume.

The probability p_1 for compound scattering that the deflexion of the particle is greater than ϕ is equal to $e^{-\phi^2/\theta_t^2}$.

Consequently				$$\phi^2 = -\frac{9\pi^3}{64}b^2 n t \log p_1.$$

Next suppose that single scattering alone is operative. We have seen (§ 3) that the probability p_2 of a deflexion greater than ϕ is given by

$$p_2 = \frac{\pi}{4} b^2 . n . t \cot^2 \phi/2.$$

By comparing these two equations

$$p_2 \log p_1 = -0\cdot 181\phi^2 \cot^2 \phi/2,$$

ϕ is sufficiently small that

$$\tan \phi/2 = \phi/2,$$

$$p_2 \log p_1 = -0\cdot 72.$$

If we suppose $p_2 = 0\cdot 5$, then $p_1 = 0\cdot 24.$

If $p_2 = 0\cdot 1,$ $p_1 = 0\cdot 0004.$

It is evident from this comparison, that the probability for any given deflexion is always greater for single than for compound scattering. The difference is especially marked when only a small fraction of the particles are scattered through any given angle. It follows from this result that the distribution of particles due to encounters with the atoms is for small thicknesses mainly governed by single scattering. No doubt compound scattering produces some effect in equalizing the distribution of the scattered particles; but its effect becomes relatively smaller, the smaller the fraction of the particles scattered through a given angle.

§ 6. *Comparison of Theory with Experiments*

On the present theory, the value of the central charge Ne is an important constant, and it is desirable to determine its value for different atoms. This can be most simply done by determining the small fraction of α or β particles of known velocity falling on a thin metal screen, which are scattered between ϕ and $\phi + d\phi$ where ϕ is the angle of deflexion. The influence of compound scattering should be small when this fraction is small.

Experiments in these directions are in progress, but it is desirable at this stage to discuss in the light of the present theory the data already published on scattering of α and β particles.

The following points will be discussed:—

(a) The 'diffuse reflexion' of α particles, *i.e.* the scattering of α particles through large angles (Geiger and Marsden).

(b) The variation of diffuse reflexion with atomic weight of the radiator (Geiger and Marsden).

(c) The average scattering of a pencil of α rays transmitted through a thin metal plate (Geiger).

(d) The experiments of Crowther on the scattering of β rays of different velocities by various metals.

(*a*) In the paper of Geiger and Marsden (*loc. cit.*) on the diffuse reflexion of α particles falling on various substances it was shown that about 1/8000 of the α particles from radium C falling on a thick plate of platinum are scattered back in the direction of the incidence. This fraction is deduced on the assumption that the α particles are uniformly scattered in all directions, the observations being made for a deflexion of about 90°. The form of experiment is not very suited for accurate calculation, but from the data available it can be shown that the scattering observed is about that to be expected on the theory if the atom of platinum has a central charge of about 100 *e*.

(*b*) In their experiments on this subject, Geiger and Marsden gave the relative number of α particles diffusely reflected from thick layers of different metals, under similar conditions. The numbers obtained by them are given in the table below, where *z* represents the relative number of scattered particles, measured by the number of scintillations per minute on a zinc sulphide screen.

Metal	Atomic weight	z	$z/A^{3/2}$
Lead	207	62	208
Gold	197	67	242
Platinum	195	63	232
Tin	119	34	226
Silver	108	27	241
Copper	64	14·5	225
Iron	56	10·2	250
Aluminium	27	3·4	243

Average   233

On the theory of single scattering, the fraction of the total number of α particles scattered through any given angle in passing through a thickness *t* is proportional to $n \,.\, A^2 t$, assuming that the central charge is proportional to the atomic weight A. In the present case, the thickness of matter from which the scattered α particles are able to emerge and affect the zinc sulphide screen depends on the metal. Since Bragg has shown that the stopping power of an atom for an α particle is proportional to the square root of its atomic weight, the value of *nt* for different elements is proportional to $1/\sqrt{A}$. In this case *t* represents the greatest depth from which the scattered α particles emerge. The number *z* of α particles scattered back from a thick layer is consequently proportional to $A^{3/2}$ or $z/A^{3/2}$ should be a constant.

To compare this deduction with experiment, the relative values of the latter quotient are given in the last column. Considering the difficulty of the experiments, the agreement between theory and experiment is reasonably good.*

* The effect of change of velocity in an atomic encounter is neglected in this calculation.

The single large scattering of α particles will obviously affect to some extent the shape of the Bragg ionization curve for a pencil of α rays. This effect of large scattering should be marked when the α rays have traversed screens of metals of high atomic weight, but should be small for atoms of light atomic weight.

(c) Geiger made a careful determination of the scattering of α particles passing through thin metal foils, by the scintillation method, and deduced the most probable angle through which the α particles are deflected in passing through known thicknesses of different kinds of matter.

A narrow pencil of homogeneous α rays was used as a source. After passing through the scattering foil, the total number of α particles deflected through different angles was directly measured. The angle for which the number of scattered particles was a maximum was taken as the most probable angle. The variation of the most probable angle with thickness of matter was determined, but calculation from these data is somewhat complicated by the variation of velocity of the α particles in their passage through the scattering material. A consideration of the curve of distribution of the α particles given in the paper (*loc. cit.* p. 496) shows that the angle through which half the particles are scattered is about 20 per cent greater than the most probable angle.

We have already seen that compound scattering may become important when about half the particles are scattered through a given angle, and it is difficult to disentangle in such cases the relative effects due to the two kinds of scattering. An approximate estimate can be made in the following way:— From (§ 5) the relation between the probabilities p_1 and p_2 for compound and single scattering respectively is given by

$$p_2 \log p_1 = -0\cdot721.$$

The probability q of the combined effects may as a first approximation be taken as

$$q = (p_1{}^2 + p_2{}^2)^{1/2}.$$

If $q = 0\cdot5$, it follows that

$$p_1 = 0\cdot2 \quad \text{and} \quad p_2 = 0\cdot46.$$

We have seen that the probability p_2 of a single deflexion greater than ϕ is given by

$$p_2 = \frac{\pi}{4}n \cdot t \cdot b^2 \cot^2 \phi/2.$$

Since in the experiments considered ϕ is comparatively small

$$\frac{\phi\sqrt{p_2}}{\sqrt{\pi n t}} = b = \frac{2NeE}{mu^2}.$$

Geiger found that the most probable angle of scattering of the α rays in passing through a thickness of gold equivalent in stopping power to about $0 \cdot 76$ cm. of air was $1° 40'$. The angle ϕ through which half the α particles are turned thus corresponds to $2°$ nearly.

$$t = 0 \cdot 00017 \text{ cm.}; \; n = 6 \cdot 07 \times 10^{22};$$

$$u \text{ (average value)} = 1 \cdot 8 \times 10^9.$$

$$E/m = 1 \cdot 5 \times 10^{14} \text{ E.S. units}; \; e = 4 \cdot 65 \times 10^{-10}.$$

Taking the probability of single scattering $=0 \cdot 46$ and substituting the above values in the formula, the value of N for gold comes out to be 97.

For a thickness of gold equivalent in stopping power to $2 \cdot 12$ cms. of air, Geiger found the most probable angle to be $3° 40'$. In this case $t = 0 \cdot 00047$, $\phi = 4° \cdot 4$, and average $u = 1 \cdot 7 \times 10^9$, and N comes out to be 114.

Geiger showed that the most probable angle of deflexion for an atom was nearly proportional to its atomic weight. It consequently follows that the value of N for different atoms should be nearly proportional to their atomic weights, at any rate for atomic weights between gold and aluminium.

Since the atomic weight of platinum is nearly equal to that of gold, it follows from these considerations that the magnitude of the diffuse reflexion of α particles through more than $90°$ from gold and the magnitude of the average small angle scattering of a pencil of rays in passing through gold-foil are both explained on the hypothesis of single scattering by supposing the atom of gold has a central charge of about $100 \, e$.

(*d*) *Experiments of Crowther on scattering of β rays.*—We shall now consider how far the experimental results of Crowther on scattering of β particles of different velocities by various materials can be explained on the general theory of single scattering. On this theory, the fraction of β particles p turned through an angle greater than ϕ is given by

$$p = \frac{\pi}{4} n \cdot t \cdot b^2 \cot^2 \phi/2.$$

In most of Crowther's experiments ϕ is sufficiently small that $\tan \phi/2$ may be put equal to $\phi/2$ without much error. Consequently

$$\phi^2 = 2\pi \, n \cdot t \cdot b^2 \quad \text{if} \quad p = 1/2.$$

On the theory of compound scattering, we have already seen that the chance p_1 that the deflexion of the particles is greater than ϕ is given by

$$\phi^2/\log p_1 = -\frac{9\pi^3}{64} n \cdot t \cdot b^2.$$

Since in the experiments of Crowther the thickness t of matter was determined for which $p_1 = 1/2$,

$$\phi^2 = 0 \cdot 96\pi \, n \, t \, b^2.$$

For a probability of 1/2, the theories of single and compound scattering are thus identical in general form, but differ by a numerical constant. It is thus clear that the main relations on the theory of compound scattering of Sir J. J. Thomson, which were verified experimentally by Crowther, hold equally well on the theory of single scattering.

For example, if t_m be the thickness for which half the particles are scattered through an angle ϕ, Crowther showed that $\phi/\sqrt{t_m}$ and also $\dfrac{mu^2}{E} \cdot \sqrt{t_m}$ were constants for a given material when ϕ was fixed. These relations hold also on the theory of single scattering. Notwithstanding this apparent similarity in form, the two theories are fundamentally different. In one case, the effects observed are due to cumulative effects of small deflexions, while in the other the large deflexions are supposed to result from a single encounter. The distribution of scattered particles is entirely different on the two theories when the probability of deflexion greater than ϕ is small.

We have already seen that the distribution of scattered α particles at various angles has been found by Geiger to be in substantial agreement with the theory of single scattering, but cannot be explained on the theory of compound scattering alone. Since there is every reason to believe that the laws of scattering of α and β particles are very similar, the law of distribution of scattered β particles should be the same as for α particles for small thicknesses of matter. Since the value of mu^2/E for the β particles is in most cases much smaller than the corresponding value for the α particles, the chance of large single deflexions for β particles in passing through a given thickness of matter is much greater than for α particles. Since on the theory of single scattering the fraction of the number of particles which are deflected through a given angle is proportional to kt, where t is the thickness supposed small and k a constant, the number of particles which are undeflected through this angle is proportional to $1 - kt$. From considerations based on the theory of compound scattering, Sir J. J. Thomson deduced that the probability of deflexion less than ϕ is proportional to $1 - e^{-\mu/t}$ where μ is a constant for any given value of ϕ.

The correctness of this latter formula was tested by Crowther by measuring electrically the fraction I/I_0 of the scattered β particles which passed through a circular opening subtending an angle of 36° with the scattering material. If

$$I/I_0 = 1 - e^{-\mu/t},$$

the value of I should decrease very slowly at first with increase of t. Crowther, using aluminium as scattering material, states that the variation of I/I_0 was in good accord with this theory for small values of t. On the other hand, if single scattering be present, as it undoubtedly is for α rays, the curve showing the relation between I/I_0 and t should be nearly linear in the initial stages. The experiments of Madsen* on scattering of β rays, although not made with

* Phil. Mag. xviii. p. 909 (1909).

quite so small a thickness of aluminium as that used by Crowther, certainly support such a conclusion. Considering the importance of the point at issue, further experiments on this question are desirable.

From the table given by Crowther of the value $\phi/\sqrt{t_m}$ for different elements for β rays of velocity $2 \cdot 68 \times 10^{10}$ cms. per second, the values of the central charge Ne can be calculated on the theory of single scattering. It is supposed, as in the case of the α rays, that for the given value of $\phi/\sqrt{t_m}$ the fraction of the β particles deflected by single scattering through an angle greater than ϕ is $0 \cdot 46$ instead of $0 \cdot 5$.

The values of N calculated from Crowther's data are given below.

Element	Atomic weight	$\phi/\sqrt{t_m}$	N
Aluminium	27	$4 \cdot 25$	22
Copper	$63 \cdot 2$	$10 \cdot 0$	42
Silver	108	$15 \cdot 4$	78
Platinum	194	$29 \cdot 0$	138

It will be remembered that the values of N for gold deduced from scattering of the α rays were in two calculations 97 and 114. These numbers are somewhat smaller than the values given above for platinum (viz. 138), whose atomic weight is not very different from gold. Taking into account the uncertainties involved in the calculation from the experimental data, the agreement is sufficiently close to indicate that the same general laws of scattering hold for the α and β particles, notwithstanding the wide differences in the relative velocity and mass of these particles.

As in the case of the α rays, the value of N should be most simply determined for any given element by measuring the small fraction of the incident β particles scattered through a large angle. In this way, possible errors due to small scattering will be avoided.

The scattering data for the β rays, as well as for the α rays, indicate that the central charge in an atom is approximately proportional to its atomic weight. This falls in with the experimental deductions of Schmidt.* In his theory of absorption of β rays, he supposed that in traversing a thin sheet of matter, a small fraction α of the particles are stopped, and a small fraction β are reflected or scattered back in the direction of incidence. From comparison of the absorption curves of different elements, he deduced that the value of the constant β for different elements is proportional to nA^2 where n is the number of atoms per unit volume and A the atomic weight of the element. This is exactly the relation to be expected on the theory of single scattering if the central charge on an atom is proportional to its atomic weight.

§ 7. *General Considerations*

In comparing the theory outlined in this paper with the experimental results, it has been supposed that the atom consists of a central charge supposed

* *Annal. d. Phys.* iv. 23. p. 671 (1907).

concentrated at a point, and that the large single deflexions of the α and β particles are mainly due to their passage through the strong central field. The effect of the equal and opposite compensating charge supposed distributed uniformly throughout a sphere has been neglected. Some of the evidence in support of these assumptions will now be briefly considered. For concreteness, consider the passage of a high speed α particle through an atom having a positive central charge Ne, and surrounded by a compensating charge of N electrons. Remembering that the mass, momentum, and kinetic energy of the α particle are very large compared with the corresponding values for an electron in rapid motion, it does not seem possible from dynamic considerations that an α particle can be deflected through a large angle by a close approach to an electron, even if the latter be in rapid motion and constrained by strong electrical forces. It seems reasonable to suppose that the chance of single deflexions through a large angle due to this cause, if not zero, must be exceedingly small compared with that due to the central charge.

It is of interest to examine how far the experimental evidence throws light on the question of the extent of the distribution of the central charge. Suppose, for example, the central charge to be composed of N unit charges distributed over such a volume that the large single deflexions are mainly due to the constituent charges and not to the external field produced by the distribution. It has been shown (§ 3) that the fraction of the α particles scattered through a large angle is proportional to $(NeE)^2$, where Ne is the central charge concentrated at a point and E the charge on the deflected particle. If, however, this charge is distributed in single units, the fraction of the α particles scattered through a given angle is proportional to Ne^2 instead of N^2e^2. In this calculation, the influence of mass of the constituent particle has been neglected, and account has only been taken of its electric field. Since it has been shown that the value of the central point charge for gold must be about 100, the value of the distributed charge required to produce the same proportion of single deflexions through a large angle should be at least 10,000. Under these conditions the mass of the constituent particle would be small compared with that of the α particle, and the difficulty arises of the production of large single deflexions at all. In addition, with such a large distributed charge, the effect of compound scattering is relatively more important than that of single scattering. For example, the probable small angle of deflexion of a pencil of α particles passing through a thin gold foil would be much greater than that experimentally observed by Geiger (§ *b–c*). The large and small angle scattering could not then be explained by the assumption of a central charge of the same value. Considering the evidence as a whole, it seems simplest to suppose that the atom contains a central charge distributed through a very small volume, and that the large single deflexions are due to the central charge as a whole, and not to its constituents. At the same time, the experimental evidence is not precise enough to negative the possibility that a small fraction of the positive charge may be carried by satellites extending some distance from the centre. Evidence on this point could be obtained by examining whether the same central charge is required to explain the large single deflexions of α and β particles; for the α particle must approach much closer to the centre of the atom than the β particle of average speed to suffer the same large deflexion.

The general data available indicate that the value of this central charge for different atoms is approximately proportional to their atomic weights, at any rate for atoms heavier than aluminium. It will be of great interest to examine experimentally whether such a simple relation holds also for the lighter atoms. In cases where the mass of the deflecting atom (for example, hydrogen, helium, lithium) is not very different from that of the α particle, the general theory of single scattering will require modification, for it is necessary to take into account the movements of the atom itself (see § 4).

It is of interest to note that Nagaoka* has mathematically considered the properties of a 'Saturnian' atom which he supposed to consist of a central attracting mass surrounded by rings of rotating electrons. He showed that such a system was stable if the attractive force was large. From the point of view considered in this paper, the chance of large deflexion would practically be unaltered, whether the atom is considered to be a disk or a sphere. It may be remarked that the approximate value found for the central charge of the atom of gold ($100 \ e$) is about that to be expected if the atom of gold consisted of 49 atoms of helium, each carrying a charge $2 \ e$. This may be only a coincidence, but it is certainly suggestive in view of the expulsion of helium atoms carrying two unit charges from radioactive matter.

The deductions from the theory so far considered are independent of the sign of the central charge, and it has not so far been found possible to obtain definite evidence to determine whether it be positive or negative. It may be possible to settle the question of sign by consideration of the difference of the laws of absorption of the β particle to be expected on the two hypotheses, for the effect of radiation in reducing the velocity of the β particle should be far more marked with a positive than with a negative centre. If the central charge be positive, it is easily seen that a positively charged mass if released from the centre of a heavy atom, would acquire a great velocity in moving through the electric field. It may be possible in this way to account for the high velocity of expulsion of α particles without supposing that they are initially in rapid motion within the atom.

Further consideration of the application of this theory to these and other questions will be reserved for a later paper, when the main deductions of the theory have been tested experimentally. Experiments in this direction are already in progress by Geiger and Marsden.

University of Manchester
April 1911

---

* Nagaoka, Phil. Mag. vii. p. 445 (1904).

Transformation and Nomenclature of the Radioactive Emanations

by PROFESSOR E. RUTHERFORD, F.R.S., and DR. H. GEIGER,
University of Manchester

From the *Philosophical Magazine* for October 1911, ser. 6, xxii, pp. 621–9

IN a recent paper, H. Geiger* has described experiments which show that the emanation of actinium contains a product of very quick transformation, which emits α rays of long range, 6·5 cm. in air, while the range of the α particles from the emanation itself is 5·7 cm. Immediately after its formation, this new substance has a positive charge, and travels to the negative electrode in an electric field. By assuming that this positive carrier has the same mobility in air as a positive ion produced in air, it was deduced that the quick product was half transformed in about 1/500 of a second. The proof of the existence of this product at once explained the observation made some time before by Geiger and Marsden that the emanation of actinium apparently emitted two α ray particles at nearly the same time. Since the new product is almost completely transformed in 1/50 of a second after its formation, the α ray particle from the emanation itself would be followed within this interval by one from the new product, and the interval between them could not be detected by the eye using the scintillation method. In a previous paper, Geiger and Marsden† have also shown that the emanation of thorium emits two α particles in rapid succession. In this case there was on the average a distinct interval between the appearance of two scintillations on the zinc sulphide screen, indicating that an α ray product was present in the emanation which had an average life of transformation not longer than 1/5 of a second. These conclusions have been confirmed by further experiment. As in the case of the actinium emanation the new product has a positive charge, and in an electric field is deposited on the negative electrode. The period of transformation, however, is much longer than that for the corresponding actinium product. It is, in consequence, not practicable to determine the period of transformation by the method previously employed by Geiger for the short-lived product of the actinium emanation. In order to determine the period, weak electric fields are required, and under these conditions the loss of charge of the carriers by recombination with the ions seriously complicates the deductions from the experiments. In an accompanying paper, the times of transformation of these new products have been determined by a new and

* Phil. Mag. July 1911.   † *Phys. Zeit.* xi. p. 7 (1910).

more direct method by H. Moseley and K. Fajans. It will be seen that the half-value period of transformation of the product in actinium is 0·002 second, a value in good accord deduced indirectly by Geiger. The corresponding half-value period for the thorium product is 0·14 second, a value within the limit indicated by the scintillation experiments of Geiger and Marsden.

The presence of these new products in the emanations of thorium and actinium can be simply illustrated by experiments in a dark room. The emanation from a strong preparation of actinium or of thorium is allowed to diffuse into a small cylindrical vessel in the centre of which is fixed a metal rod insulated from the cylinder by an ebonite cork. One terminal of a battery of about 1000 volts is connected to the rod and the other to the outside of the vessel. About one centimetre of the end of the rod is coated with a layer of zinc sulphide. In a dark room without an electric field, the zinc sulphide is seen to glow faintly due to the α rays from the emanation. When the rod is connected to the negative pole of the battery, the end of the rod is seen to light brightly on the instant. This is due to the concentration of the short-lived product from the whole volume of the emanation space on to the small negatively charged rod. On disconnecting the battery the luminosity sinks instantly. By alternate application and removal of the electric field at rapid intervals, the luminosity is seen to rise and fall with it. In the case of actinium, the effect of the ordinary active deposit becomes evident after a few minutes by a slow but steady increase of the residual luminosity when the field is cut off. The effects are more strongly marked using the emanation of thorium instead of actinium. For this purpose a tube containing an active preparation of radiothorium and mesothorium open at one end was placed at the base of the vessel. The emanation diffused from the tube into the air space. The increase of luminosity was relatively far more marked than in the case of the actinium emanation, a result probably due to the fact that a considerable fraction of the actinium product was transformed before the electric field could transport it to the negative electrode. A similar experiment was performed with a vessel containing some radium emanation; but the increase of luminosity on applying the field was small and the decrease of luminosity after the electric field was cut off was difficult to detect. In this case, the radium A of half-value period three minutes was concentrated on the rod. No appreciable effect is observed in any of the experiments if the central rod is charged positively.

In this connexion, it is of interest to recall an experiment on the actinium emanation made many years ago by Giesel.* A narrow metal tube containing a strong actinium preparation at one end was placed with the open end downwards about 5 cm. from the surface of a zinc sulphide screen. The screen could be charged negatively to a high potential, and the cylinder connected with earth. On applying the electric field a luminous spot of light instantly appeared on the screen under the tube. Giesel suggested in explanation that

* *Ber. d. D. Chem. Ges.* xxxvi. p. 342 (1903).

a new type of radiation was emitted by the emanation which he termed the E rays. It has been generally assumed that the effect observed by Giesel was due to the concentration of the ordinary actinium deposit on the zinc sulphide screen. In the light of the experiments described in this paper there appears to be no doubt, however, that the luminosity was initially mainly due to the very short-lived product of the actinium emanation. This travelled along the lines of force to the zinc sulphide screen and was transformed *in situ*, giving rise to the luminosity observed.

The presence of the new product in the thorium emanation and its rapid decay can be simply shown in the following way. A small brass cylinder about 6 cm. long is connected with a strong source of thorium preparation, and the ends are closed by small ebonite corks. An endless wire, which can be kept in constant movement by means of a motor, passes through small holes in the ebonite stoppers. The outside of the cylinder is connected with the positive pole of the battery, and the wire with the negative pole. Under these conditions, the short-lived product of the thorium emanation is concentrated on the wire. The activity of the wire as it passes from the tube is examined by a zinc sulphide screen placed close to it. It is seen that the luminosity of the screen falls off as it is moved along the wire from the end of the tube. By suitably adjusting the velocity of the wire and using a long zinc sulphide screen placed close to the wire the decay of the new product is very strikingly and simply shown. A band of luminosity is seen on the screen which decreases rapidly in intensity from the point nearest the end of the tube. In order to observe a similar effect with the actinium emanation, the wire must be moved far more rapidly. By counting the scintillations at various points along the wire it is obvious that the decay of the product can be simply determined when the velocity of the moving wire is known. As, however, the method is similar in general principle with that employed by Moseley and Fajans in the paper referred to, it has not been considered necessary to give the results obtained.

It is seen that the atoms of the new products derived from the thorium and actinium emanations carry a positive charge, for they are conveyed to the negative electrode in an electric field. An interesting question arises whether the concentration of the active deposit from these two emanations on the cathode depends entirely upon the transportation of the new products to the cathode and their subsequent transformation *in situ*. For example, there is at present no certain evidence of the sign or magnitude of the charge on the residual atom which will result from the transformation of the new products when present in the gas. This is a question of considerable interest, and will be examined in detail later. The experiments of Russ* and Kennedy† on the actinium emanation have shown that under some conditions as much active deposit appears on the anode as on the cathode, while in the case of thorium, the active deposit under normal conditions appears only on the cathode. These differences may possibly receive an explanation by taking into account

* Phil. Mag. xvii. p. 412 (1909).   † Phil. Mag. xviii. p. 744 (1909).

I

the difference in the period of transformation of the short-lived products in the two cases. Some experiments are in progress to examine these points in detail.

Nomenclature

The existence of two short-lived products following the emanations of thorium and actinium makes a change of nomenclature desirable, in order to give a definite name to each of these products, and to specify its position in the radioactive series. There seems to be no doubt that these short-lived products do not exist in the gaseous state, but behave as solids. This is borne out by the experimental results included in this paper, for there is no evidence that the products considered escape from the negatively charged wire except by the well-known process of radioactive recoil.

It seems fairly certain that the new products ought to be included as members of the 'active deposits' derived from the thorium and actinium emanations. They resemble the first member of the active deposit of radium, namely radium A, in carrying a positive charge, and in their concentration on the cathode in a strong electric field. If they are considered to be members of the group of products now generally included under the term 'active deposit,' there is seen to be a remarkably close analogy in the successive transformations of the three emanations. It is seen that the new products are very analogous, both as regards physical properties and nature of radiation emitted, to the first product of the radium emanation, namely radium A. It is, therefore, suggested that the new product in thorium should be called thorium A, and in actinium, actinium A. In consequence it is necessary to denote the products previously called thorium A and actinium A, thorium B and actinium B respectively, and similarly for the later products.

The scheme of nomenclature proposed, starting from the emanations, is shown in the following table, where the nature of the radiation emitted and the half-value period of transformation are added.

In this scheme, it is seen that not only are the A products similar in general character, but also the B products, for each of the latter emits easily absorbed β rays.

There is an apparent divergence in the mode of transformation of the third product of the three series. In the radium series, the product ordinarily called radium C emits not only α rays but also β and γ rays. The third product of the actinium series, now named actinium C, emits only one type of α rays, and is followed by another product to be called actinium D, emitting β and γ rays. This latter product was isolated by Hahn using the recoil method. In the case of thorium, the transformation of the second product, thorium B, is followed by a product of half-value period 55 minutes, which Hahn showed emitted two distinct types of α rays, whose ranges in air are $5\cdot0$ and $8\cdot6$ cm. In all other radioactive transformations, it has been observed that each product emits α particles of characteristic range. It has, therefore, been generally assumed that two successive products are present, one of period of

RADIUM SERIES			THORIUM SERIES			ACTINIUM SERIES		
Substance	Period	Radiation	Substance	Period	Radiation	Substance	Period	Radiation
Radium Emanation. →	3·86 days.	α rays [4·23 cm.]	Thorium Emanation. →	53 seconds.	α rays.	Actinium Emanation. →	3·9 seconds.	α rays. [5·7 cm.]
Radium A. →	3·0 minutes.	α rays [4·83 cm.]	Thorium A. →	0·14 second.	α rays.	Actinium A. →	0·002 second.	α rays. [6·5 cm.]
Radium B. →	26·7 minutes.	soft β rays.	Thorium B. →	10·6 hours.	soft β rays.	Actinium B. →	36 minutes.	soft β rays.
Radium C = C$_1$ + C$_2$ →	19·6 minutes. / 1·38 minutes.	α rays [7·06 cm.] / β rays.	Thorium C = C$_1$ + C$_2$	55 minutes. / ?	α rays. [5·0 cm.] / α rays [8·6 cm.]	Actinium C. →	2·15 minutes.	[α rays] [5·4 cm.]
Radium D.		soft β rays.	Thorium D.	3·1 minutes.	hard β rays.	Actinium D.	4·71 minutes.	hard β rays.

55 minutes emitting α particles of range 5·0 cm., followed by a product of probably rapid transformation emitting α particles of range 8·6 cm. Although a number of attempts have been made by various observers, it has not so far been found possible to separate the two products; for they always appear together and decay together with the same period, and each product seems to emit about the same number of α particles per second.

Hahn has also shown that from a plate coated with the active deposit of thorium, a product emitting β and γ rays of period three minutes can be obtained by recoil. We thus see that the analogy apparently breaks down at this point; for radium C or actinium C emits only one type of α rays. In addition, in the case of actinium and thorium, the penetrating β and γ rays arise from a distinct product following the α ray transformation. Recently, however, Hahn and Meitner* drew attention to the existence of a new product in radium C. This has been confirmed and examined in detail by Fajans,† who found that the product has a period of 1·4 minutes, and emits only β and probably γ rays. The absorption of these rays is about the same as for those ascribed to ordinary radium C. This new product is obtained by recoil from pure radium C, but in exceedingly small relative quantity, about 1/20000, measured by the β rays. In this respect, it differs markedly from the β ray products obtained by recoil from the corresponding products of thorium and actinium, where the relative quantities obtained by recoil are about 10,000 times greater. Fajans has discussed the question of the position of this new β ray product in the radium series, and concludes that radium C breaks up in two distinct ways, and that the new product is to be regarded as a lateral branch of the main radium series.

At the Radiology Congress in Brussels, the question of nomenclature was discussed, and it was generally agreed that if a product considered simple was shown to be complex, the original name should, if possible, be retained to signify the group, and that the individual components should be distinguished by numbers. For example, there are many practical advantages in retaining the name radium C to include the two or more components that may be present. So far only two components have been definitely distinguished, and these will be called radium C_1 and radium C_2. Fajans has suggested that the name radium C_2 should be given to the new product of period 1·4 minutes. No attempt has been made in the Table to differentiate between the products of the transformation of the two component substances. The main series is supposed to follow the group as a whole.

In regard to thorium, it is proposed that the matter in the active deposit, which always emits the two distinct types of α rays, should be called as a whole thorium C, and that its possible components should be called thorium C_1 and thorium C_2. This nomenclature has certain advantages, for in the first place on this scheme all the C products emit α rays, and the D products emit β rays. In the second place, there is at present no definite evidence that the

---

* *Phys. Zeit.* x. p. 697 (1909).   † *Phys. Zeit.* xii. p. 369 (1911).

components of thorium C are successive products in the ordinary sense. The question of the exact type of the transformation occurring in thorium C is a difficult one. If thorium C_1 and thorium C_2 are successive products, it is to be expected that a considerable quantity of C_2 should be obtained by recoil under suitable conditions. Hahn, however, was unable to obtain any evidence of the separation of thorium C_2 by recoil, and in consequence concluded that the period of the latter must be very short. On the other hand, Geiger and Marsden failed to observe any double scintillations from thorium C or any groupings of the scintillations small intervals of time apart.*

This evidence is difficult to reconcile with the view that the changes are successive in the ordinary sense. Additional support to this conclusion is given by the observations first made by Bronson, and afterwards confirmed by the scintillation method by Geiger and Marsden, that the thorium emanation together with its short-period product emits four α particles for two from the products thorium C_1 + thorium C_2 in equilibrium with it. This is difficult to account for on the view of successive products, unless it be supposed that both the emanation and the new product emit two α particles for one from thorium C_1 and one from thorium C_2. The evidence as a whole points strongly to the conclusion that the products are not successive, but are connected in some unusual way. There are several modes of transformation possible to account for the observed facts; but it is desirable to delay the discussion of these points until more experimental data are available.

As in the case of radium C, thorium D is supposed to be a successive product of the group thorium C without any assumption whether it originates from only one of the components or both. The analogy between the general modes of the three emanations is undoubtedly very close. It appears very probable that the group radium C in reality corresponds to the group thorium C + D, and actinium C + D, although the type and order of the transformations appear to be different in each case. There are, however, still many questions to be closely examined before any decisive conclusion can be reached. The analogy between the products of the emanations has been discussed in some detail partly to give reasons for the scheme of nomenclature adopted, and partly to bring out the numerous points of similarity.

University of Manchester
August 1911

* These experiments made some time ago by Geiger and Marsden have not so far been published. A more detailed account will be published by them in a later paper.

The Transformation of Radium

by PROFESSOR E. RUTHERFORD, F.R.S.

From the *Journal of the Society of Chemical Industry*, June 15, 1911, No. 11, xxx, pp. 3–14 (Presented on 10th March 1911 before the Manchester Section of the Society of Chemical Industry.)

THE subject of the transformation of radium is now a very large one, and I cannot hope to give more than a brief survey of the main results; but I shall concentrate my attention as far as possible on the more chemical side. In 1896 Becquerel discovered that the ordinary uranium compounds possessed the property of emitting a penetrating radiation of a type that would pass through matter opaque to ordinary light. This radiation darkens a photographic plate, but the effect due to uranium and its compounds is exceedingly feeble, at least a day's exposure being required to produce a sensible impression. A little later thorium and its compounds were found to possess a similar property and to about an equal degree to that of uranium. Some time later, Madame Curie made a systematic examination of the radiating power of uranium and thorium minerals, and found that the former showed a radiating power of activity between four and five times greater than that to be expected from their content of uranium. Since, in previous experiments, she had found that the radioactivity of uranium was an atomic property, she concluded that there must be a new active substance present in the mineral. Following out this line of reasoning, she was able chemically to separate two very active constituents, radium and polonium. The former in a pure state possesses a radiating power to an intense degree. While the photographic method has proved useful in many investigations on radioactivity, the electric method of measurement is far simpler and more reliable. The rays from active matter, like the Röntgen rays, possess the property of causing a discharge of electricity from a body, whether electrified positively or negatively. The ordinary gold leaf electroscope has proved very useful for such measurements. Under ordinary conditions, the insulated gold leaf system of the electroscope loses its charge very slowly. When a radioactive substance is brought near, the gas becomes a temporary conductor of electricity, and the electrified body loses its charge. This is shown by the increased rate of movement of the gold leaf. Under suitable conditions, the rate of movement of the gold leaf can be used as a relative measure of the activity of substances. The electrical effect due to uranium and thorium and the corresponding minerals, is marked and easily measured. A pure radium compound has an activity several million times greater than that of uranium, so that an insignificant quantity of radium produces a rapid discharge of the electroscope. The time is too short for me to discuss here the methods of separation and concentration of radium from uranium minerals. It suffices to say that radium has chemical properties very similar to those of barium, and is ordinarily separated with it. Finally, by the method of fractional crystallisation of the chloride or bromide, the radium is at last obtained free from barium. Mme. Curie has recently shown that radium can be obtained in the metallic state by methods similar to those employed to isolate metallic barium. The atomic weight of radium is 226·5; it has a characteristic spectrum, and from the chemical point of view is a new element with distinctive chemical and physical properties.

It is now necessary to consider the types of radiation emitted by radium and its compounds. It must be borne in mind that there are a number of active substances which possess very similar radiating properties. There are three distinct types of radiation emitted, which are known as the α, β, and γ rays. These radiations differ widely in their power of penetrating matter, and can also be distinguished by the effect on them of a strong magnetic or electric field. The α rays are of comparatively insignificant penetrating power, for they are completely stopped in passing through a few centimetres of air, or through a thin sheet of metal. They are, however, in many respects, the most important radiations emitted, for, under ordinary conditions, they are responsible for most of the electrical effect and for most of the energy emitted by radioactive substances. They have been shown to consist of a stream of particles carrying a positive charge of electricity projected with velocity of about 10,000 miles per second. I have recently shown that the α rays are material in nature, and consist of charged atoms of the rare gas, helium. The β rays are far more penetrating than the α rays, and are far more easily deflected by a magnetic field. They have been shown to consist of a stream of negatively charged particles projected at a speed between 100,000 and 180,000 miles per second. The apparent mass of these particles is only about 1/1800 of the mass of the hydrogen atom. They are, in fact, identical with the corpuscles or electrons, and their apparent mass is believed to be entirely electrical in origin. The γ rays are exceedingly penetrating, and their electrical effect can be detected through nearly a foot of lead. They are not deflected by a magnetic electric field, and appear to be analogous in their properties to Röntgen rays. The true nature of this type of radiation is at present a matter of much discussion.

The difference in penetrating power of these radiations can be easily illustrated by experiment. The discharging effect due to an uncovered layer of radium is mainly due to the α rays. On placing a sheet of paper over the radium, the α rays are all absorbed and the discharging effect is then due to the β and γ rays, but mainly to the β rays. A sheet of lead, 2 mm. thick, cuts off most of the β rays, and a discharging effect is then due to the γ rays alone. It must be remembered that the electrical effect due to the γ rays is, under ordinary conditions, less than 1/1000 of that due to the α rays.

It is seen that for the most part radiations from radium consist of a stream of charged particles projected with great velocity. The α particle is an atom of helium which is initially projected with a velocity of about 10,000 miles per second. Rutherford and Geiger have devised an electrical method of detecting a single α particle, and of counting the number emitted per second by one gram of radium itself. The α particle can also be detected by its property of producing a visible flash of light, or scintillation, when it falls on a screen of phosphorescent zinc sulphide. When the eye has become accustomed to darkness, the number of α particles falling on a given area of zinc sulphide can be counted by observing with a microscope the number of scintillations. We have thus two distinct methods of detecting a single atom of matter in swift motion, one electrical and the other optical. In addition, Mr. Kinoshita, in this Laboratory, has shown that a single α particle produces a detectable photographic effect on an ordinary plate.*

We shall now consider the interpretation of the effects taking place in radium. The radiations are emitted continuously, and show no sensible diminution over periods measured by years. In addition to the form and energy of characteristic radiations, radium is also continuously evolving heat. One gram of radium emits heat at the rate of 118 gram calories per hour; in other words, the heat is sufficient to melt an equal weight of ice in about 40 minutes. The amount of energy given out in the course of a year is very large, and yet there is no sensible alteration of the radium itself, or in its rate of emission of α and β particles. The radiating power is unaffected by chemical or physical conditions, and in this respect differs notably from any known chemical substance. Simple considerations show that chemical changes of the ordinary type are insufficient to account for the continuous emission of heat energy. At the same time it is necessary to account for the characteristic radiations. In order to account for radioactive phenomena, Rutherford and Soddy put forward some years ago the transformation theory, which is now generally accepted as a reasonable explanation of the effects observed in radium and other radioactive substances. From the chemical point of view, radium is to be regarded as an ordinary element with a definite atomic weight and with specific chemical and physical properties. It differs however from an ordinary element, like copper or gold, inasmuch as the atoms of which it is composed are not stable. Each second, a small fraction of the radium atoms present becomes unstable and disintegrates with explosive violence. This atomic explosion is generally marked by the expulsion of part of its mass, the α particle, at a very high speed, and the appearance of a new element with distinctive chemical and physical properties. Since the α particle is an atom

* Since the delivery of this lecture, Mr. C. T. R. Wilson has devised a method of making visible to the eye the path of a β as well as of an α particle. By a sudden expansion of the air, the ions produced by the β particle in the gas become centres of condensation of water vapour, and the trail of the particle is rendered visible by a fine cloud of water drops. Experiments of this character bring in evidence in a striking way the objective reality of the α and β particles.

of helium, and one α particle is expelled in the explosion of the radium atom, the residue of the atom has an atomic weight about 4 units less than radium. These residual atoms form an entirely new substance called the radium emanation. This exists at ordinary temperatures as a gas of high atomic weight (222·5).

I shall now consider in some detail the properties of this remarkable gas. The name 'emanation' was first given to an emission of what appeared to be a radioactive gas from thorium, and when, somewhat later, similar effects were observed in radium and actinium, the name was retained. In order to illustrate the properties of these emanations, a small quantity of actinium preparation wrapped in thin paper is placed in a glass tube; the emanation produced in it diffuses through the paper and mixes through the air. On sending a puff of air through the tube between the plates and the electroscope the gold leaf collapses with great rapidity. On closing up the electroscope the discharging effect of the emanation introduced falls off very rapidly. It can be shown that the actinium emanation is a type of matter which has a radioactive life of less than a minute, for half of it is transformed into another type of matter in about 4 seconds. In this emanation, we have an example of a transition element with very unstable atoms, so that the average life of the atom is very short. The radium emanation will produce similar effects but with a difference, that its atoms are far more stable. It takes 3·85 days before half the atoms are transformed. The emanation from radium can be separated from it either by heating the radium or by dissolving it. The emanation can then be collected, and if necessary purified and contained in a sealed vessel. Under these conditions, the emanation is gradually transformed into other substances, and loses its activity. The law of decay of activity is very simple, and holds universally for all radioactive substances. The radiation from an active substance is proportional at any moment to the number of atoms breaking up per second, and at the same time varies directly as the number of unchanged atoms present. For example, if the activity of a substance falls to half value in a time, T, it follows that half of the initial number of atoms present remain unchanged at that time. In a time, 2T, one-quarter remain unchanged, in a time, 3T, one-eighth, and so on; in other words, the number of unchanged atoms present varies according to an exponential law with the time. The time, T, or half value period of a product, varies for different substances, but has a characteristic value for each substance. Since the emanation has a half value period of 3·85 days, the greater part of it is transformed in one month, and the activity observed after that interval is a very small fraction of the original value. The emanation is produced at a constant rate by a radium compound. Since the emanation is itself a transition substance and continuously changing into other types of matter, its amount reaches an equilibrium or maximum value when the number of new atoms of emanation supplied per second by the radium is on the average equal to the number of the emanation atoms which are transformed per second. The actual amount of emanation produced by radium is exceedingly small, but can be

I*

measured. One gram of radium in equilibrium yields 0·6 cubic millimetre of pure emanation at standard pressure and temperature. Notwithstanding its small amount this quantity of emanation emits radiations of enormous intensity. I have in a fine glass tube the emanation derived from about 1/10 of a gram of radium, so that the volume of the gas in a pure state is only 6/100 of a cubic millimetre. The walls of the glass tube are made so thin that the α particles are projected through the glass walls and travel several centimetres through the air before they are stopped. On bringing near a screen of zinc sulphide, it is seen to be brilliantly lighted up. This luminosity is due mainly to the α rays, and if the screen were examined with a microscope, it would be seen to show a multitude of scintillations. A screen of willemite is made brilliantly luminous, while the mineral kunzite gives a marked orange coloured phosphorescence. This emanation tube emits β and γ rays of great intensity. This is seen by the rapid discharge of the electroscope through a thick metal screen. This tube of emanation, if brought close to a photographic plate, would cause a marked photographic effect in a few seconds. An interesting effect produced by this tube is that the α rays passing through the air produce ozone at a rapid rate. The emanation in the tube has practically all the radiating properties of the 100 mgrms. of radium from which it was obtained.

The radium emanation can be condensed from the gases with which it is mixed by surrounding the tube containing the emanation with liquid air. The temperature of condensation in the experimental conditions is about $-150°$ C. The emanation contained in the glass tube makes the walls phosphoresce, and the luminosity due to this is easily seen in a dark room. By opening a stop cock, this purified emanation is put in connection with a glass tube containing fragments of the mineral willemite cooled down by liquid air. The emanation is seen to pass from one tube to another, and to condense in the liquid air. This is shown by the luminosity communicated to the fragments of willemite. This condensation property of the emanation has proved of great service in purifying it from other gases, and is regularly employed in the laboratory for this purpose. Although the emanation exists in such minute amount, its physical and chemical properties have been closely examined. It has an atomic weight of 222·5, and gives a characteristic spectrum of bright lines. It behaves as a chemically inert and monatomic gas, and appears to belong to the well known group of inert gases. Its boiling point has been found to be $-65°$ C. At the temperature of liquid air it condenses probably into a solid and emits an orange coloured light. The emanation is the heaviest gas known and is, in many respects, the most remarkable.

The enormous emission of energy from a radioactive substance is well illustrated by the emanation. The heating effect of radium is mainly a consequence of the expulsion of α particles. When the radium is enclosed in a vessel, the α particles are stopped either in the radium itself or in the envelope, and their great mechanical energy is converted into heat. It is known that the heating effect of any radioactive substance is in reality a measure of the kinetic energy of the expelled α particles. The emanation separated from radium emits about 80 per cent. of the heat evolved from the radium from which it is separated. Knowing the volume and density of this gas, it is a simple matter to calculate the rate of emission of heat from a known weight of emanation. Suppose, for example, it was possible to obtain a pound of this gas. Its radiations and heating effect would be so enormous that it would be difficult to construct a vessel to hold it. A short time after its introduction, the pound of emanation would emit energy at the rate of about 10,000 horse-power. The rate of emission of energy would decrease at the same rate as the emanation loses its activity; but in the course of a month it would have emitted an amount of energy corresponding to about 60,000 horse-power-days. This means that the emanation during its life emits about ten million times as much energy as the most violent explosive that can be prepared. The energy is derived from the transformation of the radioactive atoms, and is released as the result of atomic explosions.

The radium emanation is a transition element, *i.e.*, an element which has only a limited life on account of the instability of its atoms. In breaking up, the atom of the emanation expels an α particle or helium atom, and is transformed into a new substance called radium A, which behaves as a metal at ordinary temperatures. The atoms of radium A are very unstable, and break up with a half value period of 3 minutes, emitting an α particle in the process. Radium A is transformed into another metallic substance called radium B, which is half transformed in 26 minutes, accompanied by the expulsion of β particles but no α particles. This in turn changes into radium C, also a metal, which is half transformed in 19 minutes. The explosion of the atoms of radium C are very violent, for not only is an α particle expelled with great velocity but also β particles of great speed accompanied by a very penetrating type of γ rays. Radium itself emits only α rays, and the β and γ rays which arise from it when in equilibrium with its products are mainly due to the presence of radium C. These three substances, radium A, B, and C, are usually included under the name 'active deposit' of radium. They can be distinguished from each other not only by the differences in rate of transformation and in types of radiation emitted, but also by the chemical and physical properties. The presence of these substances can easily be shown by experiment. When a platinum plate is placed in a vessel containing a large quantity of radium emanation, the active matter, A, B, and C, is deposited as a solid on the plate. On removing the plate from the emanation, the presence of active matter is seen at once by the luminosity it produces in zinc sulphide, and in the rapid discharge of the electroscope. The presence of this matter on the plate can be shown by rubbing the finger lightly on the plate, when part of the active matter is removed, as shown by the fact that the finger at once causes a rapid discharge of the electroscope. It must be borne in mind that the actual weight of matter on the active plate is exceedingly small, far smaller than could be weighed by the most sensitive balance, or to be detected by a microscope. Notwithstanding this, the properties of this substance can be examined chemically. If the platinum plate is placed in dilute hydrochloric or sulphuric acid, the active matter leaves the plate and passes into solution. On evaporating the solution the active matter again remains behind. On introducing a nickel plate into the solution of the three substances, Radium A + B + C, radium C alone is deposited on the nickel. This is a very convenient and useful method of isolating radium C for experimental purposes. The difference in physical and chemical properties between these three products can also be shown by heating a platinum plate to various temperatures. Radium B volatilises at the lowest temperatures, and radium A and radium C at still higher temperatures.

We have now followed the transformation of radium through four distinct successive stages. The process of disintegration, however, does not stop with radium C; the latter changes into another substance called radium D, which is very slowly transformed with a half value period of about 16 years. This in turn changes into radium E, half value period about 5 days. Radium E changes in turn to radium F, which has a half value period of about 140 days. In its transformation, radium D emits β rays of weak penetrating power, radium E swift β rays, while radium F emits α particles. Radium D has chemical properties very similar to those of lead. For this reason it is sometimes called radio-lead. It has not so far been found possible to isolate it chemically from the lead present in radioactive minerals. Radium F is identical with the substance called polonium, the first of the new radioactive substances separated by Mme. Curie from pitchblende and is derived from the decomposition of radium. Polonium has chemical and physical properties somewhat similar to that of bismuth, and is deposited from a solution on a bismuth, or copper or silver plate. Polonium has not yet been completely isolated, although it has been sufficiently purified to obtain a few of the characteristic lines of its spectrum. It is a matter of great difficulty to obtain sufficient quantity of this active substance for an ordinary chemical examination. The process of disintegration of radium apparently ends with polonium, for after the decay of polonium no residual activity is observed. There is good reason for believing that polonium after expelling an α particle changes into helium and lead. It is very difficult to obtain direct experimental proof that lead is the end product of the transformation of radium; but the indirect evidence strongly supports such a conclusion.

Now that we have traced the descendants of radium, it is necessary to see what evidence we have of its ancestors. We have seen that radium must be regarded as an unstable element which is slowly being transformed into the radium emanation. The rate of transformation is too slow to determine by direct activity measurements over an interval of a few years, but it has been shown by Boltwood by a direct experimental method that radium is half transformed in about 2000 years. Consequently in a gram of radium about 0·3 mgrm. is transformed initially per year. There is every reason to believe that radium is transformed according to the same law as all other radioactive products. Since radium is a changing substance, its presence to-day in old minerals, which have existed unchanged for periods measured by millions of years, can only be explained by supposing that radium is continuously produced by the transformation of another substance. The search for this elusive parent of radium has been of the highest interest. Boltwood finally showed that a new radioactive substance, which he called *ionium*, was the direct parent of radium. Ionium is a substance with chemical properties closely allied to that of thorium, and emits α particles in its transformation. If ionium is completely freed from radium, it is found after a short interval that radium has again appeared in the ionium due to the transformation of the latter. In the same way it is known that ionium in turn is derived from the transformation of uranium. This element is to be regarded as the first ancestor of the radium family. Uranium itself is transformed exceedingly slowly, and it can be calculated with some certainty that it would require about 5000 million years for half of it to be transformed.

The results I have so far discussed are now generally accepted by all workers in radioactivity. The numerous radioactive substances arising from the transformation of uranium must be regarded as new elements which have a limited life and are transformed according to a definite law. Similar series of transformations occur in thorium and actinium, but time is too short to discuss these in detail. Uranium and thorium appear to be distinct radioactive elements which have no genetic connection with each other. On the other hand, there appears to be no doubt that actinium is in some way derived from the uranium family; but so far it has not been found possible to locate its position in the general scheme of transformations. There is some evidence that actinium is to be regarded as a side branch of uranium or the uranium series. The great majority of the radioactive substances break up with the expulsion of α particles; helium is consequently one of the side products of their decomposition. For example, it has been shown experimentally that radium, the radium emanation, polonium and ionium all give rise to helium in their transformation, and in amount to be expected from the number of α particles ejected. In other cases, for example, radium B, radium D and E, the transformation is accompanied by the expulsion only of β particles of comparatively slow velocity. There is reason to believe that the transformations occurring in these cases are accompanied by no sensible change in atomic weight. The atomic weights of the various radioactive elements can be simply deduced by taking into consideration the expulsion of α particles. With the exception of uranium, each member of the uranium series expels one α particle during the transformation of one atom. The experimental evidence shows that uranium must emit two. Since the atomic weight of uranium is $238 \cdot 5$, the atomic weight of the next product, uranium X, is 8 units less, or $230 \cdot 5$. Since uranium X emits only a β particle, it is transformed, without change in atomic weight, into ionium. The latter is transformed into radium with the loss of an α particle and the atomic weight of the latter should consequently be $226 \cdot 5$. This is very close to the atomic weight found experimentally. In a similar way, the atomic weight of the other members of the series can be calculated, and are included in the table below.

	Radiation	Atomic weight
Uranium	2α	238·5
Uranium X	β	230·5
Ionium	1α	230·5
Radium	1α	226·5
Emanation	1α	222·5
Radium A	1α	218·5
Radium B	β	214·5
Radium C	1α	214·5
Radium D	β	210·5
Radium E	β	210·5
Radium F	1α	210·5
End product	—	206·5
		(lead ?)

It will be seen that the atomic weight of the end product is very close to that of lead, which we have seen from other evidence is believed to be the final product of the uranium-radium series.

The very small quantity in which most of the radioactive elements can be obtained has not so far made it possible to examine their chemical and physical properties with the same detail as in the case of the ordinary elements. In the case of radium and its emanation, a sufficient quantity of material has been available to examine these properties by direct chemical methods. It appears probable that the substances like ionium, radium D and polonium, which can be obtained in reasonable amount from uranium minerals, will ultimately be chemically isolated. It is obviously not an easy matter to examine the chemical properties of a substance like radium A or radium C, which disappear in an hour or so after separation. At the same time, a considerable amount of information can be obtained as to the chemical properties of such elements by using the radiating power as a method of qualitative and quantitative analysis. There seems to be no reason why the chemical properties of these new substances should not ultimately be determined with the same certainty as in the case of ordinary elements. A certain amount of work has already been done in this direction, but a systematic examination is much wanted.

In conclusion, I would again point out that the process of transformation occurring in the radioactive bodies cannot be influenced to the slightest degree by any chemical or physical agency. We are only able to watch these atomic processes but cannot control them. As far as we know at present, there does not seem to be much hope that we shall be able eventually to alter the rate of transformation of the radioactive substances. The forces which bind the atoms together are too powerful to be influenced by the weak physical or chemical agencies at our command.

A Balance Method for Comparison of Quantities of Radium and Some of its Applications

by PROFESSOR E. RUTHERFORD, F.R.S., *and* J. CHADWICK, B.SC.,

University of Manchester

From the *Proceedings of the Physical Society*, xxiv, 1912, pp. 141–51
(Received February 21, 1912. Read February 23, 1912.)

IN many radio-active measurements it is now of great importance to determine with accuracy the quantities of radium or of radium emanation employed in various investigations. The International Committee appointed by the Congress of Radiology at Brussels, in 1910, have now in preparation an International Radium Standard, which will contain a known weight of pure radium salt. All quantities of radium will ultimately be expressed in terms of this International Standard. It is consequently of great importance to develop reliable methods of comparing quantities of radium accurately in terms of the primary standard. The most suitable method for this purpose is to compare the γ-ray activities of the two specimens. If the radium is enclosed in a sealed tube, the γ-ray activity reaches a practical maximum after two months, and the intensity of the penetrating γ-rays which are emitted serves as a definite measure of the quantity of radium present. The greater part of the γ-rays are emitted not by radium itself but by radium C, and recent investigations by Moseley and Makower* have shown that under ordinary conditions about $11 \cdot 5$ per cent. of the total γ-ray activity of radium is to be ascribed to radium B. The γ-rays from the latter are on the average much less penetrating than those from radium C, and are completely absorbed by a lead screen 2 cm. thick.

In making a comparison, it is essential to know that the radium compound contains only radium and its products and no meso-thorium or radio-thorium. Both the latter substances emit γ-rays of about the same penetrating power as those given out by radium, and without a special examination it is difficult to tell whether the thorium products are absent. Since meso-thorium and radium are always separated together, and are chemically closely allied, it is impossible to isolate pure radium compounds from minerals containing both uranium and thorium. Fortunately the uraninite deposits at Joachimsthal contain only a trace of thorium, so that the radium separated from this ore can be obtained practically free from meso-thorium.

* Moseley and Makower, 'Phil. Mag.,' January, 1912.

The γ-ray activity of radium preparations has been usually compared by electroscopic methods. The electroscope is either composed of lead plate about 3 mm. thick or is completely surrounded by a lead cover of this thickness. Under such conditions the primary β-rays are completely stopped by the lead and the ionisation in the electroscope is due to the more penetrating γ-rays and to the β-radiation to which they give rise. The rate of movement of the gold leaf of the electroscope between two fixed points of the scale of the observing microscope is taken as proportional to the intensity of the γ-radiation. With a well-constructed electroscope it is possible to compare approximately equal quantities of radium with a probable error of not more

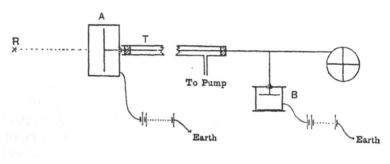

Fig. 1

than $\frac{1}{2}$ to $\frac{1}{3}$ per cent. If, however, the quantities of radium differ largely in amount—for example, in the ratio of 10 to 1—the ionisation currents in the electroscope for a given distance of the radium from the electroscope differ in this ratio, and the comparison cannot be made with the same accuracy as in the case of nearly equal quantities. The accuracy of the method obviously depends upon the certainty of complete saturation of the ionisation in the electroscope in the two cases. Unless a special investigation be made to determine the variation of current with voltage under the experimental conditions it is difficult to be certain that saturation has been reached. Any method of comparison of the intensity of the γ-rays from two preparations by the direct measurement of the ionisation they produce is subject to the same limitations.

In order to avoid this difficulty the writers have developed a balance method for accurately comparing quantities of radium by their γ-ray effects. The general principle of the method is clearly seen from Fig. 1. The γ-rays from the radium preparation at R enter the ionisation chamber A. The saturation ionisation current in A is balanced against an equal but opposite ionisation current due to a standard preparation of uranium oxide in the vessel B. The accuracy of the balance is tested by a Dolezalek electrometer. In practice, the ionisation current due to B is kept fixed, and the standard radium preparation is moved until a balance is obtained. The standard is then removed to a distance, and a balance is again obtained for the second preparation. Since the intensity of the γ-rays from a point source of radium fall off nearly inversely proportional to the square of the distance, the relative γ-ray activity can be deduced from the distances of the radium from the vessel A in the two cases.

The construction of the ionisation chamber is shown in Fig. 2A. It consists

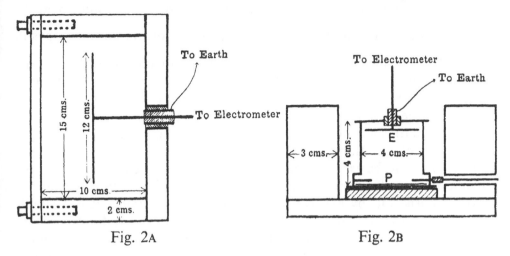

Fig. 2A   Fig. 2B

of a lead cylinder 2 cm. thick 10 cm. long and 15 cm. internal diameter. One end of the cylinder is closed by a lead plate 2 cm. thick, the other, where the γ-rays enter, by a lead plate whose thickness can be varied between 1 mm. and 2 cm. of lead. The plate ordinarily employed is of thickness 1 cm. The interior electrode consists of a circular aluminium plate 1 mm. thick and 12 cm. diameter. This is insulated from the cylinder by a sulphur stopper, provided with a guard ring in the usual manner.

The construction of the uranium vessel B is shown in Fig. 2B. It consists essentially of a brass cylinder 4 cm. high and 4 cm. diameter. On the bottom is placed a metal plate, P, which is covered with a uniform film of uranium oxide prepared by McCoy's method. Above the plate is an iris diaphragm of the type used in microscopes. The aperture of the diaphragm could be varied by means of an attachment between 0·5 cm. to 3 cm. By this means the ionisation in B can be varied over a considerable range. The base of the vessel is screwed on to an ebonite block which is fastened to a lead plate 1 cm. thick. The vessel is protected from external γ-rays by surrounding it by a lead cylinder 6 cm. high and 3 cm. thick. The interior electrode E is a circular copper plate insulated by a sulphur stopper provided with a guard ring. The electrode in A (Fig. 1) is connected with the electrometer by a wire passing through a thick lead tube, T, exhausted by means of a water pump to a pressure of about 2 cm. In this way the ionisation of the air surrounding the wire due to an external source of γ-rays was made very small. A was kept charged to −120 volts and B to +200 volts. The electrometer needle was charged through a phosphor bronze suspension to 200 volts.

The radium preparation was mounted on the carriage of a photometer bench, symmetrically in regard to the face of the ionisation chamber, and was moved along the graduated bench until a balance was obtained. Suppose r_1 is the distance of the balance point from the front surface of the air in A, and $r + a$ the distance from the surface of the back plate, a being the depth of the ionisation chamber. The intensity of the γ-rays from the radium at a distance, r, from the source within the ionisation chamber is proportional to $\dfrac{e^{-\mu_1 d} \cdot e^{-\mu r}}{r^2}$, where μ_1 is the average coefficient of absorption of the γ-rays in the lead plate of thickness d and μ the absorption of the γ-rays by the air. The actual ionisation observed in the chamber A arises in all probability mainly from the β-rays set up by the γ-rays on the front and back plates. Our knowledge of the amounts and distribution of these excited β-rays is at present too inadequate to calculate the intensity of the radiation at each point of the ionisation chamber with certainty. If, however, the distance a is small compared with r it is to be expected that the ionisation in the chamber of depth a should be very nearly proportional to $1/r(r + a)$. It will be seen later that this relation is borne out by experiment. Assuming this formula, it is a simple matter to compare the γ-ray activities of two quantities of radium by determining the distances r_1, r_2 required for a balance. Suppose the standard preparation containing S milligrammes of radium balances at a distance r_1 from the inner surface of the front plate, and the test preparation containing R milligrammes of radium balances at a distance r_2. Since when a balance is obtained, the ionisation in the chamber is equal for both preparations, and the absorption by the lead is the same in both cases—

$$\frac{R e^{-\mu r_2}}{r_2(r_2 + a)} = \frac{S e^{-\mu r_1}}{r_1(r_1 + a)}$$

or

$$R/S = \frac{r_2}{r_1} \cdot \frac{r_2 + a}{r_1 + a} e^{-\mu(r_1 - r_2)}.$$

For an accurate determination it is necessary to know the absorption of the γ-rays by air. This has been determined as accurately as possible by one of the authors, and an account of the experiments is given in an accompanying paper. If the γ-rays pass through 3 mm. of lead before entering the ionisation chamber the value of μ for air at normal temperature and pressure is given by $\mu = 0 \cdot 0000624$ (cm.)$^{-1}$, and $\mu = 0 \cdot 000059$ (cm.)$^{-1}$ at standard pressure and $15°$ C. Suppose, for example, the preparation R balances at a distance 1 metre beyond the standard preparation S at a temperature $15°$ C., then $e^{-\mu(r_1 - r_2)} = e^{0.0059} = 1 \cdot 0059$. The correction for absorption is thus $0 \cdot 59$ per cent.

Since the absorption is certainly correct within 5 per cent., the maximum error introduced by the correction in the case considered is not greater than 3 parts in 10,000.

If the standard and test preparations are contained in glass tubes of unequal thickness, it is, of course, necessary to correct for the absorption of the γ-rays in glass, and in some cases by the radio-active preparation itself. This latter correction is in general small, and can easily be applied if the weight and nature of the material are known.

It has been found that when the γ-rays have passed through 1 cm. of lead the coefficient of absorption of the γ-rays in ordinary soda glass is given by $0·117$ (cm.)$^{-1}$. The correction to be applied for $\frac{1}{10}$ mm. extra thickness of wall is consequently $0·12$ per cent., and for an extra thickness of 1 mm. $= 1·2$ per cent. In properly constructed standards contained in thin-walled glass tubes the correction is in general a fraction of a per cent.

It is, of course, necessary that the ionisation in the small vessel B, due to the γ-rays from the radium, should be negligibly small in comparison with the ionisation due to the uranium. This can readily be arranged by making the volume of B small compared with that of A. The ratio of volume of B to A used was 1 to 35. In addition, the γ-rays had to penetrate 5 cm. of lead before entering the ionisation vessel B. The relative effect of the γ-rays on the ionisation of B is clearly greater the larger the quantity of radium under comparison. In the experiments the greatest distance of the balance point from A was about 3 metres, and the distance between A and B was 4 metres. Taking the various factors into consideration, it can be simply calculated that the maximum ionisation due to the γ-rays in B is not more than $\frac{1}{3000}$ of the ionisation due to the uranium, and may consequently be neglected.

Test of Distance Law.—We have seen that the ionisation in the vessel A should vary approximately as $\dfrac{e^{-\mu r}}{r(r + a)}$, where r and a have been already defined. The accuracy of this relation has been tested experimentally in the following way: Four small glass tubes were each filled with about 20 milli-curies of emanation. After radio-active equilibrium was reached, the balance was obtained for each of the four tubes separately and then all together. Correction was made for the decay of the emanation, and consequently of the γ-ray activity during the interval of measurement. Since the time for the five observations was about one hour, the correction for the decay of the emanation was small, and could be made with accuracy. If r_1, r_2, &c., be the distance of balance for the four tubes separately, and r for the four tubes together, the values $\Sigma r_1(r_1 + a)e^{\mu r_1}$ for the four tubes should equal $r(r + a)e^{\mu r}$, if the distance law assumed is correct.

The agreement between the values is shown in the following table for a number of separate experiments. In some cases the ionisation in B was varied; in others, the ionisation was kept constant, but the distance of balance from A became smaller owing to decay of the emanation with time.

At 45 cm. from A the value corresponding to the second column was found to be $0·4$ per cent. less than in the third. At this distance, therefore, the deviation from the formula becomes appreciable. Provided the balance is not obtained for distances less than 60 cm., for the particular apparatus employed, the formula is seen to agree with experiment to 1 part in 400.

Approximate distance of single tubes from ionisation chamber	$\Sigma r_1(r_1 + a)e^{\mu r_1}$ for four tubes	$r(r + a)e^{\mu r}$ for four tubes together
120 cm.	66,860	66,740
,,	66,090	65,900
,,	65,550	65,400
110 cm.	52,140	52,170
,,	51,210	51,170
70 cm.	22,910	22,860
,,	19,550	19,580
62 cm.	18,490	18,440

Another simple method of testing the distance law is to compare the ratio of the γ-ray activity of two radium preparations which differ in considerable ratio—1 to 4, for example—for different values of the ionisation in B. If the distance law is correct the ratio of the γ-ray activities of the two preparations should agree whatever the distance of balance.

Advantages of this Method.—The balance method of comparing γ-ray activities has numerous advantages. In the first place, since the ionisation in the testing vessel is a constant when a balance is obtained, no error due to possible lack of saturation arises. There is no correction to be applied for natural leak provided it does not change during the time of observation. Since comparisons are made when the radium is at considerable distance from the ionisation chamber, the dimensions or shape of the tube containing the radium has little influence, and the correction for distribution is, in most cases, negligible.

In consequence of this the γ-ray activity of radium preparations contained in vessels of different size and shape can be readily compared. If the actual quantity of radium in the tube is to be determined in terms of the standard, correction, of course, must be made for the absorption of the γ-rays in the walls of the containing vessel and in the radio-active material itself.

The method is an ideal one for comparison of nearly equal quantities of radium, such as, for example, will be used in comparing duplicate standards in terms of the primary. Under such conditions. the balance distances are practically the same for the two preparations. Possible errors in the distance law assumed or in the amount of absorption of γ-rays by air do not appreciably affect the accuracy of the comparison. If the two standards to be compared are enclosed in tubes of equal thickness and contain radium of about equal purity, it will be a simple matter to determine the relative value of the standards to 1 part in 1,000 or even closer.

We have seen that the distance law breaks down for a distance smaller than 50 cm. If, however, the apparatus be required to determine small quantities of radium—*e.g.* $\frac{1}{10}$ milligramme—it will be necessary either to reduce the ionisation in B or to calibrate the apparatus for smaller distances. The latter can be done by observing the alteration of distance of balance for a tube containing emanation as it decays with time. Assuming the period of decay of the emanation (3·85 days) a fairly accurate calibration curve can be obtained.

The investigations of J. A. Gray* have shown that γ-rays are excited by the passage of β-rays through matter. The source of β-rays in his experiments was radium D + radium E. The authors have found by the balance method that the β-rays from radium in falling on matter also give rise to γ-rays, but in amount small compared with the primary γ-rays from the active matter itself. Since the thickness of glass tubes enclosing the radium preparation is in general not sufficient to absorb all the β-rays, the escaping β-rays no doubt produce some γ-rays in the surrounding air. This possible error in the measurements can be corrected for by surrounding the radium tube with a sufficient thickness of a substance of small density, like glass or aluminium, to absorb all the β-rays. Gray has shown that the excitation of γ-rays is small for light substances, and is approximately proportional to the atomic weight of the material.

Determination of Quantities of Emanation.—The amount of emanation in equilibrium with 1 gramme of radium is called 1 curie; from 1 milligramme of radium, 1 millicurie. It is often necessary to determine the amount of emanation stored in a tube by itself. This is simply done by comparing the γ-ray activity of the emanation tube with that due to a standard quantity of radium. Knowing the period of decay of the emanation, the amount of emanation at any time can be deduced.

In comparisons of this kind it is necessary to apply a correction. The γ-rays do not arise from the emanation itself, but from its subsequent products, radium B and radium C. Five or six hours after introduction of the emanation into a tube, a state of changing equilibrium exists between the emanation and its products, and the γ-ray activity then decays at the same rate as the emanation. It is clear, however, from general considerations that the amount of radium B or radium C present at any time must be always greater than corresponds to true equilibrium with the emanation. The theoretical correction can be easily determined. Let $\lambda_1, \lambda_2, \lambda_3, \lambda_4$ be the radio-active constants of the emanation, radium A, B and C respectively. If P_0 atoms of emanation are initially present, the amount S of radium C at a time, t, later is given† by

$$S = c_1 e^{-\lambda_1 t} + c_2 e^{-\lambda_2 t} + c_3 e^{-\lambda_3 t} + c_4 e^{-\lambda_4 t},$$

where $c_1 = \dfrac{\lambda_1 \lambda_2 \lambda_3 P_0}{(\lambda_2 - \lambda_1)(\lambda_3 - \lambda_1)(\lambda_4 - \lambda_1)}$, $c_2 = \dfrac{\lambda_1 \lambda_2 \lambda_3 P_0}{(\lambda_1 - \lambda_2)(\lambda_3 - \lambda_2)(\lambda_4 - \lambda_2)}$

and so on.

---

* Gray, 'Proc. Roy. Soc.' A. **85**, p. 131, 1911.
† *See* Bateman, 'Proc. Camb. Phil. Soc.,' 15, p. 423, 1910.

Since the values of $\lambda_2, \lambda_3, \lambda_4$ for radium A, B and C are large compared with λ_1, after five or six hours the terms in $e^{-\lambda_2 t}$, $e^{-\lambda_3 t}$, $e^{-\lambda_4 t}$ become vanishingly small, and $S = c_1 e^{-\lambda_1 t}$.

At the instant t the amount S_0 of radium C which would be in equilibrium with the amount P of the emanation, kept constant by supply from radium, is given by $S_0 = \lambda_1 P_0 e^{-\lambda_1 t}$.

Consequently
$$S/S_0 = \frac{\lambda_2 \lambda_3 \lambda_4}{(\lambda_2 - \lambda_1)(\lambda_3 - \lambda_1)(\lambda_4 - \lambda_1)},$$
$$= 1 \cdot 0089,$$

when the half value periods of emanation radium A, B and C are taken as $3 \cdot 85$ days, 3 minutes, $26 \cdot 8$ minutes, $19 \cdot 5$ minutes respectively.

This shows that the amount of radium C is always $0 \cdot 89$ per cent. *greater* than corresponds to permanent equilibrium with an equal quantity of emanation.

After the γ-radiation has passed through 2 cm. of lead, the γ-rays arise entirely from radium C. If the γ-rays from the emanation tube are compared with the γ-radiation from a radium standard through 2 cm. of lead, the amount of emanation in the tube is $0 \cdot 89$ per cent. *less* than that deduced from direct comparison with the γ-ray effect for the radium standard.

If the γ-rays pass through 3 mm. of lead, Moseley and Makower have shown that about $11 \cdot 5$ per cent. of the γ-radiation arises from radium B and $88 \cdot 5$ per cent. from radium C. The correction in this case is somewhat smaller, but can readily be shown to be about $0 \cdot 85$ per cent.

The amount of radium C present in a vessel after removal of the emanation is often determined by comparison of its γ-ray effect after a definite interval with that due to a radium standard. Correction in this case has to be made for the γ-rays emitted from radium B, unless the γ-rays from the latter are removed by using a lead screen 2 cm. thick. Numbers showing the decay with time of the γ-ray activity of the active deposit of radium have been given for one important case by Moseley and Makower (*loc. cit.*).

Experiments with Balance Method.—The decay of the γ-ray activity of a large quantity of emanation was determined over several weeks by the balance method. The decay was found to be exponential with a half value period of $3 \cdot 854$ days. This is in excellent agreement with the results of Mme. Curie and Rutherford, who found values of $3 \cdot 846$ and $3 \cdot 85$ days respectively.

Some experiments were also made to determine experimentally the time required for the γ-ray activity of a tube filled with emanation to attain a maximum. The exact time was noted when emanation was introduced into a small glass tube and sealed off. The amount of γ-ray activity was accurately measured at intervals between 2 and 6 hours later.

The activity rose rapidly at first and was within $1 \cdot 5$ per cent. of the maximum after 3 hours. Four separate determinations were made and the maximum through a lead screen 3 mm. thick was found to occur at 255, 254, 256 and 257 minutes respectively after filling the tube. In this case, about 11·5 per cent. of the γ-radiation was supplied by radium B. Taking this factor into account, it was calculated that the maximum should occur at 255 minutes, an interval agreeing with the mean experimental value.

Using this balance method experiments were made to test whether the emission of γ-rays was affected by placing a tube containing emanation between the pole pieces of a large electromagnet. No detectable effect was observed.

ABSTRACT

A balance method is described for accurately comparing quantities of radium by their γ-radiation, in which the ionisation due to the γ-rays is balanced against the constant ionisation due to uranium oxide. By observing the distance of the radium preparation from the ionisation chamber when a balance is obtained, the relative γ-ray activities of the two preparations can be determined with an accuracy of at least one part in 400. A method of calibration is given, and the corrections necessary to deduce the quantities of radium present are considered.

Calculations are given showing the correction required to determine a quantity of emanation by comparison of its γ-ray activity with that due to a radium standard.

The balance method was employed to determine the period of transformation of the radium emanation. The half value period was found to be 3·854 days. It was found experimentally that the γ-ray activity due to the radium emanation and its products reaches a maximum 255 minutes after the introduction of the emanation into a sealed tube. Calculations showed that under the experimental conditions the theoretical maximum should be reached after 255 minutes.

It was found that the γ-ray activity of a radium preparation was not appreciably altered by exposure in a strong magnetic field.

UNIVERSITY OF MANCHESTER
February, 1912

The Origin of β and γ Rays from Radioactive Substances

by E. RUTHERFORD, F.R.S.,

*Professor of Physics in the University of Manchester**

From the *Philosophical Magazine* for October 1912, ser. 6, xxiv, pp. 453–62

FROM a study of the α radiations from active matter, it has been found that each atom of a substance in disintegrating emits one α particle which is expelled with a definite velocity and with a range in air characteristic of that substance. The only exception is the product thorium C, which emits two distinct groups of α rays, each of definite but different range in air. In this case, the atom appears to break up in two distinct ways.

In many transformations β and γ rays are emitted, and, from analogy with the α ray transformations, it would be expected that one β particle of definite speed would be emitted for the disintegration of each atom. The experiments, however, of v. Baeyer, Hahn and Miss Meitner,[†] and later of Danysz,[‡] have shown that the emission of β rays from a radioactive substance is in most cases a very complicated phenomenon. The complexity of the radiation is most simply shown by observing the deflexion of a narrow pencil of β rays by a magnetic field in a vacuum. If the rays fall on a photographic plate, a number of sharply marked bands are observed, indicating that the rays are complex and consist of a number of homogeneous groups of rays, each of which is characterized by a definite velocity.

This complexity of the radiation is best shown by those products which emit penetrating β rays and intense γ rays; for example, each of the products thorium D and mesothorium 2 emits a number of well defined groups of β rays and penetrating γ rays. The complexity of the β rays is, however, most markedly exhibited in the case of the products radium B and radium C, when a very strong source of radiation is employed. Using as a β ray source a thin-walled glass tube containing a large quantity of radium emanation, Danysz found that radium B and radium C together emitted about 30 groups of homogeneous rays.

Notwithstanding this great complexity of the β rays from these products, general experiment has shown that the number of β particles emitted by them is about that to be expected if each atom in breaking up emitted only one β

* A preliminary paper on this subject was read before the Manchester Literary and Philosophical Society, 1912.

† v. Baeyer and Hahn, *Phys. Zeit.* xi. p. 488 (1910); v. Baeyer, Hahn and Meitner, *Phys. Zeit.* xii. pp. 273, 378 (1911); xiii. p. 264 (1912).

‡ Danysz, *C. R.* cliii. pp. 339, 1066 (1911); *Le Radium*, ix. p. 1 (1912).

particle. This important point has been carefully examined by H. Moseley,*
who has shown that not more than $2 \cdot 13$ β particles are emitted during the
disintegration of one atom of radium B and one atom of radium C. By
separating these two products, Moseley found that the atom of each product
contributes about half of this number. It thus seems clear that on an average
one atom in disintegrating emits about one β particle.

In addition to the well-known β ray products, Baeyer, Hahn and Meitner
have shown by the photographic method that radium itself and radium D
emit a weak β radiation which consists in each case of two definite groups of
β rays. In these cases, no evidence of the emission of a γ radiation has yet
been observed.

There appears to be no doubt that the γ rays from active matter are closely
connected with the β rays, and that both types of rays arise in the transformation of the same atom. Investigation, however, has shown that there is
not any obvious relation between the relative intensity of the β and γ rays
which are emitted from a given product. The products radium C, thorium D,
and mesothorium 2 emit β and γ rays of about the same penetrating power
and in about the same relative proportion. On the other hand, uranium X,
which emits penetrating β rays, gives relatively few γ rays. A still more
striking instance is the β ray product radium E, which, as Gray† has shown,
emits relatively an exceedingly weak γ radiation. It may prove significant that
only those products which emit well-defined groups of β rays emit also a
strong γ radiation; for, as far as observation has gone, the β rays from
uranium X and radium E give a continuous spectrum in a magnetic field.

In order to account for the emission of groups of homogeneous rays from
a single product, it is necessary to suppose either that the atom breaks up in
a number of distinct ways, each of which is characterized by the emission of
β particles of definite velocity, or that the β rays are altered in velocity in some
definite way during their escape from the disintegrating atom. On the first
hypothesis, it might be anticipated that the different modes of transformation
of a β ray product would give rise to a series of new products, but only one is
observed. In addition, the energy emitted during the transformation from one
type of matter into another would vary widely for different atoms of the same
substance, and this seems improbable. On the second hypothesis, it is
supposed that the disintegration of each atom takes place in exactly the same
way with the emission of the same amount of energy, but that the energy of
the β particle may be decreased by definite but different amounts due to
transformations of its energy in its passage through the atomic system from
which it originates. Since it is known that β rays in escaping from an atom
give rise to γ rays, it is natural to suppose that the loss of energy of the
β particle in escaping from the atomic system is connected in some way with
the excitation of γ rays.

* Moseley, 'The number of β particles emitted in the transformation of radium.' Read
Roy. Soc. June 13, 1912.

† Gray, Proc. Roy. Soc. A. lxxxv. p. 131 (1911); lxxxvi p. 513 (1912).

The work of Barkla and others on the X rays have brought out clearly that under suitable conditions of excitation, each element emits one or more definite types of X radiation which are characteristic of the element. Barkla* has determined the coefficient of absorption μ in aluminium of the characteristic X rays for elements up to atomic weight 140. The value of μ/D for aluminium, where D is the density, decreases rapidly with the atomic weight and varies between 435 for calcium of atomic weight $40\cdot1$ to $0\cdot6$ for cerium of atomic weight $140\cdot25$. Plotting the logarithms of the values of μ/D against the logarithm of the corresponding atomic weights, the points for the heavier elements are found to lie nearly on a straight line. If this straight line be produced, it can be shown by extrapolation that an element of atomic weight 214 should emit a characteristic radiation whose value of $\mu/D = 0\cdot04$ about for aluminium. This is in good accord with the value $\mu/D = 0\cdot0406$ for aluminium found by Soddy and Russell for the penetrating γ rays from radium C. It would thus appear probable that the penetrating γ radiation from radium C is to be regarded as the characteristic radiation of that element excited by the escape of β particles from the atomic system.

Whiddington† has shown that the β or cathode particle incident on matter must have a definite minimum speed for each element before the characteristic radiation of the latter is excited. Over the range examined, this velocity is directly equal to $A \times 10^8$ cm. per second, where A is the atomic weight of the element. If it be supposed that this law holds generally for all the elements, the velocity of the β particle required to excite the characteristic radiation of radium C of atomic weight 214 should be $0\cdot71$ of the velocity of light. It is not improbable, however, that the energy $\frac{1}{2}mu^2$ rather than the velocity u is the determining factor for high-speed β particles. Taking this into consideration, the velocity required to excite the characteristic radiation is given by $u = 10^8 A\sqrt{m_0/m}$, where m_0 is the mass of the β particle for slow speeds. On this hypothesis, $u = 0\cdot63$ of the velocity of light. The mean value for the two methods of calculation gives $u = 0\cdot67$. The corresponding energy of the β particle for the mean value of u is $1\cdot5 \times 10^{13}e$ ergs, where e is the charge carried by the β particle. If it be supposed that the whole energy of the β particle is converted into γ radiation, the energy absorbed in exciting the characteristic γ ray should be $1\cdot5 \times 10^{13}e$ ergs.

If the complexity of the β radiation is connected with the emission of γ rays, it is to be expected that some definite relation should exist between the energies of the β particles in each of the groups emitted. From this point of view, it is of interest to examine whether there is any evidence of such a relation for the β rays emitted by radium B and C. The results given by Danysz are included in the following table.

The various groups differ widely in photographic intensity. This is designated in column 1 by the symbols s = strong, m = mean, f = feeble,

---

* Barkla, Phil. Mag. xxii. p. 396 (1911).
† Whiddington, Proc. Roy. Soc. A. lxxxv. p. 323 (1911).

vf = very feeble. Each group is defined by a number given in the second column; for the sake of completeness, two groups of rays of low velocity A and B observed by Hahn are added to those given by Danysz. The third column gives the value of Hρ experimentally observed; the fourth column β

Intensity	No.	Hρ	β	Energy	Intensity	No.	Hρ	β	Energy
	A	Hahn	·36	$0 \cdot 353 \times 10^{13}e$.	f	11	2190	·790	$2 \cdot 595 \times 10^{13}e$.
	B		·41	0·468 ,,	s	12	2870	·862	3·711 ,,
s	1	1320	·615	1·218 ,,	f	13	3140	·882	4·155 ,,
f	2	1390	·634	1·322 ,,	f	14	3420	·897	4·60 ,,
vf	3	1490	·660	1·475 ,,	f	15	4000	·920	5·52 ,,
s	4	1580	·682	1·616 ,,	s	16	4670	·940	6·585 ,,
f	5	1680	·703	1·771 ,,	f	17	4800	·943	6·79 ,,
vf	6	1750	·718	1·885 ,,	f	18	4980	·946	7·07 ,,
s	7	1830	·735	2·017 ,,	m	19	5100	·949	7·26 ,,
m	8	1900	·748	2·132 ,,	s	20	5700	·957	8·18 ,,
f	9	1970	·760	2·246 ,,	f	21	5990	·962	8·63 ,,
s	10	2150	·786	2·535 ,,	complex	22	11200	·988	16·6 ,,
					complex	23	18100	·996	27·0 ,,

the ratio of the velocity of the β particle to the velocity of light, calculated from the Lorentz-Einstein formula. In the fifth column I have added the value of the energy of the β particle. The value of e/m for the β particle is taken as $1 \cdot 772 \times 10^7$ e.m. units.

Starting from group No. 21, the differences between the energies of the individual β particles comprising the different groups are shown in column 2 of the following table:—

Number of group	Observed difference in energy	$pE_1 + qE_2$	Calculated difference in energy
(21)—(20)	$0 \cdot 45 \times 10^{13} e$	E_1	$0 \cdot 456 \times 10^{13} e$
,, —(19)	1·37 ,,	$3E_1$	1·37 ,,
,, —(18)	1·56 ,,	E_2	1·56 ,,
,, —(17)	1·84 ,,	$4E_1$	1·82 ,,
,, —(16)	2·05 ,,	$E_1 + E_2$	2·01 ,,
,, —(15)	3·11 ,,	$2E_2$	3·11 ,,
,, —(14)	4·03 ,,	$2E_1 + 2E_2$	4·02 ,,
,, —(13)	4·48 ,,	$3E_1 + 2E_2$	4·48 ,,
,, —(12)	4·92 ,,	$4E_1 + 2E_2$	4·94 ,,
,, —(11)	6·03 ,,	$3E_1 + 3E_2$	6·03 ,,

On examining these differences, it is seen that they can be expressed closely by the relation $pE_1 + qE_2$, where $E_1 = 0 \cdot 456 \times 10^{13}e$, $E_2 = 1 \cdot 556 \times 10^{13}e$, and p and q are whole numbers which may have any values 0, 1, 2, 3, &c. The differences calculated on this hypothesis are shown in the last column, and are observed to be in close agreement for the whole series of lines from No. 21 to No. 11. This relation does not hold below line No. 11, but in all probability most of the lines, Nos. 1 to 10, belong to radium B and not to radium C. The energy of the β particle for group No. 21 is $8 \cdot 63 \times 10^{13}e$, while its velocity is $0 \cdot 962$ of the velocity of light. Groups 22 and 23 are not included in the calculation, for Danysz states that No. 22 includes from 3 to 5 groups of β rays, of which only the average velocity is given; similarly No. 23 is considered to be a complex group.

The values of the velocity of the β particles will require to be known with great accuracy before such a relation as is indicated can be definitely established; but it does not appear likely that the connexion observed is accidental. It is of interest to note that the value of $E_2 = 1 \cdot 556 \times 10^{13}e$ is in fair accord with the calculated energy of the β particle, viz., $1 \cdot 5 \times 10^{13}e$, which would be required to excite the characteristic radiation from radium C. The value of E_1 may in a similar way be connected with the energy required to excite the second type of characteristic X radiation which has been observed in a number of elements by Barkla.

It is possible that some of the groups from Nos. 10 to 1 may also belong to radium C. For example, No. 8 fits in well with a difference $4E_1 + 3E_2$ from No. 21. Hahn has determined the velocity of the stronger groups of rays from radium B and radium C separately. For radium B, the values $\beta = 0 \cdot 36, 0 \cdot 41, 0 \cdot 63, 0 \cdot 69, 0 \cdot 74$ are given. The last three no doubt correspond to groups Nos. 1, 4, and 7 respectively given by Danysz. A group for which $\beta = 0 \cdot 80$, which appears to correspond to No. 10, is ascribed to radium C. Group No. 10, however, does not fit in at all with the relation found between Nos. 11 to 21. It would be of great value to determine definitely the division of the groups of β rays observed between radium B and radium C.

If the groups Nos. 10 to 1 supposed to belong to radium B be analysed in a similar way to the groups for radium C, the differences may be approximately expressed by the relation $pE_1 + qE_2$ where $E_1 = 0 \cdot 114 \times 10^{13}e$ and $E_2 = 0 \cdot 144 \times 10^{13}e$. The agreement, however, between calculation and theory is not nearly so good as for the case previously considered, and it is doubtful whether any weight can be attached to it. For the slow velocity β rays here considered, the reduction of velocity in passing through the glass walls of the emanation tube is quite appreciable, and the correction is different for each group. Until this correction is made, it does not seem possible to draw any definite conclusions.

The simplest way of regarding this relation between the groups of β rays is to suppose that the same total energy is emitted during the disintegration of each atom, but that the energy is divided between β and γ rays in varying proportions for different atoms. For some atoms most, if not all, of the energy is emitted in the form of a high-speed β particle; in others the energy of the β particle is reduced by definite but different amounts by the conversion of part of its energy into γ rays. Suppose, for example, the total energy liberated in the form of β and γ rays during the transformation of one atom is E_0. If the β particle before it escapes from the atom passes through two regions where the energy required to excite a γ ray is E_1 and E_2 respectively, the resulting energy of the β particle is $E_0 - (pE_1 + qE_2)$, where p and q are whole numbers corresponding to the number of γ rays excited in each region. The energy emitted in the form of γ rays is $pE_1 + qE_2$, and p γ rays of energy E_1 and q of energy E_2 appear.

According to this view, the transformation of one atom gives rise to only one β ray, but to p γ rays of one kind and q of another. The groups of homogeneous β rays observed are the statistical effect due to a large number of disintegrating atoms. The relative distribution of β particles amongst the numerous groups of homogeneous rays will depend on the probability that 0, 1, 2, &c. of the units of energy E_1 and E_2 are abstracted from the β particle in traversing the atom.

This mode of regarding the connexion between β and γ rays suggests that the number of γ rays emitted from radium C is considerably greater than the number of β rays. Assuming that each γ ray from radium C was converted into one β ray, Moseley (*loc. cit.*) found that at least two γ rays appeared for the transformation of one atom of radium C. There is reason to believe that this is a minimum estimate, and that the actual number is two or three times greater.

From the results already considered, it does not necessarily follow that group No. 21 is to be regarded as the head of the β ray series. Evidence on this point can be obtained by calculating the energy E_0 liberated per atom in the form of β and γ rays during the transformation of radium C. In some recent experiments the results of which have been communicated to the Vienna Academy, the writer and Mr. H. Robinson conclude that the heating effect of the β and γ rays from 1 gram of radium is $10 \cdot 8$ gram calories, of which about $4 \cdot 3$ is due to the β rays and $6 \cdot 5$ to the γ rays. An uncertain part of this energy arises from the β and γ rays emitted by radium B; but there will not be much error if it be supposed that the energy of the β and γ rays from radium C in one gram of radium is about 8 gram calories per hour. Supposing, as Moseley found, that one β ray is on the average expelled from each atom of radium C, the energy emitted per atom of radium C in the form of β and γ rays is $17 \cdot 8 \times 10^{13}e$ ergs, assuming that $3 \cdot 4 \times 10^{10}$ β particles per second are emitted from radium C in equilibrium with one gram of radium. This approximately corresponds with the average energy of the β particle included in group No. 22, viz. $16 \cdot 6 \times 10^{13}e$.

Danysz states that No. 22 consists of 3 to 5 groups of β rays, for which only the average velocity of the group is given. In the absence of any definite information of the velocity of the components, it is impossible to ascertain whether any relationship exists between this complex group and the rest of the radium C series. The wide difference between the energies of the β rays included under No. 21 and No. 22 indicates that possibly a third region exists within the atom for which the energy required to excite γ rays is much greater than that for the other two regions considered.

Unless the energy of the β and γ rays from radium determined by experiment has been much underestimated, it does not seem possible to suppose that the swiftest β ray given by Danysz, which has the energy $27 \times 10^{13}e$, can be the head of the β ray series. The existence of such a swift group of β rays is, however, open to some doubt, as Danysz expressly states in his paper. The photographic effect of such swift β rays is very difficult to detect in the presence of a strong photographic action due to the γ rays.

We have so far confined our discussion to the connexion between the β and γ rays emitted from radium C, for in this case the necessary data are far more definite and complete than for any other product. It seems probable, however, that the same general explanation will apply to the emission of β and γ rays from mesothorium 2 and thorium D, both of which emit a number of groups of homogeneous β rays and also penetrating γ rays. In each of these products, the energy emitted in the form of γ rays is of about the same order of magnitude as the energy emitted in the form of β rays.

A difficulty arises in connexion with the β ray products like radium E and uranium X, which emit penetrating β rays but relatively weak γ rays. In the case of uranium X, some penetrating γ rays are observed, but they are weak in relative intensity compared with the γ rays from radium C. It is possible that the atomic structure of uranium X is such that only an occasional β particle loses energy by conversion into γ rays in its escape from the atom. In the case of radium E where the γ rays are very weak in intensity and of slight penetrating power, it seems probable that the β rays originate near the surface of the atom, and consequently do not traverse the regions where penetrating γ rays can be set up. There still remains the difficulty, however, of accounting for the heterogeneity of the β rays which are emitted, to which attention has been drawn by Gray and by Gray and Wilson.

There is one point of interest which has so far not been considered. Bragg has given strong evidence for believing that a β and a γ ray are mutually convertible forms of energy. The energy of a γ ray incident on matter is transformed into the energy of a β ray, and *vice versa*. On this view, it has generally been supposed that the whole of the energy of one γ ray is converted into the energy of one β ray, so that the γ ray disappears and the β particle takes its place. From the point of view outlined in this paper, it is supposed that the β ray originating in the transformation of an atom loses only part of its energy which is abstracted from it in definite units, depending on the region of the atom through which the β particle passes. A swift β ray may consequently give rise to several γ rays in escaping from the atom and yet retain a part of its initial energy.

In a previous paper* I have given reasons for believing that the atom consists of a positively charged nucleus of very small dimensions, surrounded by a distribution of electrons in rapid motion, possibly of rings of electrons rotating in one plane. The instability of the atom which leads to its disintegration may be conveniently considered to be due to two causes, although these are not mutually independent, viz., the instability of the central nucleus and

* Rutherford, Phil. Mag. xxi. p. 669 (1911). (*This vol., p. 238.*)

the instability of the electronic distribution. The former type of instability leads to the expulsion of an α particle, the latter to the appearance of β and γ rays. The instability which leads to the expulsion of a β ray may be mainly confined to one of the rings of concentric electrons, and leads to the escape of a β particle from this ring with great velocity. The β particle in escaping from the atom passes through the electronic distribution external to it, and in traversing each ring may lose part of its energy in exciting one or more γ rays which have a definite energy, which is characteristic for each ring.

At present we have no definite information of the mode in which the transformation of a β into a γ ray or a γ ray into a β ray takes place, but it is no doubt connected with the structure of the ring of electrons, and possibly with its period of free vibration. The general evidence indicates strongly that the transformation of energy from the γ ray form to the β ray form or *vice versa* takes place in definite units which are characteristic for a given ring of electrons but vary from one to the other. The transformation of energy of the β ray form into the γ ray form appears to take place far more efficiently during the disintegration of an atom than when β rays fall on the atoms of ordinary matter. This is not unexpected, for the conditions in the former case are eminently favourable to the conversion, since the β particle passes directly through the electronic distribution of the atom. It is at the same time to be expected that some of the energy of the γ rays which are formed within the atom should also be converted, in part at least, into β rays again, and should thus give rise to one or more groups of homogeneous rays. It is possible that two groups of slow velocity β rays which appear during the transformation of radium itself may arise in consequence of such a transformation of the γ rays set up by the escape of the α particles from the atom. J. Chadwick, working in the laboratory of the writer, has recently obtained evidence that α rays are able to excite γ rays in falling on ordinary matter.

If this be the case, it is to be expected that all α ray products should emit some β rays and γ rays, though probably of very weak intensity in both cases. The excitation of γ rays by α rays is not improbable if the energy of the charged particle rather than its velocity is the determining factor; for although the velocity of the α particle is small compared with that of the ordinary β particle, its energy of motion is much greater.

University of Manchester
Aug. 16, 1912

Photographic Registration of α Particles

by DR. H. GEIGER *and* PROFESSOR E. RUTHERFORD, F.R.S.,
University of Manchester

From the *Philosophical Magazine* for October 1912, ser. 6, xxiv, pp. 618–23

IN a previous paper* the authors have described an electrical method of counting the α particles from radio-active substances. For this purpose, the α particles passed through a small opening into a cylindrical vessel which contained air or another gas at a reduced pressure. The cylinder was provided with a central electrode which was connected with an electrometer, and the outer cylinder was charged to a negative potential nearly sufficient to cause the passage of a spark. Under such conditions, the small ionization produced by the α particle is magnified several thousand times by collision, and the entrance of an α particle into the testing vessel is signalized by a ballistic throw of the electrometer-needle. In order to determine the number of α particles emitted per second by one gram of radium, a source of small diameter coated with radium C was placed at some distance from the opening in an exhausted tube, so that a definite fraction of the α particles passed through the mica-covered opening of the detecting vessel.

Since the electrometer responds only slowly to the rise of potential caused by the entrance of an α particle, it was not possible to count with accuracy more than about 10 α particles per minute. Since our original experiments, several types of string electrometers have been devised. The moving part consists of a thin silvered quartz fibre placed between two charged parallel plates. The fibre follows rapidly any change of potential applied. In 1910 we used a string electrometer, devised by Laby and constructed by the Cambridge Scientific Co., for the purpose of counting α particles. The instrument was found to have sufficient sensibility, and the movement of the fibre produced by the entrance of each α particle into the detecting vessel could be simply recorded on a moving photographic film.

A number of investigations have been made from time to time in order to find the best conditions for obtaining a clear record of the individual α particles even when they enter the detecting vessel in very quick succession. The conditions for accurate counting depend on the following:—

(1) The absence of electric disturbances in the detecting vessel when no α particles enter.

(2) The equality of throws for each α particle.

* Rutherford & Geiger, Proc. Roy. Soc. A. lxxxi, p. 141 (1908). (*This vol., p. 89.*)

(3) The elimination of the effect of scattering of the α particles in the mica window and the gas in the detecting vessel.

For accurate work, it is essential that the fibre after displacement should return rapidly to its equilibrium position. To achieve this purpose, the fibre was connected through a comparatively low resistance to earth. Suitable resistances can be simply made by using capillary tubings filled with a mixture of xylol and alcohol.* The back leak so provided is ordinarily very much greater than can be used in counting with the quadrant electrometer. This large back leak has the great advantage of cutting out the great majority of slow electrical disturbances, for only rapid changes of potential make their effects visible.

The type of cylindrical vessel used in our original investigations did not fulfil condition (2) on account of scattering of the α particles by the mica-covered opening through which they entered. On account of this scattering, the α particles have different lengths of path in the gas, and produce unequal throws. The differences in the throws become less marked the smaller the length of the tube compared with its diameter. To overcome this difficulty, a detecting vessel of the type shown in fig. 1 was constructed. It consisted of a

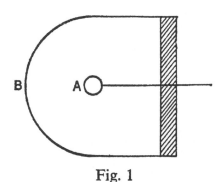

Fig. 1

metallic hemisphere, near the centre of which was a spherical electrode A supported by a metal rod. The α particles entered in the hemisphere through the opening B which was covered with a thin sheet of mica. Helium at suitable pressure was introduced into the apparatus. With this arrangement, it was found that the throws were remarkably uniform, so that it was easy to recognize by the length of the throw when two or even three α particles entered the vessel in rapid succession.

In these experiments we have used helium at a pressure of about one-third of an atmosphere as the gas in the detecting vessel. The use of helium has many advantages. In the first place, the potential difference required is much lower than with air or carbon dioxide at the same pressure. This potential was found to increase very much when small impurities were present with the

* Campbell, Phil. Mag. xxiii. p. 668 (1912).

K

helium. As is well known, however, the helium can easily be purified by passing it over charcoal cooled by liquid air. This method of purification of helium was used in all the experiments. Using helium in the detecting vessel, the necessary magnification could easily be obtained, and the magnitudes of the throws remained unchanged for hours at a time. In addition, hardly any natural disturbances were observed when the α particles did not enter. In the particular apparatus employed, the α particles travelled a distance of 3 cm. in the helium. With a pressure of 20 cm. the stopping power of the α particles in the gas corresponded to only 2 millimetres of air at atmospheric pressure. At a pressure of 20 cm. each α particle produced a large electrical effect; but even when the pressure was reduced to 5 millimetres, the entrance of each α particle could be easily detected by the movement of the fibre of the electrometer. At this latter pressure, the stopping power of the α particle corresponded to only 0·06 of a millimetre of air. In these circumstances, about 400 volts only were required to give the necessary magnification. These results bring out the enormous increase in the number of ions produced by collision under these conditions.

It is not possible to count with certainty by eye observations more than about 50 throws of the fibre per minute. The fibre moves so rapidly that it is difficult to recognize when two or more α particles enter the detecting vessel at very short intervals. It is far more certain to use a photographic method for recording the rapid movements of the fibre. For this purpose, we have found the registration apparatus constructed by Professor Edelmann very serviceable. The speed of the film could be regulated over a wide range according to the number of α particles to be registered per minute. The types of record obtained for the different rates of entrance of the α particles are illustrated in

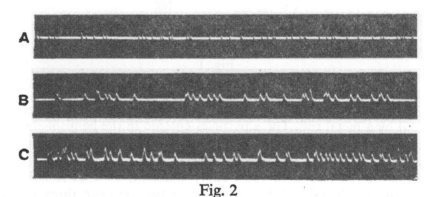

Fig. 2

fig. 2. The line A shows the part of the record obtained when the film moved at a rate of 13 cm. per minute, and for an average of 50 α particles per minute. The length shown on the photograph corresponds to an interval of about 1 minute. The line B corresponds to a movement of the film of 180 cm. per minute and for about 600 α particles per minute. The line C corresponds to an average of 900 α particles per minute.

It is seen that a clear record is obtained of the movements of the fibre even when on an average 15 α particles enter per second. With eye observations, the movements of the fibre appeared blurred and confused, but the photograph brings out clearly each individual movement. With a little experience, it is not difficult to count the number of α particles that have entered the detecting vessel in a given interval even at the rate of 1000 per minute.

The experiments described in this paper have been made with a view to determining the best conditions for measuring with accuracy the number of α particles expelled from radio-active substances, and in particular, for re-determining the number of α particles expelled per second from one gram of radium. A lengthy series of observations will be required before the value of this important constant can be definitely settled.

Recoil atoms.—It is well known from the experiments of Hahn* and of Russ and Makower,† that the emission of α particles from radioactive substances is accompanied by a vigorous recoil of the residual atoms. The recoiling atom is able to penetrate about 1/10 of a millimetre of air and of about 4/10 of a millimetre of hydrogen at atmospheric pressure, before being stopped. Wertenstein‡ has shown that these recoil atoms produce a strong ionization in the gas they traverse. This has been observed by determining the ionization due to recoil atoms and α particles at low pressures. Under these conditions, the recoil atom appears to cause several times the ionization due to an α particle over the same range.

We have seen above that it is not difficult to detect the electrical effect of a single α particle when it traverses a gas at such a low pressure that its range is only reduced by 1/20 of a millimetre. This is less than the range of the recoil atom, and it thus seems probable that the electric method could be employed to detect a single recoil atom. Some experiments were made as follows:—the mica window over the opening of the detecting vessel was removed and the radiation from a plate coated with the active deposit of actinium was allowed to enter the detecting vessel, which contained helium at a pressure of about 8 millimetres. Such an active plate causes a vigorous recoil of the atoms of actinium D. Some throws were observed large compared with those of the α particle; but the interpretation of the experiments is complicated by the natural disturbances due to the uncovered opening and to the strong ionization produced by the radiating source outside the opening. Further experiments are in progress to test whether this method can be employed to count the recoil atoms with certainty.

University of Manchester
August 16, 1912

* Hahn, *Verh. d. D. phys. Ges.* xi. p. 55 (1909).
† Russ & Makower, Proc. Roy. Soc. A. lxxxii. p. 205 (1909).
‡ Wertenstein, *C. R.* clii. p. 1657 (1911).

On the Energy of the Groups of β Rays from Radium

From the *Philosophical Magazine* for December 1912, ser. 6, xxiv, pp. 893–4

To the Editors of the Philosophical Magazine

GENTLEMEN,—

IN my paper entitled 'The Origin of the β and γ Rays from Radioactive Substances' in the October number of this Journal,* the energy E of the electron has been calculated from the formula $E = \frac{1}{2}m_0c^2 \frac{\beta^2}{\sqrt{1-\beta^2}}$, where m_0 is the mass of the electron at slow speeds, c the velocity of light, and β the ratio of the velocity of the electron to the velocity of light. This is equivalent to multiplying the 'transverse' mass of the electron by half the square of the velocity given by the Lorentz-Einstein formula. Mr. Moseley drew my attention to the fact, which I had overlooked, that according to the Lorentz-Einstein theory the total energy E of the electron is not given by the above formula but by $E = m_0c^2 \left(\frac{1}{\sqrt{1-\beta^2}} - 1 \right)$. The latter formula agrees nearly with the former for small values of β, but departs widely from it when β approaches unity. Calculating on the latter formula the energy of the electrons comprising the different groups of homogeneous rays given in the paper of Danysz, the difference between the energies of successive groups can still be expressed by a relation of the same form as that given in the paper, viz. $pE_1 + qE_2$, only E_1 and E_2 have new values. The values which fit in best with the data are $E_1 = 1 \cdot 12 \times 10^{13}\, e$ and $E_2 = 0 \cdot 356 \times 10^{13}\, e$; p has values between 0 and 9 and q between 0 and 2. All the twelve lines from Nos. 21 to 9 fit in with this relation except No. 11, which is nearly equal in energy to No. 10.

It may be significant that all the lines Nos. 8 to 1, which presumably belong to radium B, show an approximately constant difference of energy $E_3 = 0 \cdot 173 \times 10^{13}e$, which is very nearly one half of E_2. The observed and calculated energies from Nos. 8 to 1 agree within the probable limit of experimental error in the determination of the velocities. Line No. 9 fits in equally well with the radium B or radium C series, and it is not known to which series it belongs.

If the relation found by Whiddington (Proc. Roy. Soc. lxxxv. p. 323, 1911) between the velocity required to excite the characteristic radiations and the atomic weight depends on the energy of the electron rather than on its velocity, the energy of the electron to excite the characteristic radiation of the 'K'

* *This vol., p. 280.*

series in radium C of atomic weight 214 is $1 \cdot 27 \times 10^{13} e$. Chapman (Proc. Roy. Soc. lxxxvi. p. 439, 1912) has shown that the characteristic radiation of the 'L' series emitted by the heavier elements corresponds in penetrating power to the 'K' type of radiation from an element of atomic weight $\frac{1}{2}(A - 48)$. The corresponding energy to excite this characteristic radiation in radium C is $0 \cdot 190 \times 10^{13} e$ ergs. It is of interest to note that these two values do not differ much from the values deduced for E_1 and E_3.

As I stated in my paper, the values of the velocities of the different groups of homogeneous rays from a product must be known with great accuracy before the correctness of such difference relations between the energies can be adequately tested. It is of great importance to know accurately the distribution of the β rays from active products both as regards velocity and number, for it is only with the help of such data that we can hope to explain the origin of the remarkable complex β radiation from active substances and its connexion with the γ rays.

E. RUTHERFORD

University of Manchester
Nov. 4, 1912

Wärmeentwicklung durch Radium und Radiumemanation

von PROFESSOR E. RUTHERFORD *und* H. ROBINSON

Aus den Sitzungsberichten der kaiserl. Akademie der Wissenschaften in Wien. Mathem.-naturw. Klasse; Bd. CXXI. Abt. IIa. Oktober 1912, pp. 1491–516. (Vorgelegt in der Sitzung am 4. Juli 1912.)

THIS paper is noted in title only, because the later version, published in the *Philosophical Magazine*, has some additions. The later version follows on p. 312, with the papers published in 1913.—Ed.

Some Reminiscences of Rutherford during his time in Manchester

1

by H. GEIGER*

1907 trat Professor Schuster, bei dem ich Assistent war, nach 25 jähriger Tätigkeit von seinem Amt zurück. Noch im selben Herbst kam Rutherford als der neue Institutschef. Niemand von uns jungen Leuten im Institut kannte ihn aus früherer Zeit, als er aber da war, fühlten wir schnell genug, daß wir einer großen Zeit entgegengingen.

Rutherford's Arbeiten kamen 1908 dadurch rasch in Fluß, daß die Akademie der Wissenschaften in Wien ihm 250 mg. Radium als Leihgabe überließ. Das war in der damaligen Zeit eine recht erhebliche Menge. Das Radiumsalz war wenig konzentriert und konnte im allgemeinen nicht unmittelbar als Strahlenquelle benutzt werden. Rutherford hielt es für das Beste, eine Einrichtung zu treffen, die eine regelmäßige Entnahme der Emanation ermöglichte. Mit der Emanation sollte dann gearbeitet werden, wenn intensive Strahlungen gebraucht wurden. Zuerst wurde versucht, die Emanation durch Erhitzen aus dem Salz auszutreiben, was aber nicht befriedigte. Rutherford löste dann das Salz auf und pumpte die Emanation mit den sich entwickelnden Gasen ab. Anschließend wurde die Emanation durch Ausfrieren in flüssiger Luft konzentriert und in geeignete Behälter, meist feine Glasröhrchen, eingefüllt. Apparatur und Methodik hat sich sehr bewährt und wurde in vielen Instituten nachgebildet. Bei diesen Arbeiten war Rutherford hauptsächlich von Royds unterstützt. Aber auch die anderen Angehörigen des Instituts haben diese ganze Phase der Entwicklung im einzelnen miterlebt; denn Rutherford fand immer Freude daran, uns alles zu zeigen.

Nachdem die Verfahren zur Gewinnung und Reinigung der Emanation genügend sicher arbeiteten, ging Rutherford daran, eine Reihe von wichtigen Untersuchungen an hochkonzentrierter Emanation durchzuführen.

1) Die Volumenbestimmung der Emanation, d.h. Gleichgewichtsbetrag mit 1 g Radium (Phil. Mag., 16. 300. 1908).

2) Spektrum der Emanation; rund 70 Linien werden ausgemessen (mit Royds, Phil. Mag., 16. 313. 1908).

* This note was written shortly after Rutherford's death in 1937. It was sent to me by Geiger, with his permission to use it as I saw fit. A few sentences were quoted in the Obituary Notice written for the Royal Society by A. S. Eve and myself. It seems proper to print here the note in full. Geiger himself died in 1945.—Ed.

3) Kondensation der Emanation, d.h. Dampfdrucke bei verschiedenen Temperaturen (Phil. Mag., 17. 723. 1909).
4) Nachweis der Heliumnatur der α-Teilchen (mit Royds, Phil. Mag., 17. 281. 1909). Diese Arbeit hat wegen der Anschaulichkeit und Unmittelbarkeit der Experimente besonderes Aufsehen erregt. Niemand hätte damals geglaubt, daß man ein Glasröhrchen herstellen könne, das für α-Strahlen durchlässig ist, gleichzeitig aber dem vollen Luftdruck widerstehen kann. Die dünnwandigen Glasröhrchen wurden bald zu einem wichtigen Bestandteil der radioaktiven Experimentiertechnik. Fast täglich wurden solche Röhrchen fur die verschiedensten im Institut laufenden Versuche hergestellt und gefüllt.
5) Eine quantitative Fortsetzung dieser Arbeit bildete die später zusammen mit Boltwood ausgeführte Untersuchung über die Bildungsgeschwindigkeit von Helium aus Radium (Phil. Mag., 22. 586. 1911). Prof. Boltwood war ein Freund von Rutherford und verbrachte sein Urlaubsjahr in seinem Institut in Manchester.
6) Hierher gehört auch die 1913 durchgeführte Messung der Wärmeentwicklung von Radium (mit Robinson, Phil. Mag., 25. 312. 1913). Gewisse Unstimmigkeiten zwischen beobachteter und berechneter Wärmewirkung bestimmten dann Rutherford, eine Präzisionsmessung von e/m und v für α-Strahlen durchzuführen (mit Robinson, Wien. Ber., 122. 1855. 1913).

Schon Anfang 1908 sprach Rutherford zu mir von der Möglichkeit, den elektrischen Effekt einzelner α-Teilchen durch Stoßionisation zu vergrößern und so zu einer Zählmethode zu kommen. Es gab viele Schwierigkeiten, die man heute kaum mehr verstehen kann. Vor allem fehlte es an einem kurzperiodigen Meßinstrument. Die ersten Versuche mit einer Art Fadenelektrometer fielen nicht recht befriedigend aus und wir kehrten zu dem Quadrantelektrometer zurück. Viel Zeit ging auch darüber verloren, daß wir mit Ra B + C als Strahlenquelle arbeiteten und diese Präparate, wie erst allmählich klar wurde, Emanation abgaben und die Apparatur verseuchten. Als schließlich die Methode verläßlich arbeitete, wurde die Zahl der α-Teilchen pro Gramm Radium gemessen und gleichzeitig auch eine Bestimmung der Gesamtladung durchgeführt. Mit Hilfe beider Größen konnte eine Reihe wichtiger atomarer Konstanten berechnet werden, die damals nur sehr ungenau bekannt waren, z.B. das elektrische Elementarquant (Proc. Roy. Soc., 81. 141 und 162. 1908). Man erhält ein anschauliches Bild von dem damaligen Stand des Wissens durch die Presidential Address of Rutherford, Winnipeg 1909.

An der Verbesserung der Zählmethoden wurde auch in den folgenden Jahren dauernd weitergearbeitet. Vor allem wurde das photographische Registrierverfahren entwickelt (Phil. Mag., 24. 618. 1912).

Auch die Szintillationszählmethode, die von Regener eingeführt war, wurde auf eine Reihe wichtiger Probleme angewandt. Ich erwähne die

Zählung der α-Teilchen aus Uran und Thor (Geiger u. Rutherford, Phil. Mag., 20. 691. 1910) und die zeitliche Verteilung der α-Teilchen (R.u.G., Phil. Mag., 20. 698. 1910), für die Bateman, damals auch in Manchester, die mathematische Grundlage gab.

Schon bei der elektrischen Zählung der α-Teilchen war aufgefallen, daß geringe Gasreste in dem von den α-Teilchen durchlaufenen 4 m langen Rohr von Einfluß auf das Resultat waren. Wir schrieben dies einer geringen Streuung der α-Strahlen zu. Später habe ich in mehreren Arbeiten die Streuung quantitativ untersucht (Proc. Roy. Soc. 81. 174. 1908; 83. 492. 1910). Von Bedeutung wurde vor allem die Beobachtung, daß vereinzelt auch extrem große Ablenkungswinkel vorkommen, die weit ausserhalb der normalen Schwankungen liegen (mit Marsden, Proc. Roy. Soc. 82. 495. 1909). Dies war zunächst gar nicht zu verstehen. Rutherford hat wohl viel über diese Merkwürdigkeit nachgedacht; eines Tags (1911) kam er, offensichtlich in bester Stimmung, in mein Arbeitszimmer und sagte mir, er wisse jetzt, wie das Atom aussehe und wie die starken Streuungen zu verstehen seien. Wohl noch am selben Tage begann ich mit einem Versuch, die von Rutherford vorausgesagte Beziehung zwischen Teilchenzahl und Streuwinkel zu überprüfen. Wegen der starken Variation dieser Funktion mit dem Winkel war die Aufgabe relativ leicht und es konnte darum auch schon sehr bald wenigstens annähernd die Gültigkeit der Gleichung festgestellt werden. Später wurde die von der Rutherfordschen Theorie geforderte Abhängigkeit der Teilchenzahl vom Winkel, von der Ordnungszahl und der Geschwindigkeit mit großer Genauigkeit bestätigt (Geiger und Marsden, Phil. Mag., 25. 604. 1913).

Die Streuung der α-Teilchen an einem Atomkern wurde bekanntlich eine sehr wirksame Methode, um über Ladung und Größe des Kerns sowie über den Verlauf des Potentials in Nähe des Kernes bestimmte Aussagen machen zu können. Diese von Rutherford noch in Manchester begonnenen Arbeiten führten schließlich zu den Fragen der Kernstabilität und Atomumwandlung.

Als im Laufe der Jahre immer mehr junge Physiker bei Rutherford arbeiten wollten, entstand das dringende Bedürfnis, die Anfänger in kurzer Zeit in die Grundlagen der Atomphysik und in die radioaktiven Arbeitsmethoden einzuführen. Prof. Rutherford hat mich damals beauftragt, ein kleines Praktikum einzurichten, in dem die Anfänger eine Reihe von Elementarversuchen, z.B. eine Analyse von Zerfallskurven, Absorptionsmessungen, Szintillationszählungen ausführen mussten. Jeder Anfänger mußte zunächst durch dieses Praktikum hindurch, bevor er das Thema für die eigene Arbeit erhielt. Die in dem Praktikum durchgeführten Versuche sind in erweiterter Form beschrieben in dem Büchlein von Makower und Geiger, Practical Measurements in Radioactivity, 1912.

Als geschlossene Gruppe von Arbeiten aus der damaligen Zeit sind auch die Reichweite-Messungen an α-Strahlen zu nennen (Geiger und Nuttall). Der Zusammenhang zwischen Reichweite und Lebensdauer der emittierenden Substanz wurde fesgestellt. Eine Folge der Reichweite-Messungen war auch die Entdeckung der kurzlebigen Elemente Actinium A und Thor A. Die Nomenklatur der radioaktiven Elemente erhielt damals ihre endgültige Gestalt (Rutherford und Geiger, Phil., Mag. 22. 621. 1911).

Die Fragen des radioaktiven Rückstosses wurden eingehend untersucht von Makower und Russ, beide lecturer in Physics. Uber Reflexion und Absorption arbeiteten insbesondere Kovarik und W. Wilson. Auch Marsden war eine Reihe von Jahren bei Rutherford und hat unter anderem auch über Phosphoreszenz durch α- und β-Strahlen gearbeitet. Mehr nach der chemischen Seite arbeiteten Antonoff, Fajans, und v. Hevesy. Theoretische Fragen wurden von Bateman und Darwin behandelt.

Auch der Einfluß der Temperatur auf verschiedene radioaktive Vorgänge wurde immer wieder untersucht; R. W. Boyle, Russell, H. W. Schmidt und W. Wilson haben in dieser Richtung gearbeitet.

Auch Moseley, der wohl 1911 nach Manchester kam, hat zuerst radioaktiv gearbeitet und zwar über kurzlebige Radioelemente und über Zahl und Ladung der β-Strahlen. Erst später hat er sich dem Röntgengebiet zugewandt.

Ich erinnere mich auch noch an folgende Mitarbeiter: E. J. Evans, J. A. Gray, S. Kinoshita, R. D. Kleeman, May Leslie, W. C. Lantsberry, R. Rossi, Schrader, Walmsley und A. B. Wood.

2

by E. N. DA C. ANDRADE, F.R.S.

I FIRST met Rutherford in 1913, when I was appointed John Harling Fellow in succession to Moseley, to follow whom was a distinction not so clear at the time as it is now. Needless to say, it was for me a joy to experience for the first time that penetrating glance that seemed to enquire and to answer at the same time and to hear that great voice becoming more and more enthusiastic as he spoke of what he was planning, the words almost falling over one another. There was between Rutherford and the men of his research group a friendly comradeship and intercourse more like the relationship between a leader and the men of his band of explorers in an unknown and difficult country than that between an army commander and his subordinates, which was closer to the spirit that had prevailed at Heidelberg, where I had taken my doctor's degree two years earlier. Quite informal were Rutherford's daily rambles round the laboratory, in the course of which, if there was a difficult or interesting point to discuss, he would sit on a stool for some little time, throwing out incisive comments and suggestions. He talked to the research worker as to a friend and collaborator. Every day there was a meeting round the laboratory tea-table, at which Rutherford practically always presided in my days. Conversation ranged over a wide variety of topics, mainly, perhaps, concerned with physics, but including ordinary laboratory gossip, and here again, although there was no doubt as to who was the boss, everybody said what he liked without constraint.

Likewise at the Manchester colloquium, which met on Friday afternoons, Rutherford was, as in all his relations with the research workers, the boisterous, enthusiastic, inspiring friend, undoubtedly the leader but in close community with the led, stimulating rather than commanding, 'gingering up', to use a favourite expression of his, his team. At the first meeting of the session he gave an account of work carried out in the laboratory in the past year. He was always full of fire and infectious enthusiasm when describing work into which he had put his heart and always generous in his acknowledgment of the work of others.

There was at Manchester a great shortage of effective apparatus: according to Robinson the average annual grant for apparatus was, at the period in question, under £420 a year. For instance, much of the evacuation was done with Toepler pumps: there were one or two Gaede pumps, but they were jealously guarded. Improvisation was the order of the day, and not a bad order either. When I had been a few weeks in the laboratory William Kay, the indispensable laboratory steward, everybody's friend, said to me 'Papa says you'll do', 'Papa' being the usual name for the head of our family among certain of the laboratory workers. I was naturally curious to know what had established me as a fit member of the fellowship and accordingly asked Kay 'Do you know what made him say that?' 'He saw you making that plateholder out of cardboard and thought that you made a good job of it' replied Kay. For our work on gamma rays a plate holder made out of cardboard with a black paper front was as good as anything, and saved time and money. To 'get on with it' was Rutherford's great desire.

A word about Kay and 'Papa'. Some of us, including in particular Harold Robinson and E. J. Evans, had a great liking for the old-fashioned music hall, then well represented in Manchester. George Formby was a performer who was particularly popular and another was Harry Tate, who had a celebrated sketch, called *Motoring*, about a motor car, still something of a novelty in those days. The only other character in the sketch was his son, who called him 'Papa' with a peculiar intonation used by the laboratory initiates. Incidentally, since these are reminiscences, I may perhaps recall that this sketch contained the only scientific joke that I ever, with a considerable experience, heard on the music hall stage. The son said 'I know why your wheel won't go round, Papa'. 'Why, my boy, why?' 'Because they ought to be $2\pi r$ and yours are $4\pi r$.' This brought the house down, although I do not think that a large proportion of the audience knew what two pi R meant.

William Kay, the head laboratory steward, was an extraordinary man. Always always busy, he was always good-tempered and willing to help. He was an excellent mechanic, an electrician who knew all about the electrical connections of the laboratory, a skilled photographer, and a great hand with apparatus of all kinds, who excelled in setting up the striking lecture experiments for Rutherford's first-year course. He made, in his spare time, a large number of drawings for Rutherford's *Radioactive Substances and their Radiations*. He and I speedily became friends, because he had been very successful as an amateur runner and I had done a certain amount of running, having, for instance, won the University College and Hospital half mile in 1907. My time for the half mile was nothing like as good as his, but running gave us a common interest which made him accept me from the start.

Another laboratory character was the German glassblower Baumbach, an excellent craftsman who was responsible for, among other things, the alpha-ray tubes whose thin walls allowed the passage of the rays with but little decrease of velocity. When I had occasion to call on him shortly after the war broke out, he, who knew that I understood the language, broke out into a stream of fiery German prophecy as to what the German Army would do to Britain. I answered him in the kind of way he understood, telling him to keep his mouth shut or he would find himself in trouble. The next thing was that the Vice-Chancellor of the University sent for me and told me that Baumbach had complained that I had threatened him: it was very unworthy behaviour to threaten a poor defenceless German in our midst and he must ask me not to behave in this way! At the time this annoyed me, but nowadays I take such incidents as a matter of course. That I have not misrepresented Baumbach is demonstrated by what Niels Bohr has written 'but the man's temper, not uncommon for artisans in his field, and which released itself in violent superpatriotic utterances, eventually led to his internment by the British authorities.' But he was a very good glassblower.

I venture to interpolate here that I won a bet with Rutherford, a fact which I record because I think it possible that I am the only man who ever did so. Our apparatus for the work on the wave length of gamma rays included a large magnet for deflecting the beta particles out of the way. One day our plates began to be fogged systematically and we were very puzzled to account for this. Rutherford made a suggestion as to the cause of the trouble and I said—as one could to Rutherford—'No, I'm sure its not that.' 'I bet you a shilling it is', said Rutherford, and so the bet was concluded. Just after that I found the real cause of the trouble, which was not complex physics but simple meddling. All the electrical connections for heavy current were made in a room with the usual bars and plugs. Somebody had pulled out our plugs, probably by accident, and replaced them in the wrong bars, so that the direction of the field of our magnet was reversed, which, with our particular arrangement at the time, threw the beta rays on to the plate. When I told Rutherford this he duly paid me the shilling, which I long treasured. It was lost when my London laboratory was destroyed in the late war.

An incident comes to my mind indicative of Rutherford's inborn conviction that the matter of prime importance was his work in the laboratory, which nothing must interrupt. There was in the laboratory a foreign lady—let us call her Natasha Bauer—doing research, for which she showed no out-standing aptitude. She was an ardent feminist and something of a man-hater: she would never ask a man to do anything for her. One day she had a bottle of SO_2, the closing mechanism of which had stuck so that she could not open it. Not wishing to seek aid from a man she took the bottle into a very small room and did something in the way of wedging the opener in the door, which effected a sudden rush of the gas, with the result that she lost consciousness. Luckily a woman found her on the floor, and she was satisfactorily revived. Rutherford naturally heard of this—he heard of everything—and sent for her. It so happened that I was seated at the desk with him in his room on the first floor, discussing some point in connection with our work, when she arrived. 'What's this I hear, Miss Bauer' he said as soon as she had shut the door 'What's this I hear? You might have killed yourself!' to which she replied 'Well, if I had, nobody would have cared.' 'No, I daresay not, I daresay not' said Rutherford 'but I've no time to attend inquests.' I believe that he was in earnest, thinking of a morning wasted away from the laboratory, but from the lady's face this was not the reply she expected.

To attempt to give a picture of Rutherford's laboratory just before the First World War some reference to those working there is needed: 'Papa' and his family were one. Geiger was no longer there in my time, but his name echoed about freely: he was present in spirit. Chadwick, who had been working for two years under Rutherford on problems in radioactivity, had in 1913 gone to Germany to work with Geiger. Marsden was lecturer and research assistant. He was a responsible figure in the laboratory. In 1913 Moseley was just finishing his fundamental work on X-ray spectra. He would undoubtedly have become a great leader in physics if he had not been killed at Gallipoli in the First World War. I am reminded of what Newton said of the early death of Roger Cotes—'If Mr. Cotes had lived we might have known something.' Moseley was an outstanding character, not, perhaps, so hail-fellow-well-met as most of us. Typical is that he objected to other workers borrowing, perhaps permanently, his matches, so he bought a gross of boxes (which in those days cost him one shilling and sixpence) and made of them a heap on which he put a label worded to the effect 'Please take one of these boxes and leave my matches alone.' Charles Darwin had carried out with Moseley an experimental investigation on the reflexion of X-rays by crystals, an account of which appeared just before my time in the labora-tory, but he was more interested in theoretical work and in 1914 published his classical papers on the mathematical theory of X-ray reflexion. Walter Makower, who had also been a John Harling Research Fellow, was mainly responsible for teaching the technicalities of the subject: he had published with Geiger in 1912 a book *Practical Measurements in Radioactivity* which was in the hands of everybody experimenting on the subject. As *Nature* said of it 'That it should come from the laboratory of Prof. Rutherford and have for its authors two such distinguished workers on radioactivity, practically ensures its general adoption in advanced physical laboratories.' He was much interested in the atmospheric investigations carried out at the kite-flying station installed on the moors above Glossop in Derbyshire, and Harold Robinson's first appearance in print was as a collaborator in a paper by Makower and two others on the electrical state of the upper atmosphere. Makower was a man of keen musical interest and I well remember a party at his house, or at the house of another Makower, at which there was excellent chamber music.

Harold Robinson, who later adopted the additional name of Roper and became H. R. Robinson, was perhaps my closest comrade in the laboratory and we remained on terms of intimate friendship until his death in 1955. We had common interests in books, sport and the local music halls, as well as physics. He towered over Rutherford in physical height, being about six foot four inches, and was deliberate in movement and utterance, in contrast to Rutherford's excitable enthusiasm. It may seem strange to compare the great man with one of his students, but even in those days Rutherford seemed to have a particular liking for Robinson, while Rutherford was, as he always remained, Robinson's one hero. Their differences of temperament may have attracted them together, positive and negative. There is no need to say anything further about their relationship, since Robinson's lecture on *Rutherford: Life and Work to the Year 1919, with Personal Reminiscences of the Manchester Period*, which was the first of the Physical Society's Rutherford lectures, gives a striking picture of the laboratory, enriched with many anecdotes illustrating the terms on which they stood in those days. Robinson, in spite of his deep-seated reverence for his hero, was completely outspoken with Rutherford as with everybody else. Throughout Rutherford's life, from the time that I am considering onwards, Robinson was probably as close to him as anybody.

Robinson's early work was all done in collaboration with Rutherford. After investigating the heating effects of radium and the emanation they worked together on the magnetic analysis of the β-ray spectrum of Radium B and C, concerning which the first paper was published late in 1913. The magnetic analysis of X-ray electrons was Robinson's chief experimental interest throughout life. In 1914 he, like Darwin, Fajans and Makower, published a paper in conjunction with Moseley.

E. J. Evans was a skilled spectroscopist, a student of Alfred Fowler's, who in 1913–14 was carrying out precise measurements which showed that certain lines attributed to hydrogen were in fact due to ionized helium. Their wave numbers differed slightly from those to be anticipated for hydrogen, in a way that could be explained quantitatively, on Bohr's theory, as due to the finite mass of the nucleus. This offered support, valuable at the time, to the theory. Evans was a very efficient member of the teaching staff of the department. His research lay somewhat outside the sphere of interest of Rutherford and of his school, but he was on the friendliest terms with many of us, especially with Robinson, whose affection for the local music halls and for shipping he shared. He stayed on the staff at Manchester after the 1914 war broke out and afterwards became Professor of Physics at Swansea, where he died in 1944.

J. M. Nuttall and A. B. Wood held graduate scholarships and both published papers with Rutherford which are well-known. A familiar figure about the laboratory with whom I was on friendly terms was J. N. Pring, a one-time John Harling Fellow, who was Lecturer and Demonstrator in Electro-Chemistry. He was carrying out work concerned with the behaviour of hydrocarbons at high temperature and pressure, the pressure being maintained by heating petrol in a strong vessel and releasing the vapour by a valve if the pressure became dangerously high. There was a vertical explosion-mat near the apparatus and I remember asking Pring why he did not sit behind it: he replied, with a laugh, that, if the vessel burst, the mat, and everything else, would go through the ceiling.

Among the foreigners working in the laboratory in my time were Stanislaw Loria and Bohdan de Szyszkowski, both from Russian Poland and both rather older and more experienced than many of us. I think that they had come to revel in the Rutherford atmosphere and to get in touch with the latest work in radioactivity. Loria was a most friendly and sympathetic soul who was, as K. Mendelssohn has written, a citizen of the world both in outlook and habits. He had already carried out, in Breslau, Berlin and Göttingen, significant work in physical optics on such matters as the Kerr effect and the dispersion of light in gases. Under Rutherford he worked on certain aspects of radioactive decay and on the volatilisation of radioactive deposits. He published a paper on the branching of the thorium series from the Institut für Radiumforschung in Vienna in 1916, with frequent references to the work of Marsden and other Manchester workers. He was delighted to find that I knew something of the work of Witkowski, who had been his professor, and of Smoluchowski and Natanson, Polish physicists who were among his heroes. He was a man who could achieve a certain degree of mastery in whatever he undertook, and he afterwards became professor of theoretical physics, and later of experimental physics, at Lvov. He could discuss intelligently and with humour any subject that turned up—and a great variety of subjects turned up in the discussions in and around the laboratory in those days.

Szyszkowski was a man of some means, who liked to dine at the Midland Hotel, which was then under the distinguished management of Monsieur Colbert. The 7/6 dinner there was about as good as anything in Europe and to it Szyszkowski invited me every now and then. We had many interests in common outside physics. He was a close friend of Arrhenius, who in 1913 had invited me to Stockholm. We arranged to go together to Sweden for a holiday in the second half of July 1914, where we saw much of Arrhenius, a most friendly and entertaining character who had spent a period at Manchester learning radioactive techniques. He entertained us with many an anecdote of scientists of European fame and with excellent drink. The outbreak of war caught us there. In due course I was able to get back to England and Szyszkowski returned to Russia. He had estates in Podolia, a former government of European Russia. I never heard from him after the war and am afraid that he disappeared in the revolution.

Another man from Eastern Europe (Kiev) who was working in the laboratory was Stanislaw Kalandyk, a somewhat remote and melancholy character who always appeared to be pondering upon the ultimates of physics and upon time, death and eternity, without attaining anything definite.

A very attractive character from the Continent was the Dutchman A. D. Fokker, closely related to the designer of the famous aeroplane. He was even taller than Robinson. He was a theoretical physicist and mathematician who wrote on, among other things, relativity. He, too, was attracted by Rutherford's fame and personality. From the Empire were D. C. H. Florance and R. W. Varder. Varder was a South African, who had taken his degree at the Cape: he was working on the absorption of beta rays. He afterwards became professor at Rhodes University College, at Grahamstown. Florance, a specialist on the scattering of gamma rays, was from New Zealand, where he afterwards became professor.

There were others of note working in the laboratory, but those quoted will suffice to show the range of interest and character of the brotherhood, all united by the infectious enthusiasm of Rutherford, who knew so well how to produce results by a few directive words, apparently casual but probably the results of long having the matter in his thoughts.

Something must be said of Niels Bohr, who was only a visitor in the year of which I am speaking, but who had spent some months in Rutherford's laboratory before producing his revolutionary theory of light emission by the atom, and who, after the outbreak of war, succeeded Darwin as Schuster Reader. His name will ever be associated with the nuclear atom.

In 1911, when Rutherford first put forward his revolutionary conception, the nuclear atom aroused so little interest that even *Nature* did not comment on it. It is not always recalled that the structural scheme, as originally put forward by Rutherford, allowed the massive central particle to be either positively or negatively charged, since the angle through which an alpha particle, aimed to pass near the charged nucleus (as it was later called), is deflected, is independent of the sign of the charge. As Marsden said, in his admirable summary published at the time 'The author finds that an atom with a strong positive or negative central charge concentrated within a sphere of less than 3×10^{-12} cm radius, and surrounded by electricity of opposite sign distributed throughout the remainder of the volume of the atom of about 10^{-8} cm radius, satisfies all the known laws of scattering due to either α- or β-rays.' This is exactly what Rutherford showed. Even Rutherford himself, at the time, did not seem to regard his theory as of supreme significance. In his *Radioactive Substances and their Radiations,*

PHYSICS STAFF AND RESEARCH GROUP, MANCHESTER, 1913

T. S. Taylor   A. S. Russell
H. Richardson   J. M. Nuttall   B. Williams   W. Kay
A. B. Wood   E. Green   R. H. Wilson   S. Oba   E. Marsden   H. Gerrard   J. Chadwick   F. W. Whaley   H. G. J. Moseley
H. Robinson   D. C. H. Florance   Miss M. White   J. N. Pring   E. Rutherford   W. Makower, E. J. Evans, C. G. Darwin

which was published in 1913, there are only two references to it, both concerned with the scattering of alpha particles. The latter one, near the end of the book, is indexed under 'Atom, structure of, to explain scattering', which emphasises that the scattering of alpha particles, Rutherford's beloved playthings, was the chief concern. This part of the book must have been written at the end of 1912: here, in contrast with the earlier references to the atom model, the concentrated charge at the centre is made positive and it is surrounded by a distribution of electrons 'throughout a spherical volume or in concentric rings in one plane.' This, then, was 1912.

In 1913, however, the nucleus had become a centre of attraction at Manchester and it was in this and the following year that its scope and fundamental importance were established, chiefly by Bohr's work. Bohr's first paper on the subject appeared in the *Philosophical Magazine* in July 1913, to be quickly followed by two subsequent papers, making up the classic trio in which his views were fully developed. The nuclear atom had suddenly come to its own, and just as the foundation of the Royal Society has been celebrated as occurring both in 1660 and 1662, so the year 1913 may be associated with 1911 in celebrating the establishment of the nuclear atom. The quantum theory was somewhat remote from Rutherford's inborn way of visualising everything, a characteristic so important that I may, perhaps, be allowed to illustrate it by an anecdote. In the course of a discussion after a dinner at the Athenaeum at which I was present Eddington said something to the effect that electrons were very useful conceptions but might not have any real existence. Whereupon Rutherford got up and protested indignantly 'Not exist, not exist—why I can see the little beggars there in front of me as plainly as I can see that spoon.' Alpha particles and electrons he could see, but not quanta. He could not fail, however, to be impressed by the way in which Bohr's theory gave the correct value for atomic constants. At the British Association meeting in Melbourne in August 1914, the month of the outbreak of the war that closed an epoch, he opened a discussion on atomic structure with a speech in which he drew attention to the importance of Moseley's work and referred favourably, but without fully committing himself, to Bohr's theory of line spectra. Later, of course, he accepted the theory completely.

The fifty years that have elapsed since the birth of the nuclear atom have seen marvellous advances in physics, but have not, perhaps, produced another Rutherford. I do not think that this is because there are no fundamental discoveries left to make: of the fundamentals of the solid and liquid state we know very little. I doubt even if it is because men of his calibre are no longer born. A condition for great discoveries is undisturbed leisure to think: Isaac Newton told one enquirer that he made his discoveries 'By always thinking unto them' and on another occasion said 'I keep the subject constantly before me and wait till the first dawnings open little by little into the full light.' Such quiet respite to think Rutherford had, especially in the period at Manchester, a period so remote in spirit from the present day as to seem like another age. The physics research lab. of Rutherford's Manchester days, with the professor in closest touch with all his research men, who, with little thought for their future living, were eagerly engaged in obtaining results that seemed remote from any possible practical application, has passed, in the same way as the workshop where craftsmen took a pride in producing beautifully made objects has passed. Our universities have, inevitably and properly, become, in the mood of the age, factories for producing degrees and discoveries, which they do with great efficiency. It is only old men who regret the passing of the past, but it is well for us all to acknowledge the passing. 'Tempi passati sind vorüber', as they used to say in Vienna.

Rutherford at Manchester was free to do exactly as he liked, and physics was the consuming interest of his life. As far as I know, like Newton, he took no particular joy in music, poetry, the arts or the pleasures of the table. He 'kept the subject constantly before him': his life, his restless quest, was physics. He saw the atomic particles as clearly before him as your old cricketer, describing a match, will see the field before him. The radioactive elements were his children—some well-behaved, some troublesome, but he knew and loved them all. His favourite individual was, I think, the brisk little alpha particle—and how he made it work!

His debt to Manchester he freely acknowledged, with an unconcealed pride in his achievements there, as when, in 1931, towards the end of his life he said 'I owe a great debt to Manchester for the opportunities it gave me for carrying out my studies. I do not know whether the University is really aware that during the few years from 1911 onwards the whole foundation of the modern physical movement came from the physical department of Manchester University'. He knew what he had done and said so in his simple unsophisticated way. For simple and unsophisticated he was, but a supreme genius.

If I say that he was simple I have his support, for in a speech made and recorded in 1931 he said 'I am always a believer in simplicity, being a simple man myself.' I think, too, that he was a happy man and that the years spent at Manchester were the happiest time of his life. At any rate, years after he wrote to Geiger 'They were happy days in Manchester and we wrought better than we knew.' I think it was well for his peace of mind that he did not foresee the terrors, the threat to the human race, that would grow directly from the work of his school, for the nuclear bomb derives directly from the early experiments on the disruption of the nucleus, which in their turn derived from his Manchester work. Quite clearly he long believed that atomic energy could never be released on a large scale, for he said in 1933, only 28 years ago, 'These transformations of the atom are of extraordinary interest to scientists, but we cannot control atomic energy to an extent which would be of any value commercially, and I believe that we are not likely ever to be able to do so. A lot of nonsense has been talked about transmutation. Our interest in the matter is purely scientific, and the experiments which are being carried out will help to a better understanding of the structure of matter.' In 1936, the year before he died, he said 'While the over-all efficiency of the process rises with increase of energy of the bombarding particles, there seems to be little hope of gaining useful energy from the atoms by such methods. On the other hand, the recent discovery of the neutron and the proof of its extraordinary effectiveness in producing transformations at very low velocities opens up new possibilities, if only a method could be found for producing slow neutrons in quantity with little expenditure of energy. At the moment, however, the natural radioactive bodies are the only known sources for gaining energy from atomic nuclei, but this is on far too small a scale to be useful for technical purposes.' Thus he concluded his James Watt lecture. Even this, however, shows no foreboding of the nucleus as an agent of overwhelming destruction and disaster.

I have endeavoured to recall a vision of Rutherford, with a background of his collaborators, as he was at the time when I was working in Manchester, and to show him not only as an outstanding genius, who founded much of modern physics, but also as a great leader who commanded the confidence and affection of those who worked with him and who, when the mood was on him, was a figure of inspiration. At Manchester the cares of office sat lightly on him and he revelled in the free exercise of his full creative powers. He did work of the first importance after he left Manchester, but it is certain that he was never again so untrammelled as he was at Manchester and it is probable that, on the whole, he was never again so happy.

3

by A. B. WOOD

I. Pre-War Days at Manchester

THE years 1909–13, my student days at Manchester, have left in my mind the deepest impressions of Rutherford. His lectures, his remarks as Chairman of the Physics Colloquium, his stimulating help in my first efforts in research, and his breezy conversation recall very pleasant memories of those days. The staff and research school at this time formed an exceptionally flourishing group. Almost every corner and corridor housed Rutherford's enthusiastic research followers. The laboratory was full and every one was busy trying to solve some problem which formed a branch or a twig of Rutherford's main tree trunk—the nuclear atomic theory. Radio-activity with all that the name implies—spontaneous atomic disintegration (Radium to Lead, Thorium to Lead, Actinium to Lead), α, β and γ rays and recoil atoms, periods of decay of atoms, ranges, velocity and scattering of α-particles, magnetic spectra of β-rays, absorption coefficients of γ-rays, E.M. wave nature of γ and X rays, chemical properties of radio-active substances, these and many other such topics crowd to the mind.

The laboratory housed a very happy family, a large one, with Rutherford as 'father'. Problems and difficulties were often discussed at tea and some of these occasions leave impressions, apparently trivial but really fundamental. I well recall an occasion when Moseley raised the question of design of large electro-magnets. A serious difficulty in the use of the laboratory electro-magnets was due to overheating during the long exposures required to obtain α-ray deflection photographs. This overheating led to breakdown of insulation of the winding and Moseley suggested using *bare* aluminium wire which would oxidise on heating and the oxide coating, being an insulator, would *improve* with use. This topic turned to a general talk on the field strength of electro-magnets and after various members of the tea-party had aired their views, some of them obviously exaggerating what their particular brand of magnet could do, Rutherford, who had remained quiet with his knowing smile, told the story of a magnet at Montreal. This magnet, he said, would take a bunch of keys from a man's pocket as he came into the room— in fact 'the magnet was strong enough to draw the iron out of a man's constitution'. At this point, of course, the tea-party broke up, every man to his own job! In those days, all were keen on solving the problems in hand and many, with Rutherford in the forefront, worked very late hours in the laboratory. I well remember Moseley making liquid air for an experiment he was doing at 2 or 3 a.m., returning after a few hours' sleep to give a lecture at 9.30 a.m. After an intensive period of research, almost night and day, Rutherford would write up the paper for publication and take a few days' holiday; then back again full of fresh ideas. His energy was infectious and almost unlimited. He made no effort to conceal his joy at the conclusion of a successful experiment and it was his proud record that he 'never put a man on a "dud" research'. When his nuclear hypothesis seemed firmly established by experiment, he would say of the 'classical atomic theorists' that 'some of them would give a thousand pounds to disprove it'. The great delight in his face as he removed his pipe and made the remark was one of the characteristics of Rutherford which can never be forgotten by those who knew him.

II. The War (1914–18)

The war produced marked changes in Rutherford's laboratory in Manchester. The 'family' of research workers dispersed—joining various branches of the fighting forces. Only a skeleton staff remained and the laboratory looked deserted.

In July 1915 the Admiralty Board of Invention and Research (The B.I.R.) was created under the presidency of Lord Fisher. Rutherford (then Sir Ernest) served on the Panel of the Board and on the Sub-Committee dealing with submarine detection and location (amongst other important problems). At that time, enemy submarines were just beginning to be troublesome; later they became a menace.

It might have been supposed that a man like Rutherford could not 'switch over' from atoms to submarines. No doubt such a change required a great effort on his part, but he was equal to it. He tackled the problem with his customary energy. The large research laboratory on the ground level was rapidly transformed into an acoustics laboratory equipped with a large tank of water for the study of underwater acoustics. Up to this time, every one had assumed that acoustics was 'worked out'; it was regarded as a branch of mathematics. It was quickly realised, however, that little or nothing was known about underwater sound problems. Rutherford worked hard to remedy this state of affairs. His early work in the laboratory at Manchester with Broca tubes, diaphragms, microphones, magnetophones and various other forms of underwater sound sources and receivers soon yielded interesting and valuable results. This soon revealed the necessity for more extended work at sea. Apparatus devised in the laboratory could then be tested under 'service' conditions. For this purpose a B.I.R. research station was started at Hawkcraig, Aberdour (Fifeshire), in November, 1915, the staff consisting of two physicists, myself and H. Gerrard and a mechanic F. W. Pye—a small naval staff and a few ships (2 drifters and a submarine) were also attached to the station. The latter was conveniently situated near Rosyth on the Firth of Forth so that vessels of Beatty's Squadron were frequently passing. It is hardly necessary to state that the staff was kept very busy. Rutherford supplied a constant stream of apparatus and ideas to be tried out on ships and submarines at Aberdour. In the preliminary stages, late 1915 and early 1916, these consisted principally of methods of detecting submarines by means of hydrophones of various kinds (portable, hull fitting and towed). These hydrophones were also used in underwater signalling between ships. The detection problem having been solved, Rutherford next proposed a number of methods of locating submarines by 'direction finders'. His suggestions included the following:

(1) Long cones of wood or paraffin with the base immersed in water and a sensitive receiver at the apex.
(2) Long double-walled cylinders with the receiver at one end.
(3) Two similar receivers spaced half a wave length apart, and
(4) A diaphragm (with microphone at centre) mounted symmetrically in a short heavy cylinder, the water and sound having free access to both sides of the diaphragm.

(1) and (2) were not very successful; (3) was good for signalling but not for submarine detection, whilst (4) was developed into a direction finder of considerable service value in the location of submarines.

During the period up to May 1916, Rutherford had a very strenuous time; in Manchester attending to his University work and preparing apparatus for test at Aberdour, in London attending frequent meetings of the B.I.R., and in paying regular visits to Aberdour where he became a well known figure at the Shore Station and on the ships. He enjoyed his Aberdour visits; the sea air and work on ships acted like a tonic. He often talked on these occasions of his travels round the world and of the people he had met.

In a letter dated 22nd February, 1916, after suggesting a number of experiments, he said 'I can provide plenty of ideas but I have not time to work them out myself. . . . I am going to London on Friday'. On 10th April, 1916, after referring to Prof. W. H. Bragg coming to take control at Aberdour, he wrote 'I am leaving for a fortnight's holiday in the Lake District on Wednesday. I feel I need a little rest, as I have not had a day off for six months'. On the following Saturday, 18th April, 1916, in a letter from Howe Head Farm, Coniston, he says 'Glad to have a rest but my peace is bombarded by a deluge of stuff from the B.I.R. . . . the mere correspondence with B.I.R. in the past was sufficient to occupy one ordinary man's time'. But he was no 'ordinary man' and he was soon back again at work in Manchester. During Prof. Bragg's stay at Aberdour, Rutherford was a frequent and very welcome visitor. He was always brimming over with suggestions and sometimes became impatient of delays in obtaining ships for his experiments. For various reasons, the Admiralty Research Station, or rather its staff, moved in 1917 to Parkeston Quay, Harwich, the headquarters of Admiral Tyrwhitt's fleet of destroyers and submarines. Prof. W. H. Bragg moved to the Admiralty Anti-Submarine Department and Col. A. S. Eve became Superintendent. The staff was now much increased and good facilities were available for experiments on ships and submarines. During this period Rutherford played an important part in the early development of the piezo-electric quartz Asdic which ultimately became the most effective method of detecting and locating submarines. The thanks of the Navy are due to him for his untiring efforts in the anti-submarine war.

One can easily understand that at this period, he found the routine of University work somewhat irksome. He writes on 3rd January, 1917, after a long discussion of anti-submarine work, 'We had a quiet Christmas. . . . I have had the usual batch of exam. papers to look over and get to hate it more every year.' During the war period, he published only 2 or 3 papers on atomic physics. In his moments of relaxation, very infrequent, he occasionally made reference to experiments which 'ought to be done as soon as the war is over'.

One of the most remarkable features of his character was the manner in which he could divert all his thoughts and energies from their natural channels into intensive Naval research. To use a favourite expression of his, the work he did for the Navy in the war was 'colossal'.

For some years after the war he served on Admiralty Advisory Committees, visiting various Naval experimental and research establishments and advising on programmes of investigation. In this connection, he was always strongly in favour of fundamental research which ultimately leads to valuable applications, rather than so-called research in the design of 'gadgets'.

The return to peace once more found him eager to resume his life and work in atomic physics. His intuition had not diminished in the enforced period of 4 years 'break'. The news of his appointment to the Cavendish Professorship at Cambridge in 1919 gave him much pleasure. He left Manchester, the scene of some of his greatest achievements and the home of many of his friends, with many regrets, but he looked forward to his work at Cambridge with keen anticipation.

November 30, 1937

Heating Effect of Radium and its Emanation

by PROFESSOR E. RUTHERFORD, F.R.S., *and* H. ROBINSON, M.SC.,
Demonstrator and Assistant Lecturer in the University of Manchester

From the *Philosophical Magazine* for February 1913, ser. 6, xxv, pp. 312–30

SINCE the initial discovery of the rapid and continuous emission of heat from radium by P. Curie and Laborde in 1903, a number of investigations have been made by various methods to determine with accuracy the rate of emission of heat. Among the more important of these may be mentioned the determination of Curie and Dewar* by means of a liquid air and liquid hydrogen calorimeter; of Ångström;† and of Schweidler and Hess‡ by balancing the heating effect of radium against that due to an electric current; and of Callendar§ by a special balance method. It is difficult to compare the actual values found on account of the uncertainty as to the relative purity of the radium preparations employed by the different experimenters. The most definite value is that recently obtained by Meyer and Hess‖ using part of the material purified by Hönigschmid in his determination of the atomic weight of radium. As a result of a series of measurements, they found that 1 gram of radium in equilibrium with its short-lived products produces heat at the rate of 132·3 gram calories per hour.

Rutherford and Barnes¶ in 1904 made an analysis of the distribution of the heat emission between radium and its products. They showed that less than one quarter of the heat emission of radium in radioactive equilibrium was due to radium itself. The emanation and its products, radium A, B, and C, supplied more than three quarters of the total. The heating effect of the emanation was shown to decay exponentially with the same period as its activity, while the heating effect of the active deposit after removal of the emanation was found to decrease very approximately at the same rate as its activity measured by the α rays. The results showed clearly that the heat emission of radium was a necessary consequence of the emission of α rays, and was approximately a measure of the kinetic energy of the expelled α

---

* Curie and Dewar, *Mme. Curie, Recherches sur les substances radioactives*, p. 100, 2me Edition, Paris, 1904.

† Ångström, *Ark. f. Mat. Astr. och Fysik*, i. p. 532 (1904); ii. No. 12 (1905); *Phys. Zeit.* vi. p. 685 (1905).

‡ Schweidler and Hess, *Wiener Berichte*, 117, p. 879 (1908).

§ Callendar, Phys. Soc. Proc. xxiii. p. 1 (1910).

‖ Meyer and Hess, *Wiener Berichte*, 121, p. 603, March 1912.

¶ Rutherford and Barnes, Phil. Mag. vii. p. 202 (1904). (*Vol. I, p. 625.*)

particles. If this were the case, all radioactive substances should emit heat in amount proportional to the energy of their own radiations absorbed by the active matter or the envelope surrounding them. This general conclusion has been indirectly confirmed by measurements of the heating effect of a number of radioactive substances. Duane* showed that the heating effect of a preparation of polonium was of about the value to be expected from the energy of the α particles emitted, while the experiments of Pegram and Webb† on thorium and of Poole‡ on pitchblende showed that the heat emission in these cases was of about the magnitude to be expected theoretically from their activity.

It is of great interest to settle definitely whether the heat of radium and other radioactive substances is a direct measure of the energy of the absorbed radiations. Since the emission of the radiations accompanies the transformation of the atoms, it is not *a priori* impossible that, quite apart from the energy emitted in the form of α, β, or γ rays, heat may be emitted or absorbed in consequence of the rearrangements of the constituents to form new atoms.

The recent proof by Geiger and Nuttall§ that there appears to be a definite relation between the period of transformation of a substance and the velocity of expulsion of its α particles, suggests the possibility that the heating effect of any α-ray product might not after all be a measure of the energy of the expelled α particles. For example, it might be supposed that the slower velocity of expulsion of the α particle from a long-period product might be due to a slow and long-continued loss of energy by radiation from the α particle before it escaped from the atom. If this were the case, it might be expected that the total heating effect of an α-ray product might prove considerably greater than the energy of the expelled α particles.

In order to throw light on these points, experiments have been made to determine as accurately as possible:—

(1) The distribution of the heating effect amongst its three quick-period products, radium A, radium B, and radium C.

(2) The heating effect of the radium emanation.

(3) The agreement between the observed heat emission of the emanation and its products and the value calculated on the assumption that the heat emission is a measure of the absorbed radiations.

(4) The heating effect due to the β and γ rays.

It was also of interest to test whether the product radium B, which emits no α rays but only β and γ rays, contributed a detectable amount to the heat emission of the active deposit.

Method of Experiment

In order to test these points, it was essential to employ a method whereby

* Duane, *Comptes Rendus*, cxlviii. p. 1665 (1909).

† Pegram and Webb, Phys. Rev. xxvii. p. 18 (1908).

‡ Poole, Phil. Mag. xix. p. 314 (1910).

§ Geiger and Nuttall, Phil. Mag. xxii. p. 613 (1911); xxiii. p. 439 (1912); xxiv. p. 647 (1912).

rapidly changing heating effects could be followed with ease and accuracy. A sufficient quantity of radium emanation was available to produce comparatively large heating effects. It was consequently not necessary to employ one of the more sensitive methods for measuring small heating effects, such as have been devised by Callendar and Duane.

The general arrangement was similar to that employed in 1904 by Rutherford and Barnes for a like purpose. Two equal coils, P, P, fig. 1, about

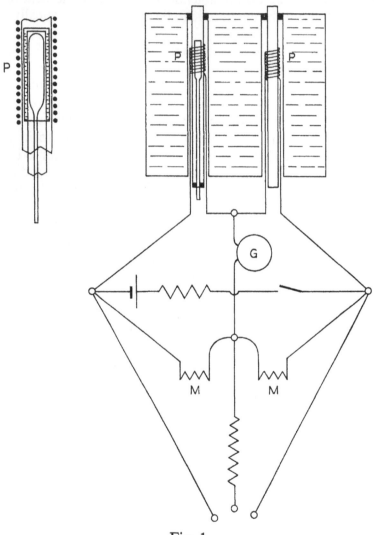

Fig. 1

2·5 cm. long, were made of covered platinum wire of diameter 0·004 cm. and length 100–300 cm., and wound on thin glass tubes of 5·5 mm. diameter. These platinum coils of nearly equal resistance formed two arms of a Wheatstone bridge, while the ratio arms consisted of two equal coils, M, M, of manganin wire, each of about the same resistance as the platinum coils, and wound together on the same spool and immersed in oil. The platinum coils had a resistance varying between 15 and 45 ohms in the various experiments. The glass tubes on which the platinum coils were wound were placed in brass tubes passing through a water-bath. When a specially steady balance was required, the water-bath was completely enclosed in a box and surrounded with lagging to reduce the changes of temperature to a minimum. In most of the experiments the correction for change of balance during the time of a complete experiment was small and easily allowed for. By means of an adjustable resistance in parallel with one of the coils, a nearly exact balance was readily obtained. A Siemens and Halske moving-coil galvanometer was employed of resistance 100 ohms. This had the sensibility required, and was found to be very steady and proved in every way suitable.

The current through the platinum coils never exceeded 1/100 ampere, and was generally about 1/200 ampere.

A calibration of the scale of the galvanometer was made by placing a heating-coil of small dimensions of covered manganin wire within one of the platinum coils, and noting the steady deflexion when known currents were sent through it. It was found that the deflexion of the galvanometer from the balance zero was very nearly proportional to the heating effect of the manganin coil, for the range of deflexion employed, viz. 400 scale-divisions. The deflexion thus served as a direct measure of the heating effect.

§ 1. *Distribution of the Heat Emission between the Emanation and its Products*

A quantity of emanation of about 50 millicuries was introduced into a thin-walled glass tube T (fig. 2) connected by a capillary tube C to a small stopcock S. The tube was attached to a mercury-pump by the aid of which the emanation could be purified and compressed into the emanation-tube. The position of the latter was adjusted to lie in the centre of one of the platinum coils.

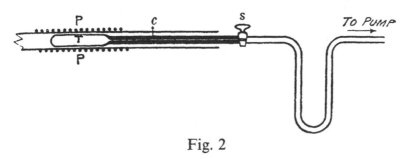

Fig. 2

As it was necessary to leave the emanation for about 5 hours in the tube for equilibrium to be reached with its successive products, it was desirable to arrange that the emanation-tube could be removed outside the water-bath at intervals to test for any change in the balance. The emanation-tube was kept fixed to the pump, but the water-bath was moved backwards along the direction of the axis of the emanation-tube. For this purpose, the water-bath was mounted on metal guides, and was moved backwards or forwards by means of a screw.

The general procedure of an experiment was as follows. The balance was adjusted as nearly as possible and the tube, which had been filled with emanation for more than five hours, was introduced within the platinum coil. In about ten minutes a steady deflexion of the galvanometer was obtained, proportional to the heating effect of the emanation. The emanation was then suddenly expanded into the exhausted pump by opening the stopcock S, and condensed in a U-tube by liquid air. The removal of the emanation caused a rapid decrease of the deflexion, followed by a slower decrease due to the decay of activity of the deposit. Observations of the deflexion were continued in some experiments for over two hours. In this time the heating effect had decayed to about 6 per cent. of the initial value. At the conclusion of the experiment, the emanation-tube was removed and the balance point again obtained. Special experiments showed that any change of balance was very slow and regular, so that a correction of the readings for the small change of balance during the observations could be made with accuracy. The readings of the galvanometer were remarkably steady, and observations of the deflexion could be made to about $\frac{1}{5}$ of a scale-division.

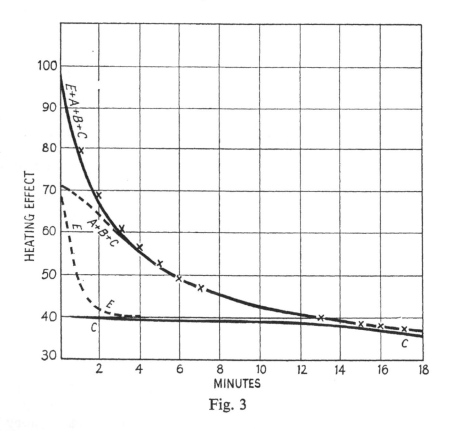

Fig. 3

A typical example showing the variation of the deflexion with time for the first 18 minutes after removal of the emanation is shown in fig. 3, curve E + A + B + C.

On account of the lag of the apparatus, the deflexion of the galvanometer at any moment is always greater than corresponds to the heat emission of the emanation-tube. A number of experiments were made to determine the amount of this lag. For this purpose a manganin coil was wound on a glass tube of the same size and thickness as the emanation-tube T and introduced into the platinum coil. A current from an accumulator was sent through this coil so as to give an effect of about the same magnitude as that of the emanation used in the experiments. The circuit was then broken and the decrease of deflexion with time was noted. For this coil the deflexion fell to half value in about 45 seconds, and after that decreased approximately according to an exponential law with a half-value period of about 30 seconds. The lag of the manganin coil was found by experiment to be slightly greater than the lag of the bare emanation-tube. It was found that for a slow decrease of heating effect the deflexion lagged about one minute behind the actual heat emission.

Analysis of the Curve

The curve given in fig. 3 is typical of a number of curves obtained which agreed closely with one another. The relative heating effects of the emanation and its products can be deduced from the observed curve by comparison with the theoretical curve of decay of the components of the active deposit.

The heating effect of the emanation itself has practically disappeared three minutes after its removal. The variation of the heating effect of the tube C resulting from the removal of the emanation alone is shown in the dotted side curve E E, where the maximum heating effect is taken as 29 per cent. of the total. This curve was deduced from a knowledge of the cooling curve of the tube under the experimental conditions when heated above its surroundings.

After subtracting the heating effect due to the emanation alone, the resulting curve A + B + C gives the heating effect due to radium A + B + C. After about 20 minutes the heating effect due to radium A has practically vanished, and the effect observed is then due to radium B + C. It was found that the curve after 20 minutes followed closely the theoretical curve to be expected if the heating effect was provided mainly by radium C. Assuming this to be the case, the heating effect due to radium C alone is shown by the curve C C, which cuts the axis of ordinates at 40. A lag of 1 minute is assumed between the observed and true heating effects after 20 minutes. The difference of the ordinates of the curves A + B + C, and C C, must be due to the heating effect supplied by radium A. After an initial lag, the heating effect of radium A should ultimately decay exponentially with its known period of transformation, viz. 3 minutes. This is shown clearly in the curve of fig. 4. The difference-curve is plotted, allowing an initial interval of 3 minutes for the emanation effect to decay. Plotting the logarithms of the deflexions as ordinates and the time as abscissæ, the curve is a straight line, showing that the heating effect due to radium A decays exponentially with a half-value period of 3 minutes. The maximum heating effect of radium A was deduced to be 31 per cent. of the total.

The variation of heating effect of the emanation-tube with time brings out clearly that about 29 per cent. of the initial heating effect of the tube is due to the emanation alone, 31 per cent. to radium A, and 40 per cent. to radium B + C together. It should be mentioned that it is difficult to deduce

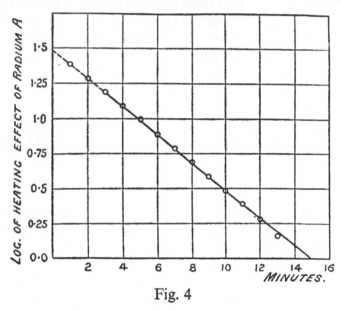

Fig. 4

with certainty the exact ratio of the heating effects due to the emanation and radium A on account of small errors in the determination of the lag. It is clear, however, from the experiments that the heating effects are nearly equal.

These deductions were verified by another method which avoided the necessity of any correction for lag or for galvanometer or scale errors. The emanation-tube was replaced by a glass tube of the same dimensions, over which was wound a layer of fine insulated manganin wire whose resistance was determined. The variation of the heating effect of the emanation-tube after removal of the emanation was then calculated from the known periods of transformation of radium A, B, and C, assuming that the emanation provided 29 per cent. of the total, radium A, 31 per cent., radium C, 40 per cent. The current through the resistance-coil to give a corresponding heating effect was then calculated, and the external resistance to be added in the battery-circuit at any moment deduced. The initial amount through the coil was adjusted to give nearly the same deflexion of the galvanometer as that due to the emanation. By means of a dial resistance-box, the resistance could be rapidly varied to give at any moment the required heating effect. For example, to imitate the removal of the emanation the resistance was suddenly increased so that the heating effect of the current changed to 71 per cent. of the initial amount.

In this way the variation of the deflexion of the galvanometer to be expected for the assumed distribution of the heating effect was directly determined. The values obtained are marked by crosses in the main curve fig. 3. It is seen that the observations lie close to the curve throughout the whole range. The lag of the resistance-coil was slightly greater than that due to the emanation-tube, and in consequence the initial points lie slightly above the curve. The agreement between the two curves shows clearly that the assumed distribution of heating effect between the emanation and its products is in close accord with direct experiment.

It is seen from the curve fig. 3 that the heating effects due to the emanation and radium A have practically vanished after 20 minutes, and the remaining heating effect is due to radium B and radium C together.

A number of experiments were made to test whether radium B provided an appreciable part of the heating effect observed. For this purpose the decay of the heating effect was carefully followed for about three hours after removal of the emanation and the results compared with those to be expected theoretically for any assumed distribution between the heating effects of radium B and radium C. The decay curve observed was found to agree closely with that to be expected if all the heating effect arose from radium C alone. The experiments were rendered difficult by the fact that a small fraction of the emanation adhered to the walls of the emanation-tube, and was gradually released during the time of the experiment. The effect of this became appreciable after two hours, when the heating effect due to radium C had decayed to about 14 per cent. of its initial value.

In addition, the method is not very sensitive, for the decay curve over the region examined is not much affected even if radium B provides 5 per cent. of the heating effect of radium C. It was concluded from the observations that radium B could not provide *more* than 5 per cent. of the heating effect due to radium C; but for the reasons mentioned the actual percentage could not be deduced with any confidence.

Agreement of Experiment with Calculation

The relative heating effects of the emanation, radium A and radium C in equilibrium can be readily calculated if it be assumed that the heating effect is a measure of the kinetic energy of the expelled α particles. Since the expulsion of an α particle causes a recoil of the residual atom, the kinetic energy of the latter should be included in the calculation. If m, M be the masses of the α particle and recoil atom respectively, and u, U the corresponding velocities, $mu = MU$, and the kinetic energy of the α particle and recoil atom is given by $\frac{1}{2}mu^2\left(1 + \dfrac{m}{M}\right)$. The ratio m/M is slightly less than 0·02, so that the heating effect due to recoil is about 2 per cent. of that due to the α particle itself.

When the emanation is in transient equilibrium with its products, it has been shown* that radium C is in excess of the true equilibrium amount by 0·89 per cent., so that a slight correction is necessary for this factor. The velocity u of the α particle is deduced from the relation found by Geiger, $u^3 = KR$, where R is the range in air. Taking the ranges of the α particles from the emanation and its products given by Bragg and correcting for the factors mentioned above, it can be simply shown that the heating effect of the emanation in transient equilibrium with its products is distributed as follows:

> Emanation 28·8 per cent.
> Radium A   30·9   „
> Radium C   40·3   „

The experimental values observed are 29, 31, and 40 per cent. respectively, and thus appear in good accord with theory.

In this comparison no account is taken of the heating effect contributed by the β rays from radium B and radium C. From experiments described later, it appears probable that the β rays from these two products contributed under the experimental conditions about 4 per cent. of the total heating effect. It follows that the percentage of the heating effect included under radium C should be 42·6 per cent. instead of 40·3.

As the result of a number of observations, the heating effect due to radium B + C was found to be certainly not greater than 40·5 per cent. and probably nearer 40·0 per cent. It thus appears that radium C provides slightly less heating effect than that to be expected theoretically. While the difference between observation and calculation is not large, it may prove to be significant; for in making the calculations no account is taken of the heating effect of the β rays or of a possible small heating effect due to radium B. If the heating effect of radium B were 5 per cent. of that contributed by C, the discrepancy between theory and experiment would be quite marked, and would indicate that the heating effect of a product was not entirely a measure of the energy of the expelled α particles and the recoil atoms.

In order to settle this point with certainty, it would be necessary to isolate radium C from radium B, and to measure accurately its heating effect. It is hoped to continue experiments in this direction, for the question to be settled is of great importance in connexion with the general theory.

§ 2. *Heating Effect of the Radium Emanation*

A series of experiments were made to determine accurately the heating effect of the radium emanation in absolute measure in order to test how far the calculated heating effect is in agreement with experiment. The general method employed was similar to that described in the earlier part of the paper. A quantity of emanation of 100 to 150 millicuries was concentrated in a small glass tube about 2·2 cm. long, 2 mm. bore, and of thickness 0·2 mm. This was attached to a long thin glass cylinder of small diameter for convenience

* Rutherford and Chadwick, Proc. Phys. Soc. xxiv. p. 141 (1912). (*This vol., p. 271.*)

of handling. In order to calibrate the heating effect observed, a coil of silk-covered manganin wire about 127 cm. long and 41·45 ohms resistance was wound uniformly for a length of 2·2 cm. on a long thin glass tube of 2·5 mm. bore. The heating-coil was of exactly the same length as the emanation-tube, but in order to make sure that the heat distribution was the same for the emanation-tube and for the heating-coil, a copper cylinder 2·7 cm. long and 0·2 mm. thick was placed over the heating-coil. The whole arrangement was placed symmetrically in the glass tube of 5 mm. bore, over which was wound one of the platinum balance coils P (fig. 1).

The procedure of an experiment was as follows. The balance of the platinum coils was accurately adjusted and the current through the coils kept constant. The emanation-tube in equilibrium with the active deposit was introduced in the platinum coil in a definite position, and the maximum deflexion of the galvanometer observed. A steady deflexion was reached in less than ten minutes. The emanation-tube was then withdrawn by the glass handle, and a known constant current from a storage-cell passed through the manganin coil to give nearly the same maximum deflexion as that due to the emanation. The current was then cut off and the emanation-tube again introduced. Alternate measurements of the heating effect of the emanation and of the current were made for a period of two hours. The emanation-tube was then removed and the change of the balance in the interval determined. The change of balance due to slight alterations of the temperature of the room was usually found to be quite regular and small, and could be easily corrected for if necessary.

In order to determine the heating effect, it was necessary to measure accurately the amount of emanation at any moment in the tube and the current through the heating-coil. The γ-ray effect of the emanation-tube at a definite time was compared in terms of the Rutherford-Boltwood standard by the electroscopic method, and also by the balance method developed by Rutherford and Chadwick. The authors are indebted to Mr. Chadwick for his kind assistance in these measurements. The results obtained by the two methods were in good agreement. A correction of 0·3 per cent. was made for the absorption of the γ-rays in the walls of the emanation-tube. The heating effect of the emanation was assumed to decrease exponentially with a half-value period of 3·85 days. This period of decay was verified on several occasions by direct measurement of the heating effect. The current through the heating-coil was determined by measurement of the E.M.F. of the accumulator by a carefully standardized voltmeter, and the total resistance of the circuit.

The measurements of the heating effect made with different quantities of emanation were in close accord, and the mean of each series of measurements agreed within 1 part in 500. In this way it was found that the heating effect of a quantity of emanation which gave the same γ-ray effect as one gram of radium (Rutherford-Boltwood standard) was 97·95 $\pm$ 0·05 gram calories per hour under the experimental conditions.

L

It is necessary, however, to correct this value to obtain the heating effect of one curie of emanation, *i.e.* of the quantity of emanation in radioactive equilibrium with one gram of radium. The amount of the products radium A, B, and C in transient equilibrium with the emanation are somewhat greater than the amounts in equilibrium with the same quantity of emanation which is maintained constant. This point has been discussed by Moseley and Makower,* and by Rutherford and Chadwick.† The amount of radium B is 0·54 per cent. and of radium C 0·89 per cent. in excess of the true equilibrium amount. Moseley and Makower showed that under ordinary experimental conditions, radium B provides about 11 per cent. of the γ-ray effect due to the emanation, and radium C 89 per cent. We have seen earlier that radium C contributes about 40 per cent. of the heating effect of the emanation. Taking these factors into account, it can be deduced that the heating effect of the emanation which gives a γ-ray effect equal to that of one gram of radium is about 0·54 per cent. less than corresponds to one curie of emanation in equilibrium with radium.

The heat emission of one curie of emanation thus reduces to 98·5 gram calories per hour in terms of the laboratory standard. By the kindness of Professor Stefan Meyer of Vienna, the laboratory standard has been compared in terms of the pure radium salt prepared by Hönigschmid. Expressed in terms of the Vienna standard, the heat emission of one curie of emanation is equal to 103·5 gram calories per hour. In the experimental arrangement the β rays traversed a thickness of glass, copper, &c., equivalent to a weight of 0·354 gram per square cm. More than 90 per cent. of the energy of the β rays was absorbed and added its heating effect to that of the α rays.

Heating Effect of the β and γ Rays

Before comparing the observed heating effect with the calculated value, it is necessary to determine how much of the heating effect observed is to be ascribed to β and γ rays. A number of experiments were made to form an estimate of the magnitude of these effects. In the experimental arrangement described the greater part of the β rays was absorbed in the glass tubes, heating-coil, and copper tube surrounding it. The heating effect under these conditions will be taken as 1. Experiments were first made to determine the alteration of the heating effect when a lead cylinder 1·2 mm. thick, which completely absorbed the β rays and some of the soft γ rays, was substituted for the copper cylinder. As a result of a series of measurements, the heating effect was found to be 1·02.

A series of measurements were then made to determine the heating effect of the γ rays. For this purpose about 4 metres of platinum wire were wound on the outside of two similar thin-walled test-tubes of 1·5 cm. diameter. Each of these was inside a metal cylinder of 6·5 cm. diameter, closed at one end and immersed in a water-bath (fig. 5). Each of the test-tubes was filled with

* Moseley and Makower, Phil. Mag. xxiii. p. 302 (1912).   † *Loc. cit.*

mercury so that the mercury extended about 5 mm. above and below the platinum coil. After the balance had been obtained the emanation-tube, surrounded by its heating-coil, as in previous experiments, was fixed in the centre of one of the mercury columns and the steady deflexion of the galvano-meter determined. The thickness of mercury traversed by the rays was 4·4 millimetres. On account of the large quantity of mercury and the distance between the platinum coils and the outside cylinder, there was a marked lag between the deflexion and the heating effect. For example, the deflexion reached half its maximum value in ten minutes.

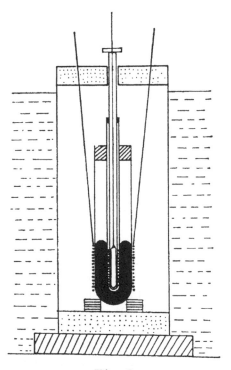

Fig. 5

It is difficult to determine the heating effect with the same accuracy as for smaller coils; but a number of fairly concordant experiments gave a heating effect of 1·034. Similar experiments were made with a thickness of mercury of 1·46 cm. The deflexion in this case reached half its maximum value in 23 minutes. On account of the large lag, the small variations in the balance point during the time of observation became more important. Suitable corrections were made for the lag of the apparatus, and also for the effect of the decay of the emanation. As a mean of several observations the heating effect was found to be about 1·05. It is, however, difficult to fix the value of the heating effect in this case closer than half of one per cent.

The heating effect due to the γ rays from radium could be determined with greater accuracy with a preparation of radium in equilibrium, for under such conditions there is no necessity, as in the case of the emanation, to take into account the decrease of the heat emission of the source.

It is now necessary to consider what fraction of the energy of the β rays was absorbed in the arrangement of the heating apparatus shown in fig. 1, by which the heating effect of the emanation was accurately determined.

Eve* has made an estimate of the relative energy emitted by the α, β, and γ rays from one gram of radium on the assumption that the energy of the radiation emitted is proportional to the total ionization produced. By this method he deduced that the β rays from radium contributed about 2 per cent., and the γ rays about 4·5 per cent. of the total heat emission. In his arrangement, however, the radium was enclosed in a glass tube which must have absorbed a considerable fraction of the soft β rays, and the correction for the ionization of these soft rays was uncertain. It was thought desirable to repeat the experiments made by Eve, using the radium emanation in place of radium, and compressing the former in a very thin glass tube, which allowed the α rays to escape freely. The stopping power of the glass tube corresponded to only 2 cm. of air. Mr. Moseley kindly assisted in these experiments, a more detailed account of which will be published later by Mr. Moseley and Mr. Robinson.

The method originally used by Eve is very suitable for the purpose for which it was designed, and was employed in these experiments. The ionization produced by the β and γ rays from the emanation-tube in a thin-walled ionization-chamber supported in the middle of a room was measured for different distances of the tube extending up to 12 metres. The ionization current in the chamber was directly measured by an electrometer using a balance method. Taking the ionization in air due to the α rays from radium in equilibrium as 100, the total ionization in air due to complete absorption of the β rays was about 3·8, and for the γ rays about 5·2. The ionization due to the γ rays is somewhat greater than that found by Eve, but the ionization due to the β rays is nearly twice as large. In Eve's experiment a large part of the soft β rays was absorbed in the radium tube. By placing over the emanation-tube the copper and glass tubes &c. used in measuring the heating effect of the emanation, it was found that about 85 per cent. of the total ionization due to the β rays was cut out. This was an unexpectedly large fraction, but was confirmed in several experiments.

Remembering that the α rays from the emanation and its products provide about 80 per cent. of the heating effect due to the α rays from radium in equilibrium, it follows that the heating effect of the β and γ rays absorbed in the experimental arrangement described in § 2 was 4·2 per cent. of that due to the α rays. This is, of course, based on the assumption that the total ionization produced by the α and β rays is a measure of their relative energy.

From the ionization measurements, it follows that the total heating effect of the γ rays from the emanation should be 6·5 per cent. of that due to the α rays. It is difficult to estimate with certainty the fraction of the γ rays

* Eve, Phil. Mag. xxii. p. 851 (1911).

absorbed by the thickness of mercury of 1·46 cm., but it was probably about 70 per cent. The heating effect of the γ rays should thus be about 4·6 per cent. of that of the emanation. This is in fair accord with the observed increase of heating effect of 5 per cent., of which probably about 0·5 per cent. was due to β rays.

It should be pointed out that the increase of the heating effect of 2 per cent. observed when the copper cylinder was replaced by a lead cylinder 1·2 mm. thick, is greater than would be expected from the ionization results. The value was undoubtedly nearly correct, for it was verified in a number of experiments. It is well known that lead shows an abnormal absorption for soft γ rays, and the heating effect observed is no doubt due partly to the absorption of the more penetrating β rays and partly to the soft γ rays. The heating effect observed for γ rays is in reasonable accord with the value calculated from the ionization, and indicates that the underlying assumption is not much in error. Since the ionization observed for γ rays is mainly, if not entirely, due to the liberation of β rays from the matter which the γ rays traverse, it seems probable that the ionization method can be used with confidence to estimate also the energy of the β rays.

In 1910 Pettersson* made a number of careful observations by balance methods of the heating effect of β and γ rays from a radium preparation. The heating effect of the radium preparation was 116·4 when the rays were absorbed in 4 mm. of lead, and 114·5 when the lead was replaced by aluminium 2 mm. thick. The rays in both cases passed through absorbing material equivalent to 4 mm. of aluminium before entering the lead or aluminium cylinder. From the measurement of ionization, it is clear that the difference is due not to the heating effect of the β rays as assumed by Pettersson, but mainly to the absorption of the γ rays by the lead. It seems certain that nearly all the energy of the β rays is absorbed in traversing aluminium 4 mm. thick.

The results obtained for the heating effect of one curie of emanation under various conditions are tabulated below.

	Heat emission of one curie of emanation in gram calories per hour			
Screen	α rays	β rays	γ rays	Total
Equivalent to 1·3 mm. of aluminium	99·2	4·2	0·1	103·5
0·7 mm. Al + 1·2 mm. of lead	99·2	4·8	1·5	105·5
,, + 4·4 mm. of mercury	99·2	4·8	3·0	107·0
,, + 14·6 mm. of mercury	99·2	4·8	4·7	108·7

These results are expressed in terms of the Vienna Radium Standard.

§ 3. *Calculation of Heating Effect of Radium and its Emanation*

The energy of the α particles and recoil atoms liberated from one gram of

* Pettersson, *Ark. f. Mat. Astr. och Fysik*, vi. p. 26, July 1910.

radium or of any of its products in equilibrium with it can readily be calculated. This energy is equal to

$$\tfrac{1}{2}mnu^2\left(1 + \frac{m}{M}\right) = \frac{1}{2}\frac{mu^2}{e} \cdot ne \cdot \left(1 + \frac{m}{M}\right)$$

for each of the α-ray products concerned. The value $\dfrac{mu^2}{2e}$ was directly determined by Rutherford for the α particle from radium C by measurement of the electrostatic deflexion of the rays, and found to be $4 \cdot 21 \times 10^{14}$. Taking the velocity of an α particle to be proportional to the cube root of its range, the corresponding values of $\dfrac{mu^2}{2e}$ for radium, emanation, and radium A, are $2 \cdot 56$, $2 \cdot 95$, $3 \cdot 13 \times 10^{14}$ respectively. The masses of the recoil atoms from radium, emanation, radium A, and radium C, are 222, 218, 214, and 210 respectively. The value of $1 + \dfrac{m}{M}$ is consequently slightly less than $1 \cdot 02$ for each product.

The value of ne, the total charge carried by the α particles from one gram of radium itself, has been found by Rutherford and Geiger to be $1 \cdot 054 \times 10^{-9}$ e.m. unit. Substituting these values, the emission of energy is for one gram of radium in equilibrium $1 \cdot 38 \times 10^6$ ergs per second, and for the emanation in equilibrium with it $1 \cdot 10 \times 10^6$ ergs.

The corresponding heating effect of one gram of radium for complete absorption of α particles is 118 gram calories per hour, and for the emanation $94 \cdot 5$ gram calories per hour.

These results are expressed in terms of the Rutherford standard, for the value ne depends on this standard. Correcting in terms of the Vienna standard, the corresponding heating effects are 124 and 100 gram calories per hour.

We have seen (§ 2) that the heating effect of one curie of emanation on the Vienna standard is $103 \cdot 5$ gram calories per hour. Since under the experimental conditions probably 5 per cent. of this is due to the β rays, it is seen that the calculated and measured values for one curie of emanation are in good agreement.

St. Meyer and Hess found that the heating effect of one gram of radium in terms of the Vienna standard was $132 \cdot 3$ gram calories per hour. This includes the heating effect of the β rays and 15 per cent. of the γ rays. This is in excellent accord with the value deduced from the observed heating effect of the emanation. The heating effect of one curie of emanation surrounded by $1 \cdot 2$ mm. of lead was $105 \cdot 5$ gram calories per hour. The heating effect due to α particles and recoil atoms is probably about $98 \cdot 5$ gram calories. The theoretical rates of the heating effects for radium in equilibrium compared with its emanation is $1 \cdot 255$. The heating effect of one gram of radium surrounded by $1 \cdot 2$ mm. of lead thus comes out to be $123 \cdot 6 + 7 = 130 \cdot 6$ gram calories per hour. Allowing a small correction for the extra absorption of γ rays in the Vienna experiments, this value is in close accord with that found by Meyer and Hess.*

Too much stress should not be laid on the agreement of the calculated value of the heating effect of radium with that deduced experimentally, for the data used in the calculations are not fixed with the accuracy required. For example, the calculation depends on the accuracy of the values $\dfrac{mu^2}{e}$ and $n.e.$ The former was determined by measuring the electrostatic deflexion of the α rays. A combination of the value so found with the value $\dfrac{mu}{e}$ found by deflexion of the same rays in a magnetic field gave a value $e/m = 5\cdot07 \times 10^3$. There is now no doubt that the α particle is a helium atom carrying two charges, and the value of e/m should be $4\cdot84 \times 10^3$. Taking the value $\dfrac{mu}{e} = 4\cdot06 \times 10^5$ found by Rutherford for the α particle from radium C as correct, and assuming $e/m = 4\cdot84 \times 10^3$, the value $\dfrac{mu^2}{2e} = 3\cdot99 \times 10^{14}$ instead of the value $4\cdot21 \times 10^{14}$ used in the calculation. Taking the new value, the calculated heating effect is reduced 5 per cent., and the agreement between calculation and experiment is not so good. In order to settle this point, experiments are now in progress to redetermine the values of u and e/m of the α particle from radium C.

From the data already given, the distribution of the heating effect between radium and its products and radiations is given below.

	Heating effect in gram calories per hour corresponding to one gram of radium			
	α rays	β rays	γ rays	Total
Radium†	25·1	..	..	25·1
Emanation	28·6	..	..	28·6
Radium A	30·5	..	..	30·5
Radium B ⎱ Radium C ⎰	39·4	4·7	6·4	50·5
Totals	123·6	4·7	6·4	134·7

It follows that the total heating effect of one gram of radium for complete absorption of the α, β, and γ rays should be about 135 gram calories per hour per gram on the Vienna standard.

University of Manchester
1912

* *Editor's footnote:* Some of the figures quoted in this paragraph are slightly different from those given in the table at the end of § 2 (p. 325). If the latter had been used, the final result for the heating effect of 1 gram of radium surrounded by 1·2 mm. of lead would have been 130·8 gram calories per hour.

† During the publication of this paper, V. F. Hess (*Wien. Ber.* cxxi. p. 1, 1912) has published the results of a direct determination of the heating effect of radium freed from all its products. The value obtained, 25·2 calories per hour per gram, is in excellent accord with the value calculated from the measurements on the emanation given above.

A New International Physical Institute

From *Nature*, **90**, 1913, pp. 545-6

IN the year 1911 an account was given in this journal (vol. lxxxviii. p. 82) of a conference of scientific men in Brussels to discuss the general theories of radiation. This meeting, which was of unusual interest and importance, was due to the initiative of Mr. Ernest Solvay, of Brussels. At the conclusion of the meeting, Mr. Solvay offered to donate a sum of money to assist scientific research in the domain of physics and chemistry. After consultation with Prof. Lorentz, of Leyden, the president of the meeting, Mr. Solvay agreed to found an International Physical Institute for a limited period of thirty years, to have its headquarters at Brussels. The resources of the institute were provided by the generous donation of a capital sum of one million francs. Part of the proceeds is to be devoted to the foundation of scholarships for the promotion of scientific research in Belgium, part to defray the expenses of international meetings to discuss scientific problems of interest, and the residue to be awarded in the form of grants to scientific investigators to assist them in their researches.

For the first year, which terminates on May 1, 1913, a sum of about 17,500 francs is available for the latter purpose. It is the intention of the committee each year to give grants for special lines of work. As the first international meeting was engaged in the discussion of the theories of radiation, it is proposed this year to assist preferentially researches on the general phenomena of radiation, comprising Röntgen rays and the rays from radioactive bodies, general molecular theory, and theories of units of energy. The grants will be awarded without distinction of nationality by the administrative committee of the institute on the recommendation of the international scientific committee.

The administrative committee is composed of Profs. P. Heger, E. Tassel, and J. E. Verschaffelt, of Brussels; the scientific committee is composed of H. A. Lorentz (Haarlem), Mme. Curie (Paris), M. Brillouin (Paris), R. B. Goldschmidt (Brussels), H. Kamerlingh-Onnes (Leyden), W. Nernst (Berlin), E. Rutherford (Manchester), E. Warburg (Berlin), and M. Knudsen, secretary (Copenhagen).

The requests for subsidies should be addressed before February 1, 1913, to Prof. H. A. Lorentz, Zijlweg 76, Haarlem, Holland. They should be accompanied by definite information on the problem to be attacked, the methods to be employed, and the sum required. Definite regulations have been drawn up for the administration of the institute and for the periodical change of the members of the international scientific committee, which are intended to be representative of the active scientific workers in physics and chemistry in Europe.

Mr. Ernest Solvay has in the past been a very generous supporter of science, and has been responsible for the endowment of several scientific institutes in Brussels. The new Solvay International Institute, which is due entirely to the generosity of Mr. Solvay, is unique in character, and promises to be of great value to science. It will offer an admirable opportunity for scientific men of all nations to meet together and to exchange views on questions connected with physics and chemistry, and to obtain a consensus of opinion as to the best direction in which grants should be given to extend or deepen our knowledge of special subjects. As the funds available for distribution are limited, the decision of the committee to restrict the grants for each year to investigations in a special department of science seems a wise one, and should be more fruitful in results than if the money were distributed in small sums over a wide field of scientific inquiry. The subjects for which grants are available will, no doubt, be changed from time to time in accordance with the decision of the international committee.

E. RUTHERFORD

L*

The Age of Pleochroic Haloes

by J. JOLY, F.R.S., *and* E. RUTHERFORD, F.R.S.

From the *Philosophical Magazine* for April 1913, ser. 6, xxv, pp. 644–57

IT is now well established that the minute circular marks seen in sections of certain coloured rock minerals—notably the coloured micas—are due to the effects produced by the alpha radiation of a central radioactive particle. The circular mark, in fact, represents the section or projection of a sphere defined by the range in the particular containing mineral of the most penetrating alpha ray emitted. If the parent radioactive substance is uranium, the fully developed halo is defined by a sphere having the radius of the furthest reaching ray, RaC. If the parent substance is thorium, the extreme radius of the sphere will be defined by the range of ThC.

Haloes are found in which the effects of other and less penetrating alpha rays of the uranium and thorium families may be clearly shown, and in some the quantity of radioactive material is so small that the halo may be described as 'under-exposed'; so that the maximum effects of RaC or ThC being distributed over a relatively large spherical surface or shell, may be faint or absent, and the containing mineral only notably darkened by those inner rays whose actions are more concentrated owing to the lesser spherical shells to which their maximum ionization effects are confined. Thus, in the case of the uranium derivatives, the halo may be limited by the range of RaA, by Ra emanation, by Ra, or even by ionium or uranium.

The halo is in every case the result of the integral actions of rays emitted since a very remote period. Haloes in the younger rocks are unknown. The quantity of radioactive material involved is very small. More especially haloes of true spherical form are necessarily formed around very minute nuclei, so that even if these were entirely composed of a parent radioactive element, it is only by the integration of effects over a very prolonged period that any results are brought about.

It is of interest to seek some estimate of the time which may have been required to generate these haloes. This can be done if the following data are available:—

(*a*) The number of alpha rays which will produce a certain intensity of staining in a particular mineral.

(*b*) The mass of the nucleus of a halo of similar or comparable intensity of staining in this same mineral, and from this an estimate of the quantity of radioactive substances which may be concerned in generating the halo.

It is evident that while we can obtain the numerical values involved in (*a*) with a considerable degree of accuracy, certain assumptions enter into the numerical values required in (*b*), which render them uncertain within particular limits. Thus, while the dimensions of the nucleus may be determined with fair accuracy, and its mineral nature inferred with considerable confidence, we have no means, at present, of ascertaining the amount of radioactive material it contains. This amount can only be stated on the basis of analyses made on specimens of the mineral large enough to be subjected to examination. Evidently, however, a probable minor limit to the age is obtained by making our assumptions exceed by a safe margin any of the ascertained results.

The following experiments and observations have been confined to the brown mica, Haughtonite, of Co. Carlow. In this mica, haloes due to the uranium family of elements are very beautifully defined, and are found in every stage of development; from the smallest, due to uranium or ionium, to fully darkened haloes completed to the range of RaC. The mica is of a rich clear brown colour, with high lustre and perfect cleavage. Particulars with reference to this mica and of its contained haloes have been given elsewhere.*

The age of the containing granite is probably late Silurian or early Devonian. The evidence for this is found in its relations with the surrounding rocks. It has upheaved and metamorphosed slates of Silurian age, and fragments of it are contained in Old Red Sandstone sediments which rest unconformably on the upturned edges of the Silurian slates, and overlap directly on to the granite. Its upheaval is generally referred to the Caledonian earth-movements.†

Some of the haloes may have been formed somewhat subsequent to the crystallization of the containing mica. This is suggested (*a*) by the fact that some of the crystallized nuclei giving rise to the haloes are oriented in the plane of cleavage; their greatest dimensions coinciding with that plane as if they had been developed *in situ*; and (*b*) by the existence of veins traversing the biotite which evidently contain radioactive substances, seeing that they are bordered by the characteristic staining, and faithfully reproduce the appearance of a tubular halo.‡ A photograph of these veins is given in 'Bedrock' for January 1913. Such conduits of radioactive substances often diverge from heavily-stained areas located on the outer margins of the crystal. Haloes are sometimes found linearly arranged along the veins, showing that they partook of the radioactive substances contained in them. It is known that towards the close of the period of consolidation of a granitic mass, mother liquors, rich in the rarer elements concentrated from the magma, make their appearance and penetrate the granite, and often the adjacent rocks. The genesis of radioactive ore deposits seems in many cases traceable

---

* Phil. Mag. April 1910; Proc. Royal Dublin Soc. xiii. p. 73; 'Bedrock,' Jan. 1913.

† Mineralogical and geological particulars respecting the granite are given by Sollas (Trans. Royal Irish Academy, xxix., Jan. 1891). The Geological Survey of Ireland Reports on Co. Dublin and Co. Carlow &c., may also be consulted.

‡ Rutherford, Phil. Mag. Jan. 1910. (*This vol., p. 178.*)

to this phenomenon. It is not impossible that the venation of the mica is referable to similar developments on a minute scale. This, however, cannot have been long subsequent to the period of consolidation. The haloes may most safely be regarded as of early Devonian age.

The experiments on the number of alpha rays required to produce notable staining of the Carlow mica were carried out in the Physical Laboratory of the University of Manchester. The mica was placed beneath and in contact with a lead plate which was perforated with a circular aperture 0·42 cm. in diameter. At a distance of 1·5 cm. vertically over the centre of the aperture a capillary alpha-ray tube, containing 25 millicuries of radium emanation to start with, was fixed. The whole was placed under a bell-glass in which a partial vacuum could be made. The thickness of the walls of the alpha-ray tube was equivalent to 1·4 cm. of air at normal density. The average range of the three sets of alpha rays falling on the mica was deduced as 3·5 cm. of air. This would ensure a penetration of about 0·016 mm. in the mica. The quantity of emanation used and the duration of the exposure enable direct determination of the number of alpha rays required to produce a particular effect.

Two cleavage flakes of the mica were dealt with. To one of these flakes three exposures on adjacent areas were given. On the second flake two exposures on adjoining areas were given. Calling the first three exposures A, B, C, and the second D, E, the number of alpha rays in each case was:—

Spot A received $3 \cdot 7 \ 10^{13}$ alpha rays per square cm.

,,	B	,,	1·6	,,	,,	,,	,,
,,	C	,,	1·5	,,	,,	,,	,,
,,	D	,,	3·7	,,	,,	,,	,,
,,	E	,,	1·5	,,	,,	,,	,,

The amount of darkening of the mica produced by these exposures was sensibly the same for A and D and for C and E. B was slightly darker than C or E. The depth of staining was, therefore, very consistently dependent on the exposure. In some places the areas overlapped a little, and this part of the mica faithfully produced an effect due to the added densities of the overlapping spots.

A subsequent measurement of the thickness of these flakes of mica, by focussing with a high power on the upper and lower faces of the flakes as they rested on the stage of the microscope, showed that the flake containing A, B, and C was where thinnest 0·022 mm. thick, and the second flake, containing D and E, was where thinnest 0·014 mm. thick. The close agreement in the effects of the staining on the two flakes shows that the whole of the alpha rays must have been absorbed in the second flake as well as in the first.

The intensity of the darkening of the mica in these experiments is such as to best admit of comparison with the less developed haloes: those darkened uniformly out to the range of Ra (0·0156 mm.), or RaF (0·0177 mm.), or

Ra Emanation (0·0196 mm.). The darkening is too faint to be measured against the more exposed haloes which extend to RaA or RaC. These are, indeed, generally almost opaque centrally. Observations involving comparison of the halo with the experimentally produced staining are, then, necessarily restricted to the less exposed haloes, whose radial dimensions may be assumed to be 0·016, 0·018, and 0·020 mm.

From the experiments it may be deduced that the numbers of rays falling upon circular areas of these radial dimensions which will give the depth of staining of spots A, B, C, will be:—

Radius of halo	0·016	0·017	0·018	0·020
Intensity of A	296×10^6	337×10^6	370×10^6	466×10^6
,, ,, B	128 ,,	146 ,,	160 ,,	202 ,,
,, ,, C	120 ,,	136 ,,	150 ,,	189 ,,

The observations on the haloes, designed to afford an estimate of their age, must be directed, as already intimated, to estimating the rate at which alpha rays are being emitted by the nucleus. We have to consider also how the

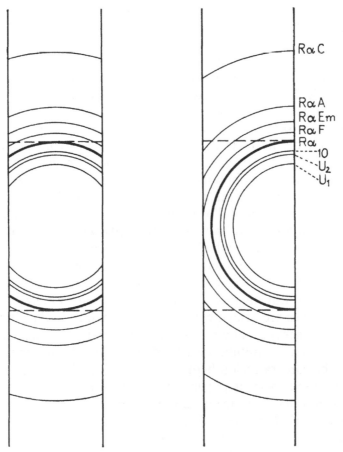

Fig. 1

staining effects produced by such radially emitted rays may be brought into comparison with the parallel rays used in the experiment. The cleavage flakes of mica used in the observations to be presently recorded average from 0·016 to 0·018 mm. in thickness, thicker flakes being too dark for convenient observation. The accompanying diagram of a flake 0·017 mm. in thickness, containing a nucleus placed centrally in the flake and again near its surface, shows that in the case of haloes extending radially 0·016 mm., the total ionization is by no means included within the flake. Only four out of the eight rays complete their ranges within the semi-halo. The maxima of RaC and RaA are wholly excluded. In the case of the centrally placed nucleus the maxima of RaEm and RaF exert but little effect. These excluded rays, however, affect the central area, where their velocities are greatest and ionization effects least. It is also seen that in cleavage flakes of this thickness as much as one-half the entire effects of the rays may be removed with the adjacent cleavage flakes. When the nucleus is at the surface the hemisphere only is contained in the flake. Any accurate allowance for these deficiencies is impossible, as they vary with the position of the nucleus in the flake. There is a small set off against them in the fact that the rays which in the experiment darken the mica are not wholly included within it. It would seem that a general allowance of 50 per cent. would not be too much to make; that is, we may consider that only half the rays actually emitted by the nucleus are effective in generating that part of the small haloes which enters into the observations.

The measurements on the dimensions of the nuclei of the haloes were made in the Iveagh Geological Laboratory of Trinity College, Dublin. As it is obviously very necessary to secure accuracy in these measurements, care was taken by attention to the best conditions of illumination and magnification and by repetition of the readings to guard against error. Professor H. H. Dixon, F.R.S., independently checked many of the observations. It was found that the best definition was obtained with an oil-immersion of Zeiss (apochromatic 3 mm. N.A. 1·4), and with a No. 7 of Leitz; but the oil-immersion gave the best definition. With careful attention to the lighting the definition could be got very perfect, and there was no uncertainty whatever as to the reliable nature of the measurements, which could be repeated again and again with always the same results. The readings were effected by two distinct methods: (1) by a micrometer eyepiece with travelling line; (2) by the camera lucida combined with the following method of measuring the diameter of the referred image. Two *fine* lines are drawn with a drawing-pen in indian-ink on smooth white paper. The lines meet at a point and very slowly diverge. The separation of the lines was about 3 millimetres at a distance of 15 cm. from the point of intersection. The camera lucida being in position the image is referred in the usual manner to the sheet of paper. The latter is now shifted till the image of the nucleus appears to fit accurately between the lines. While still looking at the object a mark is made with a pencil across the lines just where the nucleus is referred. This is, say, at a distance d_1 from the intersection. An engraved scale divided to 0·01 mm. is now substituted for the mica, and one of the subdivisions brought to fit, as before, between the lines, at a distance d_2 from the intersection. This point is marked. The diameter, x, of the nucleus is then obviously found from the ratio $x : 0\cdot01 :: d_1 : d_2$. This method gave very consistent results in successive measurements.*

In some few cases the diameter of the nucleus is the mean of two not very different diameters, the development of the nucleus not being such as would permit of the assumption that it was of approximately spherical form. In some cases the nucleus possesses a definite crystalline form, apparently dimetric. Two dimensions are then read: the axial length of the prism and its width transversely.

In comparing the density of the staining of the halo with that of the artificially darkened areas, the following procedure was found most satisfactory:—Two microscopes are placed side by side. The halo is observed in one of them, the spot in the other. The magnification is alike and the light is from the same source. Daylight only could be used. The spot covered but half its field, and the adjoining unacted on mica was, when necessary, brought to the exact same colour and luminosity as the mica around the halo by interposition of a tinted solution and manipulation of the substage fittings. In the case when the mica containing the halo is the lighter in colour these adjustments are applied to it. The depth of tint of spot and halo are now compared. These comparisons were always repeated without adjustment of the colour and luminosity of the field. The former were more consistent and more easy to be sure of. Professor H. H. Dixon quite independently checked many of these comparisons. In the majority of cases the comparison was restricted to determining whether the halo should be judged lighter or darker than, or the same as, the spots A or C. Generally there could be no doubt as to the conclusion arrived at. It will be seen presently that the interest of the results principally turns on that conclusion. However, an effort was made to evaluate the difference in staining of the halo and the spot in a few cases. For this purpose a transparent screen for producing different graded amounts of light absorption was prepared as follows:—A 'Process' plate was exposed in the dark room in such a manner that successive strips of it received quantities of light in the proportions $1 : 2 : 3 : 4$, &c. This was done by uncovering successive strips of it at intervals of one second. When developed and fixed, and a slip cut from the plate with a diamond, an absorption screen was available having successive areas darkened approximately in the proportions of the exposures. It was found that spot A was almost exactly of the darkness of the 3 second exposure; this area bringing the field of the spot into agreement with it. From this we may assume that the area receiving 6 seconds exposure is of double the density of spot A.

The mode of using this screen is evident from what has been already said. Colour and luminosity having been adjusted as before so that the fields of both spot and halo are alike, the screen is slipped in above the spot (or halo) till the darkening of these objects is brought into equality.

* Proc. R. D. S. xiii. p. 441.

The nuclei are in most cases possessed of all the optical appearances of zircon; they are transparent, colourless, and of high refractive index. An attempt to determine the index by use of refracting liquids failed. In the cases tried the nuclei were probably buried in the mica. There is no doubt that the index is considerably higher than that of the mica. Some may be orthite; some, possibly, brookite. Many of the nuclei in certain parts of the granite show no definite crystalline shape, but are granular in appearance. In other places they are all in the form of well-developed dimetric crystals placed with their long axis parallel with the cleavage of the mica. These definite crystals are beyond doubt zircons. They seem somewhat less radioactive than the irregular nuclei. Again, some nuclei are plate-like crystals, rectangular in shape and developed in the plane of cleavage. It is not practicable to ascertain the volume of these as their thickness cannot be measured. Probably their mass is often comparable with that of the smallest nuclei.

It would be very desirable to ascertain directly the radium content of these nuclei. An attempt was made by reducing a considerable quantity of the mica to a very fine powder, and then effecting separations according to density by use of heavy liquids, to isolate the zircons. But nothing of a minuteness comparable with the nuclei was extracted. It is probable that their removal from the mica is difficult. It is not safe to use solvents, for there is evidence that zircons may carry their radioactive constituents as a surface accretion and always as a purely foreign admixture. We could not then be sure that a solvent which acted on the mica did not also remove the radioactive constituents of the zircon, thus falsifying the result in the very direction which is most to be avoided.

It would appear that, for the present, we can only estimate the uranium content by analogy with known results obtained on large zircons. Strutt* has given results on 14 zircons of very different geological ages. The quantity of U_3O_8 in grams per gram of zircon varies from $1\cdot4 \times 10^{-4}$ to $75\cdot3 \times 10^{-4}$. The highest result is very exceptional, the next highest being 38×10^{-4}, and this is from the exceptionally radioactive lavas of Vesuvius. After this $13\cdot3 \times 10^{-4}$ is the highest. The mean of all is 14×10^{-4}. In other experiments Strutt found 865×10^{-12} gram of radium per gram in zircons.† This involves less than 40×10^{-4} gram per gram of U_3O_8. He also found‡ $0\cdot34$ millionths of one per cent. of radium in zircon; and in another specimen $0\cdot52$ millionths of one per cent. The latter quantity involves about $1\cdot5$ per cent. of uranium. The closely allied, and probably derived, mineral cyrtolite was found by Strutt‖ to contain $3\cdot67$ per cent. of U_3O_8 and $5\cdot05$ per cent. of ThO_2. Two analyses of cyrtolite are cited by Dana; one showing $1\cdot59$, the other $1\cdot4$ per cent of U_3O_8.

J. W. Waters showed that in gneiss of the Inner Hebrides and granite of Mourne the radioactivity was concentrated mainly in the zircons; in Cornish

* Proc. R. S., A, lxxxiii. p. 298.   † Proc. R. S., A, lxxviii. p. 150.
‡ Proc. R. S., A, lxxvi. p. 312.   ‖ Proc. R. S., A, lxxvi. p. 88.

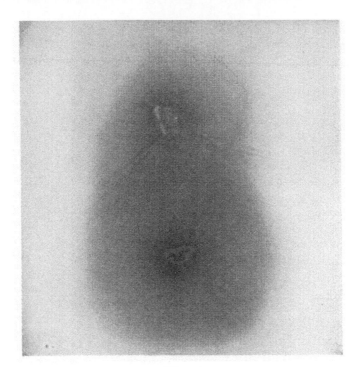

Fig. 1

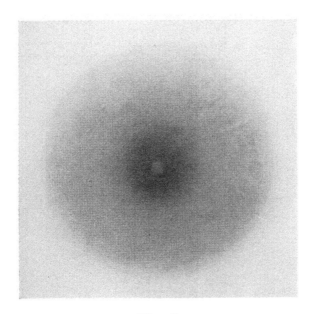

Fig. 2

granite chiefly in anatase or rutile; and in Dalbeattie granite chiefly in orthite. Quantitative results are not given.*

It will be seen that the highest of the above results is that for cyrtolite, which rises to $3 \cdot 67$ per cent. of U_3O_8. This figure does not, presumably, limit the quantity of uranium which might be present. But there is no reason to believe that radioactive constituents ever constitute a large part of zircon. On the contrary, it has been shown by Mügge† and by Gockel‡ that the radioactive matter constitutes an impurity or foreign substance in zircon.

Uranium is not recorded by Dana as a constituent of orthite or brookite. Strutt§ found in one specimen of orthite $0 \cdot 073$ per cent. of U_3O_8, and in another radium equivalent to 3 per cent. of uranium.

From the foregoing results it would seem safe to assume 10 per cent. of uranium as an upper limit to the amount now contained in the nuclei of the haloes in this mica.

The periods of time which emerge from the measurements to be presently cited are such that the wasting of the uranium should be taken into account. We must, accordingly, calculate back, deriving the mass of transformed uranium from the number of alpha rays required to form the halo according to the observations and the experimental data. The estimated existing mass of uranium in the nucleus thus becomes the amount remaining after the time t; so that if M is the mass transformed and W the existing mass, $W/(M + W)$ is the fraction remaining, or $e^{-\frac{t}{L}}$. Exponential tables give the value of t/L corresponding to $e^{-\frac{t}{L}}$, where L is the average life of uranium. From this t is found. As an example of the mode in which the figures in the ensuing table are derived:—halo No. 4 has a nuclear mass of 418×10^{-14} gram, and the uranium present is taken as 418×10^{-15} gram. The radius of the halo is $0 \cdot 016$ mm., and as it is darker than spot A we calculate that the number of alpha rays required to produce this halo is not less than $2 \cdot 95 \times 10^8$ rays, or $5 \cdot 9 \times 10^8$ for the whole halo. Now 8 rays are produced for every atom of uranium transformed. Thus the number of uranium atoms transformed has been $0 \cdot 74 \times 10^8$. Taking the mass of one atom of hydrogen as $1 \cdot 6 \times 10^{-24}$,‖ the mass of an atom of uranium is 381×10^{-24}. The mass transformed is, hence, 28×10^{-15} gram. The quantity present being assumed as 418×10^{-15}, this becomes $0 \cdot 9373$ of the original mass. The exponential tables now give the corresponding value of t/L as $0 \cdot 065$, and as L is $7 \cdot 2 \times 10^9$, the value of t is found to be $0 \cdot 47 \times 10^9$ years. The age is greater than this as the darkening of the halo exceeds that of spot A, but in this case no attempt was made to evaluate this excess.

The following table contains the results of 30 measurements of nuclei. There would have been no difficulty in adding to this number almost indefinitely, but it is thought that the number given is sufficient. Moreover, considerable

* Phil. Mag. 1909 and 1910.   † *Centralbl. f. Min.* &c., 1909, p. 148.
‡ *Chemiker Zeitung*, 1909, No. 126.   § Proc. R. S., A, lxxvi. p. 312.
‖ Rutherford and Geiger, Proc. R. S., A, lxxxi. p. 162. (*This vol., p. 109.*)

time and labour are required to complete the careful measurement of each halo. In Nos. 24, 25, and 26 the nuclei are well developed, apparently dimetric, crystals with prismatic and terminal faces. The mass is probably somewhat exaggerated, as outside dimensions are given. No allowance is made for a slight elongation of the derived halo. Both these facts will tend to lower the age.

The nuclei of Nos. 29 and 30 have been photographed, and are reproduced in the Plate, No. 29 (fig. 1) is magnified 1450 diameters; No. 30 (fig. 2) is magnified 1400 diameters. It is very difficult in such photographs to show at once the nucleus and the halo, for the intense conditions of exposure and development requisite to bring out the former are quite unsuited to reproduce the latter. The nucleus of No. 29 is characteristic of the granular type. Near it is a relatively large nucleus of well crystallized zircon. It has given rise to a less intense halo. The halo, No. 29 (which is developed round the small nucleus), is of a rich brown colour and certainly darker than spot A. Centrally it is so dark that the exposure required to take it has resulted in complete over-exposure of the surrounding halo. In calculating the volume of this nucleus it is assumed to be ellipsoidal in shape, and the mean diameter is taken as that of the equivalent sphere. In No. 30 the nucleus is rectangular in outline and is some distance down in the mica. It is apparently a flat, rectangular crystal; but its volume has been calculated on the assumption that it is cubical. It was easier to photograph this halo owing to the fact that the cleavage flake in which it is contained is so thin as only to be of a pale straw-yellow. Notwithstanding its thinness the halo is not less than twice as dark as spot A.

It must be borne in mind that the higher values of the age are those of most significance. Low values of course are deduced whenever the mass of the nucleus is large and the surrounding halo faint. Although the darkness of the halo generally varies with the size of the nucleus, there is no definite proportion of uranium in the zircon or other minerals forming nuclei, and cases of apparent low age are plentiful. They are explained if we suppose that in these cases our assumption as to their uranium content is excessive. What is significant is the fact that in many cases we can find nuclei so small surrounded by well darkened haloes.

In many examples given in the table the age must exceed that which is given in the last column. This is so because only in some of the last observations, and in those in which the halo happened to match closely the staining of one of the spots, is there any numerical evaluation of the relative darkness of spot and halo. The results are numbered in the order in which they were obtained. The highest of the measurements point to an age not less than four hundred millions of years as the time required to generate these haloes: in other words, as the age of the early Devonian. The treatment of the subject throughout this paper has been such as to render this a minor limit.

Sources of error must now be inquired into:—

(*a*) The nuclei are the *complete* nuclei. The possibility that the nuclei measured were detached fragments of larger nuclei, the greater part being cleaved off in opening the mica, was investigated by comparing the contiguous faces of cleaved flakes and making sure that the part of the halo-sphere removed contained nothing that could be regarded as part of the nucleus, or

No.	Diam. of nucleus cm. $\times 10^5$	Mass of Zircon grm. $\times 10^{14}$	Radius of Halo cm. $\times 10^4$	Comparison of Halo with intensity of spots A, B, or C	No. of α-rays for whole halo $\times 10^{-8}$	Mass of transformed U grm. $\times 10^{15}$	Age, years $\times 10^6$
1	10	243	18	Dark as C	3·0	14	400
2	38	13481	20	Nearly opaque	9·3	43	> 20
3	16	990	16	Darker than A	5·9	28	>200
4	12	418	16	,, ,,	5·9	28	>470
5	33	8800	16·5	,, ,,	6·4	30	> 30
6	37	12320	16·5	,, ,,	6·4	30	> 20
7	20	1980	16	Between A and B	4·2	20	80
8	23	3080	16	Very much darker than A	5·9	28	≫ 60
9	12	418	16	Dark as B	2·6	12	200
10	14	660	16	Darker than A	5·9	28	>270
11	21	2420	16	Much darker than A	5·9	28	≫ 90
12	22	2640	16	Darker than A	5·9	28	> 80
13	26	4400	16	Very much darker than A	5·9	28	≫ 50
14	13	506	16	Dark as A	5·9	28	390
15	30	6600	16	Very much darker than A	5·9	28	≫ 30
16	17	1210	16	Darker than A	5·9	28	>160
17	25	3740	16	Very much darker than A	5·9	28	≫ 50
18	15	792	16	Dark as A	5·9	28	250
19	27	4840	18	, ,,	7·4	35	50
20	27	4840	17	,, ,,	6·8	32	50
21	24	3388	17	Darkness = A × 2	13·6	64	140
22	26	4400	17	Darkness = A × 7/3	15·8	75	120
23	23	3080	17	Darkness = A × 7/3	15·8	75	170
24	57 × 18 × 18	8580	17	Dark as A	6·8	32	30
25	48 × 32 × 32	22000	17	Darkness = A × 2	13·6	64	20
26	40 × 22 × 22	9020	17	Dark as A	6·8	32	30
27	12·5	462	17	Dark as C	2·7	13	200
28	21	2420	18	Dark as A	7·4	35	100
29	29	5984	16	Darker than A	5·9	28	> 40
30	21 × 21 × 21	5720	19	Darkness about A × 2	16·5	78	100

that could suggest the loss of any part of it. Not a single case of a divided nucleus was met with. The fact is the cleavage of zircon is very imperfect, and even if it lay just at the surface of cleavage it would almost certainly remain bodily in the one flake or the other; or, possibly, drop out. Again, many of the observed nuclei were demonstrably beneath the surface of the flake. This was rendered quite certain by observations on thick cleavage flakes, using a very strong light.

(*b*) The crystallographic direction in which a part of the rays traverses the mica in the case of the halo, is not the same as that in which the rays move in mica when generating the experimental darkening. The fact, however, that in sections of haloes there is uniform darkening in all crystallographic directions, negatives the idea that error can arise from this source.

(*c*) Nothing is known of any mode by which the mere passage of time can intensify a halo or increase the effects of the original ionization. In the case of light-sensitive salts the ionization set up by exposure appears to weaken by the passage of time. No intensification of the latent image is known. A recombination of the products of dissociation and weakening of the halo with time would rather be expected. The halo is probably very stable.

(*d*) The rate of formation of the experimental staining is very fast compared with the rate of formation of the halo. It does not seem likely that any significance is to be ascribed to this fact. For the distribution of the alpha rays projected into the mica in the experiment is actually a very sparse one. Molecularly speaking they are finally separated by very great distances; so that the molecular groups, in which the entering rays expend their energy, remain generally isolated and undisturbed by succeeding rays. The interval of time between succeeding rays would not, therefore, appear to matter, and the resulting staining would be the same for each ray.

(*e*) Exposure to light does not appear to affect, one way or the other, the staining produced in this mica by the alpha ray. An experiment in which spots D and E were half covered by lead foil and then exposed in bright sunlight, concentrated by a lens, for several hours, revealed no effect whatever. Strong heating obliterates haloes. In the experiment strong heating was avoided.

(*f*) Thorium is absent from these nuclei. No thorium haloes have been found in this mica.

Comparing the value indicated for the age of the Devonian by the foregoing results with estimates arising from the measurement of accumulated radioactive *débris*, it will be seen that the halo indicates a somewhat excessive age. The helium ratio gave Rutherford* 241 millions of years for a specimen of fergusonite, probably from ancient rocks. Strutt, by the same method, arrived at 150 millions of years as the age of the Carboniferous Limestone and 710 millions as the age of the Archæan.† Boltwood's results with the lead ratio ranged from 246 to 1640 millions of years; the higher being referable, probably, to Archæan rocks.‡ Becker, dealing with Pre-Cambrian minerals of the Llano Group, found, by the lead ratio, from 1671 to 11,470 millions of years; regarding his figures as a 'reductio ad absurdum'.§ Boltwood criticised these results, questioning the reliability of the material dealt with. Becker urged in

---

* Phil. Mag., Oct. 1906, p. 368. (*Vol. I, p. 898.*)   † Proc. R. S., 1908–1910.
‡ Am. J. of Sc., 1907.   § Bull. Geol. Soc. Am., 1908.

reply that there was no evidence of any effect on the lead ratio arising from incipient alteration of the material. A. Holmes* investigated the age of a Devonian syenite by the lead ratio, and found this to be 370 millions of years; he rejects, however, results showing a greater age, on the ground that original lead was probably present in sufficient quantity to falsify the result.

The results founded on the accumulation of radio-active *débris* are open to the objection that addition or subtraction of the measured substances may have occurred over the long period of time involved. The results obtained from the halo are less likely to be affected by this source of error. It is of interest to find, therefore, that the results obtained from it are not very discordant with some of the higher results obtained by the lead ratio. There may be significance in this rough agreement, or errors of a nature as yet unknown may affect both determinations. We are not in a position to say. But it is certain that if the higher values of Geological Time so found are reliable, the discrepancy with estimates of the age of the ocean, based on the now well-ascertained facts of solvent denudation, raises difficulties which at present seem inexplicable. Discussion of hypotheses whereby the reliability of the one method or the other may be called in question is not within the scope of this paper.

* Proc. R. S., June 1911.

The Analysis of the γ Rays from Radium B and Radium C

by PROFESSOR E. RUTHERFORD, F.R.S., and H. RICHARDSON, B.SC.,
Graduate Scholar, University of Manchester

From the *Philosophical Magazine* for May 1913, ser. 6, xxv, pp. 722–34

IT has long been recognized that the penetrating γ rays emitted by a γ ray salt were complex in character. The examination of the radiation has been made by the electric method in two ways:

(1) by measuring the absorption of the γ rays in different materials over a wide range of thickness;

(2) by an examination of the absorption of the secondary and scattered γ rays which appear when γ rays traverse matter.

Initial experiments on the absorption of the γ rays of radium by different materials were made by Rutherford* and McClelland.† These were extended by later investigations of Eve,‡ Tuomikoski,§ Wigger,‖ and S. J. Allan.¶ The experiments showed that the absorption of the γ rays in lead rapidly decreased for the first two centimetres of thickness, but became approximately exponential for greater thicknesses. The whole question was re-examined with great detail and thoroughness by Mr. and Mrs. Soddy and A. S. Russell,** who determined the absorption of the γ rays in a number of materials and investigated the effect of different arrangements on the apparent value of the absorption coefficient. They found that the absorption of the γ rays by lead was accurately exponential for a very wide thickness, viz. from 2 to 22 cm., and concluded that over this range of thickness the γ rays were to be considered as homogeneous in type. These results were confirmed and extended by Russell, who showed that the γ rays from radium were absorbed by mercury over a range of thickness from 1 to 22·5 cm. strictly according to an exponential law. Over this range of thickness the intensity of the ionization current in the testing vessel, which served as a measure of the intensity of the γ rays, varied in the ratio of 360,000 to 1.

* Rutherford, *Phys. Zeits.* iii. p. 517 (1902). (*Vol. I., p. 410*).
† McClelland, Phil. Mag. viii. p. 67 (1904).
‡ Eve, Phil. Mag. xvi. p. 224 (1908); xviii. p. 275 (1909).
§ Tuomikoski, *Phys. Zeit.* x. p. 372 (1909).
‖ Wigger, *Jahrb. Radioakt.* ii. p. 430 (1905).
¶ Allan, Phys. Rev. xxxiv. p. 311 (1912).
** Soddy and Russell, Phil. Mag. xviii. p. 620 (1909); Mr. and Mrs. Soddy and Russell, Phil. Mag. xix. p. 725 (1910); Russell, Proc. Roy. Soc. A. lxxxvi. p. 240 (1911).

From an examination of the quality of the secondary γ rays set up in different materials by the γ rays, Kleeman[*] considered that the primary γ rays from radium could be divided into three types of widely different penetrating power. In similar experiments Madsen[†] found evidence of two types. On the other hand, Florance,[‡] who examined the character and intensity of the secondary and scattered γ rays from radiations of different materials at various angles for the primary beam, concluded that the γ rays were very complex in character and that no definite evidence could be obtained by this method of the existence of distinct groups of primary rays.

It was at first supposed that the penetrating γ rays emitted by a radium salt arose entirely from the transformation of its product radium C. Moseley and Makower,[§] however, showed in 1912 that radium B also emitted γ rays, although weak in intensity and penetrating power compared with those emitted from radium C. Even if radium C emitted only one type of radiation, it was clear from this result that the γ rays from a radium salt must contain at least two types of γ rays. In the meantime, the work of Barkla on X rays had shown conclusively that each of the elements emitted one or more types of characteristic or fluorescent radiations when X rays of suitable penetrating power traversed them. In some of the elements two types of characteristic radiations were observed. J. A. Gray[‖] extended these results to γ rays, for he found that the γ rays emitted by radium E were able to excite the characteristic radiations of certain elements. His results showed, as had long been supposed, that the γ rays were identical in general properties with X rays and possessed the fundamental property of exciting characteristic γ rays. In a paper entitled 'The origin of β and γ rays from radioactive substances,' Rutherford[¶] put forward the view that the γ rays from radioactive substances were to be regarded as the characteristic radiations of the respective elements set up by the escape of α or β rays from them. On this basis an explanation was given of the numerous groups of homogeneous β rays emitted by radium B and C, and their connexion with the γ rays was outlined. If this were the case, each type of characteristic radiation emitted should be absorbed according to an exponential law by an absorbing substance of low atomic weight like aluminium.

The present experiments were undertaken with a view of testing this hypothesis. It will be seen that this analysis brings out that the γ radiation from radium B consists of at least two and possibly of three distinct types, and from radium C of a single type, probably corresponding in penetrating power to the characteristic radiations to be expected from elements of their atomic weight.

[*] Kleeman, Phil. Mag. xv. p. 638 (1908).
[†] Madsen, Phil. Mag. xvii. p. 423 (1909).
[‡] Florance, Phil. Mag. xx. p. 921 (1910).
[§] Moseley and Makower, Phil. Mag. xxiii. p. 312 (1912).
[‖] Gray, Proc. Roy. Soc. A. lxxxvii. p. 489 (1912).
[¶] Rutherford, Phil. Mag. Oct. 1912. (*This vol., p. 280.*)

Experimental Arrangement

In the preliminary investigations the source of γ rays consisted of about 50 millicuries of radium emanation enclosed in an α ray tube with thin walls. The thickness of glass was equal in stopping power for α rays to about 2 cm. of air. In order to get rid of the effect of the primary β rays, the source was placed between the pole-pieces of a powerful electromagnet. The γ rays passed horizontally into a thin-walled ionization vessel ($3 \times 5 \times 7$ cm.) CD, fig. 1, placed at the side of the electromagnet, in which the ionization was measured in the usual way by means of an exhausted electroscope E of dimensions $5 \times 5 \times 5$ cm. The ionization produced by the γ rays in the electroscope E was negligibly small compared with that produced in the ionization vessel CD. The sides of the vessel CD, through which the γ rays passed, consisted of thin sheets of mica equivalent in stopping-power for the α particles to about 2 cm. of air. This was done in order to increase relatively the ionization due to the softer γ rays that might be present. The inside of the vessel was lined throughout with aluminium. In most experiments the pole-pieces of the electromagnet

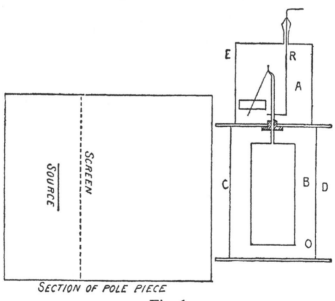

SECTION OF POLE PIECE

Fig. 1

were about 2 cm. apart. The source of γ rays was in all cases more than 9 cm. distant from the vessel CD. The absorbing metal screens were placed between the pole-pieces of the magnet close to the source. The β rays escaping from the absorbing material were removed by the magnetic field before entering the ionization vessel. The γ rays which entered the ionization vessel passed nearly normally through the absorbing screens, so that no correction for obliquity was necessary in determining the absorption coefficient of the rays.

Preliminary experiments showed that if the ionization vessel were filled with air, the effect of the penetrating γ rays from radium C was large compared with that due to the softer types of radiation that were present. When air was used, the reduction of the ionization by using absorbing screens of aluminium of different thicknesses is shown in fig. 2 A, where the logarithms of the ionization are plotted as ordinates and the thickness of aluminium as abscissæ. It will be observed that in this case there is a rapid drop of the ionization corresponding to about 10 per cent. of the whole effect.

In order to bring out prominently the effect of the softer types of γ radiation present, the vapour of methyl iodide was used instead of air. As it was impossible to exhaust the ionization vessel on account of the thin mica covering of the sides, the vapour was introduced by means of a slow current of hydrogen which bubbled through the liquid. Under these conditions, the ionization vessel was filled with a mixture of hydrogen and vapour of methyl iodide at atmospheric pressure. The ionization in the vessel was almost entirely due to the methyl iodide, and for the hard γ rays from radium C was usually about three times as great as for air at atmospheric pressure. Some difficulty was at first experienced on account of the absorption of the vapour by the wax used in fixing the mica plates and in sealing the various parts of the vessel. This effect was got rid of by reducing the amount of wax to a minimum, and covering it with a non-absorbing layer of gum. This was kindly prepared for us by Dr. Lapworth, F.R.S., by special treatment of the ordinary gum sold commercially. By this method the absorption of the vapour was so much reduced that measurements extending over several hours could be made with certainty and accuracy.

The absorption curve using methyl iodide instead of air is shown in fig. 2 B.

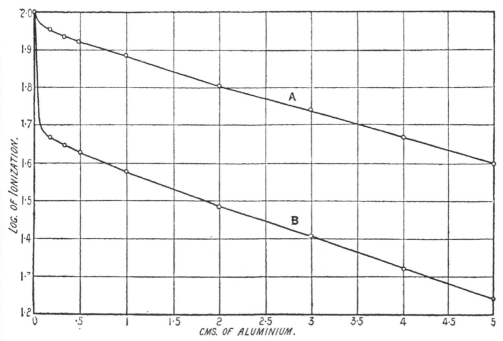

Fig. 2

An absorbing screen of 1·6 mm. of aluminium reduced the ionization to about 50 per cent. It will also be seen that the latter part of the absorption curves A and B are not linear and are not parallel to one another. This will be shown to be due to the fact that two penetrating types of radiation are present, the relative ionizations of which differ in air and in methyl iodide.

After passing through about 6 cm. of Al, the absorption curves, both for air and methyl iodide, became exponential with a value of the absorption coefficient $\mu = 0·115$, or $\mu/D = 0·0424$, where D is the density. This is practically identical with the absorption coefficient found by Russell and Soddy for the γ rays from radium C after passing through 2 cm. of lead. The absorption curve of the γ rays for thicknesses of aluminium between 0·05 and 7 cm. is shown in fig. 3, where the ionization itself is plotted as ordinates. After 6 cm. of aluminium the curve is exponential with a value of $\mu = 0·115$. It will be shown later that the absorption curve in aluminium of the γ rays from radium C is practically exponential from the beginning with a value of $\mu = 0·115$. Consequently, if the curve is produced backwards from a thick-

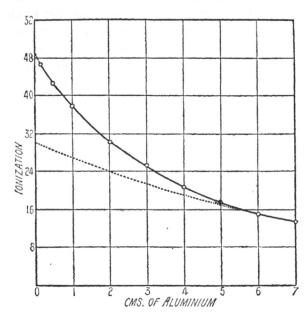

Fig. 3

ness of 6 cm. corresponding to radiations for which $\mu = 0·115$, it gives the ionization due to the γ rays from radium C alone. This is shown in the dotted curve fig. 3. If the difference between the ordinates of these curves be plotted, it is found to be an exponential curve with a value of $\mu = 0·51$ in aluminium. This radiation is undoubtedly due to radium B. A similar result was obtained when air was used instead of methyl iodide, but the effect due to this radiation is relatively smaller compared with that from radium C.

Analysis of the soft radiation

In order to find the absorption coefficient of the very soft radiation, which is shown so prominently in fig. 2 B, sheets of aluminium 0·042 and 0·084 cm. thick were used. The curve obtained is shown in fig. 4. In deducing the

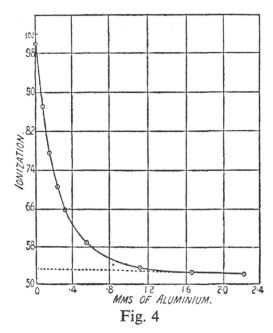

Fig. 4

Absorption of γ rays from radium B+C. Initial portion of curve $\lambda=40$.

ionization due to these soft rays, it is necessary to subtract the ionization due to the harder rays. This can easily be done since the absorption of the harder rays over the thickness of 1·51 mm. of aluminium is practically linear. The difference curve is exponential and gives an absorption coefficient in aluminium $\mu = 40$. The source of γ rays in this case was about 15 cm. from the ionization chamber. In these experiments, the pole-pieces were covered with thick cardboard in order to reduce the excitation of secondary γ rays to a minimum. With the bare pole-pieces close together, the radiation entering the ionization vessel was distinctly softer with a value of μ between 40 and 45.

Attempts to detect very soft γ radiation

Special experiments were made to examine whether radium B or radium C emitted any very soft types of γ radiation in addition to the type already discussed for which $\mu = 40$. In the experiments with the emanation tube, the γ rays before entering the ionization vessel passed through absorbing material equivalent in amount to about 14 cm. of air. Any very soft type of γ radiation which might be present would be largely absorbed in traversing this material. It is essential, however, in the experiments to use sufficient absorbing material to stop completely the α rays from radium C, which have a range of 7 cm.

in air. Since it is well known that, for equal masses, X rays pass with much less absorption through elements of small atomic weight, arrangements were made to absorb the α rays mainly by hydrogen and carbon. The source of radiation was the active deposit of radium deposited from the emanation on both sides of a very thin mica plate. This was placed in a brass vessel which was closed at one end by a thin mica plate equal in stopping power to 1·5 cm. of air. This mica plate also formed one side of the ionization chamber, shown in fig. 1. The active matter was deposited on mica to avoid the excitation of detectable characteristic X rays. For a similar reason, the inside of the brass vessel was lined entirely with thick cardboard. A continuous current of hydrogen was sent through the brass vessel. Sheets of india-paper were interposed in the path of the rays of just sufficient thickness to stop entirely the α rays. The ionization in the detecting vessel filled with methyl iodide was then carefully examined when thin aluminium screens were introduced. It was found that the ionization at first decreased more rapidly than corresponded to an exponential law of absorption of the radiation for which $\mu = 40$ in aluminium. This initial drop could be accounted for by assuming the presence of a very soft γ radiation for which $\mu = 230$ about in aluminium. Since the ionization in methyl iodide due to this radiation corresponded to only 10 per cent. of the total effect, the initial slope of the curve could not be determined with much certainty. It is difficult to decide whether this soft radiation has an independent existence, or whether it is due to an initial drop in the absorption curve of the radiation corresponding to $\mu = 40$. It will be seen later that a rapid initial drop of the absorption curve is always observed when lead is used as an absorbing material. It is possible that aluminium may show a similar effect for a comparatively soft radiation.

From the rate of decay of this very soft radiation it was clear that it arose from radium B. It was always proportional in amount to the radiation $\mu = 40$. It should be pointed out that decay of the active deposit measured under these experimental conditions is initially far more rapid than that calculated by Moseley and Makower (*loc. cit.*). This is due to the fact that radium B initially provides about 70 per cent. of the total ionization due to radium B + C instead of the 12 per cent. observed by Moseley and Makower under their experimental conditions with air in the testing vessel.

Analysis of the rays from radium C

Experiments were next made to settle which of the types of γ radiation were to be ascribed to radium B and which to radium C. It is not convenient to use radium B itself as a source, as there is a rapid growth of radium C from it. By von Lerch's method, however, it is possible to obtain a strong deposit of pure radium C on a metal plate placed in an acid solution of the active deposit radium B + C. Since radium C loses half its activity in 19·7 minutes, a large number of experiments were necessary to determine with accuracy the absorption curves for the γ rays emitted from it. The type of curve obtained with radium C on nickel is shown in fig. 5. It is seen that a very soft γ radiation is present, but after passing through two millimetres of aluminium the absorption is exponential with a value of $\mu = 0 \cdot 115$. This soft radiation was much more readily absorbed than the γ radiation, $\mu = 40$, obtained when the emanation was used as a source. It thus seemed probable that this soft radiation was excited in the nickel by the radiation from the radium C deposited on it. This conclusion was confirmed by using a deposit of radium C on silver instead of nickel. A sufficiently active preparation was obtained by using the method outlined by v. Hevesy* of placing a silver plate in a silver nitrate solution containing the radium B + C in solution. With the silver plate, no appreciable amount of soft radiation was observed, but the absorption curve in aluminium was exponential from the beginning with a value of $\mu = 0 \cdot 115$.

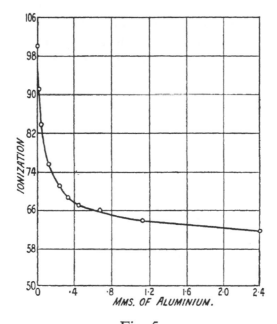

Fig. 5

Initial portion of RaC curve (on nickel).

There appears to be no doubt that this soft radiation from nickel consists mainly of the 'characteristic X radiation' of nickel excited probably by the α rays, although some rays of a more penetrating type were also present. It was observed that the amount of this soft radiation varied markedly with the orientation of the nickel plate, and was much less when the plate was parallel to the face of the pole-pieces than when it was perpendicular. Chadwick† first showed that γ rays were excited by α rays traversing different materials. The method employed by him, however, was not suitable for the detection of such a soft type of γ radiation. It is intended to make further experiments by the

* Hevesy, Phil. Mag. xxiii. p. 628 (1912).   † Chadwick, Phil. Mag. xxv. p. 193 (1913).

method outlined in this paper to examine whether the characteristic radiations of all elements are excited under similar conditions.

A number of experiments were made to test whether radium C itself emitted more than one type of radiation. For this purpose, the absorption curve in aluminium was very carefully examined over a thickness of aluminium from 0·2 to 4 cm. Over this range the absorption of the γ rays appeared to be exponential within the margin of possible experimental error with a value of $\mu = 0·115$. No evidence was obtained that a radiation for which $\mu = 0·5$ about was present. At the same time, it should be pointed out that it would be very difficult by direct measurement to detect with certainty the presence of a few per cent. of this radiation mixed with the more penetrating type for which $\mu = 0·115$.

Absorption of the γ rays by Lead

In the experiments so far described aluminium has been used as an absorbing material. Since it is well known that the absorption of γ rays in a heavy element like lead is abnormal, it was thought desirable to determine the absorption curves for this material.

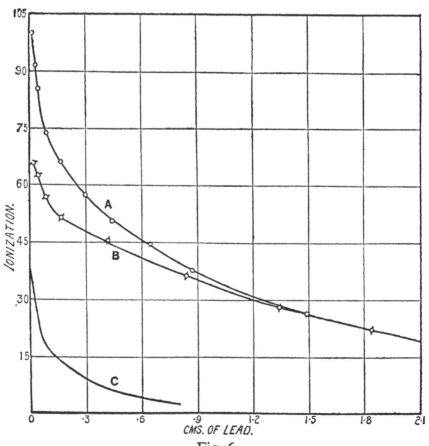

Fig. 6

The curve obtained for pure radium C on nickel is shown in fig. 6, Curve B. The soft radiation from the nickel was first cut out by a thin sheet of lead. The ionization initially fell more rapidly than was to be expected for an exponential law of absorption, but after traversing 1 cm. of lead the absorption of the rays in lead became accurately exponential with a value of $\mu = 0\cdot50$.

The absorption of the γ rays, using the emanation tube as a source, was also determined. In this case, before beginning the measurements, a thickness of lead was used sufficient to absorb completely the γ rays for which $\mu = 40$ in aluminium. The curve obtained is shown in fig. 6, Curve A. After a thickness of $1\cdot5$ cm. of lead the absorption became exponential with a value of $\mu = 0\cdot50$. Since the radiation $\mu = 0\cdot50$ comes entirely from radium C, the curve B (fig. 6) represents the part of the γ radiation due to radium C alone. The difference curve C given in fig. 6 shows the absorption in lead of the γ rays from radium B. It is seen that the curve shows a rapid initial drop, which is far more marked than in the case of radium C. The value of the absorption appears to vary from $\mu = 11$ to $\mu = 2\cdot8$ about, but it is difficult by this method to fix the values with much accuracy. These results are in general agreement with the experiments of Moseley and Makower, who showed that the absorption coefficient of the γ rays from radium B for lead varied between $\mu = 6$ and $\mu = 4$ about.

It is thus seen that the two types of γ radiation which are exponentially absorbed by aluminium both show irregular absorption curves when lead is used as absorbing material. It is intended in a later paper to discuss in more detail the relative absorption curves in lead and aluminium, and their bearing on the question of the homogeneity of the radiations concerned.

General discussion of results

The results of the analysis of the γ radiation from radium B and radium C are included in the following table. The density D of the aluminium was taken as $2\cdot71$.

	Absorption coefficient in aluminium	Mass absorption coefficient in aluminium	Absorption coefficient in lead
Radium B	$\begin{cases} 230? \\ 40 \text{ (cm.)}^{-1} \\ 0\cdot51 \quad ,, \end{cases}$	$\cdot\cdot$ $14\cdot7 \text{ (cm.)}^{-1}$ $0\cdot188 \quad ,,$	varying from 11 to $2\cdot8$ (cm.)$^{-1}$
Radium C	$0\cdot115 \quad ,,$	$0\cdot0424 \quad ,,$	$0\cdot50$ (cm.)$^{-1}$ after traversing 1 cm. of lead

It is seen that radium C emits essentially only one type of γ radiation, while radium B certainly emits two, and possibly three. In a previous paper, one of us pointed out that the rays from radium C correspond in penetrating power to the 'characteristic X radiation' of the K series to be expected from an element of atomic weight 214. The soft radiation from radium B ($\mu/D = 14\cdot7$)

undoubtedly corresponds closely in penetrating power to the radiation of the L series to be expected from an element of atomic weight 214. For example, Chapman* found that the value of μ/D in aluminium for the characteristic X radiation from bismuth (atomic weight $208 \cdot 5$) was $16 \cdot 1$, while the value of μ/D for thorium (atomic weight 232) was $8 \cdot 0$. It seems reasonable to suppose that the second type of γ radiation from radium B ($\mu/D = 0 \cdot 188$) also corresponds to a type of characteristic radiation from heavy elements which has not so far been observed with X rays on account of the difficulty of obtaining X rays of sufficient penetrating power to excite it. It is of interest to note that Chadwick and Russell† have found that three types of radiation were excited by the α rays in ionium. Two of these types, $\mu/D = 8 \cdot 35$ and $\mu/D = 0 \cdot 15$, appear to be analogous to the two types of radiation from radium B, but it is doubtful whether the very soft type ($\mu/D = 400$) observed by them in ionium is given out by radium B or radium C, although, as we have seen, careful experiments have been made to test this point. There appears to be little room for doubt that the γ rays at any rate from radium B + C are to be regarded as types of characteristic radiation from these elements. It is of interest to note that the energy of the soft γ radiation from radium B is very small compared with the energy of the more penetrating types of γ radiation from radium B and radium C. Chadwick and Russell, on the other hand, found that the soft types of γ rays excited by the α rays in ionium were relatively far more prominent. The bearing of these results on the general theory of the connexion between β and γ rays which led to these experiments will be discussed in detail in a later paper.

Summary

The γ rays from radium B consist of at least two distinct groups, each of which is absorbed exponentially in aluminium with absorption coefficients $\mu = 40$ and $\mu = 0 \cdot 51$ (cm.)$^{-1}$ respectively. The first group of γ rays is much less penetrating than the X rays excited in an ordinary focus-tube. The γ rays from radium C consist essentially of one type which are absorbed exponentially in aluminium with a value $\mu = 0 \cdot 115$. No evidence was obtained of the emission from radium C of the groups of radiation observed from radium B. The absorption of the rays by lead does not follow an exponential law.

The general evidence indicates that these radiations are to be regarded as types of characteristic radiation from the elements in question analogous to the characteristic X radiations excited in elements by X rays.

Experiments are now in progress to analyse by the methods outlined in this paper the γ radiations from all the radioactive elements which emit β and γ rays. The analysis of the γ radiation from α ray products is being undertaken by Chadwick and Russell in this Laboratory.

University of Manchester
March 6, 1913

* Chapman, Proc. Roy. Soc. A. lxxxvi. p. 439 (1912).
† Chadwick and Russell, Proc. Roy. Soc. A. lxxxviii. p. 217 (1913).

Analysis of the γ Rays from Radium D and Radium E

by PROFESSOR E. RUTHERFORD, F.R.S., *and* H. RICHARDSON, M.SC.

From the *Philosophical Magazine* for August 1913, ser. 6, xxvi, pp. 324–32

IN a previous paper* we have given the results of the analysis of the γ radiation from radium B and radium C. The radiation from the former was shown to consist of three types varying widely in penetrating power, and the latter essentially of one penetrating type. In the present paper the γ radiation from radium D and radium E has been examined by a similar method. These radio-active substances are of unusual interest and importance in considering the problem of the connexion between β and γ rays, since the γ rays emitted by radium D and radium E together are exceedingly feeble in intensity compared with the β rays. In this respect they are in marked contrast to the products radium B and radium C, where the γ rays are relatively intense and comparable in energy with the β rays.

In consequence of the short period of transformation (5 days) of radium E, it has been usual to employ radium D + E in equilibrium as a source of radiation. It has long been known that most of the β radiation emitted from such a source arises from radium E. J. A. Gray† examined the β rays from radium E in a magnetic field and found them to be very complex, giving a continuous spectrum which showed the presence of some rays of velocity close to that of light. This was confirmed by Gray and W. Wilson‡ using the electric method. Hahn, Baeyer, and Meitner§ have found that the weak β radiation from radium D consists essentially of two groups of homogeneous rays of velocity 0·33 and 0·39 compared with that of light. The question of the β radiation from radium D + E together has been re-examined by Danysz.‖ While radium D emits groups of β rays of definite velocity, radium E is anomalous in giving a continuous spectrum of β rays in which no definite evidence of groups of homogeneous rays has so far been obtained.

The only definite study of the γ radiation from radium D + E has been made by J. A. Gray.¶ He found that a comparatively soft type of γ radiation was present, and also a harder type whose mass absorption coefficient in aluminium was 0·4. Gray showed that radium D + E, in consequence of its

* Rutherford and Richardson, Phil. Mag. xxv. p. 722 (1913). (*This vol., p. 342.*)
† J. A. Gray, Proc. Roy. Soc. A. lxxxiv. p. 136 (1910).
‡ Gray and Wilson, Phil. Mag. xx. p. 870 (1910).
§ Hahn, Baeyer, and Meitner, *Phys. Zeit.* xii. p. 378 (1911).
‖ Danysz, *Le Radium*, x. (1913).
¶ J. A. Gray, Proc. Roy. Soc. A. lxxxvii. p. 489 (1912).

poverty in γ rays, was an ideal source for studying the production in ordinary matter of γ rays by β rays. The properties of the γ radiation excited in different materials have been studied by him in some detail. Gray had intended to examine the types of γ radiation emitted by radium D and radium E separately, but was unable to complete the experiments before leaving Manchester in 1912.

Preparation of Material

In order to analyse the γ radiation in detail, it was necessary to obtain very active preparations of radium D + E and to isolate the two components. From the experiments of Gray, it is known that the γ radiation excited in heavy elements by the β rays from radium D + E is greater in amount than the primary γ radiation. It is consequently of importance that the preparation of radium D and of radium E should contain a minimum amount of impurity, and should be tested either on filter-paper or a glass or aluminium vessel, for under these conditions the γ radiation excited by β rays is reduced to a small amount.

The radium D + E employed was separated from old radium preparations by adding a few milligrams of a lead salt to the solution and precipitating with hydrogen sulphide.* The separation of radium D from radium E was kindly performed for us by Dr. Russell and Mr. Chadwick. They also separated at the same time the radium F (polonium) which has been used by them in a study of the γ rays emitted by that substance. After removal of the radium F by dipping a copper plate into the solution of radium D + E + F, the radium D and radium E were precipitated together as sulphides. The precipitate was dissolved in hydrochloric acid and the radium E almost completely removed by dipping a nickel plate into the solution.

A small quantity of aluminium chloride was added to the radium D solution, and ammonia added. The radium D and lead were precipitated while any copper or nickel present was dissolved in excess of ammonia. The radium D was filtered off and the radiation from it was tested in position on the filter-paper. The amount of aluminium and lead mixed with the radium D on the filter was not more than a few milligrams, and the γ radiation excited in this material was negligible compared with the primary γ radiation.

The radium E which had been separated on the nickel plate was transferred by Dr. Russell and Mr. Chadwick on to an aluminium plate by volatilization under definite conditions in an electric furnace. As is well known, aluminium can only emit a characteristic X radiation of a very easily absorbed type, which could not be detected under the experimental conditions. On the other hand, the characteristic X radiation to be expected from nickel is of the same order of penetrating power as the primary γ rays from radium D.

The purity of the radioactive substances employed was tested by measuring

* See Boltwood and Rutherford, Wien. Ber. 120, p. 313 (1911). (*This vol., p. 221.*)

the decay of the β-ray activity of radium E, and the rise of the β-ray activity of the radium D preparation after an initial complete separation of radium E. A series of observations showed that radium D was obtained almost free from radium E, while the radium E did not contain a detectable amount of radium D.

Experimental Method

The apparatus employed to analyse the γ radiation was essentially the same in principle as that described in our former paper. An aluminium electroscope (10 × 10 × 10 cm.) was arranged as shown in fig. 1. The face E of the electro-

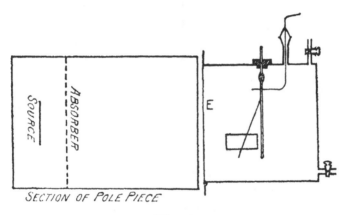

Fig. 1

scope was made of a thin sheet of mica equivalent in stopping power of the α rays to about 2 cm. of air. The electroscope was filled by displacement with a mixture of hydrogen and methyl iodide at atmospheric pressure. On account of the small ionization effect of the γ rays from the radioactive materials under examination, it was necessary to allow a comparatively wide beam of γ rays to enter the electroscope. The β rays were removed by a strong magnetic field, the pole-pieces being about 6 cm. apart. In most experiments the source was placed in the magnetic field about 12 cm. from the electroscope, and the absorbing layers were placed directly in front of the source, as shown in the figure. All exposed surfaces were covered with thick cardboard in order to reduce the possibility of excitation of γ rays by β rays to a minimum.

Analysis of the γ Rays from Radium D

The absorption curve of the γ rays from radium D was obtained immediately after its separation from radium E. No sensible change either in the intensity of the γ rays or in the shape of the absorption curve was observed with the growth of radium E in the preparation. This indicates, as is shown later by another method, that the γ radiation from radium E is very weak in intensity compared with the γ rays from radium D. In all five determinations were made with different preparations of radium D. The numbers obtained in the individual experiments under somewhat different conditions are given in the following table as an illustration of the accuracy of the individual measurements.

Thickness of Aluminium in mm.	A	B	C	D	E
0	100	100	100	100	100
0·065	74·0	74·7	75·2	77·3	77·5
0·130	57·4	57·9	58·6	58·7	60·5
0·195	..	45·2	45·7	45·4	46·4
0·260	35·4	36·8	36·9	37·3	36·5
0·390	24·5	25·2	24·8	25·5	24·3
0·56	16·1	15·9	16·3	15·4	15·8
0·82	..	10·8	10·8	10·5	10·0
1·12	8·3	8·4	8·3	7·5	7·5
1·97	6·2	..	5·7	5·1	5·6
3·94	4·3	4·7	..	4·2	4·2

A. Rather a large amount of material, consisting chiefly of aluminium. Maximum activity was 3 divs. per min. $\mu = 45$.
B. Thin preparation. Maximum activity 7 divs. per min. $\mu = 45$.
C. Same preparation as B. Measurements taken a fortnight later. $\mu = 46$.
D. Thin preparation, containing only a small amount of aluminium and lead. Preparation was placed on a filter-paper and mounted on cardboard. Maximum activity 20 divs. per min. $\mu = 44$.
E. Same preparation as D, only in this case the filter-paper was simply suspended between the poles of magnet. $\mu = 44$.

One of the several curves obtained is shown in fig. 2. It is seen that the greater part of the γ radiation is absorbed by about 1 mm. of aluminium. A small amount of a more penetrating type of radiation is also present. The curve was analysed as in previous experiments. The soft part of the radiation was found to be absorbed exponentially in aluminium (see logarithmic curve fig. 2), the value of the absorption coefficient μ being 45, 45, 46, 44, 44 in the five distinct experiments, giving a mean value $\mu = 45$ (cm.)$^{-1}$.

The absorption curve for the penetrating radiation is shown in fig. 3, the logarithm of the ionization being plotted as ordinate. It will be seen that the points lie nearly on a straight line, showing that the γ rays are absorbed according to an exponential law. The mean value of the absorption coefficient for several experiments was found to be 0·99 (cm.)$^{-1}$ in aluminium. It will be seen from the figure that only about 7 per cent. of the total γ-ray ionization is due to the more penetrating type of radiation. As the maximum ionization obtainable under the experimental conditions was about 19 divisions per

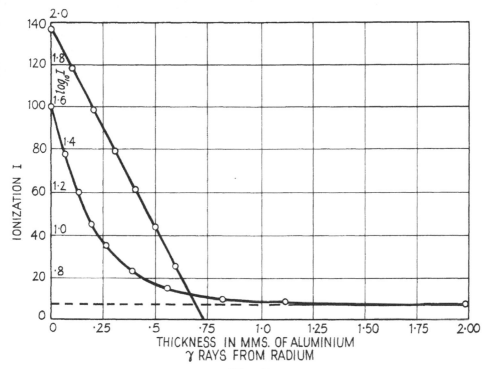

Fig. 2

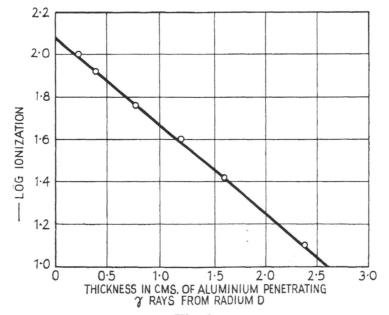

Fig. 3

minute in the electroscope, the amount due to the penetrating rays was only between 1 and 2 divisions a minute. It is consequently difficult, when dealing with such small effects, to obtain a very accurate value of the absorption coefficient of this radiation. Several experiments were made, the values of μ varying between $0\cdot97$ and $1\cdot02$. The radiation was in all cases absorbed according to an exponential law, indicating that the rays consisted of one homogeneous type. The absorption curves of both types of radiation were also determined in carbon as well as in aluminium. Allowing for the difference in the absorption coefficients, the curves for carbon were found to be identical in character with those obtained for aluminium, and the proportion of hard to soft radiation was the same in both cases. This agreement indicates that

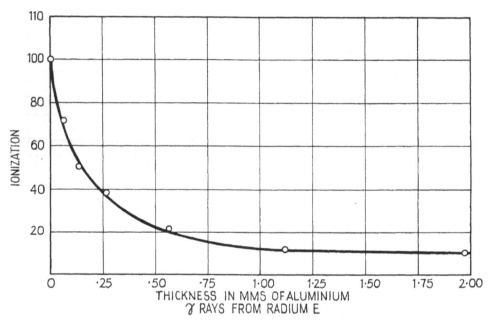

Fig. 4

the measurement of the absorption is not complicated to an appreciable extent by the excitation of γ rays in the absorbing screens by the primary γ rays from radium D. Very active preparations of radium E were obtained, and before the application of the magnetic field the β-ray activity was found to be about 2000 divisions per minute. On applying the magnetic field the amount of γ radiation observed never gave more than half a division per minute. The absorption curve of this radiation was examined in aluminium, but could not be determined with much accuracy on account of the smallness of the effects. The absorption curve obtained is shown in fig. 4. It will be seen that the curve is very similar in shape to that obtained for radium D. The results, however, are not of sufficient accuracy to analyse the constituents of the radiation with any certainty. It was of great importance to settle whether the small γ radiation observed is to be ascribed to radium E or to the presence of a small amount of radium D as impurity. For this purpose, the γ-ray activity of the radium E preparation was examined over several weeks. It was found that the amount of γ radiation decreased at the same rate as the β-ray activity, and ultimately became too small to measure. This result showed conclusively that the γ radiation must arise from radium E and not from radium D. It would appear that the emission of β rays from radium E is accompanied by an exceedingly minute γ radiation of which the soft type has about the same absorption coefficient as that from radium D. It appears fairly certain that this γ radiation is to be ascribed to the atoms of radium E, and is not excited by the β rays in the surrounding inactive matter.

Relative Intensity of the γ Rays from Radium D and Radium E

It is of importance to obtain an estimate of the relative magnitude of the β and γ radiation emitted by radium D and radium E. This was done by measuring the β-ray activity of the sources under precisely the same conditions as the γ-ray activity. The ionization in the electroscope of fig. 1 was determined with and without the magnetic field. The preparations were placed 12 cm. from the mica face E of the electroscope, and the pole-pieces were 6 cm. apart. Under these conditions, the β-ray activity of the radium D preparation was 300 divisions per minute, and the γ-ray activity 19 divisions per minute. On the other hand, the β-ray activity of the corresponding radium E preparation was 1400 divisions per minute, and the γ ray activity only 0·48 division per minute. Even without allowing for the partial absorption of the soft β rays from radium D in the air before entering the electroscope, it is obvious that there is a very marked difference between the ratios of the β- and γ-ray activity of the two preparations.

Summary

It has been shown that radium D emits two types of γ radiation for which $\mu = 45$, and 0·99 (cm.)$^{-1}$ in aluminium. Relatively to radium D, radium E emits a very weak γ radiation, and not more than two per cent. of the γ radiation given out by radium D + E in equilibrium is to be ascribed to radium E. In other words, the transformation of each atom of radium D is accompanied, on the average, by 50 times the amount of γ radiation from an atom of radium E. The soft type of radiation $\mu = 45$ from radium D and radium E is slightly less penetrating than the corresponding radiation from radium B. There appears to be little doubt that this radiation corresponds to the characteristic radiation of the L type to be expected from an element of atomic weight 210. The values of the absorption coefficients of the types of γ radiation from radium B, radium C, radium D, and radium E are collected together in the following table. The density D of aluminium was 2·72.

	Absorption coefficient μ in Aluminium	Mass absorption coefficient μ/D in Aluminium
Ra B	$\left\{\begin{array}{l} 230 \ (\text{cm.})^{-1} \\ 40 \quad\quad ,, \\ 0\cdot51 \ \ ,, \end{array}\right.$	85 $(\text{cm.})^{-1}$ 14·7 ,, 0·188 ,,
Ra C	0·115 ,,	0·0424,,
Ra D	$\left\{\begin{array}{l} 45 \quad\quad ,, \\ 0\cdot99 \ \ ,, \end{array}\right.$	16·5 ,, 0·36 ,,
Ra E	Nearly the same types of radiation as from radium D, but relatively very feeble.	

We are much indebted to Dr. Russell and Mr. Chadwick for the great care and trouble they have taken in purifying the preparations of radium D and radium E employed in these experiments.

University of Manchester
June 1913

The Reflection of γ Rays from Crystals

From *Nature*, **92**, 1913, p. 267

IN some recent investigations, Prof. Rutherford and Mr. H. Richardson have analysed the γ radiations emitted by a number of radio-active products. They have shown, for example, that radium B emits three distinct types of γ radiation, which are absorbed exponentially by aluminium with absorption coefficients $\mu = 230$, 40, and $0 \cdot 51$ (cm.)$^{-1}$ respectively. On the other hand, radium C appears to emit essentially only one type of γ radiation, the absorption coefficient of which is $\mu == 0 \cdot 115$ in aluminium.

Recently we have undertaken an examination of these types of radiation by the methods developed for X-rays by W. H. and W. L. Bragg, and by Moseley and Darwin, which consist in determining, either by the photographic or electric method, the intensity of the X-rays reflected from a crystal at different angles of incidence. In our experiments the source of γ radiation was a thin α-ray tube containing about 100 millicuries of emanation, the γ rays arising from the products of the emanation, radium B and radium C. A diverging cone of rays fell on a crystal of rock-salt, and the distribution of the reflected radiation was examined by the photographic method. The source and photographic plate were each about 10 cm. from the centre of the crystal. Suitable precautions were taken to reduce to a minimum the effect on the photographic plate of the primary and secondary β rays and penetrating γ rays. The source was first arranged so that the radiation made an average angle of about 9° with the face of the crystal.

It was calculated from the known data of the crystal that the radiation $\mu = 40$ from the radium B, if homogeneous, should be strongly reflected at about this angle. A group of fine lines comprised between the angles 8° and 10° have been observed on the photographic plate in a number of experiments. Similar results have been observed with a crystal of potassium ferrocyanide, kindly loaned to us by Mr. Moseley. On examining the reflection for an angle of 2° another series of fine lines was obtained on the plate, probably resulting from the reflection of the more penetrating radiations from radium B and radium C.

The experiments indicate that the γ radiation for which $\mu = 40$ is complex, and consists of several groups of rays of well-defined wave-length. Experiments are in progress to examine carefully the character of this reflected radiation, both by the photographic and electric method. It is hoped that in this way definite evidence will be obtained on the constitution and wave-length of each of the types of γ radiation which are emitted from radium B and radium C.

E. RUTHERFORD

E. N. DA C. ANDRADE

The University, Manchester

M*

Scattering of α Particles by Gases

by PROFESSOR E. RUTHERFORD, F.R.S., *and* J. M. NUTTALL, M.SC.

From the *Philosophical Magazine* for October 1913, ser. 6, xxvi, pp. 702–12

THE scattering of α particles by matter has been examined in detail by the scintillation method by Geiger* and by Geiger and Marsden.† In the experiments of Geiger, the most probable angle through which an α particle was scattered was determined by the scintillation method for different thicknesses of a number of elements. It was found that the most probable angle of scattering for thicknesses of different elements, equivalent in stopping power of the α particle to one centimetre of air, was proportional to the atomic weight of the scattering element. A systematic investigation was later made by Geiger and Marsden of the 'large angle scattering' by thin films of matter. The results were shown to be in complete accordance with the theory of 'single' scattering advanced by Rutherford.‡ This theory supposes that the atom consists of a charged nucleus surrounded by a compensating distribution of electrons. The large angle scattering is due to the passage of the α particle through the intense electric field of the nucleus. It was deduced by Rutherford, and Geiger and Marsden, that the charge on the nucleus for atoms between carbon and gold was approximately proportional to the atomic weight, and was equal to $\frac{1}{2}Ae$, where A is the atomic weight in terms of that of hydrogen and e is the electronic charge. On this theory it is to be anticipated that hydrogen has a nucleus of one charge, helium of two, and carbon of about six.

As this deduction is of great importance in connexion with the constitution of the simpler atoms, experiments were undertaken to determine the scattering of α particles by the simple gases. The method employed by Geiger and Marsden for solids is not altogether suitable for gases; in addition, the large angle scattering to be expected for the light elements is exceedingly small, and would be difficult to measure with accuracy.

In some preliminary experiments, the α particles were made to pass between two parallel plates, placed a small distance apart, and the number of issuing particles were observed (1) in a vacuum and (2) when the space between the plates was filled with gas at a known pressure. The number of α particles was counted photographically, using a string electrometer. On account, however, of the probability variations, a very large number would have to be counted in order to obtain a reliable result. This would involve much time and

* H. Geiger, Proc. Roy. Soc. lxxxi. p. 174 (1908).

† H. Geiger and E. Marsden, Phil. Mag. xxv. p. 604 (1913).

‡ E. Rutherford, Phil. Mag. xxi. p. 669 (1911). (*This vol., p. 238.*)

labour, and it was consequently felt desirable to use a more indirect but rapid method.

The experimental arrangement finally adopted is shown in fig. 1. A narrow pencil of α particles was obtained by placing a thin platinum wire W (see fig. 1),

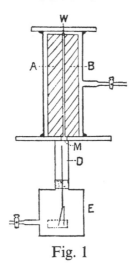

Fig. 1

coated with radium C, between two thick glass plates A and B, 14·5 cm. long and 2 cm. in width, and kept about 0·3 mm. apart by mica stops. After traversing the space between the glass plates, the beam passed through a mica window M, of stopping power equivalent to 1·9 cm. of air, and then into an ionization chamber D, consisting essentially of three parallel equidistant brass plates 5 cm. long and 3 cm. wide, 7 mm. apart. The outer plates were earthed and the central one insulated and connected to the gold-leaf system of an electroscope E. The latter was kept exhausted in order to diminish the ionization due to β and γ rays. The upper part of the apparatus, consisting of a cylindrical glass tube (with ground-glass ends) containing the plates A and B and the source of α rays W, could be completely exhausted and filled with any gas or vapour to any desired pressure, measured in the usual way by a mercury gauge. It was necessary to employ a strong source of α particles, and this was obtained by exposing about one centimetre length of thin platinum wire 0·2 mm. diameter to radium emanation for three hours as cathode in an electric field. A short time after removal from the emanation, radium C is the only α-ray product remaining on the wire, and since its curve of decay is accurately known, the amount of radium C present at any subsequent time can be readily calculated. Care was taken to remove any traces of emanation from the wire by washing it in absolute alcohol and heating it slightly.

The method of procedure in an experiment was as follows. An active wire was placed in position between the glass plates AB and the upper part of the apparatus was exhausted. The source W emitted a pencil of homogeneous α particles which passed through the mica window M, and the ionization they produced in chamber D was measured in the electroscope E. The whole issuing beam of α rays was completely absorbed in the chamber D, and the ionization due to β and γ rays was never more than 2 per cent. of the whole effect. If a gas is now introduced into the upper chamber, *i.e.* between the plates A and B, and the ionization current again measured, any decrease, when correction is made for decay of source, will be due either to partial absorption of α particles by the gas in the upper chamber or to the scattering of α particles against the faces of the plates resulting in a diminution in the number entering the ionization chamber. The decrease of ionization due to loss of range by absorption was measured in a separate experiment. For this purpose the plates AB forming the long narrow slit were removed, and the ionization current measured for various pressures of gas in the upper cylinder, the source W being placed in exactly the same position as before. In this way

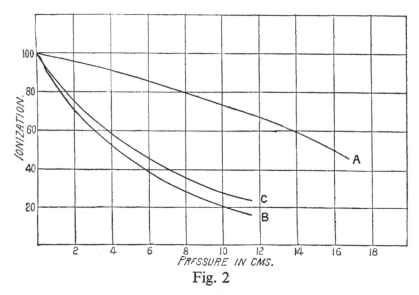

Fig. 2

a curve can be obtained showing the relation between the pressure of the gas in the upper cylinder and the ionization produced by the source in the chamber D. A typical curve is shown in the curve A of fig. 2.*

In the same figure, curve B gives the result of a typical experiment showing the effect of both scattering and absorption together. From the two curves A and B, a curve can be obtained which gives the effect of scattering alone. To obtain this, the ratios of the ordinates of the curves A and B at each pressure must be plotted. It is obvious that this ratio gives the fraction of α particles scattered by the gas at each particular pressure, and is independent of any change of ionization due to loss of range. Curve C obtained in this way indicates the amount of unscattered radiation passing between the plates at various pressures. It is seen that, over a certain range of the α particle at least, the amount of unscattered radiation passing between the plates varies nearly exponentially with the pressure of gas, *i. e.* if N_0 is the number of α particles

* Compare H. L. Bronson, Phil. Mag. xi. p. 806 (1906).

at zero pressure and N_p the number at pressure p cm., then $N_p/N_0 = e^{-\lambda p}$, where λ may be called the 'scattering coefficient' for the particular gas under the given conditions. It should be noted that this formula does not hold accurately when the pressure of the gas becomes so large that the velocity of the α particle is much reduced. For pressures of the gas, whereby the loss of range of the α particle is equivalent to more than 2 cm. of air at atmospheric pressure, the scattering will increase more rapidly. Indications of this were obtained in some of the experiments. Table I gives the results of a typical experiment, air being used as the scattering gas.

TABLE I

Pressure of Gas p	Scattering and absorption	Absorption alone	Intensity N due to scattering alone	$\dfrac{\log N_0/N_p}{p}$
0 cm.	100	100	100	
0·92	84·2	98·2	86·8	0·66
1·9	72·6	96·5	75·2	0·65
3·55	58·0	93·0	62·5	0·58
5·10	45·9	88·6	51·9	0·56
7·15	33·5	82·5	40·6	0·55
9·00	24·6	76·5	32·0	0·55

The numbers in the last column show that the radiation is scattered approximately according to an exponential law over a considerable range of pressure.

Variation of scattering with the distance between the plates

A few experiments were made to determine approximately the variation of scattering with the distance between the plates. The latter distance was varied from 0·173 mm. to 0·89 mm.; the corresponding least angles of scattering to deflect the α particles against the faces of the plates varying from 0°·069 to 0°·356.

The collected results are shown in Table II. The relative scattering coefficients are compared by observing from the curves the pressure of gas required to scatter 50 per cent. of the initial number of α particles.

TABE II

Distance d apart of plates	Relative scattering coefficient λ	$\lambda \cdot d^{5/4}$
0·172 mm.	0·562	0·618
0·325	0·256	0·627
0·586	0·126	0·643
0·890	0·0712	0·615

Within the limits employed, the scattering appears to vary rather more rapidly than the inverse of the linear distance d between the plates and to be nearly proportional to $d^{-5/4}$.

It will be shown later that the scattering coefficient increases rapidly with decrease of velocity of the α particle. A correction has consequently to be applied to reduce the amount of scattering to a standard velocity. This correction has been made in the results of Table II.

Variation of scattering with atomic weight

The gases whose scattering coefficients were compared were as follows:—air, hydrogen, helium, methane, carbon dioxide, and sulphuretted hydrogen. A few experiments were made on the scattering produced by the heavy vapours, methyl iodide and ethyl bromide, but they are not included in this paper. In all cases two sets of experiments were performed: (1) experiments on the scattering and absorption combined, and (2) experiments on the absorption alone. The latter were in agreement with the stopping powers of the gases as calculated by Bragg's* square-root law.

The distance between the plates was kept constant during the course of these experiments, and the scattering of each gas was compared with air as a standard. The scattering curve for air, which was repeated before and after

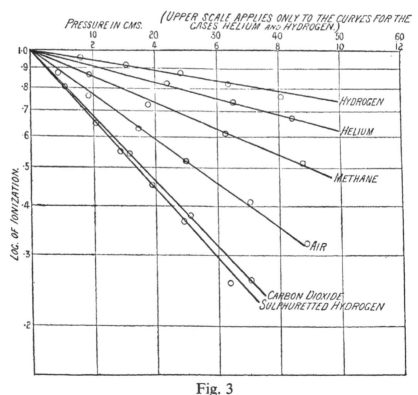

Fig. 3

* W. H. Bragg, Phil. Mag. x. p. 318 (1905).

each experiment, was found to be unchanged throughout the investigation. The logarithmic curves for the different gases are shown in fig. 3. The pressures of gas are plotted as abscissæ, and the logarithms of the intensity of the unscattered radiation passing between the plates as ordinates.

A few details of the method of preparation and purification of the gases are given.

Hydrogen. This was prepared electrolytically and dried carefully. It contained about 1 per cent. of impurity, chiefly oxygen. A correction was made for this in the curve of fig. 3, which is the mean of several experiments.

Helium. The helium was purified by passing over charcoal immersed in liquid air. This removed all traces of air and other impurities, except possibly the last traces of hydrogen.

Methane. This was prepared by acting on aluminium carbide with water and passing through cuprous sulphate to remove the acetylene. The resulting impure methane was condensed by liquid air, and pure methane obtained by fractional distillation.

Carbon dioxide was obtained from a cylinder of compressed gas, which on analysis was found to be of 99·7 per cent. purity.

Sulphuretted hydrogen was prepared by action of sulphuric acid on calcium sulphide. It was analysed and found to be of 99·8 per cent. purity.

From the results obtained in these experiments, the relative atomic coefficients of scattering were deduced for hydrogen, helium, carbon, air, oxygen, and sulphur. In the case of complex molecules, it was assumed that the scattering coefficient of the molecule was the sum of the values for each of its individual components. For example, the coefficient for carbon was deduced from the value of methane (CH_4) by subtracting from the observed value the scattering coefficient due to four atoms of hydrogen. In a similar way, the

TABLE III

Gas	Corrected pressure to scatter half of incident radiation	Relative scattering coefficient	Atomic weight	Remarks
Air	5·32 cm.	1·00	14·4	Diatomic
Carbon (from CH_4)	13·32 ,,	0·40	12·0	Monatomic
Carbon dioxide	3·64 ,,			
Oxygen	5·00 ,,	1·064	15·99	Diatomic
Sulphur	3·36 ,,	1·61	32·0	Monatomic
Hydrogen*	46·3 ,,	0·0353	1·0	Diatomic
Helium*	26·2 ,,	0·064	3·99	Monatomic

* In the cases of hydrogen and helium the pressures given are those required to cut down the incident radiation to 80 per cent. of its initial value, and they are compared with the pressure of air required to scatter to the same extent, viz. 1·68 cm. of mercury.

value for oxygen was deduced from CO_2 and sulphur from H_2S. A correction was required in each case to allow for the variation of scattering with velocity of the α particle. The reduction in velocity of the α particle in passing through a sufficient pressure of gas to scatter half the α particles was deduced from the loss of range of the α particle in the gas, using the relation between velocity and range found by Geiger. In making this correction it was assumed, as will be seen later, that the scattering coefficient varied inversely as the fourth power of the velocity. In practice, the value of the pressure was corrected so as to give the scattering observed, assuming that the velocity was equal throughout to the initial value for the α particle.

Variation of scattering with velocity

Geiger found that the most probable angle of scattering of α particles by solid matter increased rapidly with decrease of velocity of the α particle. This also holds for light gases. The initial velocity of the α particle was diminished by placing a sheet of aluminium foil of known stopping power under the active wire. The results are shown in fig. 4, where the logarithm of the corrected

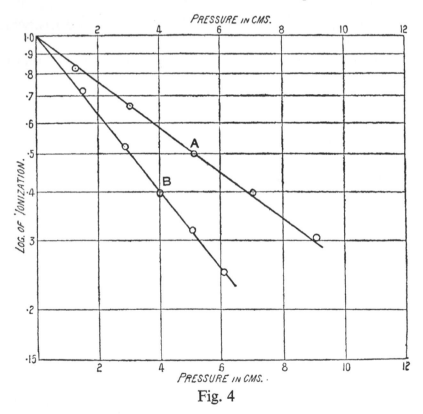

Fig. 4

ionization in the testing vessel is plotted as ordinates and pressures as abscissæ. In curve A, the initial velocity was the maximum velocity of α particles from radium C, whilst in curve B the initial velocity was 0·893 of the maximum velocity. The scattering coefficients observed in the two cases were in the ratio

1 : 1·66. The scattering thus varied approximately as the inverse fourth power of the velocity. This is the law of scattering with velocity found by Geiger and Marsden in their experiments on 'single' scattering.

Consideration of the Results

In drawing deductions from these experiments, it is of importance to decide whether the scattering coefficient observed is to be ascribed to 'single' or 'compound' scattering. If the reduction in the number of α particles in passing between the glass plates is due mainly to 'single' scattering, we should expect the scattering to vary directly as the pressure of the gas and inversely as the fourth power of the velocity—results observed experimentally. On the other hand, if the reduction in number is due mainly to 'compound' scattering, we should expect that the scattering should be proportional to the square root of the pressure, and to vary as the inverse square of the velocity. It is thus clear from the experiments that the scattering coefficient observed is a consequence mainly of 'single' scattering. This conclusion is still further strengthened by the rapid variation of the scattering with atomic weight between carbon and sulphur. No doubt 'compound' scattering produces some effect, but the main part of the scattering is to be ascribed to the scattering of individual atoms resulting from the passage of the α particle through the intense field close to the electrons and the nucleus. If we consider the atom to be composed of a nucleus with a charge ne and a compensating distribution of n electrons, the scattering due to the n electrons is proportional to n, and the scattering due to the nucleus is proportional to n^2. Mr. C. Darwin kindly examined this question mathematically for us, and concluded that if the electrons and the nucleus were a sufficient distance apart so as not to interfere seriously with the electric fields close to them, the scattering for simple atoms should be proportional to $n + n^2$ or $n(n + 1)$, where ne is the charge on the atomic nucleus. For deflexions through a small angle, such as are involved in the experimental arrangement employed in this paper, we should expect the scattering in the heavy atoms to be proportional to $n + kn^2$, where k is less than unity. This is borne out by the fact that under the experimental conditions the scattering due to heavy atoms like bromine and iodine was found to be less than the theoretical values when k is taken as unity. Assuming the simple formula found by Darwin as applicable to light atoms, the scattering coefficient λ is proportional to $n(n + 1)$ or $\lambda = cn(n + 1)$, where c is a constant.

It should be noted that Geiger and Marsden found that the 'single' scattering per atom was proportional to the square of the atomic weight. This undoubtedly holds in the case of heavy elements for large angles of scattering, where the α particle passes close to the nucleus. It is a different matter, however, when the scattering angle is only about 1/10 of a degree, as in the present experiment. It is to be anticipated that under such conditions this simple law would be widely departed from, especially in the case of heavy atoms for which the number of electrons is large.

The following table gives the observed and calculated scattering coefficient for an *atom* of each of the elements examined. The value of n, the number of electronic charges assumed for the nucleus, is given in the second column. For a comparison of the calculated with the experimental values, carbon is assumed in both to have a scattering coefficient of $0·40$ on the arbitrary scale. The scattering coefficient for nitrogen is deduced on the assumption that air is composed of 80 per cent. of N with 20 per cent. of O.

Gas	Assumed value of n	Relative scattering per atom Calculated values	Experimental values
H atom	1	0·0190	0·0176
He ,,	2	0·057	0·064
C ,,	6	0·40	0·40
N ,,	7	0·53	0·48
O ,,	8	0·69	0·53
S ,,	16	2·58	1·61

Considering the difficulty of determining with accuracy the scattering by the light gases, the agreement between the simple theory and experiment is as close as could be expected for hydrogen, helium, and carbon. The probable explanation of the divergence between theory and experiment for the heavier atoms, like sulphur, has already been outlined. From the experiments of Geiger and Marsden (*loc. cit.*) on the large angle scattering of α particles by carbon, it is clear that the carbon atom behaves as if the nucleus carries a charge of about six units. Assuming this value of n for carbon, the results indicate that the hydrogen nucleus has a charge of *one* fundamental unit and the helium nucleus of *two* units. This value for helium is to be anticipated from the observed fact that the α particle in its flight carries two unit positive charges.

The observations on the scattering of α particles by matter in general afford strong experimental evidence for the theory that the atom consists of a positively charged nucleus of minute dimensions surrounded by a compensating distribution of negative electrons. The charge on the nucleus for heavy atoms is approximately $\frac{1}{2}Ae$, where A is the atomic weight and e the electronic charge. The experiments in this paper on the scattering of simple gases indicate that the hydrogen atom has the simplest possible structure of a nucleus with one unit charge, and helium comes next with a nucleus of two unit charges. This simple structure for hydrogen and helium atoms has been assumed by Bohr* in a recent interesting paper on the constitution of atoms, and has been shown by him to yield very promising results.

We desire to express our thanks to Dr. Pring for his assistance in preparing and purifying the gases employed.

University of Manchester
July 1913

* Phil. Mag. July 1913.

The Analysis of the β Rays from Radium B and Radium C

by PROFESSOR E. RUTHERFORD, F.R.S., *and* H. ROBINSON, M.SC.,
Lecturer in the University of Manchester

From the *Philosophical Magazine* for October 1913, ser. 6, xxvi, pp. 717–29

THE problem of the distribution of the β rays from radioactive substances with regard to velocity has been the subject of much experiment and discussion. As the history of this subject has been previously treated in some detail by one of us,* it is only necessary here to draw attention to the more recent and salient facts. A great advance in our knowledge was made by the experiments of Hahn, Baeyer, and Meitner, who showed that the β rays from the majority of radioactive substances consisted of a number of separate groups, each of well-defined velocity. If a narrow pencil of β rays, falling normally on a photographic plate, were deflected by a magnetic field, the radiation was split up into its component groups and a veritable spectrum was observed. This showed that the radiation, while heterogeneous in type, was made up of a number of distinct groups of β rays, each of which comprised β particles, which were expelled from the radioactive substance with a definite and characteristic velocity. A further advance was made by Danysz† who examined, also by a photographic method, the deflexion in a magnetic field of the β rays emitted from an intense source of radium emanation enclosed in a thin-walled glass tube. The β rays in this case do not come from the emanation but from its products radium B and radium C in equilibrium with it. He showed that the β rays from radium B + C were exceedingly complex, and was able to show the presence of nearly 30 homogeneous groups of β rays from these two products together. In a later paper‡ he has used a somewhat different and more certain method for determining the velocity of each of the groups, and has given a corrected list of 25 lines.

There is a very marked distinction between the emission of α and β particles from a radioactive atom. As is well known, the atom of each α-ray product emits during its transformation only one α particle, which is expelled with a definite velocity characteristic of the substance. Since it is also known from the experiments of Makower, Moseley, Duane and Danysz that each atom of radium B and C in disintegrating does not emit more than one or two β particles, it is clear that each atom cannot contribute one β particle to each

---

* Rutherford, 'Radioactive Substances and their Radiations,' pp. 209–212 (1913).

† Danysz, *Compt. Rend.* 153. pp. 339, 1066 (1911); *Le Radium*, ix. p. 1 (1912).

‡ Danysz, *Le Radium*, x. p. 4 (1913).

group. This is also borne out by the fact that the photographic intensities of neighbouring groups of β rays differ markedly from one another. In order to explain these anomalies, Rutherford* suggested that a single β particle of definite velocity was set free by the transformation of each atom. This β particle in escaping from the atom set the electronic distribution in vibration, and in consequence lost energy in multiples of certain units depending on the system set in vibration. The decrease of energy of the issuing β particle due to these causes was variable in different atoms, depending upon the chance of passing through or close to the systems to be set in vibration. On this view, the large number of homogeneous groups of β particles emitted from a single radioactive substance was to be ascribed not to a single atom but to a statistical effect of a large number of atoms. The energy lost by the escaping β particle was converted into energy of the γ-ray form, and different types of γ rays would be emitted depending on the systems set in vibration. Taking this view, the appearance of homogeneous groups of β rays was intimately connected with the excitation of different types of γ rays in the atom. This hypothesis is strongly supported by the observation that all those radioactive substances which emit well-defined groups of β rays also give rise to marked γ radiation. In order to throw light on this question, Rutherford and Richardson† have analysed the γ radiation from several radioactive substances. They have shown that radium B emits three distinct types of γ rays, radium C one penetrating type, radium D two distinct types, while radium E, which gives a continuous spectrum of β rays, emits only a minute amount of γ radiation.

Before any theory of the origin of β and γ rays can be adequately tested, it is necessary to determine with the greatest possible precision the velocity of each of the component groups of β rays. We have consequently made a large number of experiments to determine the groups of β rays emitted by radium B and by radium C separately. A strong source of radium emanation enclosed in a thin-walled α-ray tube affords a most convenient source for determining the velocity of the swifter groups of β rays from radium B and C together. An intense source of radium C deposited on a nickel wire has allowed us to distinguish between the groups to be ascribed to radium B and radium C separately. Since it is important to determine the velocity of groups of β rays which are photographically very weak, it was necessary to devise a method of bringing out the presence of groups of β rays, the total energy of which might be only a small fraction of that distributed in the more intense groups. For this purpose, we have used a special method which appears to be very similar in principle to that employed by Danysz in his later investigation (*loc. cit.*). The great theoretical and practical advantages of this arrangement lie in the fact that the β rays of a definite velocity comprised in a comparatively wide cone of rays can be concentrated in a line of very narrow width on the photographic plate. In this way a group of β rays of very small energy can be

---

* Rutherford, Phil. Mag. Oct. and Dec. 1912. (*This vol., p. 280, p. 292..*)

† Rutherford and Richardson, Phil. Mag. May and August 1913. (*This vol., pp. 342, and 353.*)

detected even when there is a marked darkening of the photographic plate due to the γ rays and the scattered β rays which must always be present.

The method employed is shown in fig. 1. The source S is attached about 1 cm. from the end of a lead block L (9 × 4 × 3·2 cm.). A comparatively wide slit V is arranged vertically above the source, and the latter was usually 1–2 cm. below the slit. The photographic plate PP is laid on the top of the lead block. It rested against one end of a plate about 1·5 cm. from the edge of the slit. The photographic plate was kept in position by means of a small screw attached to the lead block. When the source and photographic plate were placed in position, the whole apparatus was lowered into a brass vessel B B , which was closed by an upper plate by means of screws. The vessel was then exhausted to a low vacuum in a few minutes. The brass vessel was placed between the parallel pole-pieces (17 × 10 cm.) of a large electromagnet. The pole-pieces were adjusted so that the magnetic field was very nearly uniform over the greater part of the space between them. The magnetic field was

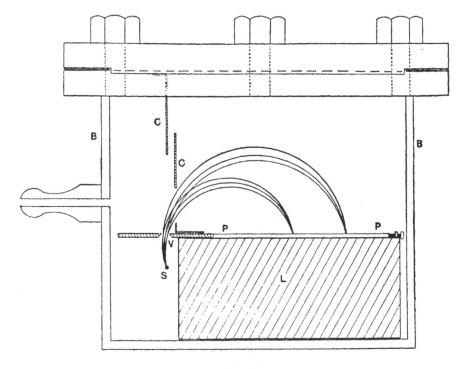

Fig. 1

parallel to the source and slit. In the uniform magnetic field the β particles of a definite velocity describe circles of the same radius. It can be shown theoretically that even with a comparatively wide slit the circles intersect the photographic plate along a curved line of comparatively narrow width. This result, however, is most clearly shown by drawing circular paths of β rays of definite velocity. It is seen from fig. 1 that if the pencil of rays comprised β rays of only one velocity, the whole pencil of rays would be concentrated along a narrow line on the photographic plate, and consequently the photographic impression for a given exposure would be far more confined and intense than if the photographic plate were placed vertically above the slit where the cone of rays is comparatively wide.

It is clear from the figure that this concentration holds for β rays of different speeds. Special precautions were taken to reduce the effect of the scattered β radiations which must necessarily be present. Apart from the question whether a continuous spectrum of β rays is superimposed on the line spectrum, β rays are excited in all parts of the vessel on which the γ rays fall; in addition, the rays falling on the photographic plate are partially scattered in all directions. These disturbing effects were reduced to a minimum by lining the vessel with thick cardboard, and placing cardboard screens CC' in the positions shown in the figure. It is also clear from the figure that the direct effect of the γ rays on the photographic plate was greatly reduced by their passage through the lead block.

The essential conditions for bringing out lines of feeble intensity depend mainly on the conditions: (1) narrow source, (2) a uniform magnetic field over the whole path of the β particles, (3) a comparatively wide slit; the width of slit employed in most of the experiments varied from 1–3 mm. The lines were comparatively sharp even with the widest slits used; and photographs could be obtained with exposures so short that the effect of the scattered radiations was relatively much reduced.

Under the experimental conditions, the *outside* edges of the lines were very sharply defined, but there was a distinct shading off towards the slit. This was especially marked in the strong groups of β rays from radium B, using the α-ray tube as a source.

The measurements were always made on the sharp outside edge of the line, and the radius ρ of the circular path could be readily determined for the rays forming this part of the trace. To a first approximation, it is given by $\rho = \sqrt{D^2 + a^2}$ where 2D is the distance of the outside of the line from the centre of the slit, and 2a the distance of the source from the slit. The position of the source relative to the slit was accurately measured in each experiment with the aid of a travelling microscope provided with a horizontal and vertical movement.

The positions of the strong lines on the plates were determined with the aid of a travelling microscope. The weak lines, however, were too faint to be seen in the microscope. It was consequently necessary to mark their positions on the plate with a fine pen. Each plate was marked several times, and the measurements of a given line in most cases were found to be in close agreement. It is obvious, however, that the position of weak lines could not be determined by this method with the same accuracy as for the stronger lines.

In order to examine the whole extent of the β-ray spectrum, photographs were taken in ten different magnetic fields, increasing with approximately constant ratio from about 200 gauss to 6300 gauss. In all, more than fifty separate photographs were taken.

The apparatus described above could be conveniently used up to fields of 3500 gauss. For the higher fields a similar apparatus to that shown in fig. 1 was used, but with an internal breadth of only 1·9 cm. instead of 3·2 cm. Another similar electromagnet was used and, with the narrower gap, constant fields of over 7000 gauss could be obtained without excessive heating of the magnet.

Each group given in the Tables I and II was observed in widely differing positions on three different plates. It was possible from this circumstance to check the accuracy of the measurements of the magnetic fields, and also the uniformity of the magnetic field over the whole region between the poles. It was usually found that when a line appeared on the plate at a distance of more than 7 cm. from the slit, the measurement gave slightly too high values of the radius of curvature, on account of the β particle entering the outer and non-uniform regions of the magnetic field. All the final measurements were confined to lines appearing on the photographic plate within 6 cm. from the slit. The uniformity of the magnetic field in both vertical and horizontal directions was carefully tested by means of a small search coil, and the positions of the poles so adjusted as to give equality of field over a considerable area. The field itself was measured by standard coils and a ballistic galvanometer in the usual way. Measurements of the intensity of the magnetic field for a given current gave results consistent to 1 part in 300, provided the iron was brought into a cyclic state by a number of reversals of the exciting current. This precaution was more especially necessary in the case of the weaker fields.

Sources of β rays employed

(1) Thin-walled emanation tube.
(2) Wires coated with radium B and C.
(3) Wires coated with radium C alone.

As already mentioned, the high velocity groups of rays were first photographed by using an α-ray tube. The thickness of the glass wall of the tube corresponded in stopping power for the α rays to 1·2 cm. of air. There was a small correction for the decrease of velocity of the β particle in traversing the glass. This was made by observing the shift of the lines when a thin mica plate of known thickness was placed immediately over the source. The corrections were found to be in good agreement with those previously tabulated by Danysz.* The α-ray tube could be used for determination of β rays from radium B and C together of velocities not less than 0·67 of that of light. For the lower velocities, a fine wire was made intensely active by exposure in a capillary tube to about 150 millicuries of purified emanation. Since here no correction for absorption of the β rays is required, the measurements afforded

---

* Danysz, *Compt. Rend.* cliv. p. 1502 (1912).

a useful check on the corrections made in some of the lower velocity lines obtained with the α-ray tube.

From the experiments of Hahn, Baeyer, and Meitner, it was known that of nine strong groups of β rays, five were due to radium B, and the remaining four to radium C. Since it was of great importance to distinguish definitely between the lines to be ascribed to radium B and radium C, a large number of experiments were made using pure radium C as a source. This was obtained by von Lerch's method, by immersing a nickel wire about 1·2 cm. long and 0·3 mm. diameter in a solution of the active deposit of radium in hot hydrochloric acid. By the use of large quantities of emanation in narrow tubes, it was found possible to make the wire so active as to equal in γ ray effect 50 milligrams of radium in equilibrium. To avoid all possibility of admixture of radium B, the nickel wire was not introduced into the solution until 20 minutes after the withdrawal of the emanation. Not a single trace of even the strongest lines of radium B was detected in any of these photographs. On the other hand, the wire was sufficiently active to bring out all but the faintest of the radium C lines observed with the relatively much stronger source afforded by the α-ray tube. The advantage of using a wide slit was especially marked in the experiment with radium C. It was found that the lines of the spectrum of radium C extended even into the regions occupied by the lowest velocity lines of radium B, although in the photographs taken with the sources containing both β-ray products, they were to a large extent masked by the strong photographic effect of the radium B rays. As a result of our experiments, we have found that radium C emits a very large number of distinct groups of β rays: in all we have observed more than 50 on the radium C plates alone. In addition, there are a number of other lines which can be seen only with great difficulty with specially arranged illumination. Only those lines are included in our list which have been definitely located on several photographs with at least two different fields. There are very great variations in the intensities of the different groups. On account of the great penetrating power of the high velocity β particles, the photographic effect for equal numbers of particles is relatively much weaker than that for the slower rays. In consequence of this, three or four times the length of exposure was required to bring out clearly the groups of high velocity β rays. In the absence of definite knowledge of the relative photographic effects of β rays of different velocities, it is desirable to classify the intensity of a given group by the strength with which it stands out above the general photographic fog due to scattered radiation &c. The following symbols, in order of decreasing intensity, have been used:—

v.s., very strong; s., strong; m.s., moderately strong; m., mean intensity; m.f., medium faint; f., faint; v.f., very faint.

In addition, it is convenient for reference to mark the strong lines by a letter.

It is evident from the nature of the problem and the large number of lines examined that this classification must be somewhat arbitrary, but it suffices for a general indication of the variation of intensity among the groups.

In the following table column I gives the distinctive number of the line, II the intensity; column III the value of $H\rho$ where H is the magnetic field in gauss and ρ the radius of curvature in cm. Column IV gives the value of β, the

TABLE I

Groups of rays from Radium B

I	II	III	IV	V
				Energy
No.	Inty	$H\rho$	β	$\div 10^{13}e$
1	f.	2450	0·823	3·852
2	m.s.	2295	0·805	3·480
3 A	s.	2235	0·797	3·332
4	v.f.	2140	0·787	3·160
5	m.s.	1990	0·762	2·758
6 B	v.s.	1925	0·751	2·610
7	m.	1815	0·731	2·366
8	m.s.	1752	0·719	2·228
9 C	v.s.	1660	0·700	2·039
10	m.s.	1470	0·656	1·650
11 D	v.s.	1392	0·635	1·519
		Groups		
12	m.	950–914	—	—
13	m.	861–836	—	—
14	m.s.	798	0·426	0·539
15 E	s.	770	0·414	0·503
16 F	v.s.	663	0·365	0·376

TABLE II

Groups of rays from Radium C

I	II	III	IV	V	VI	VII
				Observed		Calculated
No.	Inty	$H\rho$	β	Energy	Multiple	Energy
				$\div 10^{13}e$	of E_1	$\div 10^{13}e$
1	f.	9965	0·9858	25·25	59	25·30
2	f.	9605	0·9850	24·19	57	24·45
3	f.	9375	0·9840	23·51	55	23·59
4	f.	9115	0·9830	22·73	53	22·73
5	f.	8835	0·9822	21·91	51	21·88
6 A	m.f.	8530	0·9808	21·02	49	20·99
7	f.	8260	0·9797	20·22	47	20·13
8	f.	8050	0·9786	19·61	46	19·70

Continued on next page

Table II—*continued*

I	II	III	IV	V	VI	VII
No.	Inty	Hρ	β	Observed Energy ÷ $10^{13}e$	Multiple of E_1	Calculated Energy ÷ $10^{13}e$
9	f.	7820	0·9773	18·93	44	18·83
10	f.	7650	0·9764	18·43	43	18·40
11	f.	7490	0·9754	17·96	42	17·97
12 B	m.f.	7335	0·9744	17·51	41	17·56
13	f.	7200	0·9734	17·11	40	17·14
14 C	m.	7060	0·9724	16·71	39	16·71
15	f.	6910	0·9712	16·28	38	16·28
16	f.	6760	0·9700	15·83	37	15·85
17	f.	6616	0·9687	15·42	36	15·42
18	f.	6483	0·9676	15·03	35	14·99
19	f.	6310	0·9658	14·52	34	14·56
20 D	m.	6160	0·9643	14·09	33	14·13
21	f.	6000	0·9624	13·63	32	13·71
22 E	m.s.	5880	0·9610	13·28	31	13·28
23	f.	5720	0·959	12·82	30	12·85
24	f.	5520	0·956	12·24	29	12·42
25	v.f.	5386	0·954	11·86	28	11·99
26 F	m.	5255	0·952	11·49	27	11·57
27	f.	5110	0·949	11·07	26	11·14
28	f.	4990	0·947	10·73	25	10·71
29 G	m.s.	4840	0·946	10·31	24	10·28
30	m.f.	4380	0·933	8·91		
31	m.f.	4180	0·927	8·45		
32	m.f.	3900	0·917	7·68		
33	m.f.	3555	0·903	6·74		
34	m.f.	3320	0·891	6·11		
35 H	m.s.	3260	0·887	5·94		
36	m.f.	3160	0·881	5·67		
37	m.f.	3070	0·876	5·44		
38 K	s.	2960	0·868	5·16		
39	f.	2870	0·861	4·91		
40	f.	2820	0·857	4·79		
41	m.f.	2700	0·847	4·48		
42	f.	2530	0·831	4·05		
43	f.	2235	0·797	3·33		
44 L	m.	2080	0·776	2·96		
45 M	m.	1918	0·750	2·59		
46 N	m.	1550	0·675	1·81		
47	f.	1440	0·648	1·59		
48	f.	1380	0·632	1·49		

velocity of the particle in terms of the velocity of light: this is calculated from the value of $H\rho$ with the aid of the Lorentz-Einstein formula

$$m/m_0 = (1 - \beta^2)^{-\frac{1}{2}};$$

e/m_0 is taken as $1 \cdot 772 \times 10^7$ c.m. units.

Column V gives the total energy of the electron, calculated from the Lorentz-Einstein formula

$$E = \frac{m_0}{e}c^2\left(\frac{1}{\sqrt{1 - \beta^2}} - 1\right)e,$$

where c is the velocity of light.

The energy is expressed in terms of e in order to avoid any assumption with regard to the accuracy of any particular value of the electronic charge.

Radium B Groups

A diagrammatic representation of the lines from radium B is shown in fig. 2 (Plate). Using an emanation tube as a source, no lines above $H\rho = 2450$ were observed which could be definitely ascribed to radium B. It would no doubt be difficult to detect very faint lines in higher fields on account of the masking action of the strong photographic effect of the rays from the radium C groups. Both H. W. Schmidt* and Fajans and Makower† have observed that radium B emits some fairly penetrating β radiation. If this radiation be due to some groups of comparatively high velocity, it is to be expected that their photographic intensity would be small, and consequently difficult to detect in experiments with a source containing radium B and C together. In addition to the lines recorded, a number of faint groups were observed between $H\rho = 1392$ and $H\rho = 950$. These apparently consisted of two or more lines close together, but it was difficult to measure their position with accuracy.

The measurements of most of the strong lines are in fair agreement with those given by Danysz (*loc. cit.*) in the photographs obtained with radium B + C together.

It may be remarked that in a few cases lines due to radium B and radium C are exactly, or very nearly, coincident. This occurs in the case of the complex groups $H\rho$ 836–861, 914–950, in the low velocity part of the spectrum. These groups are strongly marked in the photographs obtained with wires coated with radium B + C, and fainter groups in very nearly the same positions were obtained on the radium C plates. To test this point further, photographs were taken with very short exposures of wires which had themselves only been exposed to the emanation for a short time, and thus had a much larger proportion of radium B to radium C than usual. The groups appeared of the same intensity, relative to the main radium B lines, as in the

---

* A. W. Schmidt, *Ann. d. Phys.* xxi. p. 609 (1906).

† Fajans and Makower, *Phil. Mag.* xxiii. p. 292 (1912).

other radium B + C photographs, showing that they were undoubtedly due to radium B.

Radium C Groups

A diagrammatic representation of some of the stronger lines from radium C is shown in fig. 2 (Plate). It is seen that radium C gives a spectrum in which the difference of intensity between the lines is relatively not so marked as in the case of radium B. The group of highest velocity observed in radium C, which showed up clearly on the photographs, gave the value of $H\rho = 8530$. Five other faint lines of higher velocity are recorded, and a number of others of still higher velocity could be faintly seen. Although we made a number of experiments in strong magnetic fields, no evidence of any strong line of higher velocity was observed, although lines up to $H\rho = 20,000$ would have been observed on the photographic plates.

In most of the experiments with β rays, an ordinary process plate was employed. We found it, however, advantageous for the higher fields to employ X-ray plates. These were photographically much more sensitive, and the fainter lines in the strong fields were more clearly seen. We have included a fairly complete list of the lines from $H\rho = 9965$ to $H\rho = 4840$. For still lower values of $H\rho$, the plates obtained with radium C as a source were crowded with fine lines, and only a few of the stronger are included in the table and in fig. 2 (Plate). For example, only five lines have been included between $H\rho = 4840$ and $H\rho = 3260$, although more than twenty were measured up. In a similar way, only a few of the more marked lines are recorded in the lower fields.

A complete analysis of the weaker lines would take a large amount of work, but it was not felt desirable at this stage to include them. It should be mentioned that the radium C lines extend right down to the low velocity regions, for which radium B shows several strong lines. It is of interest to note that radium C showed complex groups in about the same region as for radium B. We have already mentioned that in some cases it was very difficult to decide whether these complex groups belonged to radium B or radium C. Danysz, in his last paper, gave values for eight lines between $H\rho = 2947$ and $H\rho = 6073$ which are due to radium C. For most of these lines there is a fair agreement with the values found by him.

It should be remarked that all of the groups of β rays from radium C appear to be very closely homogeneous. The lines on the photographs were in all cases sharply marked, and there was no evidence of widening of the band to indicate that the group contained electrons of velocities varying between small limits. The strong lines due to radium B are always diffuse on the inner side when obtained with an α-ray tube. This no doubt is partly due to an alteration of the velocities in escaping through the walls of the glass tube. At the same time there is some evidence from the experiments with active wires that the velocities in the strong groups are variable over comparatively narrow limits.

General Considerations

On the theory of the origin of β and γ rays proposed by one of the authors (*loc. cit.*), it is to be expected that the differences between the energies of the β particles in different groups should be expressed by an integral number of one or more constants. These constants occurring represent the energy abstracted from the β particle in passing through certain regions of the atom, which are converted into energy of the γ-ray form. It is not proposed at this stage to enter into a discussion of the groups from radium B, as the experiments of Rutherford and Richardson (*loc. cit.*) have shown that the γ radiation from this substance is very complex, consisting of three distinct types widely differing from one another in penetrating power. Before the theory can be adequately applied in this case, the velocity of each group must be known with greater precision than it is at present. On the other hand, the same authors have observed that the γ radiation from radium C consists essentially of one penetrating type, and it is consequently to be expected that if any simple relation exists between the lines, it should be most obvious amongst the higher velocity electrons ejected from radium C. There seems to be fairly definite evidence of a relation of this character, at any rate between 29 lines given in the table for radium C. Taking the value of this difference as $E_1 = 0 \cdot 4284 \cdot 10^{13} e$, it is seen that all the lines observed from Nos. 1 to 29 fall closely into position. For the higher velocity groups, the average difference is twice this value, or $2E_1$, but all the lines between Nos. 9 and 29 are approximately expressed by the simple difference E_1, and *no lines are missing*. As the value of the energies of the fainter lines could not be determined with the same precision as the stronger lines, the agreement is probably as close as one could expect considering the difficulty of the measurements. It should, however, be pointed out that a line of medium intensity was observed before nearly all of the strong lines on the high velocity side. As far as measurements could be made, this ' outrider ' appeared to be somewhat closer to the strong line than would be expressed by a difference of the energy E_1. It is, however, difficult to decide this point definitely on account of the wide difference in intensity of the two lines.

It is also of interest to note that the energy of each of the lines is nearly an *integral* multiple of the common difference. This is clearly seen in Table II, where the value of the whole number is given in column VI and the corresponding calculated value of the energy in column VII. If this relation should hold, it would indicate that the groups of β rays observed are not directly due to the original β particle which causes the disturbance, but rather to the emission of energy in the β-ray form consequent on the vibrations of certain definite systems of electrons within the atom.

Below line No. 29, where only a few of the stronger lines are given, it was found that this simple relation no longer held, for the differences between some of the lines were much smaller than E_1. It is to be anticipated that the theoretical analysis of the slow velocity electrons for radium C would be very complex and present great difficulties. This follows from the fact that the electrons, as their velocities decrease, become more effective in exciting γ rays in the outer regions of the atoms.

The value of E_1 found by experiment is only about one-third of the value calculated from Whiddington's results on certain assumptions.* It is intended to continue experiments, and to analyse in detail the groups of β rays from other radioactive substances. It is hoped that when more data are available it will be possible to test with considerable certainty the adequacy of any theory in explanation of the origin of β and γ rays.

We are indebted to Mr. J. M. Nuttall for his kind assistance in some of the measurements.

University of Manchester
July 1913

* See Rutherford, 'Origin of β and γ rays,' Phil. Mag. Oct. and Dec. 1912. (*This vol., pp. 280, 292.*)

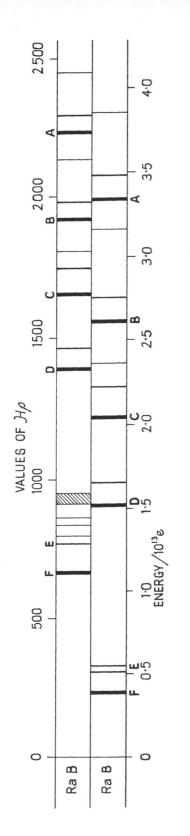

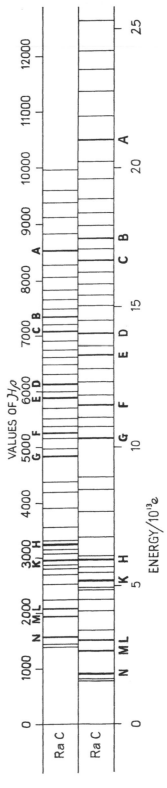

Fig. 2

Über die Masse und die Geschwindigkeiten der von den radioaktiven Substanzen ausgesendeten α-Teilchen*

von E. RUTHERFORD *und* H. ROBINSON

Aus den Sitzungsberichten der Kaiserl. Akademie der Wissenschaften in Wien.
Mathem.-naturw. Klasse; Bd. CXXII. Abt. IIa. November 1913, pp. 1855–84
(Vorgelegt in der Sitzung am 4. Dezember 1913.)

IM Jahre 1903 zeigte Rutherford,† daß die von einem Radiumpräparat ausgeschleuderten α-Strahlen aus einem Strome positiv geladener Teilchen bestehen, welche sich mit großer Geschwindigkeit bewegen und beim Durchgange durch ein magnetisches oder elektrisches Feld abgelenkt werden. Die hierbei benutzten Strahlen waren allerdings komplex, doch konnte durch Vergleich der Ablenkungen im magnetischen und elektrischen Felde von bekannter Intensität nach einer elektrischen Methode eine annähernde Schätzung ausgeführt werden, und zwar sowohl betreffs der mittleren Geschwindigkeit der α-Partikel sowie auch betreffs des Wertes $\frac{E}{m}$, des Verhältnisses der Ladung zur Masse des α-Teilchens. Es ergab sich die Geschwindigkeit zu $2 \cdot 5 . 10^9$ cm/sec und $\frac{E}{m} = 6000$ elektromagnetische Einheiten. Bald darauf wurden diese Resultate von Des Coudres‡ bei Benutzung einer photographischen Methode bestätigt. Er fand einen Wert $\frac{E}{m} = 6400$ für die α-Teilchen von Radium.

Weitere Versuche wurden von Mackenzie§ und Huff‖ angestellt, welche eine homogene α-Strahlenquelle — Polonium — verwendeten. Der Erstgenannte fand einen Wert $\frac{E}{m} = 4600$, letzterer $\frac{E}{m} = 4300$.

Im Jahre 1906 wiederholte Rutherford¶ seine Versuche, wobei er eine photographische Methode benutzte und mehrere bestimmte, voneinander

---

* *Editor's footnote:* An English translation of this paper was published under the title, 'The Mass and Velocities of the α Particles from Radioactive Substances' in *Philosophical Magazine*, xxviii, October 1914, pp. 552–72. Footnote, in the English paper, referring to new results is included here unchanged.

† Physik. Zeitschr., *4*, 235 (1903). (*English version, Vol. I, p. 549.*)

‡ Physik. Zeitschr., *4*, 483 (1903).   § Phil. Mag., *10*, p. 538 (1905).

‖ Proc. Roy. Soc. A., *78*, 77 (1906).   ¶ Phil. Mag., *12*, 348 (1906). (*Vol. I, p. 880.*)

verschiedene Quellen homogener α-Strahlung benutzte. Die erhaltenen elektrischen Ablenkungen waren bei den angewendeten Versuchsbedingungen klein und schwer mit Sicherheit zu messen. Es wurde ein Wert $\dfrac{E}{m} = 5070$ erhalten.

Ferner ergab sich der Schluß, daß die α-Teilchen von verschiedenen radioaktiven Produkten gleiche Masse besitzen, jedoch hinsichtlich der Anfangsgeschwindigkeit, mit der sie ausgeschleudert werden, differieren. Beim Durchgang durch absorbierende Schirme konnte keine Änderung der Masse der α-Partikel beobachtet werden. Der erhaltene Wert von $\dfrac{E}{m}$ für die α-Partikel führte zur Vermutung. daß diese ausgeschleuderten Heliumatome mit doppelter Elementarladung sein müßten. Dies wurde endgültig bewiesen durch die Versuche von Rutherford und Royds,* welche fanden, daß die durch ein dünnes Glasröhrchen hindurchschießenden α-Teilchen Heliumatome seien, sowie durch Rutherford und Geiger,† welche die von jedem α-Partikel getragene Ladung direkt messen konnten. Die Übereinstimmung zeischen den berechneten und den beobachteten Werten der Heliumproduktion aus Radium nach den Versuchen von Sir James Dewart‡ sowie Boltwood und Rutherford§ bestärkte noch weiter die genannten Schlüsse.

Zweck der vorliegenden Versuche war, eine möglichst genaue Bestimmung der Anfangsgeschwindigkeit sowie des Verhältnisses $\dfrac{E}{m}$ bei den α-Partikeln der radioaktiven Substanzen durchzuführen. Wenngleich, wie wir gesehen haben, die Identität des α-Partikel mit einem Heliumatom minus zwei Elektronen durch mehrere unabhängige experimentelle Methoden bewiesen ist, so sind doch die beiden obenerwähnten Größen wichtige, fundamentale radioaktive Konstanten, welche mit der größtmöglichsten Genauigkeit bestimmt sein sollten.

Überdies besteht eine beträchtliche Diskrepanz zwischen dem Werte $\dfrac{E}{m}$, der für das α-Partikel aus radioaktiven Messungen abgeleitet wurde, und dem, der für das Heliumatom aus atomistischen und elektrochemischen Daten gewonnen wurde.

Die neuesten Bestimmungen des Atomgewichtes des Heliums sind einerseits von Watson,‖ andrerseits von Heuse¶ ausgeführt worden. Ersterer fand den Wert $3 \cdot 994$ ($O = 16$), letzterer den Wert $4 \cdot 002$. Nimmt man den

---

* Phil. Mag., *17*, 281 (1909). (*This vol., p. 163.*)

† Proc. Roy. Soc. A., *81*, 141 und 162 (1908). (*This vol., p. 89 and p. 109*).

‡ Proc. Roy. Soc. A., *81*, 280 (1908), und *83*, 404 (1910).

§ Phil. Mag., *22*, 586 (1911), und diese Sitzungsber, *120*, 313 (1911). (*This vol., p. 221*).

‖ Journ. chem. Soc., *97*, 810 (1910).   ¶ Ber. d. Deutsch. physik. Ges., *15*, 518 (1913).

Mittelwert dieser zwei Zahlen, 3·998, für das Atomgewicht und den Wert von $\frac{e}{m}$ für das Wasserstoffatom zu 9570, so folgt, daß für das Heliumatom mit zwei Einheitsladungen der Wert von $\frac{E}{m}$ etwa 4826 betragen sollte, wogegen der alte experimentelle Wert für das α-Partikel 5070 beträgt.

Wir kennen keine sicheren Versuche, welche beweisen, daß der für ein mit hoher Geschwindigkeit sich bewegendes Atom gültige Wert von $\frac{e}{m}$ mit dem aus elektrochemischen Daten abgeleiteten Wert identisch ist.

Wenn auch die annähernde Übereinstimmung zwischen dem theoretischen und experimentellen Wert für das geladene Wasserstoffatom, wie er von Wien und Sir J. J. Thomson aus Versuchen mit Kanalstrahlen bestimmt worden ist, sicherlich eine solche Gleichheit vermuten lassen, so ist es doch von größter Wichtigkeit, diesen Schluß so streng als möglich zu prüfen.*

Da wir über den Ursprung der Masse und ihre Verteilung über die Bestandteile des Atoms nichts wissen, so ist es a priori nicht unmöglich, daß die scheinbare Masse des Heliumatoms, das zwei Elektronen verloren hat, merklich verschieden sein könnte von der des neutralen Atoms. Die allgemeine elektromagnetische Theorie würde ergeben, daß die elektromagnetische Masse des geladenen Heliumatoms bei einer Geschwindigkeit von $2 \cdot 10^9$ *cm*/sec von der des sich langsam bewegenden gleichen Atoms nur um $^1/_{450}$ abweichen würde. Dieser Effekt ist — selbst wenn er existierte — zu klein, um mittels den gegenwärtigen experimentellen Methoden mit Sicherheit konstatiert werden zu können.

Gleichzeitig muß daran erinnert werden, daß die Änderung des $\frac{e}{m}$ mit der Geschwindigkeit bisher nur für das negative Elektron experimentell geprüft worden ist und daß keine entsprechenden Beobachtungen über die Änderung der Masse mit der Geschwindigkeit eines positiv geladenen Atoms gemacht worden sind.

Überdies haben mehrere Beobachter aufmerksam gemacht, daß die Differenz zwischen den jetzt angenommenen Atomgewichten von Uran und Radium größer ist, als man annehmen müßte, wenn das Radiumatom aus dem Uranatom durch Ausschleuderung von drei α-Teilchen entsteht. Eine Diskrepanz im entgegengesetzten Sinne besteht zwischen den Atomgewichten von Radium und Blei, wenn das letztere — wie tatsächlich stark vermutet werden muß — das Endprodukt des Radiums ist.

* *Footnote from Philosophical Magazine:* W. Hammer has recently (*Annalen der Physik*, xliii, p. 653, March 1914) published an account of a careful determination of e/m for hydrogen Kanalstrahlen of velocity $2 \cdot 53 \times 10^5$ cm. sec.$^{-1}$, the velocity of the rays being measured directly by an oscillating condenser method. He finds $e/m = 9775$, or about 2 per cent. higher than the calculated value 9570 for the hydrogen ion in electrolysis (by an oversight Hammer quotes 9654 for the calculated value), and concludes that the difference can be accounted for by the probable experimental error.

Aus all diesen Gründen haben wir sehr sorgfältige Untersuchungen angestellt, um zu prüfen, ob der für das α-Partikel beobachtete Wert von $\frac{E}{m}$ innerhalb der Versuchsfehler identisch ist mit dem theoretischen Werte für das Heliumatom.

Seit zwei Jahren sind Versuche im Gange, die experimentelle Methode weiter auszubilden, um Resultate von der geforderten Genauigkeit — nämlich auf $^1/_{400}$ — zu erhalten.

Die Ablenkung eines engen Bündels von α-Strahlen wurde im magnetischen und im elektrischen Felde im Vakuum nach der photographischen Methode gemessen. Die Ablenkung im Magnetfeld gestattet die Bestimmung des Wertes $\frac{mv}{E}$, die im elektrischen Felde liefert den Wert $\frac{mv^2}{E}$, wobei m die Masse des α-Partikels, E seine Ladung und v die Geschwindigkeit bedeutet, mit der es ausgeschleudert wird. Durch die Kombination der beiden Beobachtungen wird dann v und $\frac{E}{m}$ einzeln bestimmt.

Die Ablenkung im magnetischen Felde

Es ist wichtig, den Wert von $\frac{mv}{E}$ für die α-Partikel, die von einer dünnen Schicht einer wohlbekannten radioaktiven Substanz ausgehen, z. B. von Ra C, zu kennen. Wenn einmal die Werte von $\frac{mv}{E}$ und $\frac{E}{m}$ für diese homogenen Strahlen von bekannter Reichweite bestimmt sind, so kann die Anfangsgeschwindigkeit der α-Partikel jeder anderen radioaktiven Substanz mit genügender Genauigkeit aus der von G e i g e r gefundenen Beziehung $v^3 = K.R$ abgeleitet werden, wobei R die Reichweite des α-Partikels, K eine Konstante bedeutet.

Die Anordnung bei den magnetischen Versuchen war die gewöhnlich angewendete: eine linienförmige homogene Strahlenquelle mit einem Schlitz parallel zur Strahlenquelle und eine photographische Platte, die parallel zum Schlitz und senkrecht zur Verbindungslinie von Strahlungsquelle und Schlitz gestellt war.

Der Schlitz war von der Strahlungsquelle und der photographischen Platte gleich weit (6·5 *cm*) entfernt. Das Magnetfeld war parallel zur Strahlungsquelle, dem Schlitz und der photographischen Platte.

Bei unseren Versuchen haben wir eine dünne Schicht von Ra C, also eine definierte und leicht reproduzierbare Quelle von homogenen α-Strahlen verwendet. Diese Strahlungsquelle bestand aus einem dünnen Platindraht, der durch mehrstündige Exposition im elektrischen Feld in einer Menge von zirka 100 Millicurie Emanation aktiviert worden war. Häufiger noch wurde ein 1 *cm* langes Stück des Platindrahtes in eine feine Kapillarröhre eingeschmolzen, in welche teilweise gereinigte Radiumemanation eingepreßt wurde. Nach zwei- oder dreistündiger Exposition wurde die Emanation mittels einer Pumpe entfernt, der Draht herausgenommen und mittels Klammern an einem Messinghalter fixiert, der dauernd an der Bodenplatte aus Messing von der Dimension $16 \times 2 \times 1$ *cm* befestigt war. Diese Grundplatte trug auch den Schlitz, der von zwei dünnen Kupferstreifen gebildet wurde, die, $^1/_{10}$ *mm* voneinander entfernt, an einem dicken Messingrahmen befestigt waren. An der Grundplatte war auch der Halter für die photographische Platte montiert. Der ganze Apparat war in seiner Konstruktion außerordentlich steif. Sobald der aktivierte Draht und die photographische Platte in ihre Stellung gebracht waren, wurde der Apparat dann in eine rechtwinklig geformte Messingbüchse geschoben, das ganze System durch Auspumpen und endlich durch in flüssiger Luft gekühlte Holzkohle evakuiert und zwischen die Polschuhe eines großen Elektromagneten gestellt. Währenddessen war das Ra *A* am Drahte praktisch vollständig verschwunden und die α-Strahlung rührte daher ausschließlich von Ra *C* her. Der Versuch bestand dann einfach darin, das Magnetfeld konstant zu halten und seine Richtung in Intervallen von einigen Minuten umzukehren.

Mit dem beschriebenen Apparate konnten befriedigende Bilder bei einer Expositionszeit von 20 Minuten erhalten werden, wenn der Draht — nach der γ-Strahlenmethode gemessen — eine anfängliche Aktivität entsprechend 30 *mg* Radium besaß.

Der Elektromagnet war einer von der großen Type, von welcher mehrere Exemplare kürzlich in unserem Institute hergestellt worden sind zu dem Zwecke, starke und gleichmäßige magnetische Felder mit großer Fläche zu erhalten.

Die Dimensionen der Polschuhe waren $10 \times 16 \times 15$ *cm*. Mit den 10×16 *cm*-Polschuhen in $2 \cdot 5$ *cm* Distanz wurde ein Feld von über 6000 Gauß mit einer Stromstärke von 5 Ampere erhalten; dabei war das Feld merklich homogen innerhalb des ganzen Zwischenraumes zwischen Strahlungsquelle und der photographischen Platte, also über zirka 13 *cm*.

Bei jedem Versuche wurde die Bestimmtheit des einem gegebenen Erregungsstrom entsprechenden Magnetfeldes noch dadurch gesichert, daß das Eisen einer Reihe von Polumkehrungen unterworfen wurde, um es in einen zyklischen Zustand zu bringen.

Die Theorie der magnetischen Ablenkung eines α-Partikels in einem Felde, welches von Punkt zu Punkt variiert, ist wie folgt:

O bedeutet in Fig. 1 die Strahlungsquelle, a den Schlitz und b die photographische Platte. Es ist angenommen, daß die Richtung des Magnetfeldes senkrecht zur Zeichnungsebene ist. Daher wird in dieser an jedem beliebigen Punkte die Bahn eines von O ausgeschleuderten α-Teilchens ein Kreisbogen vom Radius ρ sein, wobei $H\rho = \dfrac{mv}{E}$ sein muß; H bedeutet den Wert der Feldstärke an dem betreffenden Punkte.

Nehmen wir die Gerade $O\,a\,b$, welche die Strahlungsquelle mit dem Schlitze verbindet (d. h. den Weg des α-Partikels, wenn kein magnetisches Feld wirkt) als x-Achse und legen wir die y-Achse in der Zeichnungsebene senkrecht zu *O a b*.

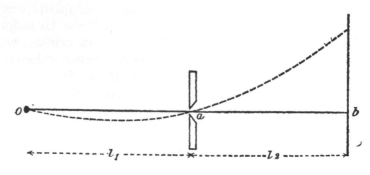

Fig. 1

Nun betrachten wir die Bahn eines α-Partikels, welches den Schlitz passiert, während das Magnetfeld wirkt; nennen wir die anfängliche Neigung $\dfrac{dy}{dx}$ dieser Bahn zur x-Achse (in O) p_0. In erster Annäherung können wir annehmen, daß die Neigung der Bahn zur x-Achse sehr klein sei, so daß deren Quadrat gegen Eins zu vernachlässigen ist und schreiben $\dfrac{d^2y}{dx^2}$ für den reziproken Wert des Krümmungsradius der Bahn.

Die Bahngleichung ist dann

$$\frac{1}{\rho} = \frac{d^2y}{dx^2} = \frac{HE}{mv}.$$

Integriert ergibt dies

$$\frac{dy}{dx} = \frac{E}{mv}\int_0^x H\,dx + p_0$$

und wenn wir setzen

$$Oa = l_1 \qquad ab = l_2,$$

so wird p_a, die Neigung der Bahn beim Schlitze a, gleich

$$p_a = \frac{E}{mv}\int_0^{l_1} H\,dx + p_0.$$

Der Wert des y beim Schlitze ergibt sich durch eine zweite Integration

$$y_{l_1} = \int_0^{l_1}\frac{dy}{dx}\,dx = p_0 l_1 + \frac{E}{mv}\int_0^{l_1}dx\int_0^x H\,dx.$$

Dieses Doppelintegral reduziert sich durch Integration per partes zu der einfachen Form

$$\int_0^{l_1} (l_1 - x) H dx.$$

Daher wird

$$y_{l_1} = p_0 l_1 + \frac{E}{mv} \int_0^{l_1} (l_1 - x) H dx.$$

Wenn aber das α-Partikel den Schlitz passieren muß, ist $y_{l_1} = 0$, d. h.

$$p_0 = - \frac{E}{mvl_1} \int_0^{l_1} (l_1 - x) H dx$$

und

$$p_a = p_0 + \frac{E}{mv} \int_0^{l_1} H dx = \frac{E}{mvl_1} \int_0^{l_1} x \cdot H dx.$$

Nehmen wir jetzt a als Koordinatenursprung von x, so ist die Gleichung der Bewegung des α-Teilchens in dem Teile des Feldes zwischen Schlitz und photographischer Platte

$$\frac{d^2 y}{dx^2} = \frac{HE}{mv}$$

und

$$\frac{dy}{dx} = p_a + \frac{E}{mv} \int_0^x H dx.$$

Die endgültige (beobachtete) Ablenkung des α-Partikels an der photographischen Platte ist dann

$$y_{l_2} = p_a l_2 + \frac{E}{mv} \int_0^{l_2} dx \int_0^x H dx$$

$$= \frac{El_2}{mvl_1} \int_0^{l_1} x H dx + \frac{E}{mv} \int_0^{l_2} (l_2 - x) H dx.$$

Dabei ist das erste Integral genommen zwischen der Strahlungsquelle und dem Schlitze, das zweite zwischen Schlitz und photographischer Platte.

Wenn wir H konstant machen, reduziert sich die letzte Gleichung zu

$$\frac{mv}{E} = H \cdot \frac{l_2(l_1 + l_2)}{2y_{l_2}}$$

und der Ausdruck $\frac{l_2(l_1 + l_2)}{2yl_2}$ repräsentiert den Näherungswert des Radius jenes Kreises, der durch O und den Schlitz hindurchgeht und einer Ablenkung y_{l_2} an der photographischen Platte entspricht.

Aus dem obigen Ausdruck ersieht man, daß der Einfluß einer etwaigen Veränderlichkeit des Feldes in Rechnung gezogen werden kann durch Ableitung einer mittleren effektiven Feldstärke $\bar{H}$, welche die gleiche Ablenkung hervorbringen würde. Dieser Mittelwert müßte ein mit Gewichtszahlen gerechnetes Mittel für die Feldstärken an allen Punkten entlang der Bahn des α-Partikels sein, wobei das einem Werte zwischen Strahlungsquelle und Schlitz zugeschriebene Gewicht der $\frac{l_2}{l_1}$ fachen Distanz x von der Strahlungsquelle proportional, in der Gegend zwischen Schlitz und Platte der Distanz $(l_2 - x)$ von der photographischen Platte proportional genommen werden müßte.

Dies zeigt, daß kleine Änderungen des magnetischen Feldes in der Gegend nahe der Strahlungsquelle und der photographischen Platte nur sehr kleine Korrekturen im Resultate involvieren würden, daß es hingegen wichtig ist, in der Gegend nahe dem Schlitze das Feld so gleichförmig als möglich zu halten.

Diese Bedingung wurde bei den vorliegenden Versuchen erfüllt, indem der Apparat so gestellt wurde, daß der Schlitz genau zwischen den Mittelpunkten der zwei großen Polschuhe des Magneten lag.

Die Werte der Integrale

$$\int_0^{l_1} x H\,dx \quad \text{und} \quad \int_0^{l_2} (l_2 - x)\,H\,dx$$

könnte man direkt erhalten nach einer von Sir J. J. Thomson* angegebenen Methode, bei welcher der magnetische Kraftfluß durch dreieckige Spulen von geeigneten Dimensionen an einem ballistischen Galvanometer oder Kraftflußmesser gemessen wird, doch bei dem gegenwärtigen Falle wurde es für bequemer und genauer gehalten, die kleinen Änderungen des Feldes von Punkt zu Punkt zu untersuchen.

Das Feld in der Mitte wurde sorgfältig gemessen mittels einer Spule in Verbindung mit einem ballistischen Galvanometer und die Veränderungen des Feldes geprüft, indem die kleine Untersuchungsspule rasch von der Mitte nach verschiedenen Punkten hin bewegt wurde.

Die Werte der Integrale erhält man dann am bequemsten dadurch, daß man die Änderung des Feldes mit der Distanz von der Mitte graphisch darstellt und aus der von den Kurven eingeschlossenen Fläche die kleine Korrektur ableitet, welche man von dem Werte des Feldes in der Mitte abziehen muß, um das mittlere wirksame Feld $\bar{H}$ zu erhalten.

Es sollen noch einige Einzelheiten angegeben werden betreffs der Konstruktion der Untersuchungs- und Standardisierungsspulen, welche speziell für die vorliegenden Versuche gewickelt worden waren.

Die angewendete Standardspirale bestand aus neun Windungen von

* Phil. Mag., *18*, 844 (1909).

seidenübersponnenem Kupferdraht Nr. 42, der auf eine Ebonitscheibe von 3·59 *cm* Durchmesser gewickelt war; die kleine Spule, welche zur Untersuchung der Änderungen der Feldstärke verwendet wurde, hatte 46 Windungen von Kupferdraht Nr. 47 auf einer Scheibe von 1·59 *cm* Durchmesser.

Die primäre und sekundäre Spule, welche zur Standardisierung der Feldstärken benötigt wurden, waren jede nur mit einfacher Lage von Drahtwindungen versehen, so daß alle erforderlichen Daten leicht genau erhalten werden konnten.

Die primäre Spule bestand aus 600 Windungen von Kupferdraht Nr. 16, die auf einem Messingrohr von 1 *m* Länge und 5·8 *cm* Durchmesser gewickelt waren. Die Spule war sehr sorgfältig auf der Drehbank hergestellt worden und die Verteilung der Windungen wurde nachher der ganzen Länge nach mit einem Ablesemikroskop untersucht.

Um die Kleinheit des Magnetfeldes, das mit natürlich schwachen Strömen in einem solchen Solenoid erhalten wurde kompensieren zu können, mußte eine sekundäre Spule von großer Fläche genommen werden; die sekundäre hatte eine Totalfläche von 10,200 *cm²*, bestehend aus 655 Windungen von Kupferdraht Nr. 42 mit einem mittleren Durchmesser von 4·45 *cm*. Der Widerstand betrug zirka 200 Ohm. Die Windungen waren aufgetragen auf einem weiten Glasrohr, das vorher bezüglich Gleichheit seines Durchmessers besonders ausgesucht worden war.

Die Änderungen in der Fläche der einzelnen Abteilungen der Röhre wurden durch Messungen in Abständen von 1 *cm* entlang ihrer Achse untersucht und nach Vollendung der Wickelung wurde die entsprechende Dichte der Windungen in jeder Abteilung durch Zählung der Windungen mit dem Ablesemikroskop ermittelt. Das sekundäre Solenoid wurde zentral innerhalb des primären gehalten und von ihm mittels zweier Ebonitringe isoliert.

Bei Benutzung von derart dimensionierten Spulen können Felder bis zu 6000 Gauß gemessen werden, ohne jede Erhitzung der primären Spule und der damit verbundenen Änderung des Widerstandes des Stromkreises.

Der Strom in der primären Spule wurde mit einem Westonamperemeter gemessen, das in der üblichen Weise mit Normalelement und Präzisionswiderständen sorgfältig geeicht worden war. Als Galvanometer wurde die Type von Siemens & Halske benutzt, welche eine sehr konstante Nullage und eine Schwingungsdauer von zirka 13 Sekunden besaß. Die Untersuchungsspulen wurden so gewählt, daß ein Ausschlag von über 400 Skalenteilen erreicht wurde und der Strom im primären Solenoid wurde so angepaßt, daß er möglichst nahe den gleichen Ausschlag gab, um etwaige Korrekturen wegen Mangel der Proportionalität der Skalen zu vermeiden.

Die Distanz zwischen den Mitten der zwei schmalen Streifen, welche auf der photographischen Platte durch Umkehrung des Magnetfeldes erhalten worden waren, wurde mittels eines Kayser'schen Komparators bestimmt. Die Einzelmessungen der Distanz zwischen den beiden Streifen stimmten innerhalb $^1/_{1000}$ überein.

Der mittlere Krümmungsradius der Bahn wurde dann berechnet nach der Formel

$$\rho^2 = \frac{1}{4d^2}(l_2^2 + d^2)[(l_1 + l_2)^2 + d^2],$$

wo l_1 und l_2 die schon erwähnte Bedeutung baben und d die Ablenkung des α-Partikels am Ende seiner Bahn, d. h. den halben Abstand der auf der Platte bei Feldumkehrung erhaltenen Streifen bezeichnet. Der Wert von $\frac{mv}{E}$ ist dann gegeben durch $\rho \cdot \bar{H}$, wo $\bar{H}$ die mittlere wirksame Feldstärke bedeutet.

Die doppelte Ablenkung des α-Strahlenbündels wurde bestimmt mit drei verschiedenen Strahlungsquellen in einem konstanten effektiven Felde von 6236 Gauß und ergab sich dabei zu 13·624, 13·634, 13·631 *mm*. Der Mittelwert ist 13·63 *mm*. Die Werte von l_1 und l_2 waren 6·450, beziehungsweise 6·618 *cm*. Der Wert von $\frac{mv}{E}$ für die α-Partikel von Ra C ergibt sich daraus zu $(3·983 \pm 0·005) . 10^5$, welchen Wert wir auf $^1/_{400}$ genau halten. E ist hierbei in absoluten elektromagnetischen Einheiten gemessen.

Es sei daran erinnert, daß die erste Bestimmung von $\frac{mv}{E}$ für die α-Strahlen von Ra C durch Rutherford* einen Wert von $3·98 . 10^5$ ergab. Bei einer späteren Bestimmung änderte sich der Wert auf $4·06 . 10^5$.

Während des Ganges unserer Versuche wurde eine unabhängige Bestimmung von $\frac{mv}{E}$ in unserem Laboratorium von Marsden und Taylor† bei ihren Messungen der Änderungen der Geschwindigkeit der α-Partikel beim Durchgang durch Materie ausgeführt. Es wurde die Szintillationsmethode benutzt und die Ablenkung der α-Strahlen in einem Magnetfeld direkt mittels eines Mikroskops gemessen. Der so erhaltene Wert von $\frac{mv}{E}$ betrug $4·00·10^5$ mit einem wahrscheinlichen Fehler von $\frac{1}{2}\%$.

Elektrostatische Ablenkung der Strahlen

Bei den ersten Versuchen von Rutherford im Jahre 1906 betrug die maximale Ablenkung der Bahn der α-Partikel, die an der photographischen Platte bei Umkehrung des elektrischen Feldes erzielt wurde, 3 *mm*. Die α-Strahlen wurden durch den evakuierten Zwischenraum zwischen zwei parallelen Platten durchgelassen, die 3·8 *cm* lang und 0·2 *mm* voneinander entfernt waren und gleichzeitig als Schlitz und zur Anlegung des elektrischen Feldes verwendet wurden. Der Nachteil der Methode liegt darin, daß es

---

* Phil. Mag., *10*, 163 (1905); *12*, 348 (1906). (*Vol. I, p. 803 and 880.*)
† Proc. Roy. Soc. A., *88*, 443 (1913).

notwendig war, die Distanz zwischen den äußeren Rändern der bei Feldumkehrung erhaltenen photographischen Eindrücke zu messen. Wie man später erkannte, hat diese Messung einige Unsicherheit an sich wegen der Zerstreuung der Strahlen an den Seiten der Metallplatten.

Bei den vorliegenden Versuchen war beabsichtigt, bei Feldumkehrung eine elektrostatische Ablenkung der α-Strahlen von mehr als 1 *cm* zu erhalten unter Versuchsbedingungen, bei denen alle in Betracht kommenden Größen mit beträchtlicher Genauigkeit gemessen werden konnten.

Mit der uns zur Verfügung stehenden konstanten Spannung von zirka 3000 Volt konnte dies nur erreicht werden, wenn man die Ablenkung des Strahlenbündels in einer Entfernung von fast 1 *m* von der Strahlungsquelle beobachtete.

In solcher Distanz ist der gesamte photographische Effekt eines sehr stark mit Ra C beschlagenen Drahtes zu klein, um beobachtet zu werden. Eine andere Schwierigkeit bietet der Abfall der Aktivität während der Zeit, die man braucht, um den großen Apparat bis zu einem solchen Grade zu evakuieren, daß beim Anlegen der hohen Spannung keine Entladung mehr eintritt. Um diese Schwierigkeiten zu vermeiden, wurde bei den vorläufigen Versuchen die Strahlungsquelle außerhalb des evakuierten Gefäßes gestellt, wobei die Strahlen durch ein Glimmerfenster von bekannter Dicke gehen mußten und die Ablenkung des Strahlenbündels durch das elektrische Feld nach der Szintillationsmethode beobachtet. Herr Marsden, der große Erfahrung und Geschicklichkeit hierin besitzt, hat uns in liebenswürdiger Weise durch Ausführung einer Zahl von Messungen unterstützt. Der Zinksulfidschirm war in einer Entfernung von zirka 1 *m* von der Strahlungsquelle; in dieser Distanz war jedoch die Linie der Szintillationen nicht sehr klar ausgeprägt. Es wurde eine Anzahl von Bestimmungen bei verschiedenen Spannungen gemacht, welche untereinander gut übereinstimmten. Doch fanden wir es sehr schwierig, mit Sicherheit die Mitte des Streifens der Szintillationen zu bezeichnen und daher ergaben sich die bei Umkehrung des Feldes beobachteten Ablenkungen in manchen Fällen unsymmetrisch in bezug auf die Mittellinie. Diese Unsicherheit bei der Szintillationsmethode führte uns schließlich dazu, eine photographische Methode anzuwenden. Bei den definitiven Versuchen war die Strahlungsquelle ein dünnwandiges α-Strahlenröhrchen mit zirka 100 Millicurie gereinigter Emanation.

Der endgültig benutzte Apparat ist in Fig. 2 dargestellt.

Das elektrostatische Feld wurde zwischen den parallelen Stirnflächen PP_1 zweier versilberter Streifen von Spiegelglasplatten, welche 35 *cm* lang, 2·5 *cm* tief und 1 *cm* dick waren, angelegt. Diese wurden 4 *mm* voneinander entfernt gehalten durch Eboniteinlagen und von Ebonitlagern getragen. Das α-Strahlenröhrchen *S*, welches an dem Punkte, wo es in die dickere Glasröhre überging, rechtwinklig gebogen und in welches die gereinigte Emanation über Quecksilber eingeschlossen war, wurde zentral nahe dem einen Ende der Glasplatten befestigt. Über dem anderen Ende der Platten war ein Schlitz montiert, der aus zwei $^1/_6$ *mm* entfernten Platten von dünnem Glimmer

N*

in einem Glimmerrahmen bestand. Dieser ganze Apparat wurde in eine
große Glasröhre gebracht, welche an ihren Enden durch eingeschliffene
Glasplatten verschlossen und mit einem seitlichen Rohr für die Evakuierung
versehen war. Der elektrische Kontakt mit den zwei Glasplatten wurde durch
Metallfedern gebildet, welche durch seitlich an dem Rohr angebrachte
eingeriebene Glasstopfen hindurchgingen. Die photographische Platte *D*,

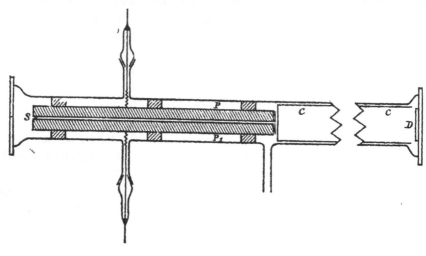

Fig. 2

welche in Aluminiumfolie eingehüllt wurde, um sie vor zerstreutem Licht und
dem Leuchten der Strahlungsquelle selbst zu schützen, war an der Platte,
welche das Rohr verschloß, in einer Entfernung von 50 *cm* vom Schlitz, d. h.
85 *cm* von der Strahlungsquelle angebracht. Der Weg der Strahlen vom
Schlitz bis zur photographischen Platte wurde von der Streuung des elektrostatischen Feldes geschützt durch einen langen Metallzylinder *CC*, der an
dem einen, dem Schlitz zugekehrten Ende geschlossen war, bis auf eine
kleine Öffnung, welche gerade den Durchgang der vom Schlitze kommenden
Strahlen gestattete.

Das flache Ende des Zylinders war in einer Entfernung von 0·5 *mm* von den
Enden der versilberten Glasplatten. Eine annähernde Berechnung des Effektes
des gestreuten Feldes wurde ausgeführt nach Methoden, ähnlich denen,
welche K a u f m a n n* angegeben hat.

Es wurde abgeleitet, daß die Korrektion wegen Streuung des Feldes unter
diesen Bedingungen fast gänzlich zu vernachlässigen war. Bei dem Ende
nächst der Strahlungsquelle war keine Korrektur erforderlich, da die Strahlungsquelle z w i s c h e n den Platten war und eine Feldstörung an diesem
Punkte einen verhältnismäßig unwichtigen Einfluß auf die Bahn hat. Die
vorläufigen Versuche mit Szintillationen hatten gezeigt, daß durch die
Einführung dieses Metallschirmes die von einem gegebenen Felde hervorgeru-

* Physik. Zeitschr., *8*, 75 (1907).

Ablenkung im
elektrostatischen
Felde

Ablenkung im
magnetischen Felde

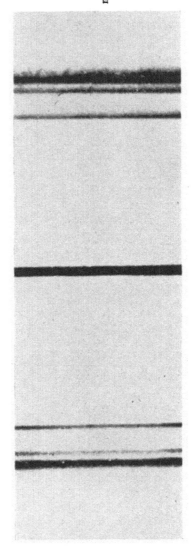

Lichtdruck v. Max Jaffé, Wien.

Bild 1

Bild 2

Tafel I

fene Ablenkung um etwa 1% verminderte. Die Potentialdifferenz zwischen den Glasplatten — im allgemeinen 1000 bis 2000 Volt — wurde erzeugt mittels einer Batterie von Akkumulatoren und kleinen Cadmium-elementen. Diese Spannung wurde mit einem Nalder-Thompsonschen elektrostatischen Voltmeter mit einem Meßbereiche von 600 bis 1500 Volt, welches eine sehr groß geteilte Skala hatte, angelegt. Dieses Instrument wurde sorgfältig geeicht durch direkten Vergleich mit zwei Weston'schen Normalelementen.

Die Batterien wurden so aufgestellt, daß praktisch jede gewünschte Spannung entnommen werden konnte, und es waren an vielen Stellen Kontakte mit isolierten Quecksilbernäpfchen angebracht, so daß jeder Teil der Batterie leicht mit dem Voltmeter geprüft werden konnte. Ein Potentiometer von hohem Widerstand, das mit einer kleinen Akkumulatorenbatterie von 40 Volt verbunden und mit einer der großen Batterien in Serie geschaltet war, diente dazu, das angelegte Potential während der Versuche bis auf weniger als 1 Volt konstant zu halten.

Die Versuche wurden in einem verdunkelten Raum ausgeführt, jedoch standen die Batterie und das Voltmeter in einem angrenzenden Raume; gut isolierte Leitungen wurden durch die Wände in Paraffinrohren geführt. Die Batterie selbst und alle Leitungen waren gut isoliert. Ein großer Kommutator mit Quecksilberkontakten auf einem Paraffinblock sowie eine Vorrichtung zum eventuellen Kurzschließen der Platten vollendeten die elektrische Apparatur. Die Evakuierung wurde mit Fleuß und Gaede-Pumpen und schließlich mit Kokosnußkohle in flüssiger Luft ausgeführt. Eine kleine, mit Induktorium betriebene Entladungsröhre, welche mit dem Apparat verbunden war, diente dazu, den Stand des Vakuums anzuzeigen und ermöglichte uns, das Risiko des Durchstoßens des α-Strahlröhrchens oder der Verschleierung der photographischen Platte bei zu frühem Anlegen des elektrischen Feldes zu vermeiden.

Im allgemeinen erforderte es zirka 1 Stunde, um die Evakuierung zu vollenden, und die darauffolgende Exposition mit dem elektrischen Felde dauerte etwa 6 Stunden. Aus der Länge dieser Expositionszeit, welche nötig war, sogar bei Verwendung der praktisch konstanten Strahlungsquelle von zirka 150 Millicuries, folgt, daß es unmöglich wäre, ein ähnliches Experiment bloß unter Benutzung eines aktivierten Drahtes als Strahlungsquelle durchzuführen.

Eine photographische Aufnahme der elektrostatischen Ablenkung der α-Strahlen ist reproduziert in Taf. I (erstes Bild). Die angelegte Spannung betrug 1435 Volt, die totale Expositionszeit 6 Stunden. Die Reproduktion ist 4·7 fach vergrößert. Der zentrale Streifen entspricht der direkten Einwirkung der α-Strahlen vor der Anlegung des Feldes; die drei Streifen an jeder Seite zeigen die bei jeder der beiden Feldrichtungen erhaltenen Eindrücke der α-Strahlen von Ra C. Ra A und Emanation. Die Linien von Ra C, dessen α-Strahlen die größte Geschwindigkeit haben, sind natürlich am wenigsten abgelenkt.

In der Photographie sieht man leicht, daß die inneren Ränder der Streifen außerordentlich scharf ausgeprägt sind. Der Grund dieser besonderen Erscheinung, welche der Ausführung genauer Messungen sehr zustatten kam, wird später besprochen werden.

Bei Benutzung des α-Strahlenröhrchens als Strahlungsquelle ist es notwendig, den Wert von $\frac{mv}{E}$ direkt für die aus dem Röhrchen kommenden Strahlen durch Messung der Ablenkung eines Strahlenbündels im Magnetfeld zu bestimmen.

Die Verminderung der Geschwindigkeit der α-Strahlen infolge ihres Durchganges durch die Wand des α-Strahlenröhrchens könnte aus dem beobachteten Absorptionsvermögen des Röhrchens zwar geschätzt werden, doch war es für unsere Zwecke vorzuziehen, die Messung direkt zu machen.

Das α-Strahlröhrchen wurde daher in den bei den magnetischen Experimenten benutzten Apparat eingelegt und nach einer Expositionszeit von 10 bis 20 Minuten die Photographien entnommen, wobei das magnetische Feld in Intervallen stets umgekehrt wurde. Dieses war von exakt derselben Stärke, wie das, welches bei der Bestimmung der Ablenkung bei einer linienförmigen Quelle homogener α-Strahlen benutzt wurde.

Eine Reproduktion eines der so erhaltenen Bilder ist in der Tafel (zweites Bild) dargestellt. Wie im vorhergehenden Falle sieht man deutlich drei getrennte Streifen, entsprechend der drei Typen von α-Strahlen, doch ist, wie zu erwarten war, die Trennung der Banden voneinander weniger scharf als im Falle der elektrostatischen Ablenkung. Die Schärfe der inneren Ränder der Linien ist wiederum, wie bei den elektrostatischen Versuchen, sehr leicht zu bemerken, ebenso das wirkliche Engerwerden der Ra C-Linie auf der Platte.

Die Korrektur wegen der Heterogenität der α-Strahlung

Der Gebrauch eines α-Strahlenröhrchens als Quelle ist nicht frei von Nachteilen, wie bei der Betrachtung der Heterogenität der herauskommenden α-Strahlen es sofort einleuchten wird. Die praktisch vollkommene Gleichförmigkeit des Röhrchens selbst war durch sorgfältige Prüfung und vorherige Auswahl garantiert, doch haben wir auch die Gegenwart von Strahlen in Betracht zu ziehen, welche die Wand des Röhrchens mit verschiedenen Graden der Neigung durchsetzen und einen Teil der erhaltenen photographischen Eindrücke ausmachen. Es ist klar, daß die einfache Theorie, welche den Fall einer linienförmigen Quelle homogener Strahlung behandelt, hier nicht mehr anwendbar ist. Jedoch ist es nicht schwer, eine Näherungsformel zu erhalten, aus welcher die Korrektur abgeleitet werden kann, um die beobachtete Ablenkung auf die entsprechende Ablenkung im idealen Falle einer homogenen Strahlungsquelle zu reduzieren.

Im letzteren Falle ist die Spur der α-Strahlen auf der photographischen Platte nach der Ablenkung das geometrische Bild der durch den Schlitz gehenden α-Strahlen. Die Entfernung zwischen den Mittellinien der bei

Feldumkehr erhaltenen Banden ist ein bestimmtes Maß für die Ablenkung der zentralen Strahlen.

Betrachten wir nun den Eindruck der von dem α-Strahlenröhrchen ausgehenden Strahlen:

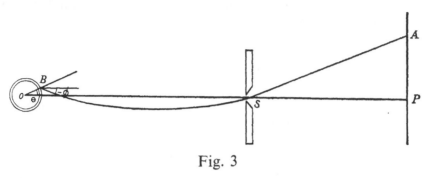

Fig. 3

In der schematischen Fig. 3 stellt *O* die α-Strahlenröhre in vergrößertem Maßstabe dar. *S* ist der Schlitz, *P* die photographische Platte. Wir wollen uns auf jene Strahlen beschränken, welche die Ablenkung kleiner als im normalen Falle machen. Bei den Strahlen, welche den oben dargestellten Eindruck *A* an der photographischen Platte hinterlassen, brauchen wir nur die obere Hälfte des α-Strahlenröhrchens zu betrachten. Nehmen wir einen Punkt in dieser Hälfte, z. B. *B* und benennen wir den Winkel, den der Radius durch *B* mit *OS* bildet, mit Θ. Wenn dann der ausgehende Strahl mit *OSP* einen Winkel −Φ bildet, so bildet er einen Winkel Θ + Φ mit der in *B* auf die Röhre errichteten Senkrechten. Wenn wir dann die Verminderung der Reichweite (z. B. in Luft) eines α-Strahles, der senkrecht durch das Röhrchen hindurchgeht, mit *D* bezeichnen, so ist die Verminderung der Reichweite eines von *B* ausgehenden Strahles näherungsweise gegeben durch *D* sec (Θ + Φ). Es sei daran erinnert, daß die Wanddicke des α-Strahlenröhrchens im Vergleich zum Radius klein ist, z. B. in einem typischen Falle mußte die Glasdicke etwa ein Zwanzigstel des Radius des Röhrchens ($\frac{1}{5}$ *mm*) betragen. Wenn wir nun als normale Ablenkung diejenige annehmen, welche bei einem α-Strahl eintritt, der gerade *D* Zentimeter seiner Reichweite in einem Röhrchen eingebüßt hat, so können wir leicht die relative Ablenkung für Strahlen, die von einem beliebigen Punkte, wie z. B. *B* ausgehen, berechnen.

Wenn *R* die anfängliche Reichweite des α-Strahles in Luft bezeichnet, so sind die Reichweiten der zwei Arten der herausdringenden Strahlen (*R*−*D*), beziehungsweise (*R*−*D*) sec (Θ + Φ) und das Verhältnis ihrer Geschwindigkeiten folglich gleich

$$\left[\frac{(R-D)}{(R-D) \sec (\Theta + \Phi)} \right]^{\frac{1}{3}},$$

wenn man der Einfachheit halber die Gültigkeit der Geigerschen Beziehung* zwischen Reichweite und Geschwindigkeit annimmt.

* Geiger, Proc. Roy. Soc. A., *83*, 505 (1910).

Nun ist im Falle der elektrostatischen Ablenkung diese verkehrt proportional zum Quadrat der Geschwindigkeit der Strahlen. Wenn daher $2d_1$ die durch Feldumkehrung erzielte doppelte Ablenkung bei einer linienförmigen α-Strahlenquelle mit einer Reichweite $R - D$ bezeichnet, so wird die doppelte Ablenkung $2d$, welche den Punkten der α-Strahlenröhre mit Winkeln $\pm\Theta$ gegen OSP entspricht, gegeben sein durch

$$2d = 2d_1 \left[\frac{R - D}{(R - D)\sec(\Theta + \Phi)} \right]^{\frac{2}{3}} - 2a \sin\Theta \times \frac{l_2}{l_1}, \qquad (1)$$

wobei a den Radius des α-Strahlenröhrchens und infolgedessen $\pm a \sin\Theta$ die Verschiebung von der Mittellinie weg für die betrachteten Punkte und l_1, l_2 die Abstände OS und SP bedeuten.

In erster Annäherung kann man Φ, welches jedenfalls sehr klein ist, als konstant und gleich dem Mittelwerte der anfänglichen Neigung der durch den Schlitz durchgehenden Strahlen betrachten, Der Wert des obigen Ausdruckes hängt dann nur von Θ ab und man findet, daß er ein Minimum erreicht für einen bestimmten Wert von Θ, welcher bei unseren Versuchsbedingungen ganz klein war; in einem besonderen Falle betrug Θ etwa 11°.

In Verbindung damit ist es erwähnenswert, daß der Wert von Θ, der die Ablenkung zu einem Minimum macht, fast direkt proportional zu $\dfrac{R - D}{D}$ ist.

Es ist daher — innerhalb gewisser Grenzen — von beträchtlichem Vorteil, diese Größe so klein als möglich zu machen, d. h. ziemlich dicke α-Strahlenröhrchen zu benutzen.

Die Wichtigkeit dieser "linsenartigen" Wirkung des α-Strahlenröhrchens, einen großen Teil seiner Strahlung in einer bestimmten Linie des Minimums der Ablenkung zu konzentrieren, drückt sich in dem Effekt aus, daß die inneren Ränder der abgelenkten Bündel außerordentlich scharf definiert sind — wie aus den Reproduktionen ersichtlich —, während die äußeren Teile der Banden, welche von Strahlen aller Geschwindigkeiten bis zur kleinsten, die überhaupt noch die Platte affiziert, herrühren, in eine allgemeine Verwaschenheit ausgehen, wie ja zu erwarten war.

Dieser Effekt ist deutlicher auf den wirklichen Photographien zu ersehen als in den Reproduktionen, welche durch Herstellung von Diapositiven der Originale und Projektion der ersteren auf „Gaslicht" -Papier hergestellt worden sind.

An den Photographien ist leicht zu sehen, daß die inneren Ränder der Streifen sehr scharf, geradezu ideal für Meßzwecke sind. Alle unsere Messungen wurden zwischen den inneren Ecken der verschiedenen Paare von Streifen ausgeführt und die normale Ablenkung dann aus diesen Messungen mit Hilfe der oben im Prinzip entwickelten Theorie berechnet.

Da die magnetischen Ablenkungen der α-Strahlen mit derselben α-Strahlenröhre bestimmt wurden, welche auch bei den elektrostatischen Versuchen verwendet wurde, so sind hier ähnliche Korrekturen notwendig, um die wahre Ablenkung der zentralen Strahlen abzuleiten. Man kann dieselbe allgemeine Theorie anwenden mit der Ausnahme, daß der Klammerausdruck in Gleichung (1) in der Potenz $^1/_3$ anstatt $^2/_3$ zu erheben ist. Der Grund hierfür ist einleuchtend, da die Ablenkung der Strahlen im Magnetfeld verkehrt proportional der ersten Potenz der Geschwindigkeit, nicht der zweiten Potenz wie im elektrostatischen Falle, ist. Die photographisch im Magnetfeld erhaltenen Streifen haben scharf begrenzte innere Ränder und die Ra C entsprechenden Streifen sind enger als das geometrische Bild.

Die Erklärung ist ähnlich wie im elektrostatischen Falle.

Die Resultate von drei getrennten Messungsreihen der Werte von $\dfrac{mv}{E}$ und $\dfrac{mv^2}{E}$ sowohl für die α-Strahlen des Ra C als auch der Emanation und des Ra A sind unten (Tabelle A) angegeben. Die Versuche Nr. 2 und 3 sind mit derselben α-Strahlenröhre, aber bei verschiedenen Spannungen ausgeführt.

Bei den elektrostatischen Experimenten betrug die angelegte Spannung bei

$$\text{Nr. 1} \ldots \ldots 1950 \text{ Volt}$$
$$\text{,, 2} \ldots \ldots 1435 \quad \text{,,}$$
$$\text{,, 3} \ldots \ldots 1958 \quad \text{,,}$$

Die erhaltenen Werte von $\dfrac{E}{m}$ sind, wie ersichtlich, sehr nahe übereinstimmend. Die an den α-Strahlen von Ra C gemachten Messungen sind wahrscheinlich am genauesten, da die photographischen Banden in diesem Falle am schärfsten definiert waren.

Unsere Endresultate für den Wert von $\dfrac{E}{m}$ für das α-Partikel liegen zwischen den Extremen $4 \cdot 813 \cdot 10^3$ und $4 \cdot 826 \cdot 10^3$, der Mittelwert ist

$$\frac{E}{m} = 4 \cdot 82 \cdot 10^3,$$

wobei E in absoluten elektromagnetischen Einheiten gemessen ist. Dies stimmt mit dem berechneten Wert innerhalb der wahrscheinlichen Fehlergrenzen gut überein, nachdem unser Wert sowohl von der elektrostatischen wie von der magnetischen Ablenkung abhängt und im letzteren Falle das Quadrat des Wertes eingeht.

Wie zu erwarten war, sind die dem Ra A und der Emanation entsprechenden photographischen Streifen nicht so scharf begrenzt wie die Ra C-Linien. Trotzdem waren einige von den späteren Aufnahmen genügend rein, um eine Ausmessung dieser Linien mit beträchtlicher Genauigkeit zu ermöglichen, und wir konnten daraus die Werte von $\dfrac{E}{m}$ für die α-Partikel von Ra A und

Tabelle *A*

Versuchnummer	Ra C			Ra A			Ra-Emanation		
	$\dfrac{mv}{E}$	$\dfrac{mv^2}{E}$	$\dfrac{E}{m}$	$\dfrac{mv}{E}$	$\dfrac{mv^2}{E}$	$\dfrac{E}{m}$	$\dfrac{mv}{E}$	$\dfrac{mv^2}{E}$	$\dfrac{E}{m}$
1 (Reichweitenverkürzung der Wand des Röhrchens 1·85 *cm*)	$3·605.10^5$	$6·269.10^{14}$	$4·824.10^3$	—	—	—	—	—	—
2 (Reichweitenverkürzung der Wand des Röhrchens 2·00 *cm*)	$3·555.10^5$	$6·083.10^{14}$	$4·813.10^3$	$2·941.10^5$	$4·174.10^{14}$	$4·824.10^3$	$2·717.10^5$	$3·560.10^{14}$	$4·822.10^3$
3 (Reichweitenverkürzung der Wand des Röhrchens 2·00 *cm*)	$3·555.10^5$	$6·100.10^{14}$	$4·826.10^3$	$2·941.10^5$	$4·185.10^{14}$	$4·837.10^3$	$2·717.10^5$	$3·563.10^{14}$	$4·826.10^3$

der Emanation berchnen. Die gefundenen Werte waren nahezu vollständig gleich wie die bei Ra *C*.

Im allgemeinen fanden wir also eine ausgezeichnete Übereinstimmung zwischen dem Werte von $\dfrac{mv}{E}$, der direkt aus der magnetischen Ablenkung abgeleitet wurde, und dem Werte, den man erhält, wenn man für die Anfangsgeschwindigkeit die Korrektur wegen Reichweitenverkürzung des α-Strahlenröhrchens einführt. Herr Marsden hat uns gütigst diese Reichweitenverkürzung nach der Szintillationsmethode experimentell bestimmt.

Zusammenfassung der Resultate

Aus den Messungen der Ablenkung der α-Strahlen in magnetischen und elektrischen Feldern haben wir für das unabgeschirmte α-Partikel von Ra *C* die Werte gefunden:

$$\frac{mv}{E} = 3 \cdot 985 \cdot 10^5 \text{ und } \frac{E}{m} = 4823 \text{ elektromagn. Einheiten.}$$

Die Kombination dieser beiden Werte ergibt für die anfängliche Geschwindigkeit der Ausschleuderung der α-Partikel von Ra *C* den Wert

$$v = 1 \cdot 922 \cdot 10^9 \text{ cm/sec.}$$

Dieser Wert ist um zirka 7 % niedriger als der früher angenommene Wert.

Nimmt man diesen Wert von v, so kann die Anfangsgeschwindigkeit der α-Partikel von den verschiedenen α-strahlenden Substanzen sofort aus der Geiger'schen Beziehung $v^3 = \kappa \cdot R$ abgeleitet werden, wo R die Reichweite in Luft bedeutet. Die Resultate sind in der folgenden Tabelle *B* enthalten.

Die Wärmeproduktion der α-Partikel des Radiums

Es ist nun von Interesse, aus den gegenwärtigen Daten die Gesamtenergie zu berechnen, welche die pro Sekunde von 1 *g* Radium samt seinen Zerfallsprodukten emittierten α-Partikel mit sich führen, und diese mit der in Form von Wärme beobachteten Energieemission zu vergleichen.

Die in Form von α-Strahlen emittierte Energie ist gegeben durch $\frac{1}{2} \Sigma nmv^2$, wo n die von jedem Produkt pro Sekunde emittierte Zahl der α-Partikel und v deren Geschwindigkeit bedeutet. Im radioaktiven Gleichgewicht wird von jedem Produkt pro Sekunde die gleiche Zahl von α-Partikeln ausgeschleudert, also

$$\frac{1}{2} \Sigma nmv^2 = \frac{1}{2} En \Sigma \frac{mv^2}{E}.$$

Den Wert $\dfrac{mv^2}{E}$ kann man für jedes Produkt aus der unten angeführten Tabelle erhalten.

Der Wert von $nE = 10 \cdot 54 \cdot 10^{-10}$ elektromagnetischen Einheiten ist von Rutherford und Geiger direkt gefunden worden, und zwar durch Messung der von den α-Partikeln transportierten Ladung. Es ist gezeigt worden, daß die so erhaltene Zahl sehr nahe mit jener übereinstimmt,

Tabelle *B*

Substanz	Reichweite der α-Strahlen in Luft bei 15° C. und 760 *mm* Druck	Geschwindigkeit *cm*/sec $v = k \sqrt[3]{R}$	$\dfrac{mv}{E}$ (in elektromagnetischen Einheiten) $^{-1}$	$\dfrac{mv^2}{E} =$ $\dfrac{\text{Energie}}{\text{Elementarquantum}}$
Uran 1	$2 \cdot 50$ *cm*	$1 \cdot 37 \cdot 10^9$	$2 \cdot 84 \cdot 10^5$	$3 \cdot 88 \cdot 10^{14}$
„ 2	$2 \cdot 90$	$1 \cdot 44 \cdot 10^9$	$2 \cdot 98 \cdot 10^5$	$4 \cdot 28 \cdot 10^{14}$
Jonium	$3 \cdot 00$	$1 \cdot 45 \cdot 10^9$	$3 \cdot 01 \cdot 10^5$	$4 \cdot 38 \cdot 10^{14}$
Radium	$3 \cdot 30$	$1 \cdot 50 \cdot 10^9$	$3 \cdot 11 \cdot 10^5$	$4 \cdot 67 \cdot 10^{14}$
Radiumemanation	$4 \cdot 16$	$1 \cdot 62 \cdot 10^9$	$3 \cdot 36 \cdot 10^5$	$5 \cdot 44 \cdot 10^{14}$
Radium *A*	$4 \cdot 75$	$1 \cdot 69 \cdot 10^9$	$3 \cdot 51 \cdot 10^5$	$5 \cdot 95 \cdot 10^{14}$
„ *C*	$6 \cdot 94$	$1 \cdot 92 \cdot 10^9$	$3 \cdot 985 \cdot 10^5$	$7 \cdot 66 \cdot 10^{14}$
„ *F*	$3 \cdot 77$	$1 \cdot 57 \cdot 10^9$	$3 \cdot 25 \cdot 10^5$	$5 \cdot 10 \cdot 10^{14}$
Thorium	$2 \cdot 72$ *cm*	$1 \cdot 41 \cdot 10^9$	$2 \cdot 92 \cdot 10^5$	$4 \cdot 10 \cdot 10^{14}$
Radiothorium	$3 \cdot 87$	$1 \cdot 58 \cdot 10^9$	$3 \cdot 28 \cdot 10^5$	$5 \cdot 19 \cdot 10^{14}$
Thorium *X*	$4 \cdot 30$	$1 \cdot 64 \cdot 10^9$	$3 \cdot 40 \cdot 10^5$	$5 \cdot 57 \cdot 10^{14}$
Thoriumemanation	$5 \cdot 00$	$1 \cdot 72 \cdot 10^9$	$3 \cdot 57 \cdot 10^5$	$6 \cdot 16 \cdot 10^{14}$
Thorium *A*	$5 \cdot 70$	$1 \cdot 80 \cdot 10^9$	$3 \cdot 73 \cdot 10^5$	$6 \cdot 72 \cdot 10^{14}$
„ C_1	$4 \cdot 80$	$1 \cdot 70 \cdot 10^9$	$3 \cdot 52 \cdot 10^5$	$5 \cdot 99 \cdot 10^{14}$
„ C_2	$8 \cdot 60$	$2 \cdot 06 \cdot 10^9$	$4 \cdot 28 \cdot 10^5$	$8 \cdot 84 \cdot 10^{14}$
Radioactinium	$4 \cdot 60$ *cm*	$1 \cdot 68 \cdot 10^9$	$3 \cdot 48 \cdot 10^5$	$5 \cdot 82 \cdot 10^{14}$
Actinium *X*	$4 \cdot 40$	$1 \cdot 65 \cdot 10^9$	$3 \cdot 42 \cdot 10^5$	$5 \cdot 65 \cdot 10^{14}$
Actiniumemanation	$5 \cdot 70$	$1 \cdot 80 \cdot 10^9$	$3 \cdot 73 \cdot 10^5$	$6 \cdot 72 \cdot 10^{14}$
Actinium *A*	$6 \cdot 50$	$1 \cdot 88 \cdot 10^9$	$3 \cdot 90 \cdot 10^5$	$7 \cdot 33 \cdot 10^{14}$
„ *C*	$5 \cdot 40$	$1 \cdot 77 \cdot 10^9$	$3 \cdot 66 \cdot 10^5$	$6 \cdot 48 \cdot 10^{14}$

welche aus den Messungen der Heliumentwicklung aus Radium* abgeleitet wurde. Vor kurzem ist der Wert von nE durch Duane und Danysz† neuerlich bestimmt worden und ergab sich ein wenig niedriger: $nE = 10 \cdot 09 \cdot 10^{-10}$. Nehmen wir den Wert $nE = 10 \cdot 54 \cdot 10^{-10}$, so ergibt sich die in Form von α-Partikeln von 1 *g* Radium im Gleichgewicht mit seinen kurzlebigen Zerfallsprodukten Emanation, Ra *A* und Ra *C* pro Zeiteinheit emittierte Energie zu $1 \cdot 250 \cdot 16^6$ Erg pro Sekunde.

Der entsprechende Wert für 1 Curie Emanation im Gleichgewicht mit Ra *A* und Ra *C* ist $1 \cdot 004 \cdot 10^6$ Erg/sec. Die bezüglichen Zahlen für die

* Boltwood und Rutherford, diese Sitzungsber., *120*, 313 (1911); Phil. Mag., *22*, 586 (1911). (*This vol., p. 221.*)   † Le Radium, *9*, 417 (1912).

Wärmeentwicklung sind dann entsprechend 107·4 und 86·3 Grammcalorien pro Stunde. Dies ist ausgedrückt nach dem alten Rutherford-Boltwood'schen Radiumstandard. Nach dem neuen internationalen Standard werden diese Zahlen 112·8 und 90·6 cal/St. Aber selbst wenn wir zu diesen Zahlen noch je 2% addieren, um die Energie der Rückstoßatome mitzuberücksichtigen, sind die Zahlen um zirka 7% niedriger als die Werte 123·6 und 99·2 cal/St, welche bei der Messung des Wärmeeffektes der α-Strahlen nach internationalem Standard erhalten wurden.*

Es ist nicht wahrscheinlich, daß der Wert von *nE* um 7% zu klein ist, und die Werte der anderen Größen sind mit einem wahrscheinlichen Fehler von weniger als 1% bekannt.

Es scheint daher vernünftig, anzunehmen, daß bei den α-Umwandlungen ein kleiner Teil der beobachteten Wärmeentwicklung aus anderen Ursachen als aus der kinetischen Energie der Partikel stammt. Zweifellos wird ein Teil der Energie in Form von γ-Strahlen frei, denn Chadwick und Russell haben gefunden, daß die Aussendung von α-Partikeln stets von schwachen γ-Strahlen begleitet wird.

Vielleicht führt die auf Emission eines α-Partikels folgende Neuordnung der Konstituenten des Atoms zu einer Freimachung von Energie in anderer Form, welche sich dann als Wärme äußert.

Es ist nicht zu erwarten, daß die ganze, bei einer Atom-explosion frei werdende Energie im Projektil und im Rückstoß enthalten ist. Das Atom selbst muß in gewaltige Schwingungen versetzt werden und sich in einen neuen temporären Gleichgewichtszustand versetzen und dies kann ein Freiwerden von Energie bedingen.

Weitere Versuche

Im Laufe der Versuche über die magnetische und elektrostatische Ablenkung der α-Strahlen wurden einige schwache Streifen beobachtet, die nicht den α-Strahlen zugeschrieben werden konnten. Diese Streifen waren besonders an den Photographien der elektrostatischen Ablenkung bemerkbar. Es wurden besondere Versuche unternommen, um zu sehen, ob diese Banden vielleicht von der Emission von Atomen von seiten des α-Strahlenröhrchens herrührten, welche entweder in ihrer Ladung oder in ihrer Masse vom α-Partikel verschieden waren. Um diesen Punkt zu prüfen haben wir den für die magnetische Ablenkung der α-Strahlen gebrauchten Apparat modifiziert. Ein mit Emanation gefülltes α-Strahlenröhrchen diente als intensive α-Strahlenquelle. Nach Durchgang durch einen engen Schlitz fielen die Strahlen auf einen Zinksulfidschirm. Die Distanzen wurden so gewählt, daß die Ablenkung der α-Strahlen von Ra *C* bei Feldumkehrung zirka 2 *cm* betrug. Es wurden besondere Vorsichtsmaßregeln getroffen, um den Effekt der Zerstreuung der α-Strahlen am Schlitz selbst zu einem Minimum herabzudrücken.

* Rutherford und Robinson, Phil. Mag., *25*, 312 (1913) und diese Sitzungber., *121*, 1491 (1912). (*This vol., p. 312.*)

Die Verteilung der Szintillationen über den Schirm wurde dann systematisch mit einem Mikroskop untersucht. Einige Szintillationen pro Sekunde wurden beobachtet auch an allen Punkten, an welchen das Hauptstrahlenbündel nicht auffiel.

Diese Zahl wuchs rapid in der Nähe des Hauptstreifens der α-Strahlen. Doch wurde keinerlei Andeutung von Maximis der Szintillationen im Gebiete zwischen den zu Ra C gehörigen Banden oder außerhalb der zur Emanation gehörigen Banden beobachtet.

Wenn irgendwelche andere Partikel von der Röhre emittiert werden, die weniger als die des Ra C abgelenkt würden, so ist sicherlich ihre Zahl kleiner als $^1/_{10000}$ der Gesamtzahl der α-Partikel.

Die wirklich beobachtete Verteilung ist zweifellos einer schwachen Zerstreuung der Strahlen an den Rändern des Spaltes zuzuschreiben. Wir können daher schließen, daß keine anderen in Masse und Geschwindigkeit mit den α-Teilchen vergleichbaren, geladenen Partikel während der Umwandlung der Radiumemanation und ihrer Folgeprodukte ausgeschleudert werden.

Zusammenfassung

1. Es wurde eine genaue Bestimmung der Ablenkung der α-Strahlen in magnetischen und elektrischen Feldern von bekannter Stärke nach der photographischen Methode ausgeführt. Ein dünnwandiges, mit Radiumemanation gefülltes Röhrchen wurde als α-Strahlenquelle benützt. Der Wert von $\frac{E}{m}$, dem Verhältnis von Ladung zur Masse des α-Partikels, wurde zu 4820 elektromagnetischen Einheiten gefunden. Dies stimmt innerhalb der Fehlergrenzen der Experimente mit dem aus elektrochemischen Daten erwarteten Werte $\frac{E}{m} = 4826$, unter der Annahme, daß das α-Partikel zwei Einheitsladungen trägt.

2. Der Wert von $\frac{E}{m}$ für das α-Partikel von Radium C ist innerhalb der Versuchsfehlergrenzen identisch mit dem entsprechenden Werte für das α-Partikel von Radium A oder Radiumemanation. Der Wert von $H\rho = \frac{mv}{E}$ für die von einer dünnen Schicht von Radium C ausgeschleuderten α-Partikel wurde genau bestimmt und ergab sich zu $3 \cdot 985 \cdot 10^5$ elektromagnetischen Einheiten.

3. Die Geschwindigkeit, mit der die α-Partikel von Radium C ausgeschleudert werden, ist $1 \cdot 922 \cdot 10^9$ *cm*/sec. Dies ist ein gegen den früher angenommen um 7% niedrigerer Wert.

4. Unter Benutzung der bekannten Beziehung zwischen Reichweite und Geschwindigkeit der α-Partikel wurden die Geschwindigkeits- und Energiewerte der α-Strahlen für alle bekannten Radioelemente bestimmt.

5. Es wurde geschlossen, daß die von der kinetischen Energie der α-Partikel des Radiums herrührende Wärmeproduktion um etwa 7 % kleiner ist als der experimentell bestimmte Wert. Daraus folgt, daß ein kleiner Teil der Wärmeentwicklung des Radiums anderen Ursachen als der Energie der ausgeschleuderten α-Partikel zuzuschreiben ist.

6. Es wurde keinerlei Anzeichen gefunden, daß Teilchen von anderer Masse oder Ladung als der der α-Partikel von dem α-Strahlenröhrchen ausgeschleudert werden.

Wir möchten nun schließlich Herrn E. Marsden für seine sehr wertvolle Beihilfe bei den Vorversuchen und seine Hilfe bei den Szintillationsversuchen unseren wärmsten Dank ausdrücken.

Universität Manchester
November 1913

(Das englische Manuskript übersetzt von V. F. Hess, Wien.)

The British Radium Standard

From *Nature*, **92**, 1913, pp. 402–3

AN account of the preparation and testing of an international radium standard was given in the issue of this journal for April 4, 1912 (vol. lxxxix, p. 115). It will be remembered that a radium standard containing $21 \cdot 99$ milligrams of pure radium chloride was prepared by Mme. Curie for the International Committee. At a meeting in Paris the standard of Mme. Curie was compared with another independent standard prepared in Vienna by Professor Hönig-schmidt, and the two were found to agree well within the limits of accuracy of measurements by the γ ray method. The preparation of Mme. Curie was accepted by the Committee as the International Standard, and was deposited in the Bureau du Poids et Mesures at Sèvres, near Paris. At the same time it was arranged that the Vienna preparation should be retained in Vienna as a secondary standard. Arrangements were made to allow Governments to obtain duplicates of the international standard. For this purpose the Austrian Government generously offered to provide the radium required at a considerable reduction in price. It was arranged that duplicate standards should be prepared and tested in Vienna in terms of their secondary standard, and then sent on to Paris to be tested again in terms of the international standard. In all six duplicate standards have now been prepared for different Governments, and the independent standardisations of the radium content in Vienna and Paris have been found to be in remarkably good agreement. The comparison of the quantities of radium is made by means of the penetrating γ rays, and it is a striking testimony to the accuracy of this method that the independent measurements have agreed so closely, although widely differing experimental arrangements have been employed in the two places.

It will be remembered that Dr. Beilby, F.R.S., very generously defrayed to Mme. Curie the cost of the radium forming the international standard, and thus relieved the International Committee of the necessity of collecting special funds for this purpose. Immediately after the fixing of the international standard, arrangements were made in this country to obtain a duplicate standard to be placed in charge of the National Physical Laboratory at Teddington. Dr. Beilby again stepped in in a very generous manner and agreed to defray the expense of acquiring the British radium standard, which was delivered to the National Physical Laboratory a few months ago. The British radium standard does not differ much in radium content from the international standard, containing about 20 milligrams of pure radium chloride.

A circular has now been issued by the National Physical Laboratory, stating that they are prepared to standardise preparations of radium and mesothorium in terms of the international standard, and a detailed list of testing charges has been issued. In the beginning, the Laboratory has very wisely confined itself to undertaking the standardisation of strong preparations of radium and mesothorium only. The comparison with the British standard will be made by γ ray methods. Tests on radio-active minerals, radio-active waters and other materials of weak activity, will not be undertaken at the moment, though, no doubt, arrangements will be made as the new radio-active department progresses to undertake some work of this character in the future. The Laboratory sends out a certificate that the active material under examination shows a γ ray activity equivalent to a certain weight of metallic radium, but no guarantee is given of whether the activity is due to radium itself, for it is well known that it is not easy to distinguish without special tests between preparations of radium and mesothorium. Preparations of the latter are standardised by expressing their γ ray activity at the time of testing in terms of a definite weight of metallic radium in radio-active equilibrium. Both the Reichsanstalt and the National Physical Laboratory express the activity of their preparations in terms of metallic radium, and not in terms of bromide or chloride. This appears to me a very wise step, for it is obviously more definite and scientific to express the results in this form. It is also very desirable that all radium should be bought and sold in terms of metallic radium, thus avoiding the uncertainty that sometimes arises as to whether the preparation is being sold as anhydrous radium bromide or radium bromide with its water of crystallisation.

The radio-active department in the Reichsanstalt has now been in operation for more than a year, under the charge of Dr. Geiger, whose radio-active researches in the University of Manchester are well known. The creation of this department has been found to fill a much-needed want, and it is not too much to say that practically all the radium and mesothorium that is bought and sold in Germany requires to-day the certificate of the Reichsanstalt. The number of standardisations required have increased very rapidly, and several assistants have been added to the department in charge of this work alone. There can be no doubt that the institution of a radio-active department in the National Physical Laboratory will prove of great service to this country, not only for scientific, but also for commercial purposes. It is well known that the buying and selling of radium in the past has been a very uncertain and risky procedure, for in most cases the radium content has not been expressed in terms of any authorised standard. This difficulty is removed by the present arrangement, and we should strongly recommend that those who wish to buy radium or mesothorium, whether for scientific or for medical purposes, should do so conditional on the certificate of standardisation from the National Physical Laboratory.

It is understood that the work of testing and standardisation will be under the supervision of Dr. W. G. C. Kaye, of the National Physical Laboratory, whose pioneer work on the production and distribution of X-rays is well known to all physicists. The ability and skill in measurements which he has shown both in his work in the Cavendish Laboratory and in the National Physical Laboratory, afford the best of guarantees that the work of the new department will be carried out in a thoroughly satisfactory manner.

E. RUTHERFORD

The Structure of the Atom

From *Nature*, **92,** 1913, p. 423

In a letter to this journal last week, Mr. Soddy has discussed the bearing of my theory of the nucleus atom on radio-active phenomena, and seems to be under the impression that I hold the view that the nucleus must consist entirely of positive electricity. As a matter of fact, I have not discussed in any detail the question of the constitution of the nucleus beyond the statement that it must have a resultant positive charge. There appears to me no doubt that the α particle does arise from the nucleus, and I have thought for some time that the evidence points to the conclusion that the β particle has a similar origin. This point has been discussed in some detail in a recent paper by Bohr (*Phil. Mag.*, September, 1913). The strongest evidence in support of this view is, to my mind, (1) that the β ray, like the α ray, transformations are independent of physical and chemical conditions, and (2) that the energy emitted in the form of β and γ rays by the transformation of an atom of radium C is much greater than could be expected to be stored up in the external electronic system. At the same time, I think it very likely that a considerable fraction of the β rays which are expelled from radio-active substances arise from the external electrons. This, however, is probably a secondary effect resulting from the primary expulsion of a β particle from the nucleus.

The original suggestion of van der Broek that the charge on the nucleus is equal to the atomic number and not to half the atomic weight seems to me very promising. This idea has already been used by Bohr in his theory of the constitution of atoms. The strongest and most convincing evidence in support of this hypothesis will be found in a paper by Moseley in *The Philosophical Magazine* of this month. He there shows that the frequency of the X radiations from a number of elements can be simply explained if the number of unit charges on the nucleus is equal to the atomic number. It would appear that the charge on the nucleus is the fundamental constant which determines the physical and chemical properties of the atom, while the atomic weight, although it approximately follows the order of the nucleus charge, is probably a complicated function of the latter depending on the detailed structure of the nucleus.

E. Rutherford

Manchester,
December 6th, 1913

Analysis of the γ Rays of the Thorium and Actinium Products

by PROFESSOR E. RUTHERFORD, F.R.S., *and* H. RICHARDSON, M.SC.,
Beyer Fellow, University of Manchester

From the *Philosophical Magazine* for December 1913, ser. 6, xxvi, pp. 937–48

IN previous papers* we have analysed the γ radiations emitted by the radium products radium B, radium C, radium D, and radium E into a number of groups of different penetrating powers. An examination of the thorium products, viz., mesothorium 2, thorium B, thorium C, thorium D, and of the actinium products has been undertaken by a similar method, and results of the same general character as in the case of radium have been obtained. The γ rays of these substances have been found to consist of distinct groups of rays absorbed exponentially by aluminium and differing widely in their absorption by matter.

As the apparatus and methods employed were the same as those previously used in the investigation of the rays from radium D and radium E, and described in a former paper, it is not necessary to discuss them further here.

Preliminary investigations of the γ rays from the thorium products were made by using the active deposit of thorium, obtained by exposure of a negatively charged wire to the emanation from a preparation of mesothorium. The effect obtained was, however, too small for accurate measurement. To overcome this difficulty, the various active products were separated from a preparation of mesothorium, which had a γ-ray activity equal to about $2 \cdot 5$ milligrams of radium. Since mesothorium always contains some radium, which is inseparable from it, it was necessary to take precautions that none of the γ-ray products of the latter should interfere in the measurements.

The mesothorium was dissolved in a very small quantity of dilute HCl and the solution made up to $0 \cdot 5$ c.c. This solution was heated on a water-bath for six hours, in order to drive off any radium emanation formed in the interval. During that time any radium B or radium C present practically disappeared. A nickel wire $0 \cdot 3$ mm. in diameter was then rotated in the solution for about 30 minutes, in order to collect the thorium C upon it. The nickel wire was found to be sufficiently active to show a strong γ-ray activity in the electroscope, and the purity of the deposit of thorium C was confirmed by examining its rate of decay. The γ-ray effect due to the nickel wire in this case arose not from the thorium C, which is an α-ray product, but from the thorium D.

* Rutherford & Richardson, Phil. Mag. xxv. p. 722 (1913), and xxvi, p. 324 (1913). (*This vol.*, p. 342 *and p. 353.*)

By suitable chemical operations the mesothorium was separated from the radiothorium. A few details of the chemical methods employed will be given, and we desire to express our thanks to Dr. Russell and Mr. Chadwick for kindly undertaking the work of chemical separation.

To the acid solution of mesothorium a few milligrams of aluminium chloride were added. Pure ammonia gas was then passed into this solution and aluminium hydroxide was precipitated. With this came down also the products mesothorium 2, radiothorium, thorium B, thorium C, thorium D, and any trace of radium B and radium C that might be present. This precipitate was filtered off and re-dissolved in a small quantity of dilute hydrochloric acid. 10 milligrams of lead chloride were then added, and H_2S was passed through the solution. The precipitate of lead sulphide was filtered off and carried with it the radium B, radium C, thorium B, thorium C, and thorium D. If any radium B or radium C were present, these would decay in the course of a few hours, and the source then consisted of thorium (B + C + D) in equilibrium. Since the radiation from thorium (C + D) was determined in the case of the deposit on the nickel wire, the γ radiation from thorium B could be deduced by a comparison of the absorption curves in the two cases.

The filtrate was evaporated to dryness and the H_2S expelled. The residue was obtained on a watch-glass and consisted of a thin layer of solid matter. This residue contained all the radiothorium and mesothorium 2. Since radiothorium itself does not emit γ rays, the examination of this source immediately after its preparation gave the γ rays belonging to mesothorium 2. The latter decays to half value in 6·2 hours, and has practically disappeared in the course of two days. In the meantime the radiothorium had produced thorium X and its subsequent products, viz. thorium emanation and the active deposit, and this source was used to test again the γ radiation from thorium (B + C + D). We obtained no definite evidence that thorium X emits a quantity of γ rays which could be detected in our experiments.*

Analysis of the γ rays from Thorium (C + D)

The γ radiation was examined in the electroscope containing methyl iodide, using, as before, a strong magnetic field to remove the β rays. The source used in these experiments consisted of the deposit of thorium (C + D) on the nickel wire mentioned above. The absorption curve showed that two types of radiation were present. About 80 per cent. of the total radiation consisted of a very hard type, which was exponentially absorbed in aluminium and had an absorption coefficient in that substance of $\mu = 0.096$ (cm.)$^{-1}$. The other radiation present was much less penetrating and was completely stopped by 2 mm. of aluminium. This soft radiation, however, was not absorbed exponentially by aluminium and would thus appear to be complex in

---

* L. Meitner and Hahn (*Phys. Zeit.* xiv. p. 873, 1913) have recently examined the products thorium X and thorium C for γ rays. Both of these products have been found to emit γ rays in small quantity. The presence of a small amount of γ radiation from these substances would, however, have no appreciable effect on the absorption curves obtained by the methods used in our experiments.

character. The general evidence indicates that this soft radiation is not emitted by thorium C or thorium D, but is excited in the nickel by the α or β rays. It will be remembered that a radiation of almost identical type was found when radium C was deposited on nickel.* It is intended to make a more detailed examination as to whether radiations of this kind are to be attributed to the excitation of one or more types of characteristic radiation by the α or β rays in the material on which the active matter is deposited.

Since thorium C does not emit any appreciable γ radiation, the results of these experiments thus indicate that thorium D emits only one penetrating type. This conclusion was confirmed by examining the γ radiation from pure thorium D obtained by recoil on an aluminium plate. Although the effects were too small and the period of transformation too short for accurate measurements, the results clearly indicated that no appreciable amount of soft radiation was present under those conditions.

Analysis of the rays from Thorium (B + C + D)

In order to determine the penetrating power of the rays given out by thorium B, it was necessary to use sources containing the active deposit in equilibrium. We have seen above that two sources were prepared for this purpose. When the radiothorium source was employed, the preparation was contained in a watch-glass and covered with goldbeater's skin to prevent the escape of the emanation. The absorption curve, in aluminium, was determined up to a thickness of 11 cm. The curves obtained are shown in figs. 1 & 2. These curves were analysed by the method described in previous papers.† It will be seen that the curves can be separated into four exponential types corresponding to groups of rays of penetrating powers $\mu = 160$, $\mu = 32$, $\mu = 0 \cdot 36$, and $\mu = 0 \cdot 096$. The last group is of course that due to thorium D, and the remaining three are given out by thorium B.

Fig. 1 shows the analysis of the soft types of radiation. The logarithm curve for the second type of radiation $\mu = 32$ is shown in the figure. The points lie on a straight line showing that this radiation is absorbed exponentially. Fig. 2 brings out in a similar way the exponential absorption of the penetrating type of radiation $\mu = 0 \cdot 096$, and also the relative effect in the experiments due to the penetrating radiation from thorium B, viz. $\mu = 0 \cdot 36$. This latter radiation probably corresponds to that described by Marsden and Wilson‡ and by Hahn and Meitner,§ and noted in papers recently published.

Analysis of the γ rays from Mesothorium 2

The mesothorium 2, prepared as described, was used for this experiment. The absorption curves obtained are shown in figs. 3 and 4. Analysis shows that the rays consist of two types, for which $\mu = 26$ and $\mu = 0 \cdot 116$ respectively. Fig. 3 shows the method of analysis of the radiation $\mu = 26$ and its

* Rutherford & Richardson, Phil. Mag. xxv. p. 730 (1913). (*This vol., p. 349.*)
† Rutherford & Richardson, Phil. Mag. xxv. p. 722 (1913). (*This vol., p. 342.*)
‡ Marsden & Wilson, Phil. Mag. xxvi. p. 354 (1913).
§ Meitner & Hahn, *Phys. Zeit.* xiv. p. 873 (1913).

exponential absorption by aluminium. The logarithm curve in fig. 4 shows
the absorption for the penetrating radiation $\mu = 0 \cdot 116$.

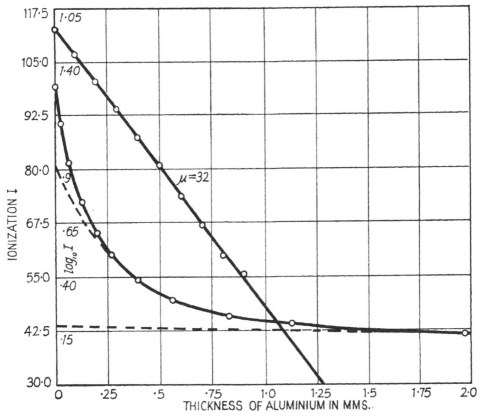

Fig. 1

Analysis of radiations from active deposit of thorium.

Analysis of the γ rays from Actinium (B + C + D)

The γ rays emitted by the actinium products were examined by the same
method as that adopted in the case of thorium. Owing to the short periods
of actinium C (2·1 min.) and of actinium D (4·71 min.) it was impossible to
complete the examination of the separate products with the small amount of
material at our disposal.*

An actinium preparation was dissolved in hydrochloric acid, and the
actinium X was separated by the ammonia method. In this manner the
actinium X was obtained on a watch-glass mixed with a thin layer of material.
The preparation was covered with goldbeater's skin as in the case of the
radiothorium preparation in order to prevent the escape of the emanation.
Shortly after the separation the actinium X is in equilibrium with its later
products actinium (B + C + D), and the whole then decays with the com-

---

* The actinium preparation used in these experiments was separated from radioactive
residues (see Boltwood, Proc. Roy Soc. A. lxxxv. 1911, p. 77) loaned to one of us by the
Royal Society.

paratively long period of actinium X (10·2 days). The γ rays in this preparation arise from the active deposit, viz. actinium (B + C + D).

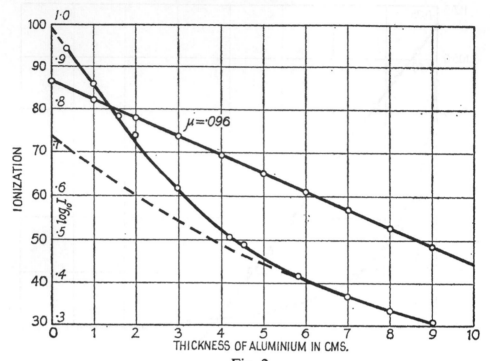

Fig. 2

Analysis of γ rays from active deposit of thorium.

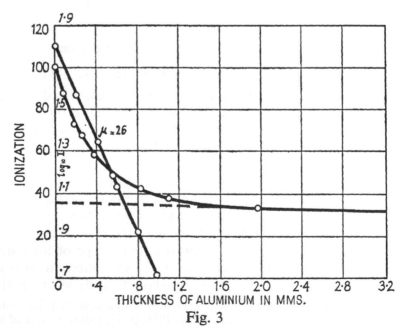

Fig. 3

Analysis of γ radiation from mesothorium 2.

The absorption curve in aluminium of this preparation was obtained up to a thickness of 9 cm. Beyond 6 cm. the absorption is exponential with an absorption coefficient $\mu = 0 \cdot 198$. The curves obtained are shown in figs. 5 and 6. Analysis as in previous cases shows that the curves can be separated

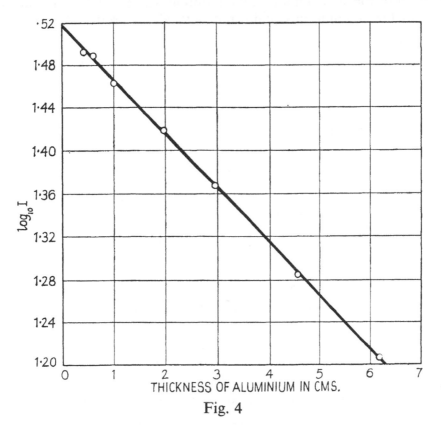

Fig. 4

Penetrating γ radiation from mesothorium 2.

into exponentials with absorption coefficients $\mu = 120$, $\mu = 31$, $\mu = 0 \cdot 46$, and $\mu = 0 \cdot 198$ respectively. Fig. 5 shows the analysis of the softer portions of the radiation, and fig. 6 the separation of the two more penetrating groups. From analogy with the thorium series, there appears to be little doubt that the radiations $\mu = 120$, $\mu = 31$, and $\mu = 0 \cdot 46$ belong to actinium B, and that for which $\mu = 0 \cdot 198$ to actinium D.

Analysis of the γ rays from Radioactinium

Dr. Russell and Mr. Chadwick recently examined in this Laboratory the γ rays from radioactinium, which they found to be far greater in intensity than could be ascribed to excitation by α rays. An approximate analysis was made by them of the types of radiation present. In the course of our later experiments with actinium, the emission of an intense γ radiation from radioactinium was evident. The actinium X was successively separated from the radioactinium preparation, and after the second operation it was found that the γ radiation from the precipitate, viz. actinium and radioactinium, reached a definite minimum of activity. Measured in the ordinary electroscope, the γ-ray effect of the actinium and radioactinium was about 25 per cent. of the activity of the actinium in equilibrium with all its products. The actinium

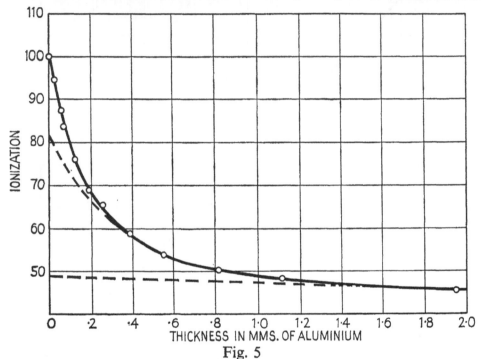

Fig. 5

Analysis of γ radiations from active deposit of actinium.

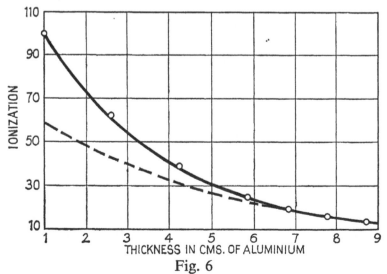

Fig. 6

Analysis of γ radiation from active deposit of actinium.

and radioactinium precipitate formed a comparatively thick deposit of several grams (about 10 grms.) in weight. Notwithstanding this comparatively thick layer, the presence of a considerable amount of soft radiation was evident. Analysis showed that the radiation from this material, which is to be ascribed to radioactinium, consists of two distinct types for which $\mu = 25$ and $\mu = 0 \cdot 190$. The absorption curves obtained are shown in figs. 7 and 8. The

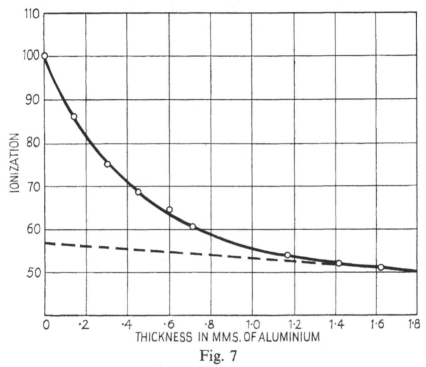

Fig. 7

Analysis of γ radiation from radioactinium.

γ-ray effect due to the more penetrating type was less than 2 divs. per min. under the experimental conditions, and it was consequently difficult to determine its absorption coefficient with as great an accuracy as was desired. An examination of the rise curve of the actinium and radioactinium preparation showed that the radiation could not be ascribed to some actinium X which had not been separated. In addition it is seen that the soft type $\mu = 25$ is quite distinct from the corresponding radiation from actinium B, viz. $\mu = 31$. It is of interest to observe that the hard type of radiation from radioactinium appears to be slightly more penetrating than that emitted from actinium D.

Summary

For convenience, the types of γ rays emitted from all the products so far examined are included in the following table. In order to complete the series it will be necessary to examine the types of γ rays emitted by uranium X. It is intended to continue the experiments in this direction.

o

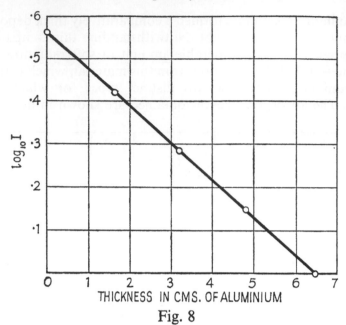

Fig. 8

Penetrating γ radiation from radioactinium.

Element	Atomic weight	Absorption coefficient μ in aluminium		Mass. absorption coefficient μ/d in aluminium	
Radium B	214	230 $\rangle$ (cm.)$^{-1}$		85 $\rangle$ (cm.)$^{-1}$	
		40 $\}$,,		14·7 $\}$,,	
		0·51 $\rfloor$,,		0·188 $\rfloor$,,	
Radium C	214	0·115 ,,		0·0424 ,,	
Radium D	210	45 $\rangle$,,		16·5 $\rangle$,,	
		0·99 $\}$,,		0·36 $\}$,,	
Radium E	210	similar types to D but very feeble			
Mesothorium 2	228	26 $\rangle$ (cm.)$^{-1}$		9·5 $\rangle$,,	
		0·116 $\}$,,		0·031 $\}$,,	
Thorium B	212	160 $\rangle$,,		59 $\rangle$,,	
		32 $\}$,,		11·8 $\}$,,	
		0·36 $\rfloor$,,		0·13 $\rfloor$,,	
Thorium D	208	0·096 ,,		0·035 ,,	
Radioactinium	..	25 $\rangle$,,		9·2 $\rangle$,,	
		0·190 $\}$,,		0·070 $\}$,,	
Actinium B	..	120 $\rangle$,,		44 $\rangle$,,	
		31 $\}$,,		11·4 $\}$,,	
		0·45 $\rfloor$,,		0·165 $\rfloor$,,	
Actinium D	..	0·198 ,,		0·073 ,,	

It will be seen that the γ rays from radioactive products can be conveniently divided into four types:—

(1) The soft radiations from the B products which vary from $\mu = 120$ to $\mu = 230$.

(2) A more penetrating type varying from $\mu = 26$ to $\mu = 45$, probably corresponding to characteristic radiations of the 'L' series.

(3) A penetrating type from the B products varying from $\mu = 0\cdot36$ to $\mu = 0\cdot51$.

(4) A very penetrating type varying from $\mu = 0\cdot115$ to $\mu = 0\cdot198$, probably corresponding to characteristic radiations of the 'K' series.

Types (1) and (2) are much less penetrating than the X rays from an average tube, although the work of Bragg* and Moseley and Darwin† has clearly shown that types of radiation similar in penetrating power to (2) are present in X rays generated with a platinum anticathode. Using a hard bulb and a consequent very high voltage (100,000 volts), S. J. Allen‡ showed that X rays could be produced of penetrating power about corresponding to type (3). Type (4) is more penetrating than any X rays produced or likely to be produced in X-ray tubes.

Thorium D emits the most penetrating type of γ rays known, while radium C and mesothorium 2 come next with radiations of nearly equal penetrating power. Actinium products, as has long been known, emit less penetrating types of radiation than radium or thorium products. The penetrating radiation from radium D has no apparent analogy with that emitted by any other product.

General Considerations

In the above table we have included the types of primary γ rays emitted by radioactive substances. In every case the emission of the γ rays observed is accompanied by a well-marked primary β radiation. As we have already pointed out in our experiments on radium D and radium E, there does not appear to be any obvious connexion between the relative intensity of the β and γ rays. For example, consider the case of radium D and radium E. On the one hand, radium D emits a comparatively feeble β radiation, shown by Hahn‖ to consist of two groups of low velocity but a well-marked γ radiation; and on the other, radium E emits an intense β radiation comprising particles projected with nearly the velocity of light, but gives a γ radiation exceedingly feeble in quantity compared with that from radium D.

The experiments of Hahn, v. Baeyer and Meitner have shown that practically all the β-ray products emit a number of distinct groups of β rays of definite speed. The only exception to this is radium E which gives a continuous

* Bragg, Proc. Roy. Soc. A. lxxxviii. p. 428 (1913).
† Moseley & Darwin, Phil. Mag. xxvi. p. 210 (1913).
‡ S. J. Allen & E. J. Lorentz, Phys. Rev. i. ser. 2, p. 35 (1913).
§ Rutherford & Richardson, Phil. Mag. xxvi. p. 324 (1913). (*This vol., p. 353.*)
‖ Hahn, Baeyer & Meitner, *Phys. Zeit.* xii. p. 378 (1911).

spectrum of β rays in which no evidence of groups has yet been observed. It should be mentioned also that no evidence of groups has so far been observed in the β rays of uranium X, but more experiments are required on this point. It may prove significant that radium E, which shows no sign of definite groups of β rays, should be the only β-ray product which emits γ rays exceedingly feeble in intensity.

In addition to these primary γ rays of marked intensity, Mr. Chadwick and Dr. Russell* have recently drawn attention to the fact that substances which emit α rays also emit a γ radiation of feeble intensity. This has been shown to be the case for the α-ray substances ionium, radiothorium, and polonium.

They have also examined the γ radiation from radium itself, which is also known to emit a weak β radiation. With the exception possibly of radium, this γ radiation observed from α-ray products has probably an entirely different origin from the intense primary γ radiation from the β-ray products. It would appear to be excited in the radioactive atoms by the escape of α particles. In their experiments on the α-ray product ionium, Chadwick and Russell have drawn attention to the fact that the softer types of radiation predominate. This is the exact opposite of the primary γ-ray products, where the penetrating types of radiation are relatively far more intense.

It is of interest to compare the types of primary γ rays emitted by different radioactive products. In the first place, it is clear from the table that each of the products radium B, thorium B, and probably actinium B, which occupy the same relative position in the radioactive series and have very similar if not identical chemical properties, emit three distinct types of γ radiation which appear to be closely analogous in relative penetrating power. Similar results appear to hold for radium C, thorium D, and actinium D, each of which emits only one penetrating type of γ radiation. Neither of the two products radium D and radium E corresponds in radioactive or chemical properties with any member of the thorium or actinium series and also shows no close analogy in their types of radiation.

In a previous paper† one of us has suggested that the types of γ rays emitted by radioactive substances should correspond to one or more of the types of characteristic radiation excited in the atom by the escape of β particles. For example, the single type of γ radiation from radium C and thorium D has about the penetrating power to be expected for a radiation of the 'K' series found by Barkla.‡

It will be observed from the table that all the B products emit a soft type of radiation which is on an average much less penetrating than ordinary X rays. From a comparison of the results given by Chapman§ it would appear that the radiations from radium B, $\mu = 40$, from thorium B, $\mu = 32$, from actinium B, $\mu = 31$, and from mesothorium 2, $\mu = 26$, belong to the 'L'

---

* Chadwick & Russell, Proc. Roy. Soc. A. lxxxviii. p. 217 (1913).

† Rutherford, Phil. Mag. xxiv. p. 453 (1912). (*This vol., p. 280.*)

‡ Barkla, Phil. Mag. xxii. p. 396 (1911).

§ Chapman, Proc. Roy. Soc. ser. A. lxxxvi. p. 439 (1912).

series of Barkla. This type of radiation seems very persistent in all except the three products radium C, thorium D, and actinium D. It should be pointed out, however, that there appears to be no very close connexion between the penetrating power of the radiation and the atomic weight. For example, radium B of atomic weight 214 gives a radiation for which $\mu = 40$, whilst thorium B of still lower atomic weight 212 gives a more penetrating radiation, $\mu = 32$. This peculiarity holds not only for the soft radiations but also for the hard types emitted by the B products. If the softer types of radiation do belong to the 'L' series, it would appear that the general rule connecting penetrating power and atomic weight may hold approximately over wide ranges but not necessarily for products whose atomic weights differ only slightly from one another. It is difficult to offer any explanation why the B products should give three types and the successive products only one type. It should be pointed out, however, that the β rays from the B products are on the average much less penetrating than the β rays from the following products, and have consequently a much better chance of exciting the softer types of γ radiation, that is, the radiations of longer wave-length. Also it should be noted that the expulsion of a long range α particle either accompanies or precedes immediately the penetrating γ radiation from these products.

During the present year, Russell,* Fajans,† and Soddy‡ have independently discussed the question of the chemical properties of the numerous radioactive products, and have shown that the sequence of chemical properties in a radioactive series follows a simple rule when the types of emitted radiation are taken into account. According to some of these views, the B products are to be regarded as chemically identical and inseparable from one another, although they may differ by about two units in their atomic weights. If we take the view that the atom consists of a positively charged nucleus of small dimensions surrounded by rings of electrons which can be set in vibration, it would seem probable that an identity of chemical nature would involve an identity in the electronic distribution and of the magnitude of the charge on the nucleus. If the γ rays set up are due to the vibration of the electronic systems, it would be anticipated that the types of γ rays emitted would be identical for such products. While, as we have previously pointed out, there is an undoubtedly close analogy between the B products, not only in the types of radiation but also in the relative penetrating power, the differences in their penetrating power are sufficiently marked to indicate a real difference in the wave-length of the radiation emitted and one which cannot be attributed to experimental error.

It is of interest to examine whether a comparison of the types and penetrating power of the radiations from the thorium and actinium products gives any indication as to the atomic weight of the latter. It has already been

* Russell, Chem. News, cvii. p. 49 (1913).
† Fajans, *Phys. Zeit.* iv. p. 136 (1913).
‡ Soddy, Chem. News, cviii. p. 168 (1913).

mentioned that the 'L' type of radiation seems to be predominant for most of the products, and for this radiation there is a close agreement between the penetrating powers of the radiations emitted by analogous products. For example, actinium B emits a radiation for which $\mu = 31$, and a corresponding radiation $\mu = 32$ is given out by thorium B. The radiation $\mu = 25$ from radioactinium also resembles closely the radiation from mesothorium 2, $\mu = 26$. Considering the very close analogy in chemical and radioactive properties of the thorium and actinium series, it is reasonable to suppose that those bodies which emit similar types of radiation of nearly equal penetrating power have nearly the same atomic weight. For example, by considering the number of α-ray products it can be estimated that thorium B has an atomic weight 212. We should expect actinium B to have an atomic weight nearly the same. In the same way we should expect radioactinium and mesothorium 2 to have nearly equal atomic weights. Calculating backwards by a consideration of the number of α-ray products, it can be deduced that the atomic weight of actinium should be (1) 228, from comparison of the B products; (2) 228 from comparison of radioactinium with mesothorium 2.

It is quite probable from other analogies, that the actual atomic weight might be two units greater or less than the above value. In the one case, the atomic weight 230 would indicate that actinium was derived from uranium after the expulsion of two α particles, and the other that it arose from ionium. On this point of view radium itself does not seem admissible as a possible origin of actinium, and this is supported by the recent experiments of Soddy.* The deduction that actinium is derived from uranium seems the more probable, and several suggestions of this possibility have already been made.†

The other more penetrating types of radiation from the actinium and thorium products do not appear to be directly comparable. For example, the hard radiations from thorium D and actinium D show wide differences in penetrating power.

When the types of γ radiation emitted not only by all β-ray products but also by all α-ray products are known, very valuable data will have been obtained for throwing light not only on the modes of vibration of the radioactive elements but also on their atomic constitution.

University of Manchester,
October, 1913

* Soddy, Nature, xci. p. 634 (1913).
† Boltwood, Amer. Journ. Sci. xxv. p. 269 (1908). Hahn & Meitner, *Phys. Zeit.* xvi. p.752 (1913). Antonoff, Phil. Mag. xxii. p. 419 (1911).

The Structure of the Atom

by SIR ERNEST RUTHERFORD, F.R.S.,

Professor of Physics, University of Manchester

From the *Philosophical Magazine* for March 1914, ser. 6, xxvii, pp. 488–98

THE present paper and the accompanying paper by Mr. C. Darwin deal with certain points in connexion with the 'nucleus' theory of the atom which were purposely omitted in my first communication on that subject (Phil. Mag. May 1911).* A brief account is given of the later investigations which have been made to test the theory and of the deductions which can be drawn from them. At the same time a brief statement is given of recent observations on the passage of α particles through hydrogen, which throw important light on the dimensions of the nucleus.

In my previous paper (*loc. cit.*) I pointed out the importance of the study of the passage of the high speed α and β particles through matter as a means of throwing light on the internal structure of the atom. Attention was drawn to the remarkable fact, first observed by Geiger and Marsden,† that a small fraction of the swift α particles from radioactive substances were able to be deflected through an angle of more than 90° as the results of an encounter with a single atom. It was shown that the type of atom devised by Lord Kelvin and worked out in great detail by Sir J. J. Thomson was unable to produce such large deflexions unless the diameter of the positive sphere was exceedingly small. In order to account for this large angle scattering of α particles, I supposed that the atom consisted of a positively charged nucleus of small dimensions in which practically all the mass of the atom was concentrated. The nucleus was supposed to be surrounded by a distribution of electrons to make the atom electrically neutral, and extending to distances from the nucleus comparable with the ordinary accepted radius of the atom. Some of the swift α particles passed through the atoms in their path and entered the intense electric field in the neighbourhood of the nucleus and were deflected from their rectilinear path. In order to suffer a deflexion of more than a few degrees, the α particle has to pass very close to the nucleus, and it was assumed that the field of force in this region was not appreciably affected by the external electronic distribution. Supposing that the forces between the nucleus and the α particle are repulsive and follow the law of inverse squares, the α particle describes a hyperbolic orbit round the nucleus and its deflexion can be simply calculated.

* *This vol., p. 238.*   †ₗProc. Roy. Soc. A. lxxxii. p. 495 (1909).

It was deduced from this theory that the number of α particles falling normally on unit area of a surface and making an angle ϕ with the direction of the incident rays is proportional to

(1) $\operatorname{cosec}^4 \phi/2$ or $1/\phi^4$ if ϕ be small;
(2) the number of atoms per unit volume of the scattering material;
(3) thickness of scattering material t provided this is small;
(4) square of the nucleus charge Ne;
(5) and is inversely proportional to $(mu^2)^2$, where m is the mass of the α particle and u its velocity.

From the data of scattering on α particles previously given by Geiger,[*] it was deduced that the value of the nucleus charge was equal to about half the atomic weight multiplied by the electronic charge. Experiments were begun by Geiger and Marsden[†] to test whether the laws of single scattering of α particles were in agreement with the theory. The general experimental method employed by them consisted in allowing a narrow pencil of α particles to fall normally on a thin film of matter, and observing by the scintillation method the number scattered through different angles. This was a very difficult and laborious piece of work involving the counting of many thousands of particles. They found that their results were in very close accord with the theory. When the thickness of the scattering film was very small, the amount of scattering was directly proportional to the thickness and varied inversely as the fourth power of the velocity of the incident α particles. A special study was made of the number of α particles scattered through angles varying between 5° and 150°. Although over this range the number decreased in the ratio 200,000 to 1, the relation between number and angle agreed with the theory within the limit of experimental error. They found that the scattering of different atoms of matter was approximately proportional to the square of the atomic weight, showing that the charge on the nucleus was nearly proportional to the atomic weight. By determining the number of α particles scattered from thin films of gold, they concluded that the nucleus charge was equal to about half the atomic weight multiplied by the electronic charge. On account of the difficulties of this experiment, the actual number could not be considered correct within more than 20 per cent.

The experimental results of Geiger and Marsden were thus in complete accord with the predictions of the theory, and indicated the essential correctness of this hypothesis of the structure of the atom.

In determining the magnitude of single scattering, I assumed in my previous paper, for simplicity of calculation, that the atom was at rest during an encounter with an α particle. In an accompanying paper, Mr. C. Darwin has worked out the relations to be expected when account is taken of the motion of the recoiling atom. He has shown that no sensible error has been introduced

[*] Proc. Roy. Soc. A. lxxxiii. p. 492 (1910).
[†] Geiger and Marsden, Phil. Mag. xxv. p. 604 (1913).

in this way even for atoms of such low atomic weight as carbon. Mr. Darwin has also worked out the scattering to be expected if the law of force is not that of the inverse square, and has shown that it is not in accord with experiment either with regard to the variation of scattering with angle or with the variation of scattering with velocity. The general evidence certainly indicates that the law of force between the α particle and the nucleus is that of the inverse square.

It is of interest to note that C. T. R. Wilson,* by photographing the trails of the α particle, later showed that the α particle occasionally suffers a sudden deflexion through a large angle. This affords convincing evidence of the correctness of the view that large deflexions do occasionally occur as a result of an encounter with a single atom.

On the theory outlined, the large deflexions of the α particle are supposed to be due to its passage close to the nucleus where the field is very intense and to be not appreciably affected by its passage through the external distribution of electrons. This assumption seems to be legitimate when we remember that the mass and energy of the α particle are very large compared with that of an electron even moving with a velocity comparable with that of light. Simple considerations show that the deflexions which an α particle would experience even in passing through the complex electronic distribution of a heavy atom like gold, must be small compared with the large deflexions actually observed. In fact, the passage of swift α particles through matter affords the most definite and straightforward method of throwing light on the gross structure of the atom, for the α particle is able to penetrate the atom without serious disturbance from the electronic distribution, and thus is only affected by the intense field associated with the nucleus of the atom.

This independence of the large angle scattering on the external distribution of electrons is only true for charged particles whose kinetic energy is very large. It is not to be expected that it will hold for particles moving at very much lower speeds and with much less energy—such, for example as the ordinary cathode particles or the recoil atoms from active matter. In such cases it is probable that the external electronic distribution plays a far more prominent part in governing the scattering than in the case under consideration.

Scattering of β particles

It is to be anticipated on the nucleus theory that swift β particles should suffer deflexions through large angles in their passage close to the nucleus. There seems to be no doubt that such large deflexions are actually produced, and I showed in my previous paper that the results of scattering of β particles found by Crowther† could be generally explained on the nucleus theory of atomic structure. It should be borne in mind, however, that there are several important points of distinction between the effects to be expected for an α particle

* C. T. R. Wilson, Proc. Roy. Soc. A. lxxxvii. p. 277 (1912).
† Crowther, Proc. Roy. Soc. A. lxxxiv. p. 226 (1910).

o*

and a β particle. Since the force between the nucleus and β particle is attractive, the β particle increases rapidly in speed in approaching the nucleus. On the ordinary electrodynamics, this entails a loss of energy by radiation, and also an increase of the apparent mass of the electron. Darwin* has worked out mathematically the result of these effects on the orbit of the electron, and has shown that, under certain conditions, the β particle does not escape from the atom but describes a spiral orbit ultimately falling into the nucleus. This result is of great interest, for it may offer an explanation of the disappearance of swift β particles in their passage through matter. In addition, it must be borne in mind that the swiftest β particle expelled from radium C possesses only about one-third of the energy of the corresponding α particle, while the average energy of the β particle is less than one-sixth of that of the α particle. It is thus to be anticipated that the large angle scattering of a β particle by the nucleus will take place in regions where the α particle will only suffer a small deflexion—regions for which the application of the simple theory may not have been accurately tested. For these reasons, it is of great importance to determine the laws of large angle scattering of β particles of different speeds in passing through matter, as it should throw light on a number of important points connected with atomic structure. Experiments are at present in progress in the laboratory to examine the scattering of such swift β particles in detail.

It is obvious that a β particle in passing close to an electron will occasionally suffer a large deflexion. The problem is mathematically similar to that for a close encounter of an α particle with a helium atom of the same mass, which is discussed by Mr. Darwin in the accompanying paper. Such large deflexions due to electronic encounter, however, should be relatively small in number compared with those due to the nucleus of a heavy atom.

Scattering in Hydrogen

Special interest attaches to the effects to be expected when α particles pass through light gases like hydrogen and helium. In a previous paper by Mr. Nuttall and the author,† it has been shown that the scattering of α particles in hydrogen and helium is in good agreement with the view that the hydrogen nucleus has one positive charge, while the α particle, or helium, has two. Mr. Darwin has worked out in detail the simple scattering to be anticipated when α particles pass through hydrogen and helium. It is only necessary here to refer to the fact that on the nucleus theory a small number of hydrogen atoms should acquire, as the result of close encounters with α particles, velocities about $1 \cdot 6$ times that of the velocity of the α particle itself. On account of the fact that the hydrogen atom carries one positive charge while the α particle carries two, it can be calculated that some of the hydrogen atoms should have a range in hydrogen of nearly four times that of the α particle which sets them in motion.

---

* Darwin, Phil. Mag. xxv. p. 201 (1913).
† Rutherford and Nuttall, Phil. Mag. xxvi. p. 702 (1913). (*This vol., p. 362.*)

Mr. Marsden has kindly made experiments for me to test whether the presence of such hydrogen atoms can be detected. A detailed account of his experiments will appear later, but it suffices to mention here that undoubted evidence has been obtained by him that some of the hydrogen atoms are set in such swift motion that they are able to produce a visible scintillation on a zinc sulphide screen and are able to travel through hydrogen a distance three or four times greater than the colliding α particle. The general method employed was to place a thin α-ray tube containing about 100 millicuries of purified emanation in a tube filled with hydrogen. The scintillations due to the α particle from the tube disappeared in air after traversing a distance of about 5 cm. When the air was displaced by hydrogen, the great majority of the scintillations disappeared at about 20 cm. from the source, which corresponds to the range of the α particle in hydrogen. A small number of scintillations, however, persisted in hydrogen up to a distance of about 90 cm. The scintillations were of less intensity than those due to the ordinary α particle. The number of scintillations observed is of the order of magnitude to be anticipated on the theory of single scattering, supposing that the nucleus in hydrogen and helium has such small dimensions, and that they behave like point charges for distances up to 10^{-13} cm.

There appears to be no doubt that the scintillations observed beyond 20 cm. are due to charged hydrogen atoms which are set in swift motion by a close encounter with an α particle. Experiments are at present in progress by Mr. Marsden to determine the number of hydrogen atoms set in motion, and the variation of the number with the scattering angle.

It does not appear possible to explain the appearance of such swift hydrogen atoms unless it be supposed that the forces of repulsion between the α particle and the hydrogen atom are exceedingly intense. Such intense forces can only arise if the positive nuclei have exceedingly small dimensions, so that a close approach between them is possible.

Dimensions and Constitution of the Nucleus

In my previous paper I showed that the nucleus must have exceedingly small dimensions, and calculated that in the case of gold its radius was not greater that 3×10^{-12} cm. In order to account for the velocity given to hydrogen atoms by the collision with α particles, it can be simply calculated (see Darwin) that the centres of nuclei of helium and hydrogen must approach within a distance of $1 \cdot 7 \times 10^{-13}$ cm. of each other. Supposing for simplicity the nuclei to have dimensions and to be spherical in shape, it is clear that the sum of the radii of the hydrogen and helium nuclei is not greater than $1 \cdot 7 \times 10^{-13}$ cm. This is an exceedingly small quantity, even *smaller* than the ordinarily accepted value of the diameter of the electron, viz. 2×10^{-13} cm. It is obvious that the method we have considered gives a maximum estimate of the dimensions of the nuclei, and it is not improbable that the hydrogen nucleus itself may have still smaller dimensions. This raises the question whether the hydrogen nucleus is so small that its mass may be accounted for in the same way as the mass of the negative electron.

It is well known from the experiments of Sir J. J. Thomson and others, that no positively charged carrier has been observed of mass less than that of the hydrogen atom. The exceedingly small dimensions found for the hydrogen nucleus add weight to the suggestion that the hydrogen nucleus is the *positive electron*, and that its mass is entirely electromagnetic in origin. According to the electromagnetic theory, the electrical mass of a charged body, supposed spherical, is $\frac{2}{3}\frac{e^2}{a}$ where e is the charge and a the radius. The hydrogen nucleus consequently must have a radius about 1/1830 of the electron if its mass is to be explained in this way. There is no experimental evidence at present contrary to such an assumption.

The helium nucleus has a mass nearly four times that of hydrogen. If one supposes that the positive electron, *i. e.* the hydrogen atom, is a unit of which all atoms are composed, it is to be anticipated that the helium atom contains four positive electrons and two negative.

It is well known that a helium atom is expelled in many cases in the transformation of radioactive matter, but no evidence has so far been obtained of the expulsion of a hydrogen atom. In conjunction with Mr. Robinson, I have examined whether any other charged atoms are expelled from radioactive matter except helium atoms, and the recoil atoms which accompany the expulsion of α particles. The examination showed that if such particles are expelled, their number is certainly less than 1 in 10,000 of the number of helium atoms. It thus follows that the helium nucleus is a very stable configuration which survives the intense disturbances resulting in its expulsion with high velocity from the radioactive atom, and is one of the units, of which possibly the great majority of the atoms are composed. The radioactive evidence indicates that the atomic weight of successive products decreases by four units consequent on the expulsion of an α particle, and it has often been pointed out that the atomic weights of many of the permanent atoms differ by about four units.

It will be seen later that the resultant positive charge on the nucleus determines the main physical and chemical properties of the atom. The mass of the atom is, however, dependent on the number and arrangement of the positive and negative electrons constituting the atom. Since the experimental evidence indicates that the nucleus has very small dimensions, the constituent positive and negative electrons must be very closely packed together. As Lorentz has pointed out, the electrical mass of a system of charged particles, if close together, will depend not only on the number of these particles, but on the way their fields interact. For the dimensions of the positive and negative electrons considered, the packing must be very close in order to produce an appreciable alteration in the mass due to this cause. This may, for example, be the explanation of the fact that the helium atom has not quite four times the mass of the hydrogen atom. Until, however, the nucleus theory has been more definitely tested, it would appear premature to discuss the possible structure of the nucleus itself. The general theory would indicate that the nucleus of a heavy atom is an exceedingly complicated system, although its dimensions are very minute.

An important question arises whether the atomic nuclei, which all carry a positive charge, contain negative electrons. This question has been discussed by Bohr,* who concluded from the radioactive evidence that the high speed β particles have their origin in the nucleus. The general radioactive evidence certainly supports such a conclusion. It is well known that the radioactive transformations which are accompanied by the expulsion of high speed β particles are, like the α ray changes, unaffected by wide ranges of temperature or by physical and chemical conditions. On the nucleus theory, there can be no doubt that the α particle has its origin in the nucleus and gains a great part, if not all, of its energy of motion in escaping from the atom. It seems reasonable, therefore, to suppose that a β ray transformation also originates from the expulsion of a negative electron from the nucleus. It is well known that the energy expelled in the form of β and γ rays during the transformation of radium C† is about one-quarter of the energy of the expelled α particle. It does not seem easy to explain this large emission of energy by supposing it to have its origin in the electronic distribution. It seems more likely that a very high speed electron is liberated from the nucleus, and in its escape from the atom sets the electronic distribution in violent vibration, giving rise to intense γ rays and also to secondary β particles. The general evidence certainly indicates that many of the high speed electrons from radioactive matter are liberated from the electronic distribution in consequence of the disturbance due to the primary electron escaping from the nucleus.

Charge on the Nucleus

We have seen that from an examination of the scattering of α particles by matter, it has been found that the positive charge on the nucleus is approximately equal to $\frac{1}{2}Ae$, when A is the atomic weight and e the unit charge. This is equivalent to the statement that the number of electrons in the external distribution is about half the atomic weight in terms of hydrogen. It is of interest to note that this is the value deduced by Barkla‡ from entirely different evidence, viz. the scattering of X rays in their passage through matter. This is founded on the theory of scattering given by Sir J. J. Thomson, which supposes that each electron in an atom scatters as an independent unit. It seems improbable that the electrons within the nucleus would contribute to this scattering for they are packed together with positive nuclei and must be held in equilibrium by forces of a different order of magnitude from those which bind the external electrons.

* Bohr, Phil. Mag. xxvi. p. 476 (1913).
† See Rutherford and Robinson, Phil. Mag. xxv. p. 312 (1913). (*This. vol., p. 312.*)
‡ Barkla, Phil. Mag. xxi. p. 648 (1911).

It is obvious from the consideration of the cases of hydrogen and helium, where hydrogen has one electron and helium two, that the number of electrons cannot be exactly half the atomic weight in all cases. This has led to an interesting suggestion by van den Broek* that the number of units of charge on the nucleus, and consequently the number of external electrons, may be equal to the number of the elements when arranged in order of increasing atomic weight. On this view, the nucleus charges of hydrogen, helium, and carbon are 1, 2, 6 respectively, and so on for the other elements, provided there is no gap due to a missing element. This view has been taken by Bohr in his theory of the constitution of simple atoms and molecules.

Recently strong evidence of two distinct kinds has been brought in support of such a contention. Soddy† has pointed out that the recent generalisation of the relation between the chemical properties of the elements and the radiations can be interpreted by supposing that the atom loses two positive charges by the expulsion of an α particle, and one negative by the expulsion of a high speed electron. From a consideration of the series of products of the three main radioactive branches of uranium, thorium, and actinium, it follows that some of the radioactive elements may be arranged so that the nucleus charge decreases by one unit as we pass from one element to another. It would thus appear that van den Broek's suggestion probably holds for some if not all of the heavy radioactive elements. Recently Moseley‡ has supplied very valuable evidence that this rule also holds for a number of the lighter elements. By examination of the wave-length of the characteristic X rays emitted by twelve elements varying in atomic weight between calcium (40) and zinc (65·4), he has shown that the variation of wave-length can be simply explained by supposing that the charge on the nucleus increases from element to element by exactly one unit. This holds true for cobalt and nickel, although it has long been known that they occupy an anomalous relative position in the periodic classification of the elements according to atomic weights.

There appears to be no reason why this new and powerful method of analysis, depending on an examination of the frequency of the characteristic X ray spectra of the elements, should not be extended to a large number of elements, so that further definite data on the point may be expected in the near future.

It is clear on the nucleus theory that the physical and chemical properties of the ordinary elements are for the most part dependent entirely on the charge of the nucleus, for the latter determines the number and distribution of the external electrons on which the chemical and physical properties must mainly depend. As Bohr has pointed out, the properties of gravitation and radioactivity, which are entirely uninfluenced by chemical or physical agencies, must be ascribed mainly if not entirely to the nucleus, while the ordinary physical and chemical properties are determined by the number and distri-

* van den Broek, *Phys. Zeit.* xiv. p. 32 (1913).
† Soddy, *Jahr. d. Rad.* x. p. 188 (1913).   ‡ Moseley, Phil. Mag. xxvi. p. 1024 (1913).

bution of the external electrons. On this view, the nucleus charge is a funda-mental constant of the atom, while the atomic mass of an atom may be a complicated function of the arrangement of the units which make up the nucleus.

It should be borne in mind that there is no inherent impossibility on the nucleus theory that atoms may differ considerably in atomic weight and yet have the same nucleus charge. This is most simply illustrated by radioactive evidence. In the following table the atomic weight and nucleus charge are given for a few of the successive elements arising from the transformation of uranium. The actual nucleus charge of uranium is unknown, but for simplicity it is assumed to be 100.

Successive Elements	$Ur_1 \rightarrow$	$UrX_1 \rightarrow$	$UrX_2 \rightarrow$	$Ur_2 \rightarrow$	$Io \rightarrow$	Ra
Atomic weights	238·5	234·5	234·5	234·5	230·5	226·5
Charge on nucleus	100	98	99	100	98	96

Following the recent theories, it is supposed that the emission of an α particle lowers the nucleus charge by two units, while the emission of a β particle raises it by one unit. It is seen that Ur_1 and Ur_2 have the same nucleus charge although they differ in atomic weight by four units.

If the nucleus is supposed to be composed of a mixture of hydrogen nuclei with one charge and of helium nuclei with two charges, it is *a priori* con-ceivable that a number of atoms may exist with the same nucleus charge but of different atomic masses. The radioactive evidence certainly supports such a view, but probably only a few of such possible atoms would be stable enough to survive for a measurable time.

Bohr* has drawn attention to the difficulties of constructing atoms on the 'nucleus' theory, and has shown that the stable positions of the external electrons cannot be deduced from the classical mechanics. By the introduction of a conception connected with Planck's quantum, he has shown that on certain assumptions it is possible to construct simple atoms and molecules out of positive and negative nuclei, *e. g.* the hydrogen atom and molecule and the helium atom, which behave in many respects like the actual atoms or molecules. While there may be much difference of opinion as to the validity and of the underlying physical meaning of the assumptions made by Bohr, there can be no doubt that the theories of Bohr are of great interest and importance to all physicists as the first definite attempt to construct simple atoms and molecules and to explain their spectra.

University of Manchester
February 1914

---

* Bohr, Phil. Mag. xxvi. pp. 476, 857 (1913).

The Wavelength of the Soft γ Rays from Radium B

by SIR ERNEST RUTHERFORD, F.R.S.,

and E. N. DA C. ANDRADE, B.SC., PH.D.,

John Harling Fellow, University of Manchester

From the *Philosophical Magazine* for May 1914, ser. 6, xxvii, pp. 854–68

DURING the last few years, a large amount of attention has been directed to the absorption of the γ rays emitted by radioactive bodies. At first, the nature of the absorption by matter of the very penetrating γ rays emitted by the products radium C, mesothorium 2, thorium D, and uranium X, was carefully examined, and it was found that all these types of radiation were absorbed by light elements very nearly according to an exponential law over a large range of thickness, but with different constants of absorption for each radiation. In order to explain the emission of homogeneous groups of β rays from a number of products, Rutherford suggested that the γ rays emitted by the radioactive products must be regarded as the 'characteristic' radiations excited in the radioelements by the escape of β particles from them. These 'characteristic' radiations were supposed to be analogous to one or more of the groups of characteristic radiations observed by Barkla to be excited in different elements by X rays. It was suggested that the emission of homogeneous groups of β rays was directly connected with the emission of different types of characteristic γ rays from each element, and that the energy of the escaping β particle was diminished by multiples of definite units depending on the energy required to set the electronic system of the atom in a definite form of vibration.

In order to test this point of view, Rutherford and Richardson* analysed in detail the γ rays emitted by a number of radioactive substances, using the absorption method to distinguish broadly between the different types of γ rays emitted. It was found that the γ radiation from the B products, viz., radium B, thorium B, and actinium B, could all be conveniently divided into three types of widely different penetrating power. For example, the absorption coefficients in aluminium for the groups of γ rays from radium B were found to be 230, 40, and 0·5. In the case of the C products, viz., radium C, thorium C, and actinium C, the γ radiation was found to be mainly of one very penetrating type exponentially absorbed in aluminium. The radiations from the various radioactive substances can be conveniently divided into three

---

* Rutherford and Richardson, Phil. Mag. May 1913, p. 722; August 1913, p. 325; Feb. 1914, p. 252. (*This vol., pp. 342, 353, 410.*)

distinct classes, viz.:—(1) a soft radiation, varying in different elements from $\mu = 24$ to $\mu = 45$, probably corresponding to characteristic radiations of the 'L' type excited in the radioatoms; (2) a very penetrating radiation with a value of μ in aluminium of about $0 \cdot 1$, probably corresponding to the 'K' characteristic radiation of these heavy atoms; (3) radiations of penetrating power intermediate between (1) and (2) corresponding to one or more types of characteristic radiations not so far observed with X rays.

In the meantime, the experiments of W. H. and W. L. Bragg* and Moseley and Darwin† had shown that the reflexion of X rays from crystals afforded a definite and reliable method of studying the wave-length of X rays. It was found that the radiations from a platinum anticathode consisted in part of a series of strong lines, no doubt corresponding to the 'L' characteristic radiation of this element. By using a number of anticathodes of different metals, the X-ray spectra of a number of elements were determined by W. H. and W. L. Bragg‡ and by Moseley.§ The latter has made a comparative study of the strong lines of the spectra emitted by the great majority of the elements. For most of the lighter elements from aluminium to silver, the spectra obtained corresponded to the 'K' characteristic radiations, while for the heavier elements the 'L' series has been determined. The simple relations which Moseley finds to hold between the spectra of successive elements has been discussed by him in his recent paper.

From the analysis of the types of γ rays, it appeared probable that each corresponded to one of the characteristic types of radiation of the element in question. It was consequently to be anticipated that each of these radiations would give definite line spectra when reflected from the surface of crystals.

In order to examine this question, experiments were begun to determine the wave-lengths of the γ radiations from the products radium B and radium C. For this purpose, a thin walled α-ray tube, filled with a large quantity of emanation, served as a source of γ rays. The rays were allowed to fall at a definite angle on a crystal, generally rocksalt, and the intensities of the 'reflected,' or rather diffracted, rays were examined by a photographic method.

The determination of the γ-ray spectra is in some respects far more difficult than similar measurements of X rays. In the first place, the photographic effect of the γ rays, even from the strongest source of emanation available, is very feeble compared with that due to the X rays from an ordinary focus tube. For example, using a source of 100 millicuries of radium emanation, an exposure of 24 hours is necessary to obtain a marked photographic effect due to the reflected γ rays. Under similar conditions, 10 minutes exposure suffices to obtain a well-marked X-ray spectrum. In the second place, special precautions have to be taken to screen the photographic plate from the effects of the very penetrating γ radiation from radium C. The greatest difficulty of

* W. H. Bragg and W. L. Bragg, Proc. Roy. Soc. A. lxxxviii. 1913, p. 428.

† H. G. J. Moseley and C. G. Darwin, Phil. Mag. July 1913, p. 210.

‡ W. H. Bragg and W. L. Bragg, Proc. Roy. Soc. A. lxxxix. 1913, p. 277, and *loc. cit.*

§ H. G. J. Moseley, Phil. Mag. Dec. 1913, p. 1024; April 1914, p. 703.

all, however, is to get rid of the disturbing effect of the very swift primary β particles emitted from the source and the swift β particles emitted from all material through which the γ rays pass. This can only be accomplished by placing the source of radiation, absorbing screens, and crystal in a strong magnetic field, so that practically all the β rays, both the primary ones and those excited by the γ rays in matter, are bent away from the photographic plate.

Method of Experiment

The straight emanation tube, A, about 0·5 mm. in diameter and 1 cm. in length (fig. 1), was fixed behind a massive block of lead BB, so that the rays from it passed through a horizontal slit in the block; a square vertical hole in the lead allowed screens SS to be interposed in the path of the rays when desired. The length of the block from back to front was 6 cm., the width of the slit 3 mm. The emergent pencil of rays fell on the crystal CC, which was

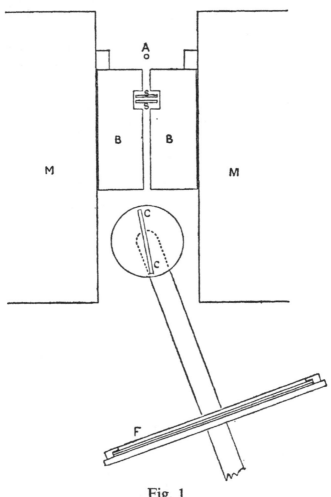

Fig. 1

mounted on a small turn-table so that the axis of revolution of the table passed accurately through the reflecting face; the angle of the crystal was measured on a metal scale by means of a glass pointer attached to the table. The emanation tube, lead block, and crystal were all placed between the rectangular pole-pieces of a powerful electromagnet MM, as indicated in the diagram; the magnetic field usually employed was 2500 gauss.

The photographic plate was held with its film towards the source of the rays in a special carrier mounted on a rotating arm, the axis of rotation of which coincided with the axis of the crystal turn-table; in front of the film there was a single thickness of black paper to protect the plate from stray light. As it was possible that the position of the plate in the holder might vary in successive experiments, it was necessary for purposes of measurement to mark its position relative to the holder. To enable this to be done, a narrow strip was removed from this black paper, and the hole thus formed covered with a slip of metal in which was a fine slit F; by means of a fixed lamp a fiducial line could then be marked on the plate, the plate-holder being always put in the same position for this purpose.

The distance of the source from the centre of the crystal was arranged so as to be exactly equal to the distance of this centre from the photographic plate (about 9 cm.). It is well known that under these conditions no correction for the length of the crystal is necessary in determining the angle of reflexion of the spectral lines, for reflexion of the same wave-length from any point of the crystal always falls at the same point on the plate. The crystal was arranged with the centre of its reflecting face as near as possible opposite the centre of the slit; the plate-holder was adjusted perpendicular to the slit for the zero reading. In making an experiment the crystal was set at a given angle with the central incident ray and the plate-holder rotated through double the angle from the zero position; the plate was inserted, the fiducial line marked on it, and an exposure of some hours (usually 24) was made, the magnetic field remaining on throughout. The crystals used were rocksalt and heavy spar. The whole apparatus was in a dark room.

Owing to the finite angle of the beam of γ rays, and to the length of the crystal, for any given setting of the crystal there are rays striking it at all angles within a certain small range. To enlarge the range and thus obtain more lines on the plate for a single exposure, the crystal was in some cases slowly rotated during the experiment, as in the experiments of M. de Broglie.* The rotation was effected by the following device. Supported by the water in a tall cylindrical vessel was a float, which subsided slowly, owing to the escape of the water, drop by drop, through a capillary tube of suitable size attached to an opening in the bottom of the vessel. The float as it sank rotated the crystal by means of a light horizontal arm, to one end of which it was fastened by means of a thread passing over a pulley; the other end was attached to the turn-table carrying the crystal. The moving end of the arm was carried without friction by means of wheels on a glass plate, and the motion

* M. de Broglie, *Journal de Physique*, Feb. 1914, p. 101.

attained was very uniform and could be adjusted by changing the length of the capillary tube. A rotation of one degree occupied from four to eight hours.

Measurement of the Plates

The positions of the lines were measured as distances from the fiducial line' which fixes the position of the plate relative to the plate-holder. From this was calculated the angle which the ray corresponding to any line made with the normal to the plate; the angle which the plate-holder made with the zero position being known, the angle of reflexion of the given ray followed at once. To correct for possible errors in the fixing of the zero, the same line was photographed twice, the crystal being rotated between the two photographs to a symmetrical position on the other side of the zero, so as to throw the line in the one case to the right, in the other case to the left, of the undeflected beam. This enables the angle of reflexion to be fixed with considerable accuracy if the zero positions are only roughly determined.

Experimental Results

In this paper an analysis will be given of the soft type of γ radiation from radium B. Evidence of lines corresponding to the more penetrating rays from radium B and the penetrating rays from radium C has been obtained on the photographs, and the spectra have been separated by the interposition of absorbing screens; lines have been found, due to radium C, with 6 mm. of lead between the radium tube and the crystal. The spectra due to the penetrating rays from radium B and radium C are faint compared with that of the soft radiation from radium B, and have not yet been fully investigated; an account of them is withheld for a future paper.

The stronger lines due to radium B appeared with great distinctness on the photographic plate, as will be seen from fig. 2 (Plate), which is reproduced from an actual photograph; they permit of accurate measurement. In the photograph B is the band made by the direct rays coming through the slit, β and α are the two strong lines formed by the reflected rays, and F is the fiducial line. The fainter lines do not appear on all the plates; however, no line is given in the table which has not been measured on at least two plates. The main feature of the spectra of the radiation reflected from rocksalt is two strong lines at almost exactly 10° and 12° respectively; they are accompanied by a number of fainter lines at angles of from 8° to 14°. There is also a large group of faint lines between 18° and 22°, which do not permit of accurate measurement, and so are omitted in the table; some of these, at least, are probably repetitions of the measured lines in the second order.

Most of the photographs were taken with crystals of rocksalt, the crystal in some cases being a slip less than a millimetre thick, in others a specimen about a centimetre thick; the results did not differ noticeably for the soft rays.

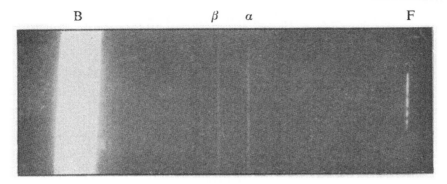

Fig. 2

Fig. 4

To check the measurements of the angles of reflexion made with the rocksalt, photographs were also taken by reflexion from the cleavage (001) face of a crystal of heavy spar (barium sulphate). Since the scattering by an atom is proportional to its atomic weight, it was thought that the heavy spar might give spectral lines of more intensity than the rocksalt for the more penetrating rays. In order to compare the constants, or 'grating space,' of the two crystals, special experiments were made with them both with X rays by the photographic method developed by Moseley, who kindly designed for us an X-ray tube with a nickel anticathode. This was provided with a side tube and slit, and the rays escaped through a thin aluminium window; the tube emitted an intense beam of soft X rays consisting mainly of the characteristic radiation of nickel. The angle of reflexion for the two strong lines in the nickel rays was directly determined for both rocksalt and heavy spar; for the (100) plane of rocksalt the angles obtained agreed closely with those determined by Moseley. The angles of reflexion from the (001) plane of heavy spar were found to be 12° 5' and 13° 23' in the first order. The ratio for the corresponding angles for the two crystals was 1·278. This enables us to compare numerically the photographs taken with the two crystals by the γ rays.

The spectra obtained with heavy spar appeared to be less intense than with rocksalt for the soft rays, and did not show the harder rays with much greater clearness. The angles of reflexion for the two strong lines of the radium B radiation were found to be 7° 52' and 9° 28'. Multiplying by the factor 1·278 to express them in terms of rocksalt, these become 10° 3' and 12° 6', agreeing closely with the values obtained directly with this crystal. This puts it beyond doubt that the lines given by rocksalt are true diffraction lines, and do not arise from irregularities in the crystal.

In Table 1 (next page) the angle of reflexion of the different homogeneous rays which make up the softer γ radiation from radium B are given for rocksalt. Their relative intensities are denoted by the letters 's.' (strong), 'm.' (medium), and 'f.' (faint), but this indication is only very rough, as the circumstances conditioning the intensity vary from photograph to photograph. The wave-lengths (in centimetres) corresponding to the different angles of reflexion are calculated from the formula $\lambda = 2d \sin \theta$, the value $d = 2·814 \times 10^{-8}$ cm. being taken from Moseley's paper. The spectrum of the characteristic radiation from platinum, suitably reduced by division by a constant factor, is added for comparison: this will be referred to again later (p. 441).

In fig. 3 the spectrum is shown diagrammatically, and below it that of platinum, the scale being adjusted so as to make the strong 10° line coincide with the corresponding platinum line. The dotted lines in the platinum spectrum are taken from a paper of de Broglie*; as his determination of the strong line differs somewhat from that of Moseley and Darwin, the whole spectrum given by him has been reduced by multiplying by a constant factor chosen so as to make the strong lines agree.

* *Journal de Physique*, loc. cit.

TABLE I

RADIUM B Soft γ-ray spectrum			PLATINUM X-ray spectrum
Angle of reflexion from rocksalt	Wave-length (in cm.) × 10⁻⁸	Intensity	Angle of reflexion 1·122
8° 6′	0·793	m.	
8° 16′	0·809	m.	
8° 34′	0·838	m.	8° 27′
8° 43′	0·853	m.	8° 43′
9° 23′	0·917	f.	
9° 45′	0·953	m.	
10° 3′	0·982	s.	10° 2′
10° 18′	1·006	m.	10° 13′
10° 32′	1·029	m.	
10° 48′	1·055	f.	
11° 0′	1·074	f.	
11° 17′	1·100	f.	
11° 42′	1·141	m.	
12° 3′	1·175	s.	12° 3′
12° 16′	1·196	m.	
12° 31′	1·219	f.	
13° 0′	1·266	f.	
13° 14′	1·286	f.	
13° 31′	1·315	f.	
13° 52′	1·349	m.	
14° 2′	1·365	m.	

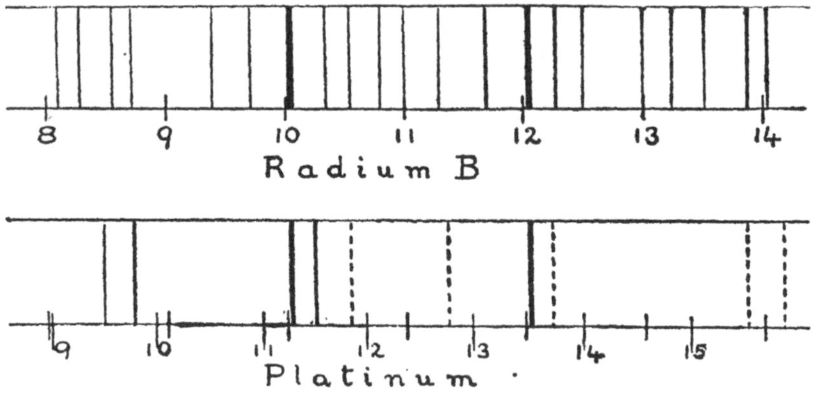

Fig. 3

Structure of the Spectral Lines

In the case of the stronger lines from rocksalt, viz. the 10° or 12° lines, the structure of the lines could be studied in some detail. They consisted of slightly curved bands about 0·5 mm. wide, the photographic intensity being greatest at the edge of the bands. A reproduction of part of one of these bands, magnified about five times, is shown in fig. 4 (Plate). With weak intensities only the outer edges of the band could be seen, and the band appeared as a close double. The spectral band appeared to be the exact mirror-image of the source, both as regards magnitude and distribution of intensity in radiation; the width of the image was the same as the diameter of the α-ray tube, viz. $0 \cdot 50$ mm. The sharp and well-marked edges of the band are due to the fact that the intensity of the radiation is least from the centre of the cylindrical α-ray tube, and increases to a maximum from the edges, owing to the active matter deposited on its inner surface. It is well known that a photograph taken of an α-ray tube by its own rays through a narrow slit parallel to the source always shows these variations of intensity. The fact that the spectral band on the photographic plate is the mirror-image of the source, indicates clearly that the scattered rays forming the band come from very near the surface of the crystal. Attention has been drawn to the completeness of the reflexion of X rays from a crystal at the proper angle by Darwin,* and shown by him to be a necessary consequence of the mathematical theory. The efficiency of the reflexion is also well shown by recent experiments of W. L. Bragg.†

Imperfection of Crystals

In most of our experiments we have employed a crystal of rocksalt, since its structure has been worked out in detail by W. H. and W. L. Bragg, and since it gives fairly strong reflexions for soft radiations. The crystals employed, however, showed many imperfections, and their behaviour was very different from that to be expected for an ideal crystal; for example, when the crystal was set at an angle of 12° to the incident beam, and with the width of the pencil such that only radiations between 11° and 13° should be strongly reflected, in addition to the lines in this region other outside lines are observed in varying positions on the photographic plate: for example, in a particular case at 2° 40′ and 13° 50′.

Special experiments showed that the position of these lines on the plate corresponded to a definite frequency of vibration in the incident beam. All our photographs showed similar peculiarities, but the outside lines which appear are very variable for different angles of the crystal. This behaviour of rocksalt led us to make many fruitless experiments to obtain a more perfect crystal, but all the crystals of rocksalt we have examined show similar imperfections, though in varying degrees. The crystal of heavy spar employed, for which the face appeared very plane and perfect, also behaved similarly. There appears to be no doubt that many crystals, and especially those of rocksalt, have a contorted or undulating surface, and that the orientation of the planes varies within certain limits from point to point of the crystal. At the same time, these irregularities may lead to the absence of a line in the photograph, although the crystal is set at the correct theoretical angle. To avoid this

* C. G. Darwin, Phil. Mag. Feb. 1914, p. 315.
† W. L. Bragg, 'Nature,' March 1914, p. 31.

difficulty, it is desirable to keep the crystal in rotation during the experiment. Darwin has examined the consequences of such imperfections in a crystalline structure,* and considers that they offer an explanation of the fact that the intensity of the reflected beam is in general greater than the theoretical value to be expected for an ideal crystal.

Connexion of Radium B with Lead

In recent papers,† Moseley has examined the X-ray spectra of a number of the ordinary elements. For this purpose, each element either in the state of metal or compound is exposed as anticathode in a focus tube, and the resulting X-ray spectra are obtained photographically by the crystal method. He has shown that the 'K' characteristic radiation of all the elements between aluminium and silver shows a similar type of spectrum, and the frequency of the corresponding lines changes by definite steps in passing from one element to the next. The frequency of the strongest spectrum line has been shown to vary as $(N - a)^2$ where N is a whole number and a a constant (about unity) for all this group of elements. N changes by unity in passing from one element to the next, and is supposed to represent the number of fundamental units of positive charge carried by the atomic nucleus and may for convenience be called the 'atomic number,' since it represents the number of the element when arranged in order of increasing atomic weight supposing that no elements are missing.

It is well known from the work of Barkla that the heavier atoms emit a second type of characteristic radiation known as the 'L' radiation. Moseley has examined the X-ray spectra of this type for elements of atomic weight from silver to gold, and finds that the spectra of all these elements are similar, but as in the case of the 'K' type, the frequency increases by definite steps as we pass from one element to the next. He has shown that the frequency of the chief line of the spectra is nearly proportional to $(N - b)^2$, where N as before is the atomic number (or nucleus charge) and b a constant (about $7 \cdot 4$) for the whole groups of elements.

On the general theory of the nucleus atom, the nucleus charge determines the chemical and physical properties of the atom, and it is consequently of great importance to determine the value of this constant for the radioactive atoms. Before the publication of this paper, Mr. Moseley kindly informed us of his experimental results, and it became of great interest to determine the nucleus charge of radium B. As we have already seen, the soft radiation from radium B, whose absorption coefficient is $\mu = 40$ in aluminium, was believed to be the 'L' type of characteristic radiation of radium B, and this is completely borne out by the comparison of the γ-ray spectrum of the soft radiations of radium B with that of platinum (see page 438). Using Moseley's formula, and assuming for the atomic numbers the values to be given in a

---

* C. G. Darwin, Phil. Mag. April 1914, p. 975.
† Phil. Mag. *loc. cit.*

following paragraph, the factor by which the angle of the strong platinum line must be divided to give the angle of the corresponding line of radium B is 1·118: the value 1·122 used in Table I. was chosen so as to make the experimental lines agree exactly.

A determination of the nucleus charge of radium B is for another reason of the highest importance, for this radioactive element has been shown by Fleck to have the chemical properties of lead and to be chemically inseparable from it. As is well known, a very comprehensive and far reaching theory of the relation between the chemical and physical properties of the radioelements has been advanced by Fajans and Soddy. From the point of view of the nucleus theory of the atom, their conclusions may be expressed by the simple relation that the expulsion of an α particle (carrying two positive charges) from an atom lowers its nucleus charge by two units, and the expulsion of a β particle (carrying one negative charge) raises its nucleus charge by one unit. Soddy has pointed out that the products radium B, actinium B, thorium B, and radium D are the 'isotopes' of lead, *i. e.* they show identical chemical properties with those of lead, from which they are inseparable by chemical methods. If this view is correct. the atoms of these elements should have the same nucleus charge, although they may differ slightly in atomic weight.

If radium B has the same nucleus charge as lead, it must give an X-ray spectra almost identical with that of lead. It should, however, be pointed out that a very small variation in the frequency of the vibrations may be possible if the nuclear masses are different. In his recent paper (*loc. cit.*) Moseley has not determined the X-ray spectra of lead, but he kindly pointed out to us that on his results its atomic number or nucleus charge should be 82. He found gold had the nucleus charge 79; the two intervening elements, mercury and thallium, should have a nucleus charge of 80 and 81 respectively. From the relations found by him, it followed by calculation that the strongest line of lead should be reflected at 12·07° from rocksalt. The strongest line from radium B found by us was 12·05°—a very close agreement.

As it was possible, however, that there might be a small error in comparing the reflexion angles of rocksalt with different crystals and with such different experimental arrangements, it was decided to test by a straightforward method whether the X-ray spectra of radium B and lead were identical within the limits of experimental error. For this purpose it was arranged that the γ-ray spectra of radium B and of lead should be compared, using the same apparatus and under as nearly as possible identical conditions. H. Richardson, working in this laboratory, has found that the β rays expelled from radium B and radium C excite strongly the characteristic 'L' type of radiation when they fall on heavy elements. In order to take advantage of this result, the β rays from an emanation tube of the kind already described were used to excite the characteristic radiation in a strip of lead, 1 mm. thick and 5 mm. broad, which was then used as the source in place of the tube itself, the rest of the apparatus being disposed much as before. The arrangement was as shown in fig. 5; the slit was narrowed down to about 0·8 mm. by means of aluminium strips N,

N, and the radiator R placed opposite it in the position indicated. The emanation tube A was fixed to one side, so that no direct rays from it could

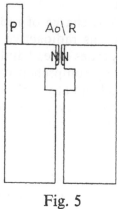

Fig. 5

strike the photographic plate. A second block of lead P was placed behind it, to increase the intensity by means of successive 'reflexions' of the β rays between this block and R.

The spectrum of the radiation excited in the lead plate L was then determined under as nearly as possible the same conditions as for the γ rays from the emanation tube. For a given source, the photographic effect of the spectrum lines from lead only showed up faintly against the general blackening of the plate, but was sufficiently clear to admit of measurements of some of the angles of reflexion. Only a few lines of lead could be measured; two of these gave reflexion angles of 10° 2' and 12° 0' in good agreement with the strong lines of the radium B spectrum. Other faint lines were also observed but were difficult to measure. There was, however, a possibility of error in such an experiment. It was conceivable that the spectrum lines observed were not due to the characteristic radiations from the lead but were to be ascribed to some of the soft γ radiations from radium B scattered by the lead plate. To test this point, the lead plate was replaced by one of platinum of the same dimensions and the spectrum again measured. The positions of the lines were quite distinct from those observed with the lead radiator, and the measurements of the reflexion angles of two of the strongest lines were in fair agreement with those given by Moseley and Darwin for platinum.

According to Moseley's results, the frequency of a reference line for the X-ray spectra of successive elements changes by well-marked steps. For example, on the formula given by him, the reflexion angles from rocksalt of the strongest line of the spectrum from an element of nucleus charge 81 is 12·41°, for 82, 12·07°, for 83, 11·77°. In order to make a mistake of one unit in the nucleus charge, an error of 2 per cent. is necessary in measuring the angle of reflexion of the reference line, while the experimental determination of the reflexion angles of the γ rays from radium B is believed to be correct within 0·3 per cent.

It thus appears that the nucleus charge of radium B is the same as that of lead, for the atomic number of radium B, deduced by Moseley's formula from the γ-ray spectrum, is that to be expected for lead, and the strong lines of the γ-ray spectrum of radium B seem to be coincident with those of lead. According to radioactive calculation, the atomic weight of radium B is 214, while that of lead is 207. Provided the difference in atomic mass has not a large influence on the vibration frequencies of the outer distribution of electrons, it is to be anticipated that the ordinary light spectra of radium B and lead should be nearly identical, while we already know that these two elements have apparently identical chemical properties.

These results confirm in an unexpected way the correctness of this deduction of Soddy and Fajans, and also give a definite verification of the hypothesis that two elements of different atomic weights may have identical spectra and identical chemical properties. A similar result has been recorded by Sir J. J. Thomson and Aston in their work indicating that neon consists of a mixture of two gases of atomic weights about 20 and 22. The theory of the nucleus atom affords a simple explanation of such a result; for the chemical and physical properties are for the most part determined by the charge on the nucleus, and are practically independent of the mass of the nucleus. The properties of radioactivity and gravitation belong mainly to the nucleus. The fact that radium B is radioactive while lead is not, shows that the constitution of the nucleus is different in the two cases, and this is borne out by the known difference in atomic weights.

Taking the nucleus charge of radium B as 82, the nucleus charge of all the elements in the uranium-radium family can be deduced at once from the generalization already referred to. The numbers are given in the following table: an α radiation gives a decrease of 2 in the nucleus charge, a β radiation an increase of 1.

TABLE II

Element	Radiation	Atomic Number
Uranium I	α	92
Uranium X_1	β	90
Uranium X_2	β	91
Uranium II	α	92
Ionium	α	90
Radium	α	88
Emanation	α	86
Radium A	α	84
Radium B	β	82
Radium C	$\alpha + \beta$	83
Radium D	β	82
Radium E	β	83
Radium F	α	84
End product (Lead)	..	82

If the general formula of Moseley holds throughout, the frequencies of vibration of the 'L' type of radiation for each of these elements can be simply calculated.

Summary

(1) The γ-ray spectrum of the soft radiations from radium B has been examined by reflexion from the cleavage faces of crystals, and found to consist of a number of well-marked lines.

(2) The γ-ray spectrum of radium B is found to be of the same general type as that found for platinum and other heavy elements when bombarded by cathode rays.

(3) Attention is directed to the structure of the spectral lines using an emanation tube as source of radiation, and also to the imperfections of the crystals employed.

(4) Evidence is given indicating that the spectrum of the soft γ-rays spontaneously emitted from radium B, is identical within the limits of experimental error with the spectrum given by lead when the 'L' characteristic radiation is excited by the bombardments of β rays.

(5) The bearing of these results on the structure of the atom is discussed.

The Structure of the Atom

by ERNEST RUTHERFORD

From *Scientia*, xvi, 1914, pp. 337–51

THE discovery of Röntgen rays in 1895 marks the beginning of a great epoch in the history of Physical science, not only from the intrinsic interest and importance of the discovery itself, but from the remarkable impetus it gave to investigations in a variety of directions. It was followed a few months later as a direct consequence by the discovery made by Becquerel of a new and unexpected property of matter, 'Radioactivity', which has led to the opening up of such a wide and fertile territory of research, resulting in a veritable revolution in our conceptions of the possibilities of matter. The investigation of the origin of Röntgen rays from a vacuum tube soon led to the discovery of the corpuscular nature of the cathode rays, and the proof that the particles constituting the cathode stream consisted of negatively charged particles whose mass was exceedingly minute, about $^1/_{1800}$ of the mass of the hydrogen atom — the lightest atom known to science. There soon followed definite evidence that these 'corpuscles' or 'electrons' as they have been termed, can be liberated from matter by a variety of agencies, and are to be considered as definite constituents of all atoms of matter. It was found that radium, and other radioactive bodies, emits such electrons, known as β rays, with enormous speed, approaching very closely that of light itself. At such high speeds, the mass of the electron is a function of its velocity, and it was shown by Kaufmann that the variation of mass with speed could be explained by supposing that the mass of the electron was entirely electrical in origin. In other words, the electron is to be regarded as a condensed charge of negative electricity existing independently of matter as ordinarily understood.

The position of the negative electron as one of the fundamental units of atomic structure is thus definitely established, but the rôle of positive electricity appears to be very different. Although a very exhaustive search has been made by Sir J. J. Thomson and others, no evidence has been obtained of the existence of a corresponding positive electron whose mass is small compared with that of the lightest atom. Positive electricity appears to be always associated with the atom of matter, and no positively charged particle has been found whose mass is less than that of the atom of hydrogen. This distinction between positive and negative electricity appears to be fundamental and has to be taken into account in all theories of the constitution of the atom.

From investigations on the passage of electricity through solutions, Faraday long ago showed that there was a definite connection between the atoms of matter and the electrical charges they carry in electrolysis. In 1887 Helmholtz definitely put forward the view that electricity is atomic in character, or in other words that there is a natural indivisible unit of electricity, and that all electrical charges must be an integral multiple of this unit. The work of the last few years has given a definite proof of this fundamental conception. The unit of positive electricity is believed to be the charge carried by the hydrogen atom in the electrolysis of water. This is equal in value but opposite in sign to the charge carried by the negative electron. The actual value of this fundamental unit has been determined by a great variety of methods with very concordant values. The researches of Millikan in particular have provided not only an accurate value of this unit but also definite evidence of the correctness of the unitary theory. The atomic nature of electricity is one of the fundamental facts on which all modern theories of the structure of the atom are based.

In the meantime great advances had been made in our knowledge of radioactivity. The separation of highly radioactive substances like radium made it possible to examine the radioactive processes by chemical as well as by physical methods. It was shown by Rutherford and Soddy that the phenomena of radioactivity received a complete explanation by supposing that atoms of a radioactive substance were undergoing spontaneous transformation. At each instant a small fraction of the total number of atoms become unstable and break up with explosive violence. In most cases, a fragment of the atom — the α particle — is ejected with very high velocity, while in a few cases the explosion is accompanied by the expulsion of a high speed electron and the appearance of a very penetrating type of Röntgen rays known as the gamma rays. The radiations accompany the transformations of atoms and serve as a direct measure of the rate of their disintegration. It was found that the transformation of an atom led to the production of an entirely new type of matter, differing completely in chemical and physical properties from the parent substance. This new substance was in turn unstable, and was transformed with the emission of characteristic types of radiations. The process once started continued through a number of definite stages, which have been traced out and analysed in great detail. As an example of such transformations, we may take the case of radium. This is a comparatively stable substance, for it is half transformed in about 2000 years. The transformation of the radium atom is accompanied by the expulsion of an α particle, which is now known to be a charged atom of helium. The resulting product, the emanation, is a heavy gas which is half transformed in $3 \cdot 85$ days. The emanation after expelling an α particle changes into radium A which is a solid, and the process of successive transformations continues through a number of further stages. From a purely chemical point of view, each one of these series of substances is to be regarded as a new element. It is distinguished, however, from the ordinary elements by the instability of its atoms.

It was thus definitely established that the atoms of some elements suffer spontaneous disintegration accompanied by the emission of an amount of energy enormous compared with that liberated in ordinary molecular changes. The processes occurring in the radioactive substances were found to be entirely unaffected by physical or chemical agencies. We may consequently conclude that the atom contains a very large store of energy either in the kinetic or potential form, but that its stability cannot be changed by the application of the most intense laboratory forces at our command. In a large number of radioactive changes, the radiation is emitted in the form of α rays which are known to consist of positively charged helium atoms projected with very great velocity. Consequently all the radioactive substances which emit α rays give rise to helium as a product of their transformation. It is thus reasonable to conclude that the atoms of the radioactive elements are complex structures consisting, in part at least, of helium atoms, and that one of the constituent helium atoms is released in the atomic explosion. For example, the element uranium, which in its transformation gives rise to ionium and radium, emits in all eight helium atoms. It is thus natural to conclude that the atom of uranium is a composite structure containing at least eight helium atoms. The study of radioactive phenomena has thus clearly indicated not only that the atoms of the heavy elements are very complex structures but that such atoms are not permanent and indestructible, but can suffer spontaneous transformation with the appearance of a number of new elements. It must be borne in mind that so far the transformation of the atoms of matter has only been observed in the case of the radioactive substances uranium, thorium and actinium and their products, all of which are composed of heavy atoms. With the exception of rubidium and potassium, no light elements have been found to possess any intrinsic radioactivity. The activity of the two latter elements is relatively very feeble compared with radioactive bodies proper, and the radiation emitted is entirely of the β ray or electronic type. No evidence, however, has so far been obtained that the atoms of these elements are undergoing a definite transformation with the appearance of new types of matter akin to that observed in the radioactive elements proper.

While the study of radioactivity has profoundly modified the older ideas of the atom, and has brought to light the existence of nearly thirty new types of matter, the atoms of which have a limited life and carry within themselves the seeds of their final destruction, it has at the same time provided us with powerful methods for showing in a very direct way the individual existence of the atom as a definite physical and chemical unit in the structure of matter.

Until a few years ago, it would have been considered impossible that we should ever be able to devise methods to detect the presence of a single atom of matter. We have seen that the α particle from radioactive substances is a charged atom of helium projected with great velocity. By suitable methods, Rutherford and Geiger have shown that each α particle in entering a vessel may be made to show an electrical effect that is easily measurable, and the number of α particles entering in a given time can be accurately counted. Each of the scintillations observed in crystalline zinc sulphide, when the α rays fall on it, have been shown to be due to the impact of a single α particle. In this case, part of the energy of the flying atom is transformed into light of sufficient intensity to be easily visible to the eye in a dark room. The application of these results provides an exceedingly direct method for determining the number of atoms in a cubic centimetre of helium, for the actual number of α particles expelled from a known quantity of radium have been counted and the volume of helium gas resulting from the α particles has been directly measured. In this way, it has been shown that there are $2 \cdot 75 \times 10^{19}$ atoms of helium in a cubic centimetre at standard temperature and pressure. From other data, we know that this value represents the number of molecules in a cubic centimetre of any gas under the same conditions.

Not only are we able to detect a single atom of matter but also an electron in swift motion although its mass and energy is small compared with that of the α particle. This has been shown to be possible by a simple electrical method devised recently by Dr. H. Geiger.

A still further advance has been made by C. T. R. Wilson who has perfected a beautiful device for showing the track of an α particle or electron through a gas. This method depends upon the fact discovered by him that the charged ions, produced by the impact of these flying particles with the molecules of the gas, become nuclei for the condensation of water upon them, when a gas saturated with water is suddenly cooled by a rapid expansion. Each ion then becomes the centre of a visible globule of water. A photograph of these minute drops taken immediately after the expansion records in a striking and perfect way the position of the ions in the gas which mark the path of the particle. Experiments of this kind bring out in a vivid fashion the individual existence of these flying particles and the processes occurring in their passage through matter.

There is another important field of work which has thrown much light on the structure of the atom. The α particle and the β particle are expelled from radioactive matter with such great energy of motion that they are able to pass actually through the atoms in their path. Many of these particles are deflected from their rectilinear paths by their passage through the strong electric field within the atom, and from a study of the amount of deflection, we are able to form an idea of the strength and distribution of the electric field inside the atom.

From a brief review given in this article, it will be seen that the development of modern physics is closely connected with the problem of the structure of the chemical atom. On the one side, the examination of radioactive phenomena has thrown much light on the processes of the transformation of heavy atoms, and the products of their disintegration. On the other, a study of the effects produced by the new types of penetrating radiation in their passage through matter has yielded information of great importance in regard to the structure of the atoms themselves. As a result of these important additions to our knowledge, there is a general agreement amongst physicists that the atom of a heavy element like gold for example must be regarded as a complex structure consisting of negatively and positively charged particles, which are held in equilibrium by electrical forces. It is supposed that the negative electricity is distributed in discrete units in the form of negative electrons but the exact number of such electrons in each atom is still to some extent *sub judice*. There is, however, great difference of opinion with regard to the nature and distribution of the positive electricity and of the positively charged particles within the atom. The differences in point of view can be most simply illustrated by the consideration of two distinct types of atomic models which have been advocated in recent years. The first is the well-known type of atom, first suggested by Lord Kelvin, and afterwards modified and examined in great detail by Sir J. J. Thomson. In this type of atom, the positive electricity is supposed to be uniformly distributed throughout a spherical volume of dimensions comparable with that of the atom itself as ordinarily understood. Negative electrons are supposed to be distributed through this volume either in one plane in concentric rings or in concentric spherical shells like the coats of an onion. For the atom to be electrically neutral, it is, of course, necessary that the charge carried by the negative electrons should be equal in magnitude to the positive charge. Sir J. J. Thomson has examined mathematically in detail the possible stable distributions of electrons in one plane and has deduced the possible arrangements of the electrons for a number of different values of the positive charge.

The 'Thomson' atom has undoubtedly served a very useful purpose in giving a simple and easily understood idea of atomic structure. It has the great advantage that the law of force involved admits readily of mathematical calculation and the position and number of the electrons in different rings can be directly calculated. The ultimate test of any atomic model lies, however, in its ability to explain the experimental facts, and we shall see reasons for believing that this type of atom must be much modified before it is capable of accounting for some unexpected phenomena which have been brought to light in recent years.

It is now well known that swift α and β particles in passing through matter, whether solid, liquid or gaseous, are occasionally deflected from their rectilinear paths. A narrow pencil of rays in traversing a sheet of matter is always 'scattered' into a diffuse beam. Such a result is to be anticipated if there are strong electrical fields within the atoms which the particles occasionally penetrate. Geiger and Marsden, however, drew attention to the remarkable fact that in addition to the general scattering of a beam of α rays, a small fraction of the total number of α particles were diffusely reflected from a thin metal sheet. When account is taken of the enormous energy possessed by an α particle from radium, it is impossible to explain such a result unless it be supposed that the α particle occasionally passes through such an intense electric field within the atom that it is deflected from its path through an angle of more than a right angle at a single atomic encounter. Simple calculation shows that the fields inside the Thomson atom are much too feeble to account for this result unless it be supposed that the positive electricity is

P

much more concentrated than was originally assumed. In order to account for this 'large angle' scattering of α particles, the writer was led to suppose that the positive electricity within the atom was concentrated within an exceedingly small region. In order to make the atom electrically neutral, the positive charge was supposed to be surrounded at a distance by a suitable distribution of electrons. It was also necessary to assume that the positive nucleus was the seat of most of the mass of the atom. The deflections of the α particle are supposed to occur only when the α particle passes through the intense field close to the nucleus where the magnitude of the deflections are but little affected by the external electrons. Assuming that the positively charged α particle and the positive nucleus repel each other according to the law of the inverse square of the distance, it can simply be shown that the α particle describes a hyperbolic orbit round the nucleus and at the same time sets the nucleus in motion. The relation between the number of α particles and the angle through which they are deflected at a single encounter can be simply calculated, and also its dependence on the velocity of the α particle. It is obvious that on this point of view the chance of a large angle deflection due to the close approach of the α particle to the nucleus, is much less than for small angle deflections. The large angle scattering of α particles for thin films of metal was examined in detail in an important investigation by Geiger and Marsden, and their results were in very satisfactory agreement on all points with the theory. Darwin showed that any other law of force but the inverse square was inconsistent with the experimental data. The idea that the atom contains a positively charged nucleus of small dimensions has thus a strong basis of experimental support.

From the agreement of experiment with theory, it was calculated that the radius of the nucleus of a heavy atom like gold was not greater than 3×10^{-12} cms. This is very small compared with the ordinarily accepted idea of the radius of the sphere of action of an atom, which is about 10^{-8} cms. It thus seems certain that the nucleus of the atom has dimensions very small compared with the electronic distribution which surrounds it.

Special interest attaches to the effect of a collision of a swift α particle with a light atom like that of hydrogen. According to the views discussed later, the α particle, which is to be regarded as the helium nucleus, carries two unit charges, and the hydrogen nucleus one charge. It can be simply calculated that in a close encounter between the α particle and the hydrogen atom, the latter should in rare cases be set in motion with a velocity about 1·6 times that of the α particle and should travel about four times as far through a gas as the α particle itself. Under these conditions, these swift hydrogen atoms should have sufficient energy to produce visible scintillations in zinc sulphide. In some recent experiments, Marsden has found in this way definite evidence of the existence of such swift hydrogen atoms when an intense beam of α rays passes through hydrogen. A small number of scintillations can be detected at a distance of about four times the maximum range of the colliding α particle. The number and range of these penetrating particles are in agreement with the view that they consist of hydrogen atoms, or rather hydrogen nuclei set in swift motion by close collisions with an α particle. From these experiments, a still lower limit may be given for the dimensions of the hydrogen and helium nuclei. In order to give the observed velocity to the hydrogen nucleus, the α particle must approach within a distance $1 \cdot 7 \times 10^{-13}$ cms. of the hydrogen nucleus. This is an exceedingly small distance, even smaller than the calculated diameter of the electron itself, viz. $3 \cdot 8 \times 10^{-13}$ cms. According to the electromagnetic theory, the mass of a moving charge depends on the degree of condensation of the charge. If it were supposed that the linear dimensions of the hydrogen nucleus were about $1/_{1800}$ of that of the electron, its mass could be about 1800 times greater than the electron.

It thus appears possible that the hydrogen nucleus of unit charge may prove to be the positive electron, and that its large mass compared with the negative electron may be due to the minuteness of the volume over which the charge is distributed. It is, of course, difficult to give a definite proof of such an hypothesis, but at present there are no experimental facts to contradict it. It would be natural on this view to suppose that the positive and the negative electrons are the two fundamental units of which all the elements are composed.

On the theory of the nucleus atom it must be supposed that the α particle has its origin in the nucleus and acquires part of its energy of motion in escaping from it. It is well known that the majority of radioactive transformations are accompanied by the expulsion of α particles. There are, however, a small number of changes in which only β particles, i. e. swift electrons, are expelled. The general evidence indicates that these transformations owe their origin to the escape of a high speed electron from the nucleus. The excitation of γ rays is due to the passage of the electrons through the external electronic distribution, which is set in violent vibration. If this be the case, the nucleus, though of minute dimensions, is in itself a very complex system consisting of a number of positively and negatively charged bodies bound closely together by intense electrical forces. The fact that helium nuclei are expelled so frequently in radioactive processes suggests that the helium nucleus is one of the units, possibly secondary, of which the nucleus of the heavy atom is built up. The fact that the helium nucleus survives unchanged such a violent atomic explosion indicates that it must be a very stable structure.

We have so far not considered the magnitude of the excess positive charge carried by the atomic nucleus. From the experiments of scattering, it was deduced that the positive charge on the nucleus was equal to about $\frac{1}{2} A e$ where A was the atomic weight in terms of hydrogen and e the unit charge. This would mean that the number of external electrons in an atom is represented numerically by about half its atomic weight. This result is strongly substantiated by an entirely distinct line of evidence. It is well known that when the X rays pass through matter, they are scattered in all directions. Assuming that this scattering is due to the external electrons composing the neutral atom, Barkla deduced from the theory of Sir J. J. Thomson that the number of electrons in an atom was equal to about half its atomic weight. On account of the experimental and theoretical difficulties, too much stress cannot be placed on the accuracy of this deduction.

It is obvious that this rule in regard of the number of external electrons cannot apply strictly to light atoms. The radioactive evidence suggests that the helium nucleus contains two unit charges. If this be the case, the hydrogen nucleus of mass 1 should contain one unit charge. Van den Broek made the interesting suggestion that the nucleus charge might prove to be equal to the atomic number of the element, i. e. the number of the element when all the atoms are arranged in order of increasing atomic weight. Supposing that all the elements are known, this means that hydrogen has a nucleus charge 1, helium 2, lithium 3, ... carbon 6, nitrogen 7, oxygen 8, and so on. Such a view agrees quite well with the experimental data of scattering, and its simplicity has much to commend it. It has formed a basis of the recent theories of Bohr to explain the structure and spectra of elements.

On the nucleus theory, the properties of the atom are mainly determined by the magnitude of the nucleus charge, which can only vary by integral units. This must necessarily be the case, for the number and distribution and mode of vibration of the external electrons in the neutral atom are controlled by the forces arising from the charge at the centre of the atom. It is thus possible on this view that elements may exist of almost if not completely identical properties but whose atomic weight may be sensibly different. It should be pointed out that the mass of the nucleus as well as its charge may have a small effect on the equilibrium of the external electrons, and thus modify to a slight degree the mode of vibration of the external system of electrons. During the last year or so some very striking evidence has been obtained in support of these views by investigations in a number of different fields. In the first place, a wide and important generalisation due to Fajans, Soddy and Russell has been put forward to account for the change of properties of the radioactive elements as a result of successive transformations of the parent elements. A detailed account of this generalisation has been given by Soddy last year in this journal. It is found that the expulsion of an α particle from a radioactive atom gives rise to a new element which occupies a place in the periodic table two places lower in the direction of decreasing atomic weight; the expulsion of a β particle causes the resulting element to change its position by one group in the opposite direction. This simple rule is found to embrace all the radioactive elements and led to the prediction of the position and properties of an undiscovered radioactive element. This missing element was a few weeks later isolated by Fajans and Gohring, when it was found that its chemical and physical properties were in close accord with the predictions of the theory.

This generalisation can be expressed in a very simple way on the nucleus theory. The expulsion of an α particle lowers the nucleus charge of the atom by two units, while the expulsion of a β particle, carrying one negative charge from the nucleus, raises the charge of the latter by one unit. It follows at once from this point of view that a number of the radioactive elements are defined by nuclear charges varying by unity. For example, uranium 1 of nucleus charge taken as 92 units, expels an α particle and gives rise to uranium X_1 of charge 90. This in turn expels a β particle and gives rise to uranium X_2 of charge 91. This again expels a β particle and gives rise to uranium 2 of charge 92. By the loss of an α particle, this changes to ionium 90, and this in turn to radium 88 and so on.

Important evidence on this point has been obtained by Moseley for the non-radioactive elements using a novel method. Following the important discovery of Laue that X rays give marked interference effects in their passage through crystals, Professor Bragg and W. L. Bragg and also Moseley and Darwin have developed methods for determining the frequency of vibration of X rays by examining the reflection, or rather diffraction, of X rays at the surface of crystals. It is to be anticipated that the X rays correspond to vibrations of the electrons close to the nucleus where their rate of vibration is very rapid under the intense electrical forces, and it is of great interest to examine how the X ray spectrum of one element differs from another. Moseley has recently investigated the X ray spectra of a large number of elements of atomic weights between aluminium and gold. He found that the lighter elements all give rise to similar spectra, known as the K series, consisting of two strong lines, and that the frequency of vibration of the corresponding line is markedly altered in passing from one element to the next. For the elements of atomic weight above silver, the spectra of the 'L' series was determined. Here again the spectra were all similar in type and the frequency of each corresponding line changed by well marked steps in passing from one element to another. He found that the frequency of vibration of the 'K' spectra is proportional to $(N - a)^2$ from aluminium to silver; for the 'L' spectra, the frequency is proportional to $(N - b)^2$. Here a and b are constants, a being about unity and $b = 7·4$. N is a whole number which is considered to correspond to the atomic number of the element i. e. it represents the number of units in the nucleus charge. For example, N varies from 13 for aluminium to 79 for gold. Quite apart from any theory of the origin of these spectra, it is evident that we have here a very simple relation between the X ray spectra of the elements and their atomic numbers; for the frequency of vibration of corresponding lines is proportional to the square of a number which changes by unity in passing from one element to the next. These conclusions are of great practical as well as of great theoretical importance.

If it be supposed that elements can exist representing each atomic number, from 1 for hydrogen to 92 for uranium, the X ray spectra supplies a simple method for determining missing elements. Moseley has examined this question, and has shown that only three possible elements are missing between aluminium of number 17 and gold number 79. A new and powerful method of chemical analysis has thus been placed in our hands, for it is possible to predict with certainty the X ray spectra of these missing elements, and thus afford an important aid to their discovery. This method is especially applicable to fixing the number and position of elements in the rare earth group.

The results of Moseley show conclusively that there is a quantity in the atom, viz. its nucleus charge, which is more fundamental and varies more regularly than the atomic weights. The chemical and physical properties of an element are defined by a whole number corresponding to the nucleus charge which varies by unity from one atom to another. It is well known, however, that the atomic weight varies very irregularly for successive elements. It would thus appear that the atomic weight of an atom is a complicated function of the structure of the nucleus, while the physical and chemical properties are largely independent of the structure of the nucleus but depend mainly on its charge.

We have already mentioned that it is possible for atoms to exist of practically identical physical and chemical properties but of different atomic mass. This possibility was first suggested by radioactive evidence, for it was found that there were a number of radioactive elements which appeared to be identical in ordinary physical and chemical properties, and inseparable from each other but which showed distinctive radioactive properties. Examples of this connection are the elements ionium and thorium and also radium and mesothorium. No evidence could be obtained that ionium gave a different light spectra from thorium. Recently also this idea has been strengthened by experiments in another direction. In his investigation on the positive rays, Sir J. J. Thomson found that the rare gas neon appeared to consist of two elements of atomic weight about 20 and 22. Unsuccessful attempts were made by Aston to isolate these two components by fractional distillation of the gas in charcoal cooled by liquid air. He, however, found that it was possible to effect a partial separation of neon into two gases of unequal density by the methods of diffusion. Such a result suggests that neon consists of a mixture of two gases of the same nucleus charge but of different atomic weights.

We have already referred to the fact that a number of the radioactive elements of different atomic weights appear to be chemically identical. Such *isotopes*, as they have been called by Soddy, are fairly numerous. For example, the elements radium *B*, thorium *B*, and actinium *B*, appear to be chemically identical with one another and with the element lead. This suggests that these elements have the same nucleus charge and should consequently give identical X ray spectra. In conjunction with Dr. Andrade the writer has recently made experiments to determine the spectra of the γ rays from radium *B*. The spectra of the soft γ rays was found to be identical in general form with that given by the 'L' series of the heavy elements, and special experiments showed that the two strong lines in this spectrum agreed with the corresponding lines of the X ray spectra of lead. Here we have a case of two elements of the same nucleus charge which give identical X ray spectra but whose atomic weights differ considerably. For example, the atomic weight of radium *B* is believed to be 214 while that of lead is 207. Evidence of the same general character is rapidly accumulating. Soddy and Fajans have suggested that lead is the end inactive product of both radium and thorium, and that these two types of lead differ in atomic weight by two units. The question whether lead obtained from radioactive sources has different atomic weight from ordinary lead is now under examination by a number of investigators in different countries. Already preliminary announcements have been made that a considerable difference in atomic weight has been detected, and it is probable that we soon shall have a large amount of evidence to settle this point definitely. It has been suggested that many of the ordinary elements may consist of a mixture of isotopes of the same nucleus charge but of different atomic weights. These elements would be inseparable by ordinary chemical methods, but might be conceivably separated by diffusion. The atomic weight of such elements would depend on the relative proportion of the two constituents, and this might be anticipated to vary when the materials were obtained from widely different sources. We see from such results that the study of the atomic weights of the atom is entering on a new phase, and it is probable that important conclusions are likely to follow in the near future.

We have so far not discussed the distribution of the external electrons which make the atom electrically neutral. This is an exceedingly involved problem and any attempt at its solution at once raises many difficulties. If we adopt the ordinary theory that an accelerated electron radiates energy, it is to be anticipated that the external electrons surrounding the nucleus should ultimately lose energy by radiation and fall into the nucleus. In order to get over this difficulty, Bohr has assumed that the radiation of energy from an atom can only take place in certain definite ways. By the introduction of a conception connected by Planck's quantum, he has suggested a possible structure of the simple atoms. The great difficulties involved in such an attempt can be appreciated when it is remembered that the complex light spectra of hydrogen must arise from the motion of one electron about a nucleus of unit charge. The deductions drawn by Bohr from his theory are of great interest and importance as the first attempt to define the structure of the actual atoms. There no doubt will be much difference of opinion as to the validity of the assumptions made by Bohr in his theory of the constitution of atoms and molecules, but a very promising beginning has been made on the attack of this most fundamental of problems, which lies at the basis of Physics and Chemistry.

Manchester University

The Spectrum of the Penetrating γ Rays from Radium B and Radium C

by SIR ERNEST RUTHERFORD, F.R.S.,

and E. N. DA C. ANDRADE, B.SC., PH.D.,

John Harling Fellow, University of Manchester

From the *Philosophical Magazine* for August 1914, ser. 6, xxviii, pp. 263–73

IN a previous paper,* we have given the results of an examination of the wave-lengths of the soft γ rays from radium B, for angles of reflexion from rock-salt between 8° and 16°. It was shown that the two strong lines at 10° and 12° correspond to the two characteristic lines always present in the spectra of the 'L' series for heavy elements. It was deduced from the experiments of Moseley, that the spectrum of radium B corresponded to an element of atomic number or nucleus charge 82. Direct evidence was obtained that the strong lines of the γ ray spectrum of radium B were identical with the corresponding lines in the X-ray spectrum of lead—thus confirming the hypothesis that radium B and lead have in general identical physical and chemical properties, although their atomic weights differ probably by seven units.

In the present paper an account is given of further experiments to determine the γ-ray spectra of the very penetrating rays from radium B and radium C. The strong lines from radium B, which are reflected from rock-salt at angles of 10° and 12°, undoubtedly supply the greater part of the soft radiation for which $\mu = 40$ (cm.)$^{-1}$ in aluminium. There still remained the analysis of the frequency of the lines included in the penetrating radiations from radium B, for which $\mu = 0 \cdot 5$, and from radium C, for which $\mu = 0 \cdot 115$. It may be mentioned at once that there is undoubted evidence that a large part, if not all, of these penetrating radiations give definite line spectra and correspond to groups of rays of very high frequency; but it has been a difficult task to determine the wave-lengths of the lines with the accuracy desired. We have been much aided by the development of a new method for finding the wave-length, which depends on the measurement of absorption as well as of reflexion lines.

In our first experiments the same general method was employed as in the previous work. A fine glass tube containing about 100 millicuries of emanation was used as a source. The distances between the source and crystal and between the crystal and the photographic plate were equal, and, as in the

---

* Phil. Mag. May 1914, p. 854. (*This vol., p. 432.*)

previous experiments, about 9 cm. A beam of γ rays passing through a narrow opening in a lead block fell on the crystal, the arrangement being that shown in fig. 1 of our previous paper. The width of the direct photographic impression on the plate was in general about 3 mm. It was important in these experiments with penetrating γ rays to use a thin crystal, since the rays pass right through the crystal and the exact plane of reflexion is consequently uncertain. The crystal employed was a slip of rocksalt about 3 cm. long, 2 cm. wide, and about 1 mm. thick. For a radiation which is reflected at about 1° from the (100) planes of rocksalt, the diffraction line is displaced on the photographic plate little more than 3 mm. from the centre of the dark band. The fact that some lines may be reflected from near the front face of the crystal, and others from further back, introduces a possible source of error in the determination of the correct angle of reflexion. This difficulty could have been avoided if the lines could have been measured in the second order as well as in the first, but the second order lines were too faint to pick out with certainty in the presence of a number of other faint lines. A large number of photographs were taken and measured up, and the mean of the deflexions of the strongest and easily observed lines should not be much in error.

The crystal was kept in slow rotation by the method described in the previous paper. This prevented the appearance of apparent lines due to crystal imperfections. The 'centre' of the crystal was in most cases obtained by observing, on a special photograph taken for the purpose, the position of the strong 10° and 12° lines. This was checked also by obtaining the same line on both sides of the zero by rotating the crystal through about 180°, as described in the previous paper.

Table 1 (next page) gives the results of these experiments. The wave-lengths have been deduced from the data given in the previous paper.

Of these lines, those marked A and B showed up strongly on the photographic plate, A being the more intense. The other lines were comparatively faint. In addition to those recorded, which have been observed on several plates, a number of very faint lines have also been observed, but it has not been thought desirable to include them.

It is of importance to decide which of these lines belong to radium B and which to radium C. To bring out this, photographs taken with a screen of 6 mm. of lead between the source and the crystal were compared with similar photographs either with no absorbing screen, or with a screen of aluminium 2 mm. thick between the source and crystal. The latter absorbed the soft γ rays from radium B but did not stop more than a small percentage of the penetrating rays. The lead plate, on the other hand, practically cuts out the greater part of the radiations from radium B, and no doubt also some of the softer possible constituents of radium C,* but does not stop half of the very penetrating rays from radium C. With the lead plate the strong line A disappeared, but the group of lines reflected in the neighbourhood of B still remained. We may consequently conclude that the radiations with reflexion

---

* See Rutherford and Richardson, Phil. Mag. May 1913, p. 722. (*This vol., p. 342.*)

P*

angles greater than 1° 24′ belong mainly to radium B, and the smaller reflexion angles mainly to the penetrating rays from radium C.

TABLE I

Penetrating γ rays from radium B and radium C

	Angle of reflexion from rocksalt		Wave-length in cms. $\times 10^{-9}$
		44′	0·72
B	1°	0′	0·99
	1°	11′	1·16
	1°	24′	1·37
A	$\left\{\begin{array}{l}1° \quad 37′\\1° \quad 43′\end{array}\right\}$ *		1·59
			1·69
	2°	0′	1·96
	2°	28′	2·42
	2°	40′	2·62
	3°	0′	2·96
	3°	18′†	3·24
	4°	0′†	3·93
	4°	22′	4·28

Transmission method of determining wave-lengths

While the reflexion method described above left no doubt as to the approximate wave-lengths of the penetrating radiations, the uncertainty in regard to the plane or point of reflexion of each type of radiation of the crystal might lead to a possible error of some minutes in the angle of reflexion of a line. This is a considerable percentage error for an angle of reflexion of 1° or 2°.

A number of experiments were made to test whether a more reliable method could be devised for determining the wave-lengths of these very penetrating rays.

For this purpose, the α-ray tube R was placed behind a rocksalt crystal C, and a photographic plate PP placed at D, the whole apparatus being placed between the poles of the electromagnet to get rid of the effect of β rays.

Suppose, for simplicity, that the source R emits a radiation of one definite frequency which would be reflected at an angle θ with the surface (cleavage plane) of the crystal. If RD is the normal to one face (and hence parallel to another face of the cubic crystal), the ray RA making an angle θ with the normal is in position to be strongly reflected in passing through the cubic

---

* The lines at 1° 37′ and 1° 43′ may possibly be one broad line, having a mean reflexion angle of 1° 40′. The evidence, however, seems to be in favour of the double nature of the lines, which was by far the strongest line on most photographs. We were for some time in doubt as to the existence of a line reflected at 44′, but its presence has been confirmed in several reflexion and transmission (see p. 459 et seq.) photographs.

† Possibly second order.

crystal; similarly also the corresponding ray RA'. The reflected rays from each point of EF cut the normal in the neighbourhood of O. A photographic plate placed at O normal to RD should thus show a narrow dark band on the general background, while if placed at D it should show dark bands at B and B'. If a considerable part of the rays in the direction RA, RA' are reflected, the lines A and A' should be positions of minimum intensity of radiation, and thus appear as *white* lines on the general dark background. In other words, reflexion lines appear at B and B' and absorption lines at A and A'.*

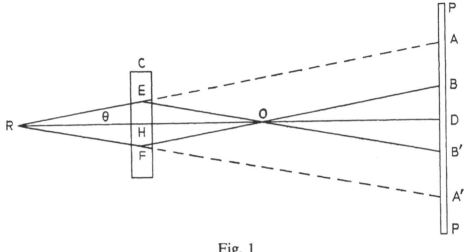

Fig. 1

These conclusions have been completely confirmed by experiment. When the photographic plate is placed at O, a central dark band appears with two absorption lines symmetrically placed on either side of it. An actual untouched photograph of this kind is shown in fig. 3 (Plate), and brings out the main points quite clearly.†

If an emanation-tube, with the emanation compressed into a small length of the tube, is placed normal to the face of the cubic crystal, it corresponds nearly to a point source. Under such conditions, the photographs obtained show a beautifully symmetrical pattern. The two sets of strong absorption bands cut one another at right angles, while the crystal planes at 45° also give bands cutting the main system of bands at an angle of 45° and passing through their points of intersection. The 45° bands are relatively less marked than the bands from the (100) planes. The results obtained are in exact agreement with those to be expected from the geometry of a cubic crystal. Reproductions of two photographs of this kind made with different crystals are shown in figs.

* Professor W. H. Bragg has found evidence of a similar absorption in passing X rays through a thin crystal of diamond at the reflexion angle, using the electric method to detect the rays. See 'Nature,' March 12th, 1914, p. 31.

† The three reproductions (on the Plate) are of *prints* from the original negatives: in consequence the reflexion lines show white, and the absorption lines black.

1 and 2, Plate, fig. 1 showing the 45° lines well. The nearly monochromatic radiation producing the pattern is the 'A' line, or doublet; the lines due to the other radiations are too faint to show in the reproduction.

Considering the fact that no attempt was made to cut off the effect of the general γ radiation on the photographic plate, the sharpness with which the bands show up is very remarkable. Results of this kind bring out clearly the large amount of the radiation that is 'scattered' or 'absorbed' when passing through the crystal at the proper reflexion angle. The distance A, A' between the absorption bands can be easily measured, and the angle of reflexion is given by $\tan^{-1} \dfrac{\frac{1}{2}AA'}{RD}$ and can be determined with accuracy.

The dark lines are fairly sharp, as might be expected for such small angles of reflexion (1° to 2°), even if the rays are reflected throughout the thickness of the crystal. To estimate the angle from them, the rays forming them must be considered as proceeding from the point O, whose distance from the photographic plate was obtained by subtracting twice RH, the mean distance of the emanation-tube from the crystal, from RD, the distance of the tube from the plate. When this was done, the angles of reflexion as given by the narrow black bands, or lines, agreed excellently with those given by the white absorption lines.

It is seen that the precision of these photographs depends upon the theoretical perfection of the crystal planes. If, for example, the planes of the crystal happened to be distorted into a curved shape, the distance apart of the absorption lines would be different if the crystal were reversed so as to have the other side towards the source of the rays. With some crystals the lines were slightly bent, but with those selected for the final work the lines were straight, and no measurable differences in the distances apart of the lines were observed when the crystal was reversed in the manner indicated.

In the actual experiments, it was found that the A line gave a very marked absorption band for a thickness of crystal of 8 mm. and also a visible band in the second order. The first order of this band was observed on a crystal only two millimetres thick. It is obvious, however, that if the spectrum of the rays is complicated, a dark line may in some cases fall on an absorption band and mask it. For this reason, it is sometimes desirable to take photographs at O (fig. 1) where there is no chance of confusion due to these causes.

In order to overcome this difficulty, recourse was also had to another method involving the measurements of the dark bands arising from reflexion. The general arrangement of apparatus is clearly seen in fig. 2. The emanation-tube was placed close to the crystal, and a lead screen LL, with a narrow slit to allow the passage of the reflected rays, was placed at the focus O. The photographic plate was placed some distance away at D. A dark central band was observed in the photograph produced by the direct radiation through the slit, and the dark lines due to the reflected radiations appeared symmetrically on either side. The experiments were not complicated by the presence of absorption bands, which were cut out by the lead screen. Since the dark bands appeared on the photographic plate at points outside the central beam, they were relatively more intense than in the experiments where no lead screen

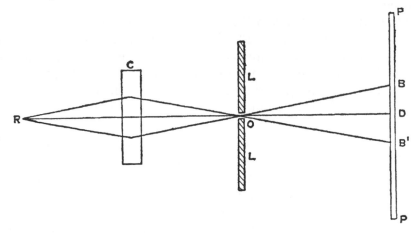

Fig. 2

was used to cut down the main radiation. In this way, we were able to determine the wave-length of the stronger lines due to the very penetrating rays.*

TABLE II

Penetrating γ rays from radium B and radium C

Transmission method		Reflexion method	Wave-length (mean) × 10^{-9} cms.
Absorption	Reflexion		
42′	43′	44′	0·71
1° 0′	1° 0′	1° 0′	0·99
1° 10′	1° 10′	1° 11′	1·15
1° 24′	1° 25′	1° 24′	1·37
⌠ 1° 37′†	⌠ 1° 36′	⌠ 1° 37′	⌠ 1·59
⌡ 1° 44′	⌡ 1° 44′	⌡ 1° 44′	⌡ 1·69
	2° 0′	2° 0′	1·96
2° 20′	2° 20′		2·29
		2° 28′	2·42
		2° 40′	2·62
	3° 0′	3° 0′	2·96
3° 18′‡		3° 18′‡	3·24
		4° 0′‡	3·93
		4° 22′	4·28

* At each edge of the dark band corresponding to the beam directly transmitted through the lead slit appear close lines not due to beams diffracted by the crystal, since they appear equally whether the crystal is there or not. These 'false' lines render this method unsuitable for investigating the lines reflected at the smallest angles. Their cause is uncertain, and is now being investigated by one of us.

† See note to Table I.   ‡ Possibly second order.

This method has obvious advantages in determining wave-lengths where only a small fraction of the total intensity of the radiation is due to the waves under examination.

In Table II (p. 461) are given the wave-lengths of the stronger lines determined by the transmission method. Column 1 gives the measurements made by means of absorption bands, and column 2 by means of the reflexion bands. The results obtained by the two methods are seen to be in very fair agreement. In column 3 the lines determined by the first reflexion method are given again. The wave-lengths given in column 4 correspond to the mean value of the reflexion angle.

It will be seen that there is also a very good agreement between the values obtained by the direct reflexion and by the transmission method, but for the very penetrating rays under examination, the results obtained by the transmission method were more definite and reliable, while the exposures required for the photographs were relatively much less.

On p. 457 we gave evidence that the lines reflected in the neighbourhood of 1° belong mainly to radium C, and those at greater angles mainly to radium B. In examining the transmission photographs, strong evidence was obtained that the absorption band at 1° 40′ must consist of a close double. In experiments with some crystals of rock-salt, this was more marked than with others. The absorption line was considerably wider than would be expected from the dimensions of the source if only a single frequency were involved. This observation has been indirectly confirmed by some experiments now being carried out in the Manchester Laboratory by Mr H. Richardson, M.Sc.

In examining the absorption by lead of the γ rays from radium B and radium C separately, he has found evidence that each of these substances emits two similar types of γ radiation, which are absorbed far more easily in lead than the very penetrating radiation from radium C. We should thus anticipate that the spectra of the soft radiation from radium C and the corresponding radiation from radium B should partly overlap. It is, however, impossible from the photographs to determine differences in frequency of possible close doubles at these small angles.

Some experiments were made to try to obtain transmission photographs with pure radium C deposited on a nickel wire as a source. The photographic effect, however, was far too faint for measurement. It seems not unlikely that the radiations in the neighbourhood of 1° 10′, 1° 24′, and 1° 40′, are to be ascribed to both radium B and radium C, although the actual frequencies of the corresponding lines may be distinct in the two cases.

Discussion of Spectra

It will be seen that the wave-lengths of the penetrating γ rays from radium B and radium C are much shorter than any previously determined. Moseley has determined the 'K' spectra of silver and found the wave-length of the strong line $0 \cdot 56 \times 10^{-8}$ cm. The wave-length of the most penetrating γ ray observed is $0 \cdot 7 \times 10^{-9}$, or eight times shorter. When the great penetrating power of the radiations from radium C—half absorbed in 6 cm. of aluminium—is considered, and the shortness of its wave-length, it is surprising that the architecture of the crystal is sufficiently definite to resolve such short waves. This is especially the case when we consider that owing to the heat agitation of the atoms, the distance between the atoms must be continually varying over a range comparable with the wave-length of the radiation. One photograph was taken with the crystal immersed in liquid air, but no obvious improvement in definition was observed.

The appearance of these high frequency vibrations from radium B and radium C is accompanied by the expulsion of very high speed β particles from the atom. It does not, however, follow that it will be necessary to bombard the material with such very high speed β rays to excite the corresponding radiation. If we may assume, as seems probable, that Planck's relation $E = h\nu$ holds for the energy of the β particle required to excite radiation of frequency ν, it can be deduced that the electron to excite this radiation in radium C must fall freely through a difference of potential of 180,000 volts, which is equivalent to a velocity of about $0 \cdot 7$ that of light. This is much smaller than the velocity of the swift β particles from radium B or C, and is not beyond the range of possible experiment. With the tube recently designed by Coolidge* there should be no inherent difficulty in exciting the corresponding radiation in a heavy element like platinum or uranium.

We have seen that the soft γ rays defined by the absorption coefficient $\mu = 40$ in aluminium correspond to the 'L' series of characteristic radiations for an element of atomic number 82. Moseley has examined the spectra of the K series for elements from aluminium to silver and finds them all similar, consisting of two well-marked lines differing in frequency by about 11 per cent. The frequency of the more intense line (α) is approximately proportional to $(N - 1)^2$ where N is the atomic number of the element. Supposing this relation to hold for all the elements of higher atomic weights, the angle of reflexion for the strong line of the K series for an element of number 82 (radium B) should be $1° 46'$. The observed value of the strong line is about $1° 40'$—a very fair agreement, considering the wide range of extrapolation.

We may consequently conclude that the penetrating γ rays from radium B, correspond to the characteristic radiation of the K series of this element. It had been previously supposed that the very penetrating rays from radium C belong to the K series of characteristic radiations for that substance, but if the relation found by Moseley holds even approximately for the heavy elements, this cannot be the case.

Radium C corresponds to an element of atomic number 83, and the frequency of its 'K' radiation should be only a few per cent. higher than that for radium B. Actually the average frequency of the main radiations from radium C is roughly twice that for the average frequency of the penetrating rays from radium B. We are thus driven to conclude that in the case of radium

* W. D. Coolidge, 'Physical Review,' Dec. 1913, p. 409.

C, and probably also thorium D, which emits an even more penetrating γ radiation than radium C, another type of characteristic radiation is emitted which is of higher mean frequency than for the 'K' series. In other words, it is possible, at any rate in heavy elements, to obtain a line spectrum which is of still higher frequency than the 'K' type. This may for convenience be named the 'H' series, for no doubt evidence of a similar radiation will be found in other elements when bombarded by high speed cathode rays.

Connexion of Absorption with Frequency

Owen has pointed out that the relation between the absorption coefficient μ in aluminium of a characteristic radiation and the atomic weight A can be expressed approximately by $A = k\mu^{-1/5}$ where k is a constant. Taking the numbers given by Moseley for the frequencies of corresponding lines of the 'K' series for the elements, it can be simply shown that the frequency v is connected with μ by the formula $v = k_1\mu^{-2/5}$ where k_1 is a constant.

It is of interest to examine whether this formula holds approximately for the penetrating radiations from radium B and radium C. The soft γ rays for which $\mu = 40$ give rise to two strong lines at $10°$ and $12°$ or a mean reflexion angle of $11°$. It follows from the above formula, that the strong line or lines due to the penetrating radiations from radium B, for which $\mu = 0 \cdot 51$ in aluminium, should give a mean reflexion angle of $1° 54'$. The actual reflexion angle observed for the strongest line is $1° 40'$ about—a fair agreement.

Similarly, the penetrating rays from radium C, for which $\mu = 0 \cdot 115$ in aluminium, should give an average reflexion angle of $63'$. This is in good accord with the average $58'$ of the reflexion angles of the stronger lines, viz. $43'$, $60'$, $70'$. The results thus show that this simple method of calculating frequencies from absorption coefficients holds approximately over a very wide range. It is possible, for example, by this method to calculate approximately the mean position of the strongest line or group of strong lines of the various types of γ rays from radioactive substances whose absorption coefficients have been given by Rutherford and Richardson.

Hardening of γ Rays

It has long been known that the γ radiation from radium C becomes more penetrating for aluminium after it has passed through a centimetre or more of lead. This hardening effect has been recently examined in detail by Oba,* who has shown that after passing through a considerable thickness of lead the radiations become permanently more penetrating for aluminium. He has explained these results on the supposition that the spectrum of radium C is complex. We have seen that the main reflexion lines from radium C are at $44'$, $1° 0'$, and $1° 10'$. It is probable that the radiations of frequencies corresponding to the two latter are absorbed more rapidly than that corresponding

* S. Oba, Phil. Mag. April 1914, p. 601.

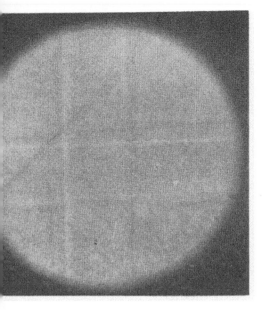

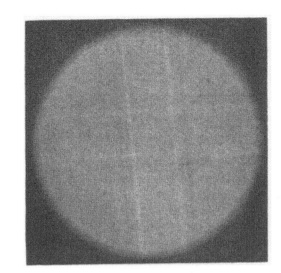

Fig. 1   Fig. 2

Fig. 3

PLATE

Facing page 464

to the former. On such a view it would be anticipated that the penetrating power of the γ radiations from radium C would increase until the softer rays are all absorbed, and then remain constant, corresponding to a homogeneous radiation reflected at 44′ from rock-salt.

Our best thanks are due to Dr. E. S. Kitchen for a large number of crystals of rock-salt.

Spectrum of the β Rays excited by γ Rays

by SIR ERNEST RUTHERFORD, F.R.S., H. ROBINSON, M.SC.,
and W. F. RAWLINSON, B.SC., *University of Manchester*

From the *Philosophical Magazine* for August 1914, ser. 6, xxviii, pp. 281–6

IT is well known that the primary β radiation emitted by the great majority of radioactive substances contains a number of groups of rays expelled with definite velocities. When the β rays are analysed by their passage through a magnetic field, and received on a photographic plate, they give a veritable spectrum, consisting of a number of well-marked bands. In a previous paper,* Rutherford and Robinson have analysed in some detail the magnetic spectra of the β rays of both radium B and radium C, and have measured the velocities and energies of a large number of groups from each element.

It is also well known that the γ rays of radium B and C in passing through matter give rise to high-speed β rays. The penetrating power of these rays has been examined by the electrical method, and has been shown to correspond approximately with that of the primary β rays from the same radioactive substance. It is of great importance to determine accurately the distribution of velocities of these excited β rays, and to trace their connexion, if any, with the primary β rays emitted by the radioactive substance.

For this purpose, we have made a number of experiments by a method similar to that described in the paper mentioned above. The diagram, fig. 1, is reproduced from that paper. The whole apparatus is placed in a uniform magnetic field between the pole-pieces of a large electromagnet. The β rays issuing from the radioactive source S describe circular orbits and after passing through a wide slit fall on the photographic plate PP. The great advantage of this method is that the β rays in a comparatively wide cone are concentrated into a narrow band on the photographic plate, thus making it possible to detect the presence of groups of β rays of feeble intensity and to deduce their velocities with accuracy. The source of γ radiation S was a fine-bore glass tube containing about 50 millicuries of radium emanation. The glass walls were sufficiently thick to absorb all the α rays and the low velocity primary β rays. This source was surrounded by a cylinder of the absorbing substance which in different experiments varied between 0·1 and 1 millimetre in thickness.

From the work of Baeyer and Danysz it is known that the velocities of the groups of β rays are reduced by their passage through a thin sheet of

---

* Rutherford and Robinson, Phil. Mag. xxvi. p. 717 (1913). (*This vol., p. 371.*)

matter, and that the β radiation at the same time ceases to be homogeneous. After passing through a small thickness of matter, the sharp bands initially observed on the photographic plate become broader and more diffuse, and with increasing thickness become undetectable against the general background of continuous radiation. Even the swiftest groups from radium C are undetectable after passing through the glass tube and an additional thickness of 0·14 mm. of lead or gold. All the low velocity β rays escaping under these conditions are those liberated by the passage of the γ rays through the absorbing screen. The magnetic spectrum of these excited β rays was examined in exactly the same manner as that of the primary β rays.

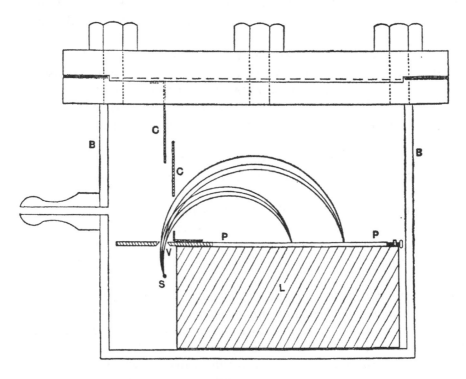

Fig. 1

A number of fairly broad bands were clearly visible on the photographic plates, above the general fog due to the penetrating γ radiation. The outside edges of the bands, corresponding to the swiftest rays of the groups, were sharply defined, but the bands gradually decreased in intensity on the low velocity side. The appearance of the bands is easily explained on general considerations. If the passage through matter of γ rays of definite frequency gives rise to a β particle of definite speed, then the β rays excited at the outer surface of the absorbing screen should escape with no loss of velocity, but those generated some distance below decrease in velocity before they are able to escape. Consequently we should expect that the bands would show a well-marked outer edge corresponding to the swiftest rays, and a gradual fading of intensity on the inner side due to the effect of particles with lower velocities. In this respect the spectrum of the β rays excited by γ rays differs from that of the primary rays, where each group consists of rays of the same speed, but is very similar to the spectrum observed when the primary β rays have passed through an absorbing screen.

With a lead-absorbing screen, the most prominent groups of excited β rays were observed in that part of the spectrum which corresponded to the strong groups of primary rays from radium B, viz. those having velocities between $0 \cdot 6$ and $0 \cdot 8$ of the velocity of light. The photographs were practically identical in appearance, whether the tube was covered with $0 \cdot 1$ or 1 mm. of lead, showing that a thickness of $0 \cdot 1$ mm. was sufficient to cut out entirely the effect of the primary β rays. When the tube was covered with thin lead, evidence was obtained of groups, of very feeble intensity, corresponding approximately in position with the strong groups of primary rays from radium B marked E and F in the list of lines given by Rutherford and Robinson. These are rays having velocities approximately $0 \cdot 4$ of that of light.

The values of the velocities of the excited β rays were determined by direct comparison of the positions of the groups with those of the stronger lines due to the primary rays from radium B. For this purpose, an α-ray tube containing about 10 millicuries of radium emanation was put in the place of the first tube, and the primary β-ray spectrum obtained with the same magnetic fields. Small corrections were necessary for the reduction in velocity of the primary β rays in passing through the walls of the α-ray tube, and also for the increase in dimensions of the radiating source when surrounded by a metal cylinder. With thicknesses of lead of about 1 mm., the latter correction is difficult to make with certainty, for account has to be taken of the fact that the excited β rays are mainly projected in the direction of the exciting γ rays, and are thus most concentrated in the direction normal to the absorbing cylinder.

The experiments with lead as the absorbing screen are of special interest, for it has been shown by Rutherford and Andrade in a previous paper that the spectrum of the soft γ rays from radium B corresponds to that of the 'L' characteristic radiation of lead. It would thus appear very probable that radium B and lead have the same nucleus charge and the same external distribution of electrons although their atomic weights are different. If the primary β rays from radium B result from the conversion of γ rays into β rays in their escape from the radioactive atom, it is to be anticipated that the spectrum of the β rays excited in lead should be identical with that given by the primary β rays. It was obvious on inspection of the plates that the stronger groups of excited β rays corresponded very nearly in position with the primary β-ray spectrum of radium B. The values of Hρ for the stronger groups of β rays excited in lead are given in the following table; for the purpose of comparison, we have added a list of some of the primary β rays of radium B, taken from the paper of Rutherford and Robinson. It will be seen that within the limits of experimental error, the velocities of the rays are the same in each case.

Table of β rays excited in Lead

Hρ of excited β ray	Intensity	Hρ of primary β ray	Intensity
3610	f.		
3250	f.		
2990	s.		
2735	m.f.		
2225	m.	2235	s.
2130	f.	2140	f.
2000	f.	1990	m.s.
1935	v.s.	1925	v.s.
1825	m.f.	1815	m.
1750	m.f.	1752	m.s.
1670	v.s.	1660	v.s.
1560	f.		
1400	s.	1392	v.s.
1240	m.		
1150	m.f.		
1080	m.		
1010	m.s.		
950	s.	950–914 (group)	m.
870	v.s.		
800	s.	798	m.s.

[v.s. = very strong; s. = strong; m.s. = moderately strong; m. = moderate intensity; m.f. = medium faint; f. = faint.]

Dependence of intensity and velocity of the excited β rays upon the material

It is of great interest to determine whether the velocity of the excited β rays depends on the material through which the γ rays pass. For this purpose, comparative measurements were made with screens of aluminium, silver, gold, and lead surrounding the emanation tube. With the screen of aluminium 0·6 mm. thick, it was impossible to detect any groups of β rays against the general fog of the plates: with silver the lines were faint and difficult to measure, but the lines from lead and gold were well marked and of about equal intensity. It appears certain that the groups of β rays observed with the two latter substances are due to the conversion into β rays of some of the more penetrating γ rays from radium B. It is well known that the greater part of the γ radiation from this substance is rapidly absorbed in heavy elements like lead and gold, and no doubt the greater intensity of the groups of β rays from these substances is a result of the rapid absorption of the γ rays and their conversion into β rays, since the absorption of γ rays of this type increases with atomic weight more rapidly than the absorption of β rays. In consequence, the β rays are more intense from heavy than from light elements.

The lines obtained with a silver screen were compared with those obtained with a lead screen for the same field, and no certain difference in the velocity of the three strong groups of rays could be detected.

On the other hand, the velocities of a number of the β rays excited in gold were distinctly higher than the corresponding velocities of the β rays excited in lead, the difference being of the order of two per cent. In view of the method of comparison it seems scarcely likely that the difference is due to experimental error.

Experiments on this subject will be continued by Robinson and Rawlinson. It is intended to continue the investigation into the region of the groups of swift β rays of radium C and to examine the velocities of the β rays excited in a number of different metals. Preliminary photographs have shown that the β rays excited by the very penetrating γ rays from radium C are difficult to observe, as would be anticipated from the considerations already advanced.

University of Manchester
June 29, 1914

The Structure of Atoms and Molecules*

(Abstract of a Discussion at The British Association, August 18, 1914)

Sir ERNEST RUTHERFORD (abstract of remarks): In recent times there has been an accumulation of convincing evidence of the independent existence of the chemical atom. The atomic theory is no longer merely an hypothesis introduced to explain the laws of chemical combination; we are able to detect and count the individual atoms. We can determine the actual mass of an atom in various ways, and know its value with considerable accuracy. The idea that the atom is an electrical structure received a great impetus by the detection of the electron by J. J. Thomson; and, moreover, the Zeeman effect showed that all atoms must contain electrons. The atomic character of negative electricity is well established; we always find the negative electron, however produced, carrying a definite charge. We have, unfortunately, not the same certainty with regard to the behaviour of positive electricity, for it cannot be obtained except associated with a mass comparable with that of a hydrogen atom. In J. J. Thomson's model of the atom the positive electricity was supposed (for mathematical reasons) to be distributed throughout a large sphere with the negative corpuscles moving inside it. This hypothesis has played a useful part in indicating possible lines of advance; but it does not fit in with more recent discoveries, which point to a concentrated positive nucleus.

We have now two powerful methods that aid us in determining the inner structure of the atom—the scattering of high-speed particles in transit through matter, and the vibrations of the interior parts of the atom. In C. T. R. Wilson's photographs of the tracks of the α particles through a gas we notice many sudden bends in the paths. In order to account for these deflections I have found it necessary to believe that there is a concentrated nucleus in the atom (having a certain number of units of charge), in which the main part of the mass resides; outside this there are a corresponding number of electrons. The whole dimensions of the nucleus are very small indeed compared with the distance of the outer electrons. From the scattering experiments it appears that the law of force right up to the nucleus is the inverse square law; no other formula would give accordance with the observations. The radius of the nucleus is of the order 10^{-12} cm. in the case of gold, and for a lighter element it is smaller still. The approach of the α particle to the nucleus of the hydrogen atom when the latter is set into very swift motion is exceedingly close—a distance even less than the diameter of an electron. From this it is probable that the hydrogen nucleus is simply the positive electron with a large electrical

* The report and discussion will be published in full in Vol. IV [Ed.].

mass due to the great concentration of the positive charge. Another fact that appears from the scattering experiments is that the number of electrons (outside the nucleus) is about half the atomic weight. There is now fairly good evidence that, if the elements are numbered in order of atomic weight, the numbers will actually express the charge on the nucleus. The rate of vibration of the inner parts of the nucleus can now be measured by means of the characteristic X-rays emitted. Each substance has several strong lines in its X-ray spectrum, and as we pass from element to element in order of atomic weight the frequencies of these change by regular jumps. H. G. J. Moseley has investigated all the known elements in this way, and he is even able to show at what points elements are missing, because at such points the X-ray frequencies make a double jump. In this way he has found that between aluminium and gold only three elements are now missing. It is deduced from these considerations that there is something more fundamental in the atom than its atomic weight, viz., the charge on the nucleus, and that this is the main factor which controls the frequency of the interior vibrations, the mass having only a slight influence.

There are certain elements with identical chemical properties, but different atomic weights. Thus Radium-B (atomic weight 214) and lead (207) are chemically inseparable and have the same γ-ray spectrum. It is quite clear that some new conception is required to explain how the atoms, having the structure we have supposed, can hold together. N. Bohr has faced the difficulty by bringing in the idea of the quantum in a novel way. At all events, there is something going on in the atom which is inexplicable by the older mechanics.

Sir E. RUTHERFORD (replying to the discussion which followed) said that the chemical inseparability of certain isotopes was, indeed, derived from experiments with small quantities, but the methods used were very delicate. The separation of Radium D from lead was a most important problem; there seems evidence that different leads exist, having different atomic weights. The difficulty of stability is common to all theories of the atom; but what it points to is that there is something wrong with the theory of electromagnetic radiation—not of the atom.

The Connexion between the β and γ Ray Spectra

by SIR ERNEST RUTHERFORD, F.R.S.,
Professor of Physics, University of Manchester

From the *Philosophical Magazine* for September 1914, ser. 6, xxviii, pp. 305–19

THE problem of the nature of emission and absorption of radiation has occupied a very prominent position in modern Physics, both on account of its outstanding importance and of the great difficulties involved. It is clear that the question of the excitation of X rays and their conversion into β rays, and also the spontaneous emission of β and γ rays from radioactive substances, must be included in any general theory of radiation. A study of the β and γ rays from radioactive matter is of especial interest in this connexion, since the β rays are expelled with a very high velocity and a considerable fraction of the energy is emitted in the form of γ rays of very short wave-length. It is to be anticipated that a close study of the emission of these radiations from radioactive bodies should throw light of a fundamental character on the radiation problem on the high frequency side.

During the last few years a number of careful investigations have been made in this Laboratory bearing on this problem, and it may prove of interest to discuss briefly the evidence that has so far been obtained and to indicate the general conclusions that can be drawn from it. The problem is much too large and involved to hope for an immediate and definite solution, but the experimental results are sufficiently complete to afford some data for drawing some tentative conclusions.

In a paper published two years ago* I discussed the possible connexion between β and γ rays emitted from radioactive substances and outlined a general theory in explanation of the magnetic 'spectrum' observed when the β rays are analysed by their passage through a magnetic field. It was pointed out that the emission of a large number of groups of the β rays of definite velocity from a single substance could be most simply explained by supposing that it is a statistical effect due to a large number of atoms each of which gave rise to a few only of the groups of β rays observed.

In a transformation where primary β and γ rays appear, it was supposed that each atom broke up with the emission of a β particle of definite speed. The latter in passing through the external electronic system set it into vibration, and energy was abstracted from the β particle in definite *integral* units depending on the vibrating system. If, for example, the β particle passed

* Phil. Mag. xxiv. p. 453 (1912). (*This vol., p. 280.*)

through two distinct vibrating systems A_1 and A_2, the final energy of the escaping β particle was given by $E_0 - (pE_1 + qE_2)$,* where E_0 was the initial energy of the β particle, p and q whole numbers which might have any values 0, 1, 2, 3, etc., and E_1, E_2 the units of energy abstracted in passing through A_1 and A_2 respectively. It was supposed that the energy $pE_1 + qE_2$ which was abstracted from the β particle appeared in the form of p gamma rays each of energy E_1, and of q gamma rays each of energy E_2. It was suggested that the γ rays so excited corresponded to one or more of the types of characteristic radiations brought to light by the experiments of Prof. Barkla on X rays. This theory has formed a starting point for a number of subsequent researches. In the first place, in order to test the theory, it was necessary to know the energy of the β particles comprising the different groups with the greatest possible accuracy. The initial experiments made by Baeyer, Hahn and Meitner, and by Danysz on the groups of β rays emitted from radium B and radium C were repeated with great care by Mr. H. Robinson and myself.† By the adoption of a modified method, the magnetic spectra due to radium B and radium C were separately determined. The spectrum of radium C was greatly extended and found to consist of a great number of lines, about 50 of which were measured. It was pointed out that there appeared to be certain simple numerical relations between a number of the groups of β rays from radium C. In the meantime, the problem had been attacked from another direction. According to the theory, the γ rays emitted from a radioactive substance should consist of types of characteristic X radiations which should be exponentially absorbed by a light substance like aluminium. This question has been examined in detail by Mr. H. Richardson and myself,‡ and the results obtained have fully confirmed this point of view. The γ rays from each radioactive substance can be analysed into a number of distinct groups. Some of these groups undoubtedly correspond to the characteristic radiations to be expected from elements of their atomic weight; but attention was drawn to the evidence of the existence of other types of characteristic radiation not previously observed by workers with X rays. It was found that the different radioactive substances showed great variety in the types of γ rays emitted, but they could be classified by their power of penetration as belonging to certain general types of characteristic radiations. Mr. H. Richardson has continued these investigations and has recently obtained evidence of the excitation of characteristic radiations in a large number of elements when the β rays of active matter fall upon them.

The discovery of Laue of the diffraction of X rays and the subsequent work of W. H. and W. L. Bragg and of Moseley and Darwin and others, have placed into our hands a powerful and simple method for determining the

---

* The particular point of view of which this formula is an expression has been modified subsequently.

† Rutherford & Robinson, Phil. Mag. xxvi. p. 717 (1913). (*This vol., p. 371.*)

‡ Rutherford & Richardson, Phil. Mag. xxv. p. 722; xxvi. p. 324; xxvi. p. 937 (1913). Richardson, Phil. Mag. xxvii. p. 252 (1914). (*This vol., pp. 342, 353, 410.*)

wave-lengths of the X rays. If the γ rays from radioactive matter consisted of groups of characteristic rays, it was to be anticipated that the rays would show a line spectrum when reflected from a crystal surface. This point of view has been completely confirmed by subsequent researches of Dr. Andrade and myself. In the first paper* we gave an account of the examination of the spectrum of the soft γ rays from radium B, and adduced evidence that the strong lines of the spectrum of this substance were identical with the characteristic 'L' spectrum of lead. In a subsequent research† we have determined the wave-lengths of the penetrating γ rays from radium B and radium C and verified the results by the adoption of a new experimental method.

Distribution of energy between β and γ rays

It was initially supposed that a large fraction of the β radiation from substances like radium B and radium C, which give a marked β-ray spectrum, appeared in the form of the homogeneous groups of β rays observed. J. Chadwick‡ has shown, however, in a recent paper that even the intense lines in the magnetic spectrum of radium B represent only a small fraction of the total number of β rays emitted. This result was obtained by direct counting of the β particles and confirmed by showing that an increase of intensity of only a few per cent. of the β radiation falling at a given part of the photographic plate gave the impression on development of a strongly marked band.

We may conclude from these results that the magnetic spectrum of the β rays from radium B or radium C consists of a *continuous* spectrum of β rays on which is superimposed a *line* spectrum corresponding to groups of rays expelled at definite speeds. A satisfactory explanation of these results and also of other marked differences in the distribution of β and γ rays from radioactive substances can, I think, be given on the following lines. Suppose—as seems probable—that the disintegration of the atom leads to an expulsion of a high speed β particle from or near the nucleus. This β particle in passing through the outer distribution of electrons will, on the average, suffer several collisions of an ordinary type with the electrons, and will share its energy with them. As a statistical result of a large number of atoms, the velocity of the escaping β particles will, on the average, be continuously distributed within certain limits of velocity. This would give rise to the continuous spectrum of β rays which is most typically illustrated by the β rays from radium E. Next suppose that there are certain well-defined regions in the electronic distribution which can be set into definite vibration by the escaping β particle. These regions are to be identified as containing the particular structures which give rise to the 'characteristic' γ radiations from the atom. If some of the β particles in escaping from the atom pass through one or more of these regions,

---

* Rutherford and Andrade, Phil. Mag. xxvii. p. 854 (1914). (*This vol., p. 432.*)

† Rutherford and Andrade, Phil. Mag. August 1914. (*This vol., p. 456.*)

‡ J. Chadwick, *Ber. d. D. Phys. Ges.* xvi. p. 383 (1914).

they give rise to a line-spectrum of γ rays and at the same time to one or more groups of β rays of definite speed. The connexion between the energy of the γ ray and of the β ray will be discussed later.

On this view, the appearance of homogeneous groups of β rays and the line-spectrum of γ rays are to be ascribed to certain definite regions of vibration within the atom. It is to be anticipated that characteristic γ rays will always accompany a line-spectrum of β rays. This seems to be in harmony with radioactive data. Radium E, which gives rise to a *continuous* β-ray spectrum, emits exceedingly little γ radiation in comparison with typical β and γ ray products like radium B and radium C, which give well marked spectra for both β and γ rays.

While it would appear probable that the greater part of the γ radiation from radium B and radium C is composed of several groups of rays of definite frequencies, no doubt a small part of the γ radiation gives a continuous spectrum. Such a result is to be anticipated from analogy with X rays. This general radiation probably has its origin in the electronic collisions of an ordinary type when the β particle is escaping from the atom or to the passage of the β particle close to the nucleus.

Importance of direction of escape of a β particle

There is another very interesting point that arises in consideration of this question. Why does radium E, which emits β rays of great intensity and over a wide range of velocity, not emit γ rays at all, or at any rate in very small amount compared with radium B or radium C? There appears to be no reason to suppose that radium E would not give rise to characteristic radiations of a frequency corresponding to its atomic weight or atomic number when bombarded by cathode rays of suitable speed. In order to explain this anomaly, it appears necessary to assume that the primary β particle from a given radio-element is always expelled in a *fixed* position with regard to the structure of the atom itself. Considering the remarkably definite way in which the atom of the same substance disintegrates, this assumption does not seem improbable. On this view, the absence of γ rays from radium E is due to the fact that the direction of escape of the β particle does not pass near or through the definite regions where characteristic radiations are set up, and in consequence only a continuous spectrum of β rays is observed. An explanation may be given on similar lines of many remarkable anomalies in the types and relative intensities of γ rays emitted from radioactive substances. This is well illustrated by a comparison of the γ rays from radium B and radium C which are nearly of the same atomic weight. Radium B emits a very soft radiation which is almost entirely absent in radium C, while radium B does not emit the very penetrating radiation observed from radium C. We must suppose that the β particle from radium B passes through one or more distinct regions which give rise to the corresponding characteristic radiations, while the β particle from radium C escapes in such a direction that it does not pass through the corresponding regions but does pass through a new region not involved in the case of the expulsion of the β particle from radium B.

The general evidence available indicates that radium B and radium D have identical general physical and chemical properties with those of lead, although differing from the latter in atomic weight. If this be correct, we should anticipate that these elements should give identical X ray spectra when bombarded by cathode rays. On the other hand, from the work of Rutherford and Richardson, radium B and radium D are known to emit types of γ rays which are widely different in relative amount and penetrating power. Such results are, however, at once intelligible if it be supposed that the β particles from these two elements are expelled in different directions with regard to the atomic structure.

It is to be anticipated that some of the lines of the γ-ray spectra of these two radioactive elements should be coincident, but some may be absent or very faint in one spectrum and strong in another. In fact, the relative intensities of the spectral lines of the γ rays from radioactive substances may in all cases be very different from those which would be observed when the element is bombarded by cathode rays. In the latter case, all types of characteristic radiation have a chance of excitation—supposing of course account is taken of the speed of the incident cathode rays—since the β particles enter the atom on an average equally in all directions.

Theory of the origin of the β and γ rays

One fundamental fact that has to be taken into account in considering the origin of the β and γ rays is the conversion of the energy of a γ ray into the form of a high speed electron or β ray, and *vice versa*. This point has been emphasized by Bragg in his papers on the nature of X rays. He supposed that the energy of a single X ray could be converted by its passage through matter into the energy of a single β ray of appropriate speed, and that no loss of energy occurred in the process. It would appear, however, necessary to generalize this conception, for it will be seen later that there is considerable evidence from a study of the β and γ rays from radioactive matter that a train of X rays of the same frequency may be given out each of definite energy, and that the whole energy of this train of waves may, under suitable conditions, appear in the form of a swift β ray. There is now strong evidence from a variety of directions that the energy emitted by a source of radiation of frequency v is in definite quanta E where $E = hv$, h being Planck's constant. The general idea that the energy of the γ rays is emitted in definite units or quanta appears to be necessary to explain the origin of homogeneous groups of β rays expelled from radioactive matter. We shall first assume that the energy in a single γ ray of frequency v is given by Planck's formula, and then discuss how far this particular relation is supported by the experimental evidence.

We shall suppose certain regions in the atom are set in vibration by the escape of the β particle during the atomic explosion. If v_1, v_2 ... are the frequencies of vibration, the energy emitted in the form of γ rays is phv_1, $qhv_2 \ldots$, where p and q may have any integral values. For example, one atom may emit one γ ray of frequency ν and energy $h\nu$, another may emit a train of two γ rays of the same frequency but of energy $2h\nu$, another three, and so on. There is no method at present of deciding the most probable value of p for a single atom, nor to fix an upper limit to its value. This may depend on the intensity of the disturbance communicated to a vibrating system by the escaping β particle. The energy of these γ rays is supposed to be partially or wholly converted into the β ray form in their escape from the radioactive atom. Unfortunately there is no evidence available at present of how this conversion occurs or whether any energy is absorbed in the process. If the conversion takes place without loss of energy, a train of γ rays of frequency ν_1 will give rise to a β particle of energy given by $E = phv_1$.

It is possible the conversion may occur in one of the regions of the atom which give rise to characteristic radiations, and is accompanied by the appearance of a new type of γ rays of frequency ν_2 etc. In such a case, we should anticipate that the energy of the escaping β particle is given by $E = phv_1 - qhv_2$, where p and q are whole numbers which may have all possible values consistent with $p\nu_1$ being greater than $q\nu_2$.

The value of E here refers to the energy of the β particle at the point in the atom where the conversion of energy occurs. It seems possible that there may be a further change of the energy of the β particle in escaping from the atom. In this case, the energy of the β particle after it has escaped from the atom is given by $phv_1 - qhv_2 - A$, where A may have a negative value and is at present indeterminate.

It will be seen that the present theory of the origin of the β rays differs somewhat from that advanced in the earlier papers (*loc. cit.*). I there supposed that homogeneous groups of β rays were due to the decrease of energy in definite units of the primary β particle in exciting the vibrations in the atom. The present theory supposes that the homogeneous groups of β rays arise from the conversion of the energy of the γ rays into the β ray form. In other words, the primary effect in the atom is the excitation of γ rays by the escape of the β particle from the nucleus. The appearance of groups of homogeneous β rays is a secondary effect due to the partial conversion of the γ rays into β rays in their passage through the radioactive atom. On the other hand, the continuous β radiation is ascribed mainly to the effect of the primary β particles escaping from the nucleus which have lost energy, though not in definite quanta, in setting the electronic system of the atom into vibration.

Consideration of the experimental evidence

We shall now consider briefly some recent experimental evidence which has thrown light on this question. Robinson, Rawlinson, and the writer* have shown that the β rays excited by the γ rays in their passage through matter

* Phil. Mag. August 1914. (*This vol., p. 466.*)

consist of definite groups which, no doubt, would be homogeneous if the layer of matter in which the β rays were excited was exceedingly thin. As far as experiment has gone, the velocities of these groups of 'excited' β rays are in close if not complete agreement with the velocities of the stronger groups of primary β particles from the source of radiation. The velocities of the corresponding groups appear to vary slightly when the β rays are excited in different metals, but the differences, though no doubt real, are not very marked. We may conclude from these results that the primary β particles, for example from radium B, like the β rays excited by the γ rays in traversing absorbing material, are due to the conversion of γ rays into β rays in their escape from the radioactive atom. The observed variations of velocity between the primary groups of β rays from the radioactive atom and the β rays excited by the γ rays in different substances, may be due to the variation of the part $- (qhv_2 + A)$ in the expression for the energy $E = phv_1 - qhv_2 - A$. The experiments, however, show that the variation due to this cause is small for the swift groups of β rays excited by γ rays in different kinds of matter, and for simplicity it may be assumed as a first approximation that the value of $qhv_2 + A$ is very small compared with phv_1 for the actual groups of swift β rays under consideration.

We shall now consider whether the experimental evidence supports the view that the energy of the homogeneous groups of β rays from radium B and radium C are connected with the values of hv, where v is the frequency of one or more of the stronger lines in the β ray spectrum.

For convenience of discussion, the frequencies of the stronger lines of the γ ray spectrum of the penetrating γ rays from radium B and radium C are included in the following table. The value of hv is calculated from the observed frequency, assuming $h = 6·55 \times 10^{-27}$, and expressed in the same form as the energies of the groups of β rays from radium B and radium C in the previous paper of Rutherford and Robinson. The value of e is taken as $4·69 \times 10^{-10}$ e.s. units—the value deduced directly by Planck from his theory of radiation. The energy hv of the strong line, which is reflected at 1° from rocksalt, is $1·25 \times 10^{13} e$, where e is in electromagnetic units. If Millikan's value of e is taken, viz. $4·77 \times 10^{-10}$, the corresponding value of hv is $1·23 \times 10^{13} e$.

Angles of reflexion and energies of lines in γ-ray spectra of radium B and radium C

Angle of reflexion from rocksalt.	43'	1° 0'	1° 10'	1° 24'	1° 37'	1° 43'	2° 0'	2° 20'	2° 28'	2° 40'	3° 0'	3° 18'	4° 0'	4° 22'
$(hv) \div 10^{13} e$	1·74	1·25	1·07	0·893	0·773	0·728	0·625	0·537	0·507	0·469	0·417	0·379	0·312	0·286

β-ray spectrum of radium C

In the previous paper by Rutherford and Robinson (*loc. cit.*), it was shown that the energy of the lines of the β-ray spectrum of radium C from No. 1 to No. 29 could be expressed approximately by an integral multiple of a unit of energy $E = 0.4284 \times 10^{13} e$. It was pointed out, however, that in several cases the adjacent lines appeared to be closer together than would be indicated by this difference relation. Since one of the strong lines of the γ ray spectrum of radium C, reflected at 1°, has an energy of $1.25 \times 10^{13} e$ on Planck's relation, it is to be anticipated on the theory that the energy of a number of lines should show differences of this unit or integral multiples of it. As a special case, the energy of the β particle itself might be expected to be integral multiples of this unit. The original difference unit $0.4284 \times 10^{13} e$, when multiplied by three gives $1.285 \times 10^{13} e$, which does not differ from $1.25 \times 10^{13} e$ by more than the experimental error. It will be seen in the table below that a number of the lines, including some of the stronger lines, can be expressed very closely as integral multiples of the unit $E = 1.285 \times 10^{13} e$.

The numbers of the lines refer to those given in the previous paper by Rutherford and Robinson.

Index number of line	Intensity	Velocity as a fraction of light	Observed energy $\div 10^{13} e$	Energy $\div 1.285 \times 10^{13} e$	Integral multiple
M 45	m.	0·750	2·59	2·015	2
K 38	s.	0·868	5·16	4·016	4
32	m.f.	0·917	7·68	5·98	6
G 29	m.s.	0·946	10·31	8·02	8
F 26	m.	0·952	11·49	8·94	9
23	f.	0·959	12·82	9·97	10
D 20	m.	0·9643	14·09	10·96	11
17	f.	0·9687	15·42	12·00	12
C 14	m.	0·9724	16·71	13·01	13
11	f.	0·9754	17·96	13·97	14

Of the twelve strong lines in the table marked with index letters, six are seen to be expressed very nearly as integral multiples of E_1. In picking out possible other units, it appears significant that another series of lines can be expressed as a multiple of $E_2 = 0.74 \times 10^{13} e$. This is seen in the following table.

Only two of these lines, viz. of energies 10·31 and 14·09, are included in the first table. These lines may possibly be close doubles.

It has been pointed out in the paper of Rutherford and Andrade that the strong γ ray line reflected at 1° 40′ is probably a close double which it is difficult to separate under the experimental conditions. One component is believed to belong to radium B and the other to radium C. This appears very

Multiple ..	2	4	6	8	9	12	14	15	16	18	19	22	23
Calculated energy ÷ 10^{13} e.	1·48	2·96	4·44	5·92	6·76	8·88	10·36	11·10	11·84	13·32	14·06	16·28	17·02
Observed energy ÷ 10^{13} e.	1·49	2·96	4·48	5·94	6·74	8·91	10·31	11·07	11·86	13·28	14·09	16·28	17·11
Intensity ..	f.	m.	m.f.	m.s.	m.f.	m.f.	m.s.	f.	v.f.	m.s.	m.	f.	f.

probable, for Richardson has found recently that radium C emits some γ radiation which has about the same absorption in lead as the soft γ rays from radium B. We should consequently anticipate that radium C should give a strong line near 1° 40′ of energy about $0·75 \times 10^{13}$ e—in good agreement with the unit found above.

Only two strong lines remain which are not included as multiples of these two units $0·74$ and $1·29 \times 10^{13}$ e. These are the two lines A and B of energies $21·02$ and $17·5 \times 10^{13}$ e. These can be expressed as 12 E and 10 E where $E = 1·75 \times 10^{13}$ e. It appears more than a coincidence that this value is very close to the energy of the shortest wave-length observed for the γ rays from radium C, viz. 43′ of energy $1·74 \times 10^{13}$ e. It was also found that several lines in the spectrum were multiples of the energy of the strong β-ray line of energy $1·81 \times 10^{13}$ e. This corresponds to nearly twice the energy of the line 1° 24′ of energy $0·89 \times 10^{13}$ e.

There are a number of other faint lines not included in the above tables. It is probable these may owe their origin to other frequencies not observed in the γ-ray experiments.

It is seen from the above that there is strong evidence that the energies of many of the lines in the β-ray spectrum may be expressed as *integral* values of certain definite units which correspond with the frequencies of vibration of the γ rays, assuming Planck's relation between frequency and energy. It hardly seems possible that the numerous close agreements observed are accidental; for it must be remembered that the units are chosen to fit in with three strong lines of comparatively slow velocity observed in the β-ray spectrum of radium C, whose energies are $1·81$, $2·59$, $2·96 \times 10^{13}$ e, where the multiple is 2 for the first two units and 4 for the last.

If these deductions are correct, it follows that not a single γ ray but a train of γ rays of definite frequency can be emitted from each vibrating centre of the atom. The number of waves composing the train varies for different atoms and for the different frequencies, but evidence is obtained that the number of complete waves may in some cases be ten or more. The train of waves on passing through matter may under suitable conditions give all its energy to a single electron, and thus give rise to the high speed electrons when γ rays pass through matter. We have already drawn attention to the fact that the primary β rays emitted from a radioactive atom have a very similar spectrum

Q

to that observed for the β rays excited by γ rays, indicating that both have a similar origin.

β-ray spectrum from radium B

We shall now consider the β-ray spectrum of radium B for which the energies of the groups of β rays have been carefully determined by Rutherford and Robinson. The results are included in the following table. Column I gives the number of the line, II the intensity, III the ratio β of velocity of the β particle to the velocity of light, IV the energy of the β particle composing each group, and V the suggested unit and its multiple.

I	II	III	IV	V
			Energy	Unit of energy
No.	Intensity	β	$\div 10^{13} e$	$\div 10^{13} e$
1	f.	0·823	3·852	$5 \times 0·770$
2	m.s.	0·805	3·480	$4 \times 0·870$
3	s.	0·797	3·332	$7 \times 0·476$ or $6 \times 0·555$
4	v.f.	0·787	3·160	$5 \times 0·632$
5	m.s.	0·762	2·758	$5 \times 0·552$
6	v.s.	0·751	2·610	$3 \times 0·870$
7	m.	0·731	2·366	$5 \times 0·473$
8	m.s.	0·719	2·228	$3 \times 0·743$ or $4 \times 0·557$
9	v.s.	0·700	2·039	$4 \times 0·510$ or $6 \times 0·373$
10	m.s.	0·656	1·650	$4 \times 0·412$ or $3 \times 0·550$
11	v.s.	0·635	1·519	$2 \times 0·760$ or $3 \times 0·506$
12	m.	—	—	—
13	m.	—	—	—
14	m.s.	0·426	0·539	$1 \times 0·539$
15	s.	0·414	0·503	$1 \times 0·503$
16	v.s.	0·365	0·376	$1 \times 0·376$

It is certainly striking that the energies of the three strong groups of β rays 14–16 agree very closely with the calculated values for the reflexion angles of the γ rays for the lines at 2° 20′, 2° 28′, and 3° 18′, which gave energies 0·537, 0·507, 0·379 $\times 10^{13}$ e respectively. When there are so many lines as in the γ-ray spectrum of radium B, and uncertainty as to whether some of the lines are to be ascribed to radium B or to radium C, it is obvious that it is difficult to fix with certainty the unit in which to express the energies of the groups. It is noticeable, however, that the strong lines 2 and 6 are expressed by small integral multiples of $0·870 \times 10^{13}$ e, which agrees fairly well with the energy to be expected for the γ radiation reflected at 1° 24′. With the exception of lines 1 and 11, there appears very little evidence of the unit of energy corresponding to the strong reflexion line at about 1° 40′, the energy of which should be about 0·750. There is another unit, however, 0·553 which agrees fairly well with a number of lines. The observed γ-ray line nearest to this value is reflected at 2° 20′ of energy 0·537, but the difference in the observed and calculated energies is more than two per cent. Possibly the unit may be 1·10 corresponding to the 1° 10′ line.

It will be observed that several of the lines are in approximate accord with multiples of different units calculated from the frequencies of observed lines. This will be seen by comparison of the above table with the energies for frequencies given on p. 479.

Apart from the group of slow velocity lines and lines 2 and 6, the agreement between observed and calculated energies is fair, but is not sufficiently definite to draw certain deductions as to the origin of the individual lines in the spectrum.

There is one interesting point which should be mentioned here. Danysz* determined the velocities of the slow velocity groups of β rays from radium D, and found them to be quite different from the corresponding groups of β rays (14–16) in the spectrum of radium B, although the actual numerical differences between the energies of the corresponding lines were approximately the same for the two substances. Now the general evidence indicates that radium B and radium D, although of different atomic weights, have identical general chemical properties, and would be expected to give the same γ-ray spectra, and also the same type of primary β rays. It has been pointed out earlier in this paper that the observed differences in β and γ ray spectra may possibly be ascribed to the different conditions of excitation of the γ rays in the atom in the two cases due to the expulsion of the β particle from the nucleus in a different direction with regard to the structure of the atom. If the latter point of view is correct, we should anticipate that the energies of the groups of β rays from radium D should be expressed in terms of some of the frequencies in the γ rays of radium B.

The energies observed by Dansyz are 0·309, 0·311, 0·435, 0·468 × 10^{13} e. Two of these agree well with the values of the energies 0·312, 0·469 × 10^{13} e, calculated for the lines of radium B reflected from rocksalt at angles of 4° and 2° 40′. The line 2° 51′ of energy 0·438 was observed on one or two plates but was not recorded in the list of lines published by Rutherford and Andrade. The corresponding lines involved in the radium B spectrum are at 3° 18′, 2° 28′, and 2° 20′.

It is of interest to note that the soft γ rays of radium B which give two strong spectral lines reflected at 10° or 12° from rocksalt and have values hv, 0·125, 0·104 × 10^{13} e, respectively, do not appear to be responsible for any of the observed β-ray lines in radium B. The fact that the energy of the slowest group of β rays 0·376 is about 3 × 0·125 is probably merely a coincidence.

Some recent measurements of Miss Szmidt in this Laboratory have brought out the surprisingly small amount of energy emitted in the form of these soft γ rays, the amount being only about 2 per cent. of the energy of the more penetrating γ rays from radium B and only about 0·014 per cent. of the total energy of the γ radiation from radium B and radium C together. Although

* *Le Radium*, x. p. 5 (1913).

the energy of these soft rays is only about 1/1000 of the total γ-ray energy emitted from an emanation tube, they yet give the most intense spectrum lines under the experimental conditions.

The small relative amount of energy from the soft γ rays shows that either these rays are emitted from only a small fraction of the disintegrating atoms, or, what is more probable that they do not consist of trains of waves but mostly of single waves.

The general comparison of the soft β and γ ray spectra of radium B with radium C indicates that the higher the frequency of the radiation, the greater the tendency to excite long trains of waves. It does not seem possible at this stage to decide whether this is a definite property of the vibrating systems, or whether it is due to the violence of the disturbance in the case of the higher frequencies of vibration.

Excitation of β-ray spectra by characteristic X rays

It is known that when a comparatively light element like nickel is bombarded by cathode rays, the X rays initially emitted consist mainly of the 'K' characteristic radiation of that element. Rawlinson* has found that with increasing voltage, more and more penetrating types of radiation appear in addition, but no certain evidence has been found that these penetrating rays give a line spectrum on reflexion from crystals. Moseley showed that the characteristic radiation of nickel consists mainly of two strong lines reflected from rocksalt at angles of $15°\cdot5$ and $17°\cdot15$. From analogy with the case of β rays from a radio-element, it is to be anticipated that with a high voltage discharge, the nickel radiation would consist of a train of one or more waves each of energy $h\nu$ corresponding to the frequency ν of each of the strong lines of nickel. When this radiation falls on another element, it is absorbed and partially converted into β rays. For a given frequency ν_1, we may suppose as before that the excited β rays would fall into groups of energy $ph\nu_1 - qh\nu_2 - A$, where q is an integer and ν_2 is the frequency of the characteristic radiations excited by the X rays in the element in question. The general evidence indicates that the constant A is negligible. If no characteristic radiation is excited, it is to be expected that the energy of the groups of β rays should be given by $ph\nu_1$, where p is an integer. If the conversion of the X rays into β rays is accompanied by the characteristic radiation of the element, another series of lines should make their appearance of energy given by $ph\nu_1 - qh\nu_2$. At the same time, the characteristic radiation of the frequency ν_2 would be partially converted into β rays in escaping from the substance and would thus be expected to give rise to groups of β rays of energy $qh\nu_2$, where q is an integer, and so on.

It is obvious, that on these views, the spectrum of the β rays excited by the passage of only one frequency of X radiation through matter may show a very complex β-ray spectrum, consisting of a number of distinct groups of

* Phil. Mag. August 1914.

β rays. If the primary radiation consists of a number of frequencies, the resulting β-ray spectrum will probably prove as complex as the β-ray spectrum of radium B or radium C.

A preliminary examination of the β-ray spectra excited in different elements by the characteristic radiation of nickel has been made by Robinson and Rawlinson.* They have found that the magnetic spectrum of the β rays does show the presence of a number of well-marked groups of β rays, but the experimental evidence is not yet complete enough to test the correctness of the above point of view.

University of Manchester
June 30, 1914

* Phil. Mag. August 1914.

Radium Constants on the International Standard

by SIR ERNEST RUTHERFORD, F.R.S.,
Professor of Physics, University of Manchester

From the *Philosophical Magazine* for September 1914, ser. 6, xxviii, pp. 320–7

DURING the past year, the National Physical Laboratory has obtained a radium preparation certified by the γ ray method in terms of the International Radium Standard preserved in the Bureau des poids et mesures at Sèvres. It has consequently been possible to compare the standards in use for many years in the Laboratories in Montreal and Manchester with the International Standard.

It may be of interest to mention briefly the history of the preparation which for ten years has served as a laboratory standard and in terms of which a number of important radioactive magnitudes have been measured.

In connexion with the radioactive work in Montreal, it became important in 1903 to adopt a radium standard in which to express the results of various measurements. For this purpose, Professor A. S. Eve weighed out for me $3 \cdot 69$ milligrams of a preparation of radium bromide bought from Dr. Giesel, and enclosed it in a sealed tube. This preparation has served as the primary laboratory standard, and was assumed to contain $3 \cdot 69$ milligrams of pure radium bromide or $2 \cdot 16$ milligrams of radium element. At the same time about one milligram of the same material, calibrated in terms of the larger quantity by the γ-ray method, was dissolved and a number of standard radium solutions were prepared. Part of this standard solution was sent to Professor Boltwood in New Haven in order to make a direct determination by the emanation method of the quantity of radium in a mineral per gram of uranium. The preliminary measurements of Professor Boltwood gave a result much higher than that measured by Professor Eve by direct comparison of the γ-ray effect of a uranium mineral with the primary standard. It thus appeared that some error had crept in the preparation of the standard solution. An investigation by Eve showed that the radium in the standard solution had partly deposited out on the walls of the glass vessels, and that the solutions were consequently unreliable.

Another standard solution was then prepared in which the precaution was taken of adding a considerable quantity of hydrochloric acid to keep the radium in solution. These standard solutions are believed to have kept their strength unaltered during the past ten years. With the aid of these standard solutions, the quantity of radium per gram of uranium was determined by

Boltwood. The result was expressed in terms of the 3·69 milligram standard, which has generally been referred to as the 'Rutherford-Boltwood' standard.

A radium standard in use in the Laboratory in Manchester was accurately compared for me by Professor Stefan Meyer, Secretary of the International Radium Committee, in terms of the Vienna standard, which has been set aside as a secondary International standard. At the same time, Mr. Chadwick, in the Manchester Laboratory, compared by a balance method the laboratory standards with a secondary standard kindly lent to me by the Radium Institute of Vienna. Recently Dr. Kaye has compared two of the Laboratory standards with the radium standard of the National Physical Laboratory. The numerous cross measurements at different institutions have all been found to be in excellent agreement, and bring out the reliability of the γ-ray method of measurement.

The original Rutherford-Boltwood standard was assumed to contain 3·69 milligrams pure radium bromide. The actual γ-ray activity of the standard as employed in actual use was found to correspond to 3·51 mg. $RaBr_2$ in terms of the International Standard, and was thus 4·9 per cent. too low. Considering the time of preparation of this standard, the choice of material has turned out to be very fortunate. No doubt part of the radium bromide from which the standard was prepared had been converted before weighing into carbonate by exposure to the air, and this would account for the high radium content, considering that no allowance was made for the water of crystallization. In any case, the result brings out the purity of the radium preparations sold commercially by Dr. Giesel more than eleven years ago.

It has been suggested to me that it would be a convenience to many workers if the various radioactive magnitudes determined in the Laboratory at Manchester were re-calculated in terms of the International Standard. In my work 'Radioactive Substances and their Radiations,' most of the results were expressed in terms of the laboratory standard, but the heating effect, which was determined while the book was passing through the press, was given in terms of the Vienna Standard, which had been compared with the International Standard. Confusion may consequently arise in regard to the standard in which some of the results have been expressed.

In the following table the various magnitudes involving the radium standard are expressed in terms of the International Standard.

It may be of interest at this stage to discuss briefly in some cases the agreement between the observed and calculated values, and to consider the fundamental data involved in the calculations.

In the first place, it can be easily shown that the calculated values of (1) the production of helium by radium, (2) the volume of the emanation, (3) the heating effect, (4) the life of radium, all involve the quantity nE, wheie n is the number of α particles expelled per second per gram of radium itself, and E the charge carried by the α particle. In other words, the agreement of calculation with experiment in these cases does not involve the accuracy of the actual value of n or of E (which is twice the unit charge), but the product of

Original paper	'Radio-active Substances'	Quantity	Rutherford & Boltwood Standard	International Radium Standard
Rutherford & Boltwood, Amer. Jour. Sci. xxii. p. 1 (1906); Boltwood, Amer. Jour. Sci. xxv. p. 269 (1908).	pp. 16, 462	Amount of radium in equilibrium with one gram of uranium.	$3·4 \times 10^{-7}$ gr. Ra	$3·23 \times 10^{-7}$ gr. Ra
Boltwood & Rutherford, *Wien. Ber.* cxx. p. 313 (1911); Phil. Mag. xxii. p. 586 (1911).	p. 557	Production of helium per gram of radium per year.	156 c.mms.	164 c.mms.
		Calculated value ..	155 c.mms.	163 c.mms.
Rutherford & H. Robinson, *Wien. Ber.* Oct. 1912; Phil. Mag. xxv. p. 312 (1913).	pp. 578–581	Total heating effect of one gram of radium and its products in equilibrium with it.		134·7 gr. cals. per hour.
		Radium alone	....	25·1 gr. cals. per hour.
		Emanation „	....	28·6 „ „
		Radium A „	....	30·5 „ „
		Radium B ⎱ „ Radium C ⎰	....	50·5 „ „
Rutherford, Phil. Mag. xvi. p. 300 (1908).	p. 480	Volume of the emanation from one gram of radium in equilibrium.	0·60 c.mm.	0·63 c.mm.
		Calculated value ..	0·59 „	0·62 c.mm.
Rutherford & Geiger, Proc. Roy. Soc. A. lxxxi. p. 141 (1908).	pp. 128–133	Number of α particles expelled per second per gram of radium itself.	$3·4 \times 10^{10}$	$3·57 \times 10^{10}$
		Number per second from one gram of radium in equilibrium.	$13·6 \times 10^{10}$	$14·3 \times 10^{10}$
Rutherford & Geiger, Proc. Roy. Soc. A. lxxxi. p. 162 (1908).	pp. 135–137	Total charge carried by the α particles per sec. from one gram of radium itself and from each of its products in equilibrium with it.	31·6 e.s. units. $1·05 \times 10^{-9}$ e.m. units.	33·2 e.s. units $1·11 \times 10^{-9}$ e.m. units.

Original paper	'Radio-active Substances'	Quantity	Rutherford & Boltwood Standard	International Radium Standard
Geiger, Proc. Roy. Soc. A. lxxxii. p. 486 (1909).	p. 502	Total current due to the α rays from one curie of emanation. (1) by itself (2) with its α ray products.	$2 \cdot 75 \times 10^6$ e.s. units. $9 \cdot 46 \times 10^6$ e.s. units.	$2 \cdot 89 \times 10^6$ e.s. units. $9 \cdot 94 \times 10^6$ e.s. units.
H. Moseley, Proc. Roy. Soc. A. lxxxvii. p. 230 (1912).	p. 204	Total charge carried by the β particles emitted per second by radium B or radium C in equilibrium with one gram of radium.	$17 \cdot 4$ e.s. units.	$18 \cdot 3$ e.s. units.
	p. 459	Calculated half value period of transformation of radium.	1780 years.	1690 years.

these numbers, which is a measure of the charge carried by the α particles expelled per second per gram of radium.

The value nE was directly determined by Rutherford and Geiger, and is given by nE $= 1 \cdot 11 \times 10^{-9}$ e.m. units on the International Standard. This value has been used in making the calculations of the four quantities mentioned above.

The value of this quantity deduced from the rate of production of helium by radium (viz. 164 c.mm. per gram per year) is $1 \cdot 12 \times 10^{-9}$—a close agreement with the direct experimental value.

In a similar way, the value of nE deduced from the observed volume of the emanation is in close accord with the experimental value.

There is thus a satisfactory agreement between theory and experiment in the above cases, but the agreement is not nearly so good for the value nE deduced from the heating effect and the life of radium. These points will consequently be considered in more detail.

Heating Effect of Radium Emanation

In a recent paper, Mr. H. Robinson and myself* have re-determined the

* *Wien. Ber.* cxxii. Abt. IIa, Nov. 1913. (*This vol.*, p. 383.)

Q*

velocity and value E/m for the α particles expelled from radium. From these data we have compared the calculated heating effect due to one curie of radium in equilibrium *due to α rays alone*, with the observed heating effect due to these radiations. On the International Standard, the observed heating effect due to the α rays from one curie of emanation is 99·2 gr. cals per hour, and the calculated is 92·4, assuming $n\mathrm{E} = 11·1 \times 10^{-10}$ e.m. units, and adding 2 per cent. for the energy of the recoil atoms. The calculated heating effect is thus 7 per cent. lower than the observed. If all the heating effect of the α ray products is due to the energy of the expelled α particles, it would follow that the value of $n\mathrm{E}$ is 7 per cent. too small. This seems improbable, so we must look for an explanation of this apparent discrepancy in another direction. The general radioactive evidence indicates that the loss of an α particle from an atom lowers the positive charge of the atomic nucleus by two units, and the expulsion of a β particle from the nucleus raises it by one unit. Without entering into a discussion of the possible distribution and velocities of the electrons external to the nucleus, it is to be anticipated on general grounds that the kinetic energy of the total electronic distribution external to the nucleus should *increase* with increase of charge on the nucleus. The expulsion of an α particle should thus result in a lowering of the total kinetic energy of the electrons, and the expulsion of a β particle to an increase. Suppose, for simplicity, that this change of energy is proportional to the variation of charge on the nucleus, and is the same for each α ray transformation. If A be the energy per atom liberated from the electrons resulting from the expulsion of an α particle, then A/2 is the energy absorbed in consequence of the expulsion of a β particle. If E_1, E_2, E_3 be the kinetic energies of the expelled α particles from the three products, emanation, radium A and radium C respectively, then the energy per atom released during the transformation of one atom is

$E_1 + A$ for emanation, $E_2 + A$ for radium A, $-\dfrac{A}{2}$ for radium B, $E_3 + \dfrac{A}{2}$

for radium C, since the latter expels both an α and β particle, and thus its nucleus charge is finally lowered by one unit.

The resultant energy due to a transformation of one atom of all of these substances is $E_1 + E_2 + E_3 + 2A$, and the ratio of the energy liberated from radium B and radium C together to the total is

$$\frac{E_3}{E_1 + E_2 + E_3 + 2A} \quad \cdot \quad \cdot \quad \cdot \quad \cdot \quad \cdot \quad \cdot \quad (1)$$

In this calculation, no account is taken of the energy expelled in the form of primary β rays and γ rays, but only of the energy from the α rays and from the electronic distribution.

If the above point of view is correct, the ratio of the heating effect of radium B and radium C together (subtracting the energy due to β and γ rays from these products), should be *less* than the value calculated from the energy of the α particles. Now this point was carefully examined by Mr. Robinson and myself (*loc. cit.*) some years ago, and we drew attention to the fact that the heating effect of radium C was distinctly *less*, compared with that due to the emanation and radium A, than the theoretical ratio calculated from the energy of the expelled α particles.

It was found that the observed heating effect due to radium (B + C) together in equilibrium with one curie of emanation was actually in very nearly the theoretical ratio with the calculated heating effect due to the emanation and its products, viz. $0 \cdot 403$. Of the observed heating effect, however, $4 \cdot 3$ gr. cals. out of a total of $103 \cdot 5$ per curie of emanation per hour were to be ascribed to the absorption of part of the β and γ rays emitted by radium B and radium C.

From these data, the value of A (equation 1) can be deduced, and is found to be about $3 \cdot 2$ gr. cals. per hour corresponding to one curie of emanation. The heating effect due to the α rays alone should be consequently $6 \cdot 4$ gr. cals. less than the observed value $99 \cdot 2$, that is $92 \cdot 8$. The actual calculated heating effect comes out $92 \cdot 4$, assuming $n\mathrm{E} = 11 \cdot 1 \times 10^{-10}$ e. m. units. When this additional factor is taken into account, the observed heating effect is thus in complete accord with the value of $n\mathrm{E}$ deduced by direct measurement and by determining the volume of helium produced from radium.

We have in the above made no assumptions as to the form in which the energy is emitted from the external distribution of electrons, but have supposed that it ultimately appears as heat. It is, however, implicitly assumed in the calculation that this energy is not emitted in the form of radiations so penetrating that they are able to escape through the absorbing material employed in the actual measurements. No doubt part of the energy may be emitted (as possibly in the case of radium itself) in the form of slow β rays and soft γ rays, but no certain evidence is available on this point.

It follows from these considerations that the heating effect of all α ray products should be *greater* (on the average about 10 per cent.) than the kinetic energy of the expelled α particles. Similarly the heating effect due to a β ray transformation may in some cases be *less* than the value calculated from the energy of the expelled β particles, but it is difficult to be certain how far the energy of the latter may be affected by the change of nucleus charge.

Life of Radium

The half-value period of transformation of radium can be calculated at once from the value $n\mathrm{E}$, without any assumption of the actual value of n or E except that E is twice the unit charge. On the International Standard, the half-value period comes out to be 1690 years,* taking $n\mathrm{E} = 11 \cdot 1 \times 10^{-10}$ e. m. units. This is much less than the experimental value found by Boltwood,†

---

* By an oversight, the period of radium was calculated as 1850 years instead of 1620 years in 'Radioactive Substances' p. 459. The error arose in the correction in terms of the International Standard.

† Boltwood, Amer. Journ. Sci. xxv. p. 493 (1908).

viz. about 2000 years, but is in better accord with the value 1800 years given by Keetman,* and 1730 years found by Stefan Meyer.†

Unless the determinations of nE from the charge carried by the α particles, the production of helium and the heating effect, are all seriously overestimated and to the same extent, the value 1690 years cannot be far from the truth. It is desirable that this important constant should be re-determined, and I understand that this is being undertaken by Professor Boltwood and Mlle. Gleditsch.

It should be mentioned that the accuracy of the original determination of the period by Boltwood is quite independent of the correctness of the radium standard, since it merely involves the comparison of two quantities of emanation.

There is no radioactive method of checking the accuracy of the value of n, the number of α particles expelled per second per gram, except by comparison of the value of E, which is deduced from the measurement of the total charge nE carried by a counted number of α particles with other measurements of the unit charge. Taking the original determination of Rutherford and Geiger, the electronic charge comes out to be $4 \cdot 65 \times 10^{-10}$ e.s. units. If, however, we substitute the recent value of e found by Millikan,‡ viz. $4 \cdot 77 \times 10^{-10}$, the value of n reduces to $3 \cdot 48 \times 10^{10}$ instead of $3 \cdot 57 \times 10^{10}$.

An accurate re-determination of the value of n and of nE for radium is much to be desired; for both of these quantities are fundamental constants which should be known with the greatest possible precision.

University of Manchester
June 1914

* Keetman, *Jahrb. d. Radioakt.* vi. p. 265 (1909).
† *Wien. Ber.* cxxii. p. 1086 (1913).
‡ Millikan, Physical Review. ii. p. 109 (1913).

Origin of the Spectra given by β and γ Rays of Radium

by PROFESSOR SIR ERNEST RUTHERFORD, M.A., D.SC., F.R.S.

From the *Proceedings* of the Manchester Literary and Philosophical Society, 1915, IV, vol. lix, pp. 17–19

AN account was given of recent experiments of Sir Ernest Rutherford and Dr. Andrade to determine the wave length of the very penetrating gamma rays which are emitted from radium. The spectrum of the gamma rays was obtained by a photographic method by reflecting the rays from a thin slip of rock salt. The radioactive source consisted of a fine glass tube containing a large quantity of radium emanation. Special precautions were necessary to get rid of the effect of the penetrating beta rays which are emitted with gamma rays. A large number of lines were observed in the spectrum over a wide range of wave length. Two well-marked lines are reflected from rock salt at 10° and 12°, and correspond to some soft gamma rays. There were other strong lines of 1° and 1·7°, corresponding to the very penetrating rays. The shortest wave length observed was 0·07 Ångström unit, which is about 1/50000 of the wave length of visible light. This radiation has much the shortest wave length at present known.

An account was also given of the methods for determining the magnetic spectrum of the beta rays. For this purpose, the rays from a fine source, passing normally in a strong magnetic field, describe a circular path and fall on a photographic plate. Under these conditions, a number of well-marked lines are observed on the photographic plate, which correspond to groups of rays of definite velocity. The speed and energy of the beta particle comprising each of these groups of rays from radium products have been accurately determined by Rutherford and Robinson. The general evidence indicates that there is a very close connection between the emission of beta and gamma rays from radioactive bodies, and that the energy of the groups of beta rays are intimately related with the frequency of the gamma radiation from which they arise.

An account was given of a general theory to explain the connection between the beta and gamma rays. It is supposed that the breaking up of an atom is accompanied by the expulsion of a swift beta particle from the nucleus. This beta particle in escaping from the atom sets the external electrons in rapid vibration, and gives rise to the gamma rays observed. The radiation from the disturbed electrons may be given out either in the wave form or as swift electrons, and these two forms of energy are mutually convertible. Evidence was also given that during the disintegration of an atom, the beta particle is always expelled in a certain definite direction with regard to the structure of the atom. It was pointed out that the emission of beta and gamma rays from radioactive substances must be regarded as part of the general radiation problem, and its importance in this connection was discussed

Radiations from Exploding Atoms

From *Nature*, **95**, 1915, pp. 494–8
(Discourse delivered at the Royal Institution on Friday, June 4, 1915)

IT is now well established that the radio-active substances are undergoing spontaneous transformation, and that their characteristic radiations—the α, β, and γ rays—accompany the actual disintegration of the atoms. The transformation of each atom results from an atomic explosion of an exceedingly violent character, and in general results in a liberation of energy many million times greater than from an equal mass of matter in the most vigorous chemical reaction.

In the majority of cases the atomic explosion is accompanied by the expulsion of an actual atom of matter—an α particle—with a very high speed. It is known that the α particle is an atom of helium which carries two unit positive charges, and which leaves the atom with a velocity of about 10,000 miles per second. In some transformations no α particle is ejected, but its place is taken by a swift β particle or electron. These β rays carry with them a large amount of energy, for in some cases they are expelled very close to the velocity of light, which is the limiting velocity possible for such particles. The expulsion of high-speed β particles is usually accompanied by the appearance of γ rays, which correspond to X-rays, only of greater penetrating power than has so far been obtained from an X-ray tube even when a high voltage is employed. The emission of energy in the form of γ rays is not negligible, for in some cases it is even greater than the energy emitted in the form of high-speed β particles, and may amount per atom to as much as 20 per cent. of the energy released in the form of a swift α particle.

By the application of a high voltage to a vacuum tube it is quite possible to produce types of radiation analogous to those spontaneously arising from radium. For example, if helium were one of the residual gases in the tube, some of its atoms would become charged, and would be set into swift motion in the strong electric field. In order, however, to acquire a velocity equal to the velocity of expulsion of an α particle, say, from radium C, even in the most favourable case nearly four million volts would have to be applied to the tube.

In a similar way, in order to set an electron in motion with a velocity of 98 per cent. the velocity of light, at least two million volts would be necessary. As we have seen, it has not so far been found possible to produce X-rays from a vacuum tube as penetrating as the γ rays. The study of the radiations from radio-active substances is thus of especial interest, not only for the information obtained on the structure of the atoms themselves, but also in providing for investigation special types of radiation of greater individual intensity than can be obtained by ordinary experimental methods. The enormous energy of motion of swift α and β particles must exist in the atom before its disintegration, either in a potential or a kinetic form, and may arise either from the passage of the charged particles through the intense electric fields within the atom, or from the very swift motion of these particles within the atom before their release. In any case, there can be no doubt that electric fields, and possibly magnetic fields, of enormous intensity exist within the very small volume occupied by the essential structure of the atom — fields many million times greater in intensity than we can hope to produce in laboratory experiments.

In order to explain certain experimental results, I have suggested that the main mass of the atom is concentrated within a minute volume or nucleus, which has a positive charge, and is of dimensions exceedingly minute compared with the diameter of the atom. This charged nucleus is surrounded by a distribution of electrons which may extend to distances comparable with the diameter of the atom, as ordinarily understood. The general evidence indicates that the α and primary β particles are expelled from the nucleus, and not from the outer structure of the atom. If this be the case, the α particle which carries a positive charge would have its velocity increased in passing through the strong repulsive field surrounding the nucleus; on the other hand, the β particle which carries a negative charge must be retarded in its escape from the nucleus, and must possess great initial energy of motion to escape at all. There appears to be no doubt that the penetrating γ rays have their origin in some sort of disturbance in the rings of electrons nearest to the nucleus, but do not represent, as some have supposed, the vibrations of the nucleus itself.

α Rays

A brief account was given of the recent work of Rutherford and Robinson in determining with accuracy the velocity of expulsion of the α particles from certain radio-active substances. This was done by measuring the deflection of a pencil of α rays in strong magnetic and electric fields. With the aid of intense sources of radiation, it was found that the value of the E/M—the ratio of the charge carried by the α particle carried to its mass—was 4820 units, a value to be expected if helium has an atomic weight 4 and carries two unit charges. This experiment also shows that the mass of the flying positive particle is not affected appreciably by its swift motion. From known data the initial velocity of the expulsion of the α particles from all other radio-active substances can be deduced with accuracy.

If the expulsion of an α particle from an atom is the result of an internal explosion, we should anticipate, from the analogy of a shot from a gun, that the residual atom would recoil in a direction opposite to the escaping β particle. The existence of these 'recoil' atoms can be shown in a variety of ways, for the velocity of recoil is sufficient to cause the atoms to leave the surface on which they are deposited and to pass through a considerable distance in air at a pressure of one millimetre before they are stopped. It is to be antici-pated that the momentum of a recoiling atom should be equal and opposite to that of the escaping α particle. Since the deflection of a charged particle in motion in a magnetic field is inversely proportional to its momentum, the deflection of a stream of recoiling atoms should be the same as for the α particles if the atoms carry the same charge. Dr. Makower has examined the deflection of a pencil of recoil atoms in a magnetic field, and found it to be exactly half of that due to the α particle, proving definitely that the recoiling atom carries only one unit of positive charge in place of two for the α particle.

We thus see that the simple application of momentum enables us to deduce the mass and energy of the recoiling atoms. Since the mass of the radio-active atoms is about fifty times that of the α particle, the velocity, and also the energy, of recoil is only about one-fiftieth of that of the escaping α particle. In a similar way, it can be shown that the ejection of a swift β particle should cause a vigorous recoil of the atom, though not so marked as in the case of the more massive α particle.

β Rays

During the last few years notable advances have been made in our know-ledge of the mode of emission of β particles from radio-active atoms. The work of Baeyer, Hahn, and Meitner, and of Danysz, has shown that the β rays from a radio-active substance like radium B or radium C contain a number of definite groups of rays which are expelled with definite velocities. This is best shown photographically by examining the deflection of a pencil of β rays in a magnetic field. In a uniform field, each of the groups of rays describes a circular path the radius of which is inversely proportional to the momentum of the β particle. By the application of special methods it has been found possible to obtain a veritable spectrum of the β rays. The spectrum of the β rays from radium B and radium C has been very carefully examined by the writer and Mr. Robinson, and found to give a large number of well-marked bands, each of which represents a group of β rays, all of which are expelled with identical speed. It was at first thought that most of the energy of the β rays was comprised in these groups, as some of the bands on the photographic plate were very marked. Chadwick, however, has recently shown that the fraction of the rays which give a line spectrum is only a few per cent. of the total radiation. The general evidence shows that the β radiation from these substances gives a *continuous* spectrum due to β rays of all possible velocities, on which is superimposed a *line* spectrum due to a small number of β particles of definite velocity comprising each group.

Lines in the β-ray spectrum have been observed for particles which have a velocity not far from that of light, but the photographic effect of the particles becomes relatively feeble for such high speeds.

It is known from direct measurement that each atom of radium B or of radium C in its distintegration emits *on an average* one β particle. In the β-ray spectrum of radium C at least fifty definite bands are observed, differing widely in intensity. It is thus clear that a single atom in disintegrating cannot provide one β particle for *each* of these numerous groups. It is thus necessary to conclude that each atom does not emit an identical β radiation. The results are best explained by supposing that the β-ray spectrum is the statistical effect due to a large number of atoms, each of which may only give one or two of the groups in its disintegration. In this respect a β-ray transformation is distinguished from an α-ray transformation, for in the latter case each atom emits one α particle of characteristic speed. It will be seen later that there is undoubtedly a very close connection between the emission of β and γ rays from radio-active atoms, and the probable explanation of the remarkably complex β-ray spectrum will be discussed later.

With the exception of one element, radium E, and possibly uranium X, all the radio-active substances which emit primary β rays give a line spectrum. For the majority of elements the strong lines in the β-ray spectrum have been determined by Baeyer, Hahn, and Meitner, but more intense sources of radiation will be necessary to map accurately the weaker lines.

γ Rays

The earlier experiments on the γ rays were mainly confined to a determination of the absorption of the more penetrating radiations by different kinds of matter. It was early observed, however, that some of the radiations appeared to be complex. This was shown by anomalies in the initial part of the absorption curve. In the meantime, a notable advance in our knowledge of X-rays had been made by the work of Barkla. He found that under certain conditions each element when bombarded by X-rays of suitable penetrating power gave rise to a strong radiation which was characteristic for that element, *e.g.*, the lighter elements from aluminium to silver emitted characteristic radiations called the 'K' series, which increased rapidly in penetrating power with the atomic weight of the radiator. It was found that the heavier elements emitted in addition another characteristic radiation of softer type, which was called the 'L' series. These results showed clearly that there must be definite structures within the atom which gave rise to a definite radiation under suitable conditions of excitation. From these results it seemed probable that the γ rays from radio-active matter must consist of the characteristic radiations of these heavy elements, analogous in type to the corresponding radiations observed in ordinary elements when excited by X-rays or kathode rays. These conclusions were confirmed by a series of investigations made by Rutherford and Richardson. The γ rays were analysed by means of their absorption by aluminium and by lead, the disturbing effects of the primary β rays being eliminated by means of a strong magnetic field. It was found, for example, that the γ rays from radium B, when examined by their absorption in aluminium, consisted of at least two types, one easily absorbed, and the other eighty times more penetrating. By further observations of the absorption of the γ rays by lead, Richardson found that the rays from radium B could be divided into at least four definite types, each of which was absorbed exponentially by lead. Similar results were obtained for all the radio-active elements which emitted γ rays. In some cases the soft γ rays, *e.g.*, those from radium B, corresponded to the characteristic radiation of the 'L' series, and others to the 'K' series. The general results, however, indicated that several additional series of characteristic radiations are present in some cases. It was clear from these experiments that the γ rays corresponded to the natural modes of vibration of the inner structure of the radio-active atoms. In the meantime the experiments of W. H. Bragg and W. L. Bragg, and of Moseley and Darwin, had shown that the characteristic X-radiations of the elements gave definite and well-marked line spectra. These spectra were simply determined by reflecting the rays from crystals. If this were the case, it seemed probable that the γ rays from the radio-active atoms would also give line spectra, and thus allow the natural frequencies of vibration of these atoms to be determined. During the past year, a number of experiments have been made to test this point by Rutherford and Andrade, using radium B and radium C as the source of γ radiation. As was anticipated, it was found that the γ rays from radium B and radium C gave well-marked line spectra. The general method employed was to use an α-ray tube containing a large quantity of emanation as a source of radiation. The γ rays were reflected from a crystal of rock-salt, and the position of the spectrum lines determined photographically. Usually twenty-four hours were necessary to obtain a marked photographic effect. Special difficulties arose in these experiments which are absent in an investigation of a similar kind with X-rays. In addition to γ rays, the radio-active matter emits very penetrating β rays which have a strong photographic action; while the γ rays in their passage through matter themselves give rise to high-speed β rays. The disturbing effect of these radiations has to be eliminated by placing the whole apparatus between the poles of a powerful electromagnet. In this way it was found that the spectrum of radium B consisted of a large number of lines, of which the most intense were deflected at angles of 1° 46', 10°, and 12°. The more penetrating radiation from radium C gave a strong line of 1° and a fainter line at 43'. The strong lines at 10° and 12° are due to easily absorbed γ rays, and undoubtedly correspond to the 'L' radiation of radium B. The line at 1° corresponds to a very penetrating radiation which has a wave-length less than one-tenth of an Angström unit. The penetrating γ rays from radium C have by far the shortest wave-length so far observed. It does not seem probable that such short waves can be produced artificially in an X-ray tube unless possibly an exceedingly high voltage be applied.

There is one interesting result of these investigations that should be mentioned. The two strong lines of the radium B spectrum deflected at 10° and 12° were found to correspond exactly in position to the X-ray spectrum of lead. These experiments thus confirmed the view based on chemical evidence that radium B and lead were isotopic, *i.e.*, they were elements of practically identical chemical and physical properties, although their atomic weight differed by seven units.

Connection between β and γ Rays

Before considering in detail the difficult problem of the connection between β and γ rays, it is desirable to summarise the main facts that have been established in regard to the relations between kathode rays and X-rays:—

(1) A small part of the energy of kathode rays falling on a radiator is converted into X-rays, the average frequency of the latter increasing with the velocity of the kathode particle.

(2) X-rays in passing through matter give rise to a β radiation. The initial energy of the escape of the electrons increases with the frequency, and is probably proportional to it.

(3) Electrons or X-rays of appropriate energy are equally able to excite the characteristic radiations in an atom.

The results which have been shown to hold for the X-rays hold equally for the β and γ rays, which have much greater individual energies, *e.g.*, Gray and Richardson have shown that the β rays from radio-active matter are able to excite the characteristic radiations of the elements in a number of substances, while γ rays in passing through matter give rise to high-speed electrons. It was long ago suggested by Bragg that β rays and X-rays are mutually convertible forms of energy, *e.g.*, a β particle falling on matter may be converted into an X-ray of the same energy, and the latter in passing through matter may in turn be converted into an electron of identical energy. This assumes that the energy of an X-ray and an electron are mutually convertible, and the energy may appear under suitable conditions in either of the two forms. While the general evidence indicates that this point of view may hold closely for the conversion of the energy of a single X-ray into that of a swift electron, it is very doubtful whether it holds for the converse case of the excitation of an X-ray by an electron. We shall see later from experimental evidence that in general the energy of the electron required to excite an X-ray of definite frequency is always greater than the corresponding energy carried off in the form of an X-ray.

It was early observed that there appeared to be a close connection between the emission of β and γ rays from radio-active matter. In all cases, the two types of radiation appeared together. A closer examination, however, showed that there were very marked differences between the relative energies of the β and γ rays from different radio-active elements. For example, radium C emits intense β rays and also intense γ rays; on the other hand, radium E emits intense β rays over a wide range of velocity, but exceedingly weak γ rays. Differences of a similar kind were observed amongst a number of the radio-active elements. One striking distinction, however, was to be noted. All the radio-active substances which give a marked line spectrum of β rays also emitted intense γ rays. On the other hand, a substance like radium E, which gave scarcely any γ rays at all, gave a continuous spectrum of β rays in which no lines have so far been observed. It thus appeared probable that the line spectrum of the β rays was intimately connected with the emission of γ rays, and this conclusion has been completely established by recent experiments. As we have seen, γ rays in passing through matter give rise to high-speed β rays. Using radium B and radium C as a source of γ rays, the β radiation excited in a number of metals by the passage of γ rays was analysed in a magnetic field by Messrs. Robinson and Rawlinson and the writer, and was found to consist in part of definite groups of β rays. When lead was the absorbing material, the magnetic spectrum of the β rays excited by the γ rays was found to be nearly identical with the primary β-ray spectrum of radium B. This striking result shows that those β rays escaping from the radio-active atom which give rise to a line spectrum must result from the conversion of γ rays into β rays in the radio-active atom. The slight differences observed in the spectrum for different metals is probably connected with the energy required to excite one of the characteristic radiations of the element used as absorber.

An explanation of the marked differences in the character of the β and γ radiation from different radio-active atoms can, I think, be given on the following lines. Some of the γ rays are broken up in their escape from the atoms, and the energy of each converted γ ray is transferred to an electron which escapes with a definite velocity dependent on the frequency of the γ radiation. Taking into account a large collection of disintegrating atoms, each of the possible modes of characteristic vibration of the atom gives rise to an electron of definite speed. In this general way we may account for the line spectrum of the β rays which is so commonly observed. On this view, we should expect to obtain a well-marked line spectrum of β rays when a substance emits strong γ rays—a result in accord with observation.

In order to account for the marked differences in the types and intensity of γ rays from different radio-active substances, it seems necessary to suppose in addition that the primary β particle always escapes from the nucleus in a *fixed* direction with regard to the structure of the atoms under consideration. For example, we have already pointed out that radium E, although it emits intense β rays which give a continuous spectrum over a wide range of velocity, emits very weak γ rays. Since there can be no doubt that the β rays have sufficient speed to excite the characteristic modes of vibration which must be present in the atom, we are driven to the conclusion that the β particle escapes in such a direction that it does not pass through these vibrating centres. On this view, the type of characteristic γ rays which are excited, and consequently also the corresponding speed of the β rays which arise from the converted γ rays, will depend entirely on the direction of escape of the primary β particle. The definite direction of escape of the primary β particle, which varies for atoms of different substances, also suffices to explain a number of other differences observed in the mode of release of energy from various radio-active atoms. It is supported by many other observations which indicate that the atoms of a particular radio-active substance break up in an identical fashion.

We have so far considered only in a qualitative way the relation between the groups of rays in a β-ray spectrum and the emission of characteristic γ rays. During the last few years there has been a growing body of evidence that the energy E carried off in an X-ray of frequency ν is proportional to this frequency, and is given by E $= h\nu$ where h is Planck's fundamental constant. If the whole of the energy of an X-ray can be given directly to an electron, the energy communicated to the latter should be $h\nu$. There is no doubt that in many cases this simple relation holds very approximately, but the measurements so far available are not sufficiently precise to settle definitely whether a part of the energy may not appear in another form.

Assuming that the transfer of the energy from an X-ray to an electron is complete, we should expect to find groups of β rays of energy corresponding to $h\nu$ where ν is the frequency of the γ rays found experimentally. Such a relation is found to hold within the limit of experimental error for three marked groups of low-velocity β rays emitted from radium B. On the other hand, it is found that many of the high-velocity groups of β rays both from radium B and radium C have energies many times greater than correspond to any observed frequency. Not the slightest evidence, however, has been obtained that the corresponding high frequencies of vibration exist in the radio-active atom; in fact, all the evidence points to the fact that these high-speed electrons arise from one or more of the observed frequencies in the γ-ray spectrum.

In order to account for such results, it seems necessary to suppose that the γ rays of high frequency are not necessarily emitted as single pulses, but consist of a train of pulses either produced simultaneously or following one another at very short intervals. Each of these pulses has an energy $h\nu$ corresponding to the frequency ν, but the total energy in the train of waves is $ph\nu$ where p is a whole number, which may have possible integral values 0, 1, 2, 3, . . . etc., depending on the structure of the atom and the conditions of excitation. The penetrating power of such a train of waves corresponds to that of a single wave of frequency ν, but on passing through matter the energy of the whole train of p waves occasionally may be transferred to an electron which consequently is expelled with an energy $ph\nu$. There is very strong evidence of the general correctness of this point of view, for most of the stronger lines in the β-ray spectrum of radium C have energies which correspond to an *integral* multiple of the energy corresponding to the strong lines actually observed in the γ-ray spectrum. It seems probable that under the ordinary conditions of excitation by kathode rays in a vacuum tube, the X-ray contains only one pulse or wave, but under the far more powerful stimulus of the very swift β particle escaping from the atom, a long train of waves, each of the same frequency, is produced. The energy of the whole train of waves may under suitable conditions be given to an electron, which consequently has a speed very much greater than that impressed upon it by a single wave of the same frequency.

Limit to the Frequency of Vibration of the Atom

There is one question of fundamental importance which arises in considering the modes of vibration of the atom, viz. whether there is a definite limit to the frequency of the radiation which can be excited in a given atom. Theory does not provide us with an answer to this problem, since little is known about the conditions of excitation, nor even of the nature of such high-frequency vibrations. A study of the frequency of the γ rays from radio-active substances is of great importance, as it throws much light on this problem.

As we have seen, the energy of the β particle escaping from the nucleus of radium C is equivalent to that acquired by an electron moving in an exhausted space under a potential difference of several million volts. This high-speed electron passes through the electronic distribution in its escape from the atom. Notwithstanding such ideal conditions for the excitation of high-frequency radiations of the atom, the highest frequency in the radiation emitted by radium C is only about twice that obtainable from an ordinary hard X-ray tube excited by 100,000 volts. It thus appears probable that there is a definite limit to the frequency of the radiation obtainable from a given atom, however high the speed of the disturbing electron. This limiting frequency is determined not by the speed of the electron but by the actual structure of the atom. Since the γ radiation from radium C gives a line spectrum, it would appear that the highest frequency obtainable is due to a definite system of electrons which is set into characteristic vibration by the escape of a β particle. In order to throw further light on this point, Prof. Barnes, Mr. H. Richardson and myself have recently made experiments to determine the maximum frequency obtainable from an X-ray tube for different constant voltages. The Coolidge tube, which has recently been put on the market, is ideal for this purpose, as it provides powerful radiation at any desired voltage. The anti-kathode is of tungsten of atomic weight 184, so that we are dealing in this case with the possible modes of vibration of a heavy atom. The maximum frequency of the radiation was deduced by measuring the absorption by aluminium of the most penetrating rays emitted at different voltages. The absorption of X-rays of different frequencies by aluminium has been examined over a very wide range, and can be expressed by simple formula. It was found that for 20,000 volts the frequency of the radiation was slightly lower than that to be expected if Planck's relation held. With increasing voltage there is a rapid departure from Planck's relation. The frequency reaches a maximum at about 145,000 volts, and no increase was observable up to the maximum voltage employed, viz. 175,000 volts. The experiments thus show that the frequency of radiation reaches a definite maximum, which is no doubt dependent on the atomic weight of the particular radiator employed. It is of interest to note that the maximum penetrating power of the X-rays from the Coolidge tube in aluminium is about the same as the γ rays from radium B, but is about three-tenths of the γ rays from radium C. There is evidence which suggests that the very penetrating γ rays from radium C correspond to the octave of the 'K' characteristic radiation of that element. If this be the case, it may prove possible that a still more penetrating radiation might be obtained from tungsten, but in order to excite it a voltage of the order of a million volts would probably be required. In any case, it seems clear that Planck's relation does not hold for excitation of high frequencies by swift electrons, but may hold very approximately for lower frequencies corresponding to the radiation excited by a few hundreds or thousands of volts. On the other hand, the evidence obtained from a study of the β rays excited by X-rays or γ rays certainly indicates that the relation $E = ph\nu$ holds at any rate very approximately for the highest frequency examined. It is thus obvious that the emission of β and γ rays from the radio-active atom is clearly connected with the general theory of radiation, and it seems likely that a close study of these radiations will throw much light on the mechanism of radiation in general.

There can be little doubt that the penetrating γ rays from active matter have their origin in the vibration of electronic systems in the structure of the atom outside the nucleus. The nucleus itself, however, must be violently disturbed by the expulsion of an α or β particle. If this leads to the emission of a γ radiation, it must be of exceedingly high frequency, as the forces holding together the component parts of the nucleus must be exceedingly intense. We should anticipate that this radiation would be extraordinarily penetrating, and difficult to detect by electrical methods. So far no experimental evidence has been obtained of the existence of such very high-frequency radiations, but it may be necessary to devise special methods before we can hope to do so.

[E. R.]

Maximum Frequency of the X Rays from a Coolidge Tube for Different Voltages

by SIR ERNEST RUTHERFORD, F.R.S., PROFESSOR J. BARNES, PH.D., *and* H. RICHARDSON, M.SC.

From the *Philosophical Magazine* for September 1915, ser. 6, xxx, pp. 339–60

IN the course of last year, Mr. C. G. Darwin began an investigation in the University of Manchester to examine the relation between the velocity of cathode rays and the frequency of X rays excited by them in different radiators. The cathode rays were generated by the electric discharge in a suitable vacuum-tube, and by means of an adjustable magnetic field rays of definite speed were allowed to fall on a radiator. It was the intention of Mr. Darwin to examine the frequency of the X rays initially by measuring their absorption in aluminium, and if possible by direct reflexion from crystals. Some difficulty was experienced in obtaining a sufficiently good and constant vacuum, and measurements were interrupted by the departure of Mr. Darwin to the seat of war. The experiments were continued by Mr. H. Richardson, but in the complicated apparatus employed it was found difficult to keep the vacuum sufficiently constant when a discharge was passed.

As soon as the Coolidge tube was put on the market, it was recognized that it afforded a much more convenient method of attacking a part of the main problem and over a much wider range of voltage. As is well known, the discharge in the very perfectly exhausted Coolidge tube is mainly carried by the negative electrons liberated from a tungsten spiral heated to a high temperature by means of the electric current. The anticathode is of tungsten, and the exhaustion in the tube employed in the present experiments was so perfect that the tube maintained, with suitable precautions, 175,000 volts across its terminals without any obvious breakdown of the vacuum. Since tungsten is of atomic weight 184, and atomic number 74, the X radiation from the Coolidge tube corresponds to that emitted from a heavy element; but it is to be expected that the frequency of the X radiation for a high voltage should be somewhat less than from the ordinary X-ray tube with a platinum or platinum-iridium anticathode, since the atomic weights of iridium and platinum are 193 and 195, and atomic numbers 77 and 78 respectively.*

General method of the Experiment

The primary object of the experiment was to determine the maximum

* See Moseley, Phil. Mag. xxvii. p. 703 (1914).

frequency of the X rays emitted from a Coolidge tube excited by different *constant* voltages. Since the penetrating power of X rays in a light element like aluminium increases regularly and rapidly with the frequency, an estimate of the maximum frequency of the radiation can be made by determining the absorption in that metal of the 'end' radiation from the tube, *i. e.* the absorption of the most penetrating rays present when the rays of smaller frequency have been almost completely absorbed.

It has been known for some time that the absorption coefficient μ in aluminium of the characteristic X radiation from different radiators is given approximately by $\mu = k\lambda^{5/2}$ where λ is the average wave-length of the radiation.* This relation has been recently examined in detail by W. H. Bragg and Pierce† by determining the value of μ in aluminium for individual spectrum lines of definite frequency, and found to hold fairly accurately over the limited range employed, viz. for wave-lengths between $0 \cdot 49 \times 10^{-8}$ and $0 \cdot 615 \times 10^{-8}$ cm., *i. e.* for radiations which are reflected from rock-salt between angles of $5°$ and $6°$. There seems to be no doubt that this relation will hold very approximately for the much shorter wave-lengths contained in the more penetrating γ radiations emitted by radium B and radium C. It can be deduced from Bragg's results as a mean of the measurements on the silver β and palladium β lines that a radiation of wave-length $\lambda = 5 \times 10^{-9}$ cm. has an absorption coefficient $\mu = 5 \cdot 6$ (cm.)$^{-1}$ in aluminium. If the above relation between absorption and frequency for aluminium holds, the γ radiation from radium C, which has the value $\mu = 0 \cdot 115$ in aluminium, corresponds to a radiation of $\lambda = 1 \cdot 06 \times 10^{-9}$ cm., which should be reflected from rock-salt at an angle of $1° 5'$. By far the strongest line in the γ-ray spectrum of radium C was found by Rutherford and Andrade‡ to be reflected at an angle of $1°$ from rock-salt. There is thus a very fair accord between experiment and calculation when the relation is extrapolated over a wide range in the value of μ, viz. nearly 50 times. We may consequently assume with confidence that the relation $\mu = k\lambda^{5/2}$ holds very approximately over the whole range of wave-lengths employed in the experiment, viz. from $\lambda = 14 \times 10^{-9}$ cm. to $\lambda = 1 \times 10^{-9}$ cm.

It was hoped at the same time to make a systematic examination of the X-ray spectrum of the radiations, and to determine if possible the voltage at which the spectrum-lines appeared. Barnes in an accompanying paper§ has given the wave-lengths of the spectrum-lines observed with the Coolidge tube. On account of the thickness of the glass of the vacuum-tube, the intensity of the issuing 'L' characteristic radiation was weak when examined electrically or photographically, but permitted of the determination of the wave-lengths of the stronger lines. No evidence of other well marked lines was noted in the region of higher frequencies, but the main part of the

* See Owen, Proc. Roy. Soc. lxxxvi. p. 426 (1912).
† 'X rays and Crystal Structure,' by W. H. & W. L. Bragg, pp. 180, 181.
‡ Rutherford and Andrade, Phil. Mag. xxviii. p. 263 (1914). (*This vol., p. 456.*)
§ See Barnes, Phil. Mag. xxx, p. 368 (1915).

spectrum appeared to be continuous with the crystals employed. As will be seen later, the experiments on absorption of the radiation showed that the 'end' frequency increased regularly with increase of voltage. No evidence was obtained that the maximum frequency for different voltages varied by jumps, such as might be expected if the issuing radiations were mainly confined to a few waves of definite frequency.

Arrangement of the Experiment

The energy of the electrons striking the anticathode in the high vacuum of the Coolidge tube depends only on the voltage applied. In order to investigate the effect of electrons at definite speed, it was necessary to employ a constant voltage delivered by an influence-machine in place of the variable voltage due to an induction-coil or transformer. For this purpose, the only machine available was a large Wimshurst of 12 plates of diameter 71 cm. This had been presented to the Department about 15 years previously, and was in some respects not nearly so well suited for the experiment as one of the more modern types of high-speed machines. The Wimshurst, of which 10 plates had survived, was run by a motor, and after a month's fairly continuous running, four more of the plates cracked and were removed. This proved fortunate, for it was found that the machine with six plates gave nearly the same maximum voltage and delivered nearly the same current as in the beginning, and in addition ran much more steadily.

In order to reduce the losses to a minimum, all the conductors consisted of light metal tubing 2·7 cm. in diameter with rounded ends. These passed through large paraffin insulators to the Coolidge tube T, which was placed in a box covered with sheet-lead 3 mm. thick. The general arrangement of the apparatus is shown in fig. 1. Special precautions were taken to prevent losses

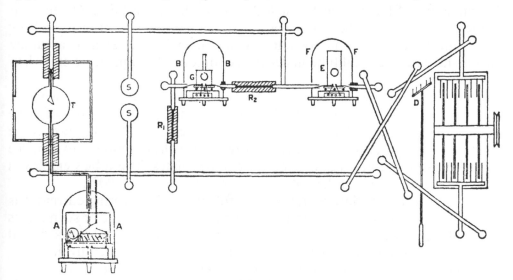

Fig. 1

from the external electrodes of the bulb and to prevent discharges over its surface. For this purpose, the dust collecting on its surface was regularly removed and the surface washed with alcohol.

As the machine after stoppage occasionally reversed its voltage, suitable cross connexions (see fig. 1) were arranged to rectify rapidly the direction of the current. The accessories for the Coolidge tube, viz., the battery, adjustable resistance. etc., were placed on an insulated stand and completely covered with a rounded metal case. The current could be controlled by an insulating handle coming through a small opening. The conductors for the heating current passed inside the hollow metal tubes.

Measurement of Voltage and Current

In order to determine the absorption curve of the radiation with accuracy, it was necessary to keep the voltage very constant, and to have some method of knowing the voltage at any moment. For this purpose, the conductors to the bulb were shunted through a high resistance in series with a galvanometer. The resistance consisted of two capillary tubes, R_1 and R_2, filled with xylol and thoroughly insulated by a thick layer of paraffin. The moving-coil galvanometer G with suitable shunts was placed between the two resistances, so that its potential was never far from zero. In order to prevent electrostatic disturbances and electrical losses, the galvanometer was placed on a rounded metal base, insulated on paraffin blocks and covered with a round metal screen, BB. The resistance in series in the galvanometer was about 5000 megohms, and was such that only about 30 per cent. of the total current from the machine passed through it even at the highest voltage employed, viz. 115,000 volts.

It was at first intended to measure the voltage across the tube directly by determination of the current through the known high resistance. On account of the charging up of the xylol tubes after the current had been passed for some time, it was not found possible to determine with the requisite accuracy the resistance of the xylol tube under the conditions of temperature etc. when in actual use.

The deflexions of the galvanometer were instead standardized by the spark method. For this purpose, we made use of a spark-gap composed of large copper spheres, SS, 20·5 cm. in diameter, constructed by Dr. Makower some years ago. Over the range of voltage employed, the voltage required to produce a spark was practically the same as for parallel plates. The tables employed were those given by C. Müller, *Ann. d. Phys.* xxviii. p. 612 (1907).

The method of standardization was as follows:—The heating current was adjusted to the required value and the machine was run for five or six minutes at about the voltage required, so that the resistance of the xylol tubes should reach a steady state. The length of the spark-gap was then carefully adjusted, and a number of observations made of the deflexion of the galvanometer at the moment the spark passed, care being taken that the voltage rose slowly to the sparking-point. In a similar way, observations were made at the end of a series of experiments, but no certain change in the deflexion was ever observed over the interval of a few hours. A gradual increase of the xylol resistance was, however, observed over the interval of several months required for the experiments. The deflexion of the galvanometer was found to be nearly proportional to the voltage over a considerable range. For voltages greater than 30,000, the spark-gap method was very suitable for calibrating the galvanometer directly. For voltages below this, it was found more convenient and accurate to take the deflexion of the galvanometer as proportional to the voltage. The correctness of this was confirmed on several occasions by determination of the deflexion of the galvanometer for a voltage read directly on a Kelvin astatic voltmeter.

As it was very important to keep the voltage constant during an absorption experiment, it was necessary to control the voltage within small limits by means of an adjustable point discharger D placed near one of the high potential conductors. The arrangement for this purpose is seen clearly in fig. 1. The voltage galvanometer worked throughout in a very satisfactory manner, and it was possible by its aid to keep the potential steady within about one per cent. over the interval of an hour or more required for a complete absorption experiment.

Since the galvanometer was highly damped, it would not indicate any rapid surges in the voltage. These surges, however, occasionally made themselves manifest at the extreme end of the absorption curve, where the intensity of the radiation had been reduced to about 1/1000 of its initial value. Even when the voltage appeared quite steady by the galvanometer, the presence of surges could be detected by the irregular rate of movement of the electrometer.

Measurement of Current

A moving-coil galvanometer E was placed in the main circuit to measure directly the current delivered by the machine, and was protected against electrostatic disturbances by a metal shield FF, as in the other case. The actual current passing through the bulb was measured in the following way. The deflexion was first observed under the experimental conditions of excitation of the radiation. The heating circuit was broken, and the voltage retained at the same value by means of the point discharge and by altering the speed of the machine. The deflexion rapidly dropped to a constant value, which was due mainly to the current taken by the voltage galvanometer, but partly also to conduction over the surface of the heated bulb. The difference between these two readings thus gave a definite measure of the current passing through the bulb, quite independently of all other losses in the circuits. The current passing through the bulb in the various experiments at different voltages varied from 1/100 to 6/100 of a milli-ampere. It was found in all experiments that the ionization for a given voltage applied to the tube was directly proportional to the current through the bulb—in other words, the intensity of the radiation was directly proportional to the number of electrons incident on the anticathode.

Determination of the Absorption Curves

The general arrangement of the apparatus for this purpose is shown in fig. 2. The X rays, passing through a rectangular opening (6 × 6 cm.) in the lead box, entered the ionization chamber A. This consisted of a lead box

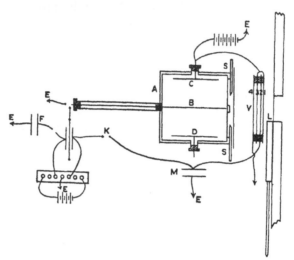

Fig. 2

(15 × 15 × 15 cm.) 3 mm. thick, divided into two equal partitions by the electrode B. The two insulated plates C and D were generally connected together and charged to a P.D. of 1000 volts, which was found to be sufficient to produce saturation under the experimental conditions. Two equal openings were cut in the front face, which were closed with a thin sheet of mica to make the vessel air-tight, and covered with thin aluminium foil to make it conducting. Two thick lead slides, SS, were constructed to control the width of the beam of rays entering either half of the ionization vessel. The current supplied to the electrode B was measured by a Lutz string electrometer, which proved very suitable in all respects for this purpose. The instrument was very easily set up, and sufficient sensibility was obtained when the plates were charged to ± 15 volts, while the zero was very steady and the deflexions read with ease. The quartz fibre was broken during the course of the experiments, and the later measurements were made with a Kaye-Wilson electroscope.

The main difficulty in these experiments lay in the fluctuations of the current through the bulb, and consequent variation in the intensity of the radiation. To correct for this, the radiation before entering the ionization chamber A passed through a 'standardizing vessel' V of thin parallel aluminium sheets. The plates 1, 3 were connected to the high potential battery and the plate 4 earthed. The plate 2 was connected with a mica condenser M of suitable capacity, which could be connected at will through the insulating key K to the string electrometer. The method of conducting an experiment was as follows. By means of the key K, the standardizing vessel was disconnected from the electrometer and the connexion of the latter with earth broken. A thick lead slide L, which completely stopped the radiation, was rapidly drawn aside and the radiation allowed to pass into the ionization vessel for a period varying from 10 to 60 seconds in different experiments. The slide was then closed, and the steady deflexion of the electrometer read. After discharge of the latter, the standardizing vessel was connected, and the deflexion again read. The corrected deflexions for the two vessels should be proportional to each other provided the fluctuations are in the current and not in the voltage; for variations of the latter alter the penetrating power of the radiation as well as the current through the bulb. In this way, provided the voltage was kept steady, it was found possible to correct for any changes in the intensity of the radiation over the long interval of an hour or more required to obtain a complete absorption curve.

It was necessary to measure the absorption curve, especially for high voltages, over a very wide range of thickness of aluminium where the current in the ionization vessel was reduced in some cases to 1/10000 of its initial value. This was done as follows. An air-condenser F of capacity about 400 e. s. units was placed parallel with the electrometer, and the width of the opening of the ionization vessel adjusted till there was a convenient deflexion of the electrometer in 10 seconds. Successive screens of aluminium were introduced, and the currents in the testing and standardizing vessels compared as the current diminished. The air condenser was removed, and before the current became too small to measure with accuracy, the opening of the ionization vessel was widened until again a convenient deflexion was obtained when the condenser was in the circuit. This process was continued two or three times, depending on the variation in range of the ionization current to be measured. For the end part of the curve, the testing vessel was always completely open to the radiation. In this way it was possible to determine the complete absorption curve over a very wide range without introducing any uncertainty in regard to saturation. The capacities of the circuits were carefully determined, and the readings corrected for change of capacity of the systems and for inequalities in the electrometer scale.

In the later experiments, an additional method was used in order to correct for changes in intensity of the radiation. This depended on the observed fact that the intensity of the radiation for a given voltage was directly proportional to the current through the bulb. The value of the latter was read by means of the galvanometer during the course of each observation by the methods already described. The values obtained by the two standardizing methods were in good agreement, and each served as a useful check on the other.

Experimental Results

Some of the absorption curves in aluminium of the radiation at different voltages are shown in fig. 3, where the logarithm of the current is ordinate, and the abscissæ the thickness of aluminium. If the absorption were exponential, the curve should be a straight line, but it is seen that this is not the case for any of the curves. The radiation at first rapidly diminishes owing to the absorption of the softer radiations, and on the average gradually becomes more penetrating with increase of thickness of absorber.* Finally, for the higher voltages, when the current is reduced to about 1/500 of its value, the absorption curve becomes very nearly a straight line until the radiation is completely absorbed. This shows that the end radiation is approximately homogeneous, and it is the absorption coefficient of this end radiation that was carefully determined.

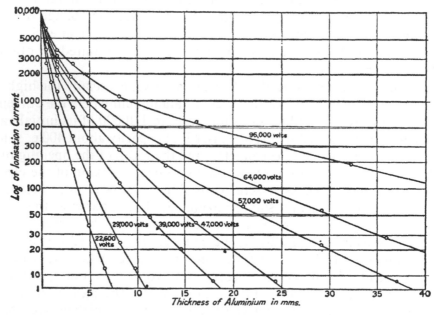

Fig. 3

From 40,000 volts upwards, the end radiation was absorbed exponentially over a considerable range of intensity, but below this voltage the penetrating power of the radiation appeared to slightly increase until it was completely absorbed. The analysis of the radiation by reflexion of crystals, as given in the following paper, shows that the 'L' characteristic radiation of tungsten escapes from the bulb. Its intensity, however, is greatly reduced by passing through the glass of the bulb, which was found by direct measurement to be about 0·5 mm. thick. Careful observations were made of the voltage for which the ionization in the testing-vessel was first measurable. In order to increase the electrical effect, the testing-vessel was filled with sulphur dioxide.

* Preliminary measurements on the absorption of the radiation from a Coolidge tube have been made by S. Russ (Journ. Röntgen Soc. April 1915), using an induction-coil to excite the rays. He observed that the radiations were not homogeneous, but tended to become so with increase of thickness of absorber.

No ionization was observed below 10,300 volts, and it then increased very rapidly with the voltage. The absorption of the main radiation by aluminium was examined at the lowest possible voltage, and was found to be $\mu = 69$ cm.[1] in aluminium. This coefficient no doubt corresponds to that of the characteristic 'L' radiation of tungsten under experimental conditions. The absorption coefficient of this radiation measured by Chapman* was found to be 81, but as in passing through the bulb the softer components were relatively cut out, the issuing radiation would be expected to be more penetrating than that observed under normal conditions with no absorber. The range over which the ionization could be measured increased rapidly with the voltage above 10,000 volts, but on account of the presence of a large proportion of softer radiations, it was difficult to determine with certainty the absorption coefficient of the end radiation for the lower voltages. In every experiment it was found, however, that the penetrating power of the radiation increased rapidly and regularly with increase of the voltage.

Attempts were made to test whether the absorption curves could be analysed into components corresponding to definite characteristic radiations. Kaye† has made numerous experiments of this kind to analyse the radiation from different anticathodes, and obtained indications that by the use of suitable absorbers the radiation could be analysed into definite components. While it was not difficult to express the absorption curve for any particular voltage with considerable accuracy by the sum of three exponentials, the values of the constants changed with voltage, and it was concluded that such an analysis, apart from showing the presence of the 'L' characteristic, had no physical meaning. It may yet be possible, as Kaye has suggested, to analyse the radiation as a mixture of two or more characteristic radiations, but before this can be accomplished it will be necessary to carry out a large number of experiments on the absorption of the radiation by different materials. It appears to us, however, unlikely that the absorption curves can be completely expressed as the result of a small number of characteristic radiations each of which are absorbed exponentially.

Experiments with Induction-coil

It was not found feasible to examine the absorption curves of the radiation excited by the Wimshurst above 115,000 volts. In order to carry the experiments still further, a large induction-coil which gave a 20-inch spark, operated by a Sanax break, was used. The gap between the sparking spheres was set to the potential required, and the current through the coil carefully adjusted so that an occasional spark passed. In many cases, a spark-gap ending in fine needle-points was used in parallel with the sparking spheres, to test whether there was any sensible alteration of the potential required to spark across the spheres, owing to possible alteration of their surface by the passage of the preliminary sparks.

---

* Proc. Roy. Soc. A. lxxxvi. p. 439 (1912).

† Kaye, 'X Rays,' pp. 121–123, Longmans, Green & Co., 1914.

The current through the heating-coil was adjusted so that the intensity of the radiation was about the same as that excited by the Wimshurst, and the absorption curves in aluminium determined as before.

It was anticipated that the radiation excited by the variable voltage of the induction-coil would on the average contain a larger proportion of softer radiation than the radiation excited by the Wimshurst. To our surprise, however, we found that the curves for equal intensities of radiation were very similar, and it was difficult to distinguish one from the other.

Russ (*loc. cit.*) had observed that the proportion of penetrating radiation increased with increase of current through the bulb. This, if true, is a very important observation, for it would indicate that the quality of the rays depends not only on the velocity of the electrons but also on their number. In our preliminary examination of this point, results were obtained in general agreement with those of Russ, but the difference between the absorption curves was finally traced to another cause. When working with very intense radiations, it was necessary to nearly close the opening in the ionization vessel by means of the lead slides. Some radiation was found to enter the vessel by scattering from one lead plate to another or by the excitation of characteristic radiations. When the front of the ionization vessel was covered with a thick lead sheet and the rays allowed to enter through a small opening, the disturbance was eliminated, and the absorption curves were found to be independent of the current through the bulb over a wide range.

Since a much greater intensity of radiation could be excited by the use of the coil, the absorption curve could be obtained over a greater range of thickness of aluminium. In such cases, the penetrating power of the 'end' radiation appeared to be slightly greater than that observed with the Wimshurst. On account, however, of the uncertainty as to the equality of the maximum potential in the two cases, not much stress can be laid on this difference; for it is probable that the minimum voltage required to produce a given length of spark is greater for the coil than for the Wimshurst, on account of the rapid variations in the potential of the former.

Since for the 'end' radiation, the coil gave about the same value as the Wimshurst, the former was employed to determine the penetrating power of the radiation between 110,000 and 175,000 volts. For such high voltages, the end radiation is absorbed nearly exponentially over a wide range. On account of the danger that the bulb might break down under such high voltages, the experiments were confined to an examination of the end radiation alone. In order to detect small variations in the absorption of the rays, the experiments for 125,000, 145,000, and 175,000 volts were made in the following way. Two sheets of lead each of thickness 0·62 mm. were placed in front of the ionization vessel, and the current determined with the fixed capacity in parallel. A thickness of aluminium 3·24 cm. was added, and the current measured with the capacity removed. An exactly similar process was carried out at 145,000 and 175,000 volts, and the variation of the ratios of the two currents allowed of an accurate measure of the small change in absorption coefficient in the three cases.

Variation of Absorption of End Radiation with Voltage

The curve showing the values of μ in aluminium for the end radiation at different voltages is shown graphically in fig. 4, where the abscissæ represent

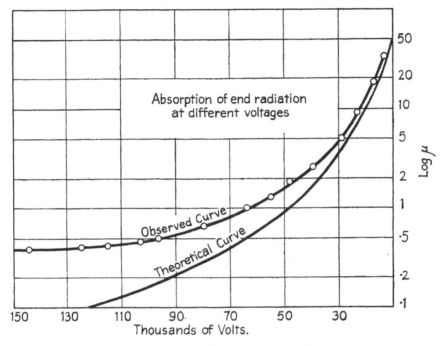

Fig. 4

volts, and the ordinates log μ. The experimental points are indicated in the figure, and from the smoothed curve the values of μ for different voltages are given in the following table. Taking $\lambda = k\mu^{2/5}$, the values of the wavelength λ are calculated, the value of k being deduced from the data of Bragg previously quoted, viz. $\mu = 5 \cdot 6$ (cm.)$^{-1}$ for $\lambda = 5 \cdot 0 \times 10^{-9}$ cm.

It will be seen that the value of μ decreases rapidly at first with voltage, but decreases slowly after 100,000 volts, and very slowly after 125,000 volts. There is no measurable change in μ between 145,000 and 175,000 volts.

During recent years, experimental evidence of various kinds has indicated that the energy of radiation is emitted in definite quanta, expressed in the simplest case by Planck's relation $E = h\nu$, when E is the energy of the ray and ν its frequency, and h Planck's fundamental constant, which has a value $h = 6 \cdot 55 \times 10^{-27}$ erg. sec. It is of great interest to see how far such a relation holds for the excitation of X rays by electrons. If all the energy E of the electron can be converted into a single X ray of definite frequency ν, then $E = h\nu$. Assuming that such a relation holds for excitation, we can at once calculate the values of μ and the wave-length λ to be expected for each voltage. The calculated tables for μ and λ are given in columns III and V of Table I. The value of $e/h = 7\cdot27 \times 10^{16}$ is taken to calculate the value of λ. This is deduced from the results given by Warburg, Leithauser, Hupka, and Müller (*Annal. d. Physik*, xl. p. 609, 1913) without any assumption of the value of e.

TABLE I

I	II	III	IV	V
		Calculated μ		Calculated
Voltage in thousands	Observed μ in aluminium	in aluminium on quantum theory	Observed wave-length	wave-length on quantum theory
13·2	33 (cm.)$^{-1}$	26·3 (cm.)$^{-1}$	$10\cdot2 \times 10^{-9}$ cm.	$9\cdot4 \times 10^{-9}$ cm.
20	12	9·4 ,,	6·8 ,,	6·19 ,,
30	4·7	3·4 ,,	4·66 ,,	4·13 ,,
40	2·46	1·66 ,,	3·60 ,,	3·10 ,,
50	1·53	0·95 ,,	2·98 ,,	2·48 ,,
60	1·07	0·60 ,,	2·58 ,,	2·06 ,,
70	0·81	0·41 ,,	2·31 ,,	1·77 ,,
80	0·66	0·29 ,,	2·13 ,,	1·55 ,,
90	0·54	0·22 ,,	1·96 ,,	1·38 ,,
100	0·48	0·17 ,,	1·87 ,,	1·24 ,,
110	0·45	0·133,,	1·82 ,,	1·13 ,,
125	0·42	0·085,,	1·77 ,,	0·99 ,,
145	0·39	0·066,,	1·72 ,,	0·85 ,,
175	0·39	0·041,,	1·72 ,,	0·71 ,,

The relation between calculated and observed frequency is simply shown in fig. 5, where the frequency is plotted as ordinate and the voltage as abscissæ. Since the energy E of the electron is proportional to the voltage, the theoretical curve on Planck's relation is a straight line. For the lowest voltages, the experimental curve is seen to fall below the theoretical, the observed frequencies being about 10 per cent. less than the calculated. The departure between theory and experiment becomes more and more marked with increase of voltage, and at about 142,000 volts the frequency reaches a maximum which is not altered by increase to 175,000 volts.

We shall discuss later the probable reason why the theory is so widely departed from in excitation of X-rays by swift electrons, but at present we shall consider whether there is any simple relation between frequency and voltage. The frequency curve is approximately parabolic in shape and can be expressed by the relation $\nu = aV - bV^2$, where a and b are constants and V is the voltage. This relation can be put into a simpler form, viz.,

$$h\nu = \text{E} - c\text{E}^2, \qquad \qquad (1)$$

where E is the energy of the electron moving through a difference of potential

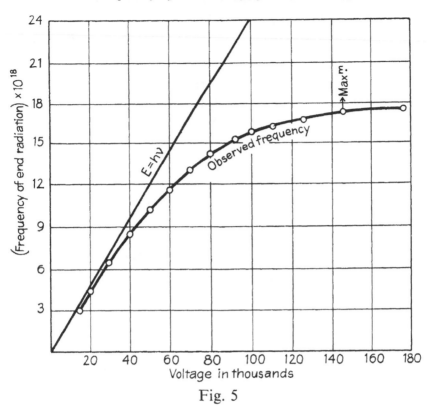

Fig. 5

E and c a constant. If Planck's relation held for excitation, $E = h\nu_p$ where ν_p is the calculated frequency. It is seen that the relation reduces to the simple form

$$\nu/\nu_p = 1 - kV.$$

By differentiating equation (1), it is seen that the frequency ν reaches a maximum when $E = 1/2c$ or $V = 1/2k$.

If V is expressed in volts, the value of $1/k$ which fits the results best is 285,000 volts. The frequency reaches a maximum at half this voltage, or at 142,500 volts.

The following table shows how closely the frequencies up to the maximum as given by this empirical relation agree with the observed values.

With the exception of the values for low voltages, where the frequency is difficult to determine accurately, the agreement between the calculated and observed values is remarkably good and within experimental error.

Absorption in Lead

Several experiments were made to find the absorption in lead of the radiation excited by the higher voltages. Preliminary experiments showed that the 'end' radiation was absorbed nearly exponentially in that substance as in aluminium. The values of the absorption coefficient μ for lead were found to be

36, 29, 23, 23, 23 for the end radiation excited by 96,000, 110,000, 125,000, 155,000, 175,000 volts respectively. Since the absorption coefficient for lead on the average decreases rapidly with increased frequency over the range

TABLE II

Volts in thousands	Calculated frequency	Observed frequency
13·2	$3 \cdot 07 \times 10^{18}$	$2 \cdot 94 \times 10^{18}$
20	4·54 ,,	4·41 ,,
30	6·55 ,,	6·44 ,,
40	8·39 ,,	8·34 ,,
50	10·06 ,,	10·1 ,,
60	11·56	11·6 ,,
70	12·88 ,,	13·0 ,,
80	14·04 ,,	14·1 ,,
90	15·02 ,,	15·3 ,,
100	15·83 ,,	16·0 ,,
110	16·47 ,,	16·5 ,,
125	17·1 ,,	16·9 ,,
142·5	17·4 ,,	17·4 ,,

considered, it was thought that the small change of frequency observed by measuring the absorption of the radiation in aluminium would show up more prominently when lead was the absorber. No certain difference in the absorption was, however, observed from 125,000 to 175,000 volts.

The final measurement of the end radiation was made after the rays had passed through 2·49 mm. of lead. The ionization observable through 3 mm. of lead is certainly less than 1/10000 of the initial value. Even with very intense radiation, a thickness of 3 mm. of lead affords practically complete protection against the rays. Through a thickness of 4 mm. it would be difficult to detect the ionization even when the bulb was strongly excited at 125,000 volts.

Discussion of the Results

It has been shown that the penetrating power in aluminium of the X rays from a Coolidge tube reaches a maximum at 142,000 volts, and that no sensible alteration has been observed when the voltage is raised to 175,000. The maximum value of the absorption coefficient μ in aluminium is 0·39. Remembering that the value of μ for the penetrating rays from radium C is 0·115 in aluminium, it is seen that the rays from the Coolidge tube have only about 3/10 of the penetrating power of the gamma rays from radium C. The radiation from the Coolidge tube is, however, slightly more penetrating than some of the rays from radium B, for which μ has been found to be 0·51. From the variation of frequency with voltage, it would appear that the frequency of the radiation reaches a natural limit, probably controlled by the frequency of the 'K' characteristic radiation of that element. This and other points are very clearly brought out by comparison of the radiations from the Coolidge tube at different voltages with the γ rays emitted by radium B. The latter has been carefully analysed by H. Richardson.* By examining the absorption in aluminium, the rays were found to consist of two components, for which $\mu = 40$ and $0\cdot5$; the former radiation, which is easily absorbed, undoubtedly corresponds to the 'L' radiation of radium B. The analysis of the radiation was carried still further by determining the absorption curve for lead. In addition to the 'L' characteristic, the rays were found to consist of three components, for which μ in lead was 46, 6, and $1\cdot5\,(\text{cm.})^{-1}$. Since the most penetrating radiation of the Coolidge tube gives a value $\mu = 23$ for lead, it is clear that the radiation from the Coolidge tube is more penetrating than one of the main components of the radiation from radium B, but is far less penetrating than two other components. In the following table are given the wave-lengths of the chief lines observed in the radium B spectrum by Rutherford and Andrade,† and the absorption coefficients in aluminium and in lead to be probably ascribed to the radiations, and also some comparative results obtained with the Coolidge tube.

TABLE III

Wave-length of strong lines from Radium B	Absorption coefficient μ of radiations from Radium B	Wave-length of radiation from Coolidge tube	Absorption coefficient μ of radiations from Coolidge tube
$1\cdot37\times10^{-9}$ cm.	$1\cdot5\,(\text{cm.})^{-1}$ in Pb		
$1\cdot59$,, ,,⎫ $1\cdot69$,, ,,⎭	6 ,, ,,		
		$1\cdot72\times10^{-9}$ cm.	23 (cm.)$^{-1}$ in Pb
$1\cdot96$,, ,,	46 ,, ,,	$1\cdot91\times10^{-9}$ cm.	36 ,, ,,
	$0\cdot51$,, in Al		$0\cdot45$,, in Al
$2\cdot29$,, ,,			
$9\cdot82$,, ,,	'L' radiation	$10\cdot82$	'L' radiation
to		to	
$11\cdot75$,, ,,	40 in Al	$14\cdot77\times10^{-9}$ cm.	81 in Al

The line in radium B, $\lambda = 1\cdot96 \times 10^{-9}$, appears to be mainly responsible for the radiation which has an absorption $\mu = 0\cdot51$ in aluminium and $\mu = 46$ in lead. This is clear from a comparison of the results with the Coolidge tube. Taking Moseley's observation that the frequency of the corres-

* H. Richardson, Proc. Roy. Soc. A. xci. p. 396 (1915); Rutherford & Richardson, Phil. Mag. xxv. p. 722 (1913). (*This vol., p. 342.*)

† Rutherford & Andrade, Phil. Mag. xxviii. p. 263 (1914). (*This vol., p. 456.*)

ponding lines in the K radiations of different elements is proportional to $(N - 1)^2$ where N is the atomic number, it follows that the line in the radium B spectrum for which $\lambda = 1 \cdot 37 \times 10^{-9}$ should have a value $\lambda = 1 \cdot 69 \times 10^{-9}$ for tungsten, since the latter has an atomic number 74 and radium B, 82. This calculated value is in good agreement with the maximum wave-length $\lambda = 1 \cdot 73 \times 10^{-9}$ found for tungsten. It thus appears probable that the radiation from tungsten is analogous to the radiation from radium B. Since the speed of the beta rays issuing from radium B corresponds to a fall of potential of at least 400,000 volts, and from radium C of 2,000,000 volts, it seems clear that we cannot expect to obtain a more penetrating radiation from tungsten unless possibly a voltage of the order of 1,000,000 volts is applied. Even with electrons corresponding in energy to over 2,000,000 volts, the wave-length of the strong line due to the penetrating gamma rays from radium C, viz. $\lambda = 0 \cdot 99 \times 10^{-9}$ is only 6/10 of the shortest wave from the Coolidge tube. The comparison of the results with the Coolidge tube with the gamma rays thus leads to the conclusion that there is a definite limit to the maximum frequency to be obtained from an element bombarded by swift electrons. This limit is probably determined by the characteristic radiation of highest frequency which exists in the atom. Since radium C has an atomic number 83 and uranium—the heaviest known element—92, we should anticipate from Moseley's relation that the shortest wave-length to be obtained with a uranium anticathode in a vacuum-tube is $\lambda = 1 \cdot 40 \times 10^{-9}$. The penetrating power of this radiation in aluminium should be $\mu = 0 \cdot 23$ instead of $= 0 \cdot 39$ from the Coolidge tube. Under possible laboratory conditions, it thus appears very improbable that we can obtain X rays as penetrating as the gamma rays from radium C.

The Excitation of X rays and the Quantum Theory

We have seen that for the Coolidge tube the connexion between the maximum frequency ν and the energy E of the exciting electron is given by

$$h\nu = E - cE^2$$

or

$$\nu/\nu_p = 1 - kV,$$

where ν_p is the frequency to be expected on Planck's theory if the whole energy of the electron is transferred into radiation; V the voltage; c and k are constants. These formulæ do not hold beyond the maximum frequency given by $E = \dfrac{1}{2c}$, $V = \dfrac{1}{2k}$. If V is expressed in volts, the value of k is 1/285000. If this formula holds for lower voltages than those actually examined, it is seen that the value of ν becomes more nearly equal to ν_p the lower the voltage. The formula suggests that, for a heavy atom like tungsten, the frequency excited by low voltages should very closely agree with that expected on the quantum theory, supposing that the whole energy of the electron is transformed into that of the X ray. It thus appears probable that the simple quantum theory holds for excitation if the voltage is sufficiently small, but that a large correction is required for high voltages.

It will be of very great interest to examine the corresponding relations between frequency and voltage for lighter atoms, and to test whether such a simple relation as that found for tungsten holds in such cases. It is to be anticipated that the maximum frequency would be reached for a voltage which diminishes in value as the atomic weight or atomic number decreases. There is one point, however, in this connexion that should be mentioned. We have seen that the maximum frequency obtainable from tungsten is about that to be expected for the component of the shortest wave-length for the 'K' characteristic radiation of that element. On the other hand, the experiments of Kaye have shown that a radiation can be obtained from aluminium at 30,000 volts, which is much more penetrating than the 'K' characteristic of that element. Results of a similar kind have been obtained by Rawlinson,* who found that at about 50,000 volts he obtained a radiation from nickel for which $\mu = 6 \cdot 9$, while the value of μ for the 'K' characteristic is 148 (cm.)$^{-1}$. Such results show that the highest frequency to be obtained from these atoms is much greater than that of the K radiation. It is possible, however, that the radiation of higher frequency may represent the octave or still higher harmonic of the fundamental mode of radiation which is represented by the 'K' characteristic. A close analysis of the frequency-voltage curves of the radiation from different elements should throw much light on this question. Arrangements have been made to continue experiments, such as have been made for tungsten, for a number of other elements.

We have already drawn attention to the fact that, even with the very high velocity of projection of the beta rays from radium B and radium C, no frequencies have been observed much higher than those to be expected for the 'K' radiation.

Some years ago, Whiddington† made a number of experiments on the minimum voltage required to excite the 'K' characteristic of a number of light elements. Assuming, as seems probable, that it is necessary to excite the *shorter* wave-length of the two main components of the 'K' radiation before the characteristic appears, the voltage required on the simple quantum theory to excite the radiations is given in the following table, taking the values for the beta component of the 'K' radiation found by Moseley.

It is seen that for all metals except aluminium the voltage required to excite the 'K' radiation is distinctly higher than that expected on the simple quantum theory. Such a result is to be anticipated if the relation between frequency and voltage is of the same general form as that observed for tungsten.

We have seen that the radiation from the Coolidge tube was first detected at 10,300 volts. It seems probable that this corresponds to the voltage required

---

* Rawlinson, Phil. Mag. xxviii. p. 274 (1914).

† Whiddington, Proc. Roy. Soc. A. lxxxv. p. 323 (1911).

R*

to excite the 'L' characteristic of tungsten. Assuming that the line $\lambda = 1 \cdot 277 \times 10^{-8}$ cm.* must be excited for the appearance of the 'L' radiation, the theoretical voltage is 9600 volts, while the voltage deduced from the equation $v/v_p = 1 - k$V is 9800. The difference in this case between the observed and computed voltage is about 2 per cent., and is not beyond the experimental error.

TABLE IV

Element	Atomic number	λ for the β line	Calculated voltage	Observed voltage	Volts Calc. / Volts obs.
Aluminium	13	$7 \cdot 91 \times 10^{-8}$	1570	1200	$1 \cdot 3$
Chromium	24	$2 \cdot 093 \times 10^{-8}$	5900	7320	$0 \cdot 80$
Iron	26	$1 \cdot 765 \times 10^{-8}$	7000	9600	$0 \cdot 73$
Nickel	28	$1 \cdot 506 \times 10^{-8}$	8200	10750	$0 \cdot 76$
Copper	29	$1 \cdot 402 \times 10^{-8}$	8800	11080	$0 \cdot 79$
Zinc	30	$1 \cdot 306 \times 10^{-8}$	9500	11280	$0 \cdot 84$

In order for the quantum theory to hold for excitation, it is necessary that the whole of the energy of the electron should be directly converted into energy of radiation. It seems probable that if an electron makes an 'end on' collision with another, the whole of the energy of the one is transferred to the other, but we have no definite evidence that radiation is emitted in such a collision. Taking, for example, the point of view proposed by Bohr,† that the radiation arises from the fall of an electron from one ring of electrons to the next, the essential antecedent to the emission of radiation is the removal of an electron from one of the rings. If the whole of the energy of the incident electron is expended in ejecting the electron from the ring, it is to be anticipated that the energy of the X radiation should equal the energy of the incident electron. This condition should approximately be fulfilled in the relatively sparse distribution of electrons in the outer rings. In order, however, to excite the higher frequencies, the electron must penetrate deep into the atom where the electrons are more closely packed, and part of its energy will be used up in setting neighbouring electrons in motion, and only a fraction will be available to eject an electron from its ring. Quite apart from any special theory of the mechanism of radiation, such a factor must always enter into the energy of the electron finally available to excite a characteristic radiation.

This point of view offers a simple and probable explanation of the reason why the quantum theory holds closely for excitation of low frequencies by slow speed electrons, but fails for high frequencies. The relation found experimentally between v/v_p for tungsten suggests that the correction for this effect increases rapidly with the frequency.

Summary

(1) The absorption curves in aluminium of the X radiation from a Coolidge tube have been examined over a wide range of constant voltages supplied

* See Barnes, Phil. Mag. xxx, p. 368 (1915).   † Bohr, Phil. Mag. July 1913.

by a Wimshurst machine. While the main radiation is complex, the 'end' radiation is found to be absorbed exponentially.

(2) The absorption curves for different voltages obtained with an induction-coil are nearly the same as for the Wimshurst machine, and the penetrating power of the end radiation is nearly the same.

The maximum frequency of the 'end' radiation for voltages between 13,000 and 175,000 volts has been deduced by examining its absorption in aluminium. The frequency and penetrating power reach a maximum value of 145,000 volts, and are not altered by increase of the voltage to 175,000.

(3) The shortest wave-length emitted by the Coolidge tube is $1 \cdot 71 \times 10^{-9}$ cm. or $0 \cdot 17$ A.U. The absorption coefficient of this radiation in aluminium is $0 \cdot 39$ (cm.)$^{-1}$, and in lead 23 (cm.)$^{-1}$. The penetrating power of this radiation is about 3/10 of the penetrating gamma rays from radium C.

(4) The relation between the frequency and the voltage is expressed by the formula $h\nu = E - cE^2$, where E is the energy of the electron, c a constant, and h Planck's fundamental constant. This relation holds up to 142,000 volts when the radiation has its maximum frequency. Evidence is given that the quantum theory is directly applicable for the excitation of waves of low frequency, but for higher frequencies requires a correcting factor, the value of which increases rapidly with the frequency.

(5) A comparison is given of the radiation from a Coolidge tube with the gamma radiations emitted by radium B and radium C.

University of Manchester
July 1915

Efficiency of Production of X Rays from a Coolidge Tube

by SIR ERNEST RUTHERFORD, F.R.S.,

and PROFESSOR J. BARNES, PH.D.

From the *Philosophical Magazine* for September 1915, ser. 6, xxx, pp. 361–7

IN the preceding paper we have given an account of experiments which have been made to determine the maximum frequency of the X radiation excited in a Coolidge tube for different constant voltages supplied by a Wimshurst machine. It was thought desirable to extend the experiments to determine the efficiency of the production of X rays in the Coolidge tube for comparatively high voltages.

The question of the efficiency of the production of X rays, *i. e.* the ratio of the energy of the generated X rays to that of the exciting cathode rays, has been the subject of several investigations. Wien[*] determined the energy of the X rays generated in a platinum anticathode for a potential difference of 58,700 volts, by measuring the heating effect of the radiation. Similar experiments have been made by Angerer[†] and Carter.[‡] The latter observed that the efficiency increased with the exciting voltage. In general, it was found that the efficiency was of the order of 1/1000.

Recently the question has been attacked under more definite conditions by R. T. Beatty.[§] Cathode rays of definite velocity sorted out by a magnetic field fell on a radiator. The generated X rays passed out through a very thin window, and were then completely absorbed in a cylinder 150 cm. long filled with the vapour of methyl iodide. The total ionization of the rays was thus measured, and the corresponding energy deduced from general data. When characteristic radiation was not excited, the energy of the X rays for an equal number of exciting electrons was found to be proportional to the fourth power of the velocity of the cathode rays. Kaye[‖] had previously observed that the energy of X rays under constant condition of excitation was approximately proportional to the atomic weight A of the radiator. Beatty finally concluded that the efficiency of production of X rays by matter in general was given by the formula

$$\frac{\text{X-ray energy}}{\text{cathode-ray energy}} = 2 \cdot 54 \times 10^{-4} A\beta^2,$$

* Wien, *Ann. d. Phys.* xviii. p. 991 (1905).   † Angerer, *Ann. d. Phys.* xxi. p. 87 (1906).
‡ Carter, *Ann. d. Phys.* xxi. p. 955 (1906).
§ Beatty, Proc. Roy. Soc. A. lxxxix. p. 314 (1913).
‖ Kaye, Phil. Trans. Roy. Soc. A. cciv. p. 123 (1908); Proc. Roy. Soc. A. lxxxi. p. 337 (1908).

where β is the velocity of the cathode as a fraction of the velocity of light. From the curves given by Beatty, it would appear that the maximum speed of the cathode rays employed by him corresponded to 23,000 volts.

In the experiments with the Coolidge tube we have examined the efficiency of production of the radiation escaping from the bulb for voltages 48,000, 64,000, and 96,000 volts. The rays were excited by a Wimshurst machine, and the current through the bulb measured by a galvanometer. The general arrangements for controlling and calibrating the voltage were the same as those described in the accompanying paper. For the high voltages employed, the radiation is too penetrating for complete absorption in a reasonable length of air or other gas. In order to overcome this difficulty, we have measured the ionization due to a definite length of a beam of X rays in air, and deduced the absorption in air indirectly by determining the absorption curve of the radiation in *water*. It is known that the absorption of X rays by complex molecules is additive. The absorption of the radiation by water is mainly due to the oxygen atoms, whose atomic weight differs only slightly from that of the average atomic weight of air. For the relatively penetrating rays employed, we can assume with very little error that the absorption of the X rays by water is very nearly equal to that of a column of air of the same thickness compressed to the same density. The absorption by 1 cm. thickness of water is thus equivalent to that by $8 \cdot 2$ metres of air at laboratory temperature.

The ionization due to the X rays was measured in air without the X rays impinging on the electrodes in order to avoid the introduction of surface effects due to scattering or excitation of characteristic radiations. The general

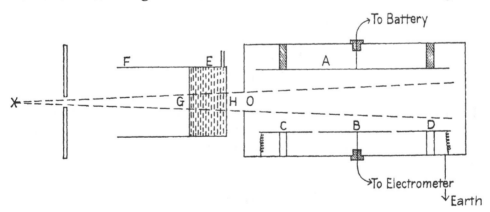

Fig. 1

arrangement of the apparatus is shown in fig. 1. The ionization vessel consisted of a rectangular box ($20 \times 12 \times 12$ cm.) lined with aluminium plate. The ionization was measured between the aluminium plate A and the central plate B. The former was connected to a 1000-volt battery and the latter to the electrometer. Two additional plates C, D connected with earth extended on both sides of the plate B. The cone of rays entering the ionization vessel was fixed by the circular opening O, 2 cm. in diameter, in a lead plate, the size of opening being adjusted so that no radiation fell on the electrodes. The disturbing effect due to the X rays impinging on the aluminium end of the box was small, and could be neglected. In order to determine the absorption curve for water, a vessel was constructed consisting of two closely fitting brass tubes E and F. Openings, G, H, cut in the ends of these tubes, were covered with a thin sheet of mica. The length of the column of water in the path of the rays could be simply adjusted, and care was taken that the radiation entering the ionization vessel did not strike the brass ends of the vessel. The maximum length of the column of water in the experiment was 5 cm. It is seen from fig. 2 that this is sufficient to reduce the intensity of the radiation to a small fraction of its initial value.

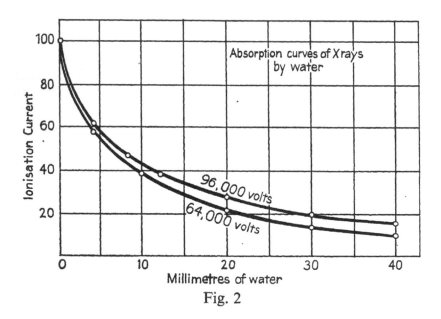

Fig. 2

In order to determine the absorption of the end part of the radiation by water, the vessel EF was removed, and a sufficient thickness of aluminium introduced in the path of the rays to reduce the ionization to an equal degree. The water column was again introduced, and the experiments continued up to a thickness of 15 cm. of water.

The capacity of the circuit and of the condensers in parallel was carefully determined. Changes in the intensity of the radiation were controlled by means of the 'standardizing vessel' described in a previous paper, and also by the current passing through the Coolidge tube. This current was determined by the method described in a previous paper. The deflexion of the galvanometer in the circuit was observed; the current through the tungsten spiral broken, and the voltage kept at the same value by varying the speed of the machine and by means of adjustable point discharge. The deflexion observed under the latter conditions was due to the current through the xylol resistances and voltage galvanometer in parallel with the Coolidge tube, and to electrical losses in the leads or over the surface of the bulb. The difference between the two readings served as a measure of the actual current conveyed by the electrons from the heated spiral.

The earlier parts of the absorption curves in water for 64,000 and 96,000 volts are shown in fig. 2. The absorption is not exponential, but decreases steadily with increase of thickness of water. Knowing the initial saturation current through the air with no absorber, the total ionization current due to complete absorption of the radiation can be at once deduced by determining the area included between the curve and the two axes, assuming that 1 cm. of water is equivalent in absorbing power to $8 \cdot 2$ metres of air at $15°$ C.

Supposing that the X radiation from the tungsten anticathode is emitted equally in all directions, the fraction of the total radiation entering the ionization vessel was $8 \cdot 6 \times 10^{-5}$.

The intensity of the radiation was found to be directly proportional to the current through the bulb. In the following table the total ionization current in air due to complete absorption of the whole radiation is expressed for each voltage in terms of 100 divisions of the current galvanometer which correspond to $2 \cdot 92 \times 10^{-5}$ amp.

Voltage	Current i_2 due Cathode Rays	Total Ionization Current i_1 for Complete Absorption of the Radiation	$X = i_1/i_2$	β	X/β^4
48,000	$2 \cdot 92 \times 10^{-5}$ amp.	$2 \cdot 5 \times 10^{-5}$ amp.	$0 \cdot 86$	$0 \cdot 406$	32
64,000	,, ,,	$4 \cdot 2$,, ,,	$1 \cdot 44$	$0 \cdot 459$	32
96,000	,, ,,	$8 \cdot 8$,, ,,	$3 \cdot 01$	$0 \cdot 540$	35

Beatty (*loc. cit.*) concluded from his experiments that $X = 0 \cdot 58 \, A\beta^4$, where A is the atomic weight of radiator and β the velocity of the cathode rays as a fraction of the velocity of light. It is seen from the last column that X/β^4 is sensibly constant. Substituting the value of $A = 184$ for tungsten, the values of X to be expected from this equation are $2 \cdot 9$, $4 \cdot 8$, $9 \cdot 1$ for 48,000, 64,000, and 96,000 volts respectively in place of $0 \cdot 86$, $1 \cdot 4$, $3 \cdot 0$ observed experimentally. The observed values are about one third of the values calculated on Beatty's relation.

Correction for Absorption in Bulb

In our calculations, however, we have measured the total ionization produced *outside* the bulb, and have not corrected for the absorption of the rays by the wall of the bulb and by the air and other absorbers in the path of the rays. Special measurements showed that the wall of the bulb where the rays issued was $0 \cdot 5$ mm. in thickness. The absorption in glass for equal thicknesses is about the same as for aluminium, and there will not be much error in taking the absorption of the rays before entering the ionization vessel as equivalent to 0·6 mm. of aluminium. Comparing the relative absorption of air and aluminium for soft radiations, 0·6 mm. of aluminium is equivalent to about 8 mm. of water. Until experiments are made of the total radiation from tungsten under conditions that the absorption of radiation in escaping from the tube is a minimum, it is difficult to make more than a rough estimate of this correcting factor for absorption. From a consideration of the absorption curves in water, it seems probable that the correcting factor for the total ionization is at least 2 for 48,000 volts, and may be somewhat greater. It is to be expected that this factor would be somewhat less for the higher voltages. Making this correction, it is seen that the results for tungsten are in very fair agreement with Beatty's relation, even though it is extrapolated over a wide range of atomic weight and voltage.

Energy of the X Rays

Knowing the current due to complete absorption of the radiation, the energy of the radiation can at once be deduced if the average energy required to produce a pair of ions in air is known. The value of this important quantity can best be deduced from the total ionization current produced by the absorption of a single alpha particle of known energy. Geiger found that a single alpha particle from radium C gave rise to $2·37 \times 10^5$ ions in air each of charge $4·65 \times 10^{-10}$ e.s. units, *i. e.* a quantity of electricity $3·67 \times 10^{-15}$ e.m. units. On the latest data,* the initial energy of the alpha particle from radium C is $7·66 \times 10^{14}e$, where e is the charge on the ions in electromagnetic units. From this it can be deduced that the energy required to produce a pair of ions in air is equal to the energy acquired by the unit charge in moving freely through a potential difference of 33 volts. If i_1 is the total ionization current and i_2 the electronic current

$$\text{efficiency} = \frac{\text{X-ray energy}}{\text{cathode-ray energy}} = \frac{i_1 v}{i_2 \text{V}},$$

where $v = 33$ volts and V = the voltage applied to the tube.

Voltage	Ratio, i_1/i_2	Efficiency	β^2	Efficiency/β^2
48,000	0·86	$0·59 \times 10^{-3}$	0·165	3·6
64,000	1·44	0·74 ,,	0·211	3·5
96,000	3·01	1·04 ,,	0·292	3·6

The efficiency deduced in the above table is for ordinary working conditions when no correction is made for absorption in the glass walls, &c. Under these limitations, the efficiency is seen to be proportional to β^2. No doubt the closeness of the agreement is accidental, for the numbers would be changed if corrections were made for absorption of the radiation.

* Rutherford and Robinson, Phil. Mag. xxviii. p. 552 (1914). (*This vol., p. 383.*)

ADMIRALTY PHYSICS BOARD AT THE MINING SCHOOL, PORTSMOUTH, 1921

Middle Row: Col. Stevenson, F. E. Smith, Sir E. Rutherford, Sir J. J. Thomson, Sir W. H. Bragg, Col. Fleming, H. T. Tizard

We have seen earlier that about half the energy of the radiation is probably absorbed in the bulb. We should consequently expect the efficiency under ideal conditions to be about 1/500 for 96,000 volts, and 1/800 for 48,000 volts. It is of interest to note that Wien (*loc. cit.*) found an efficiency of $1 \cdot 09 \times 10^{-3}$ for a platinum anticathode, using a bolometer method to measure the energy of the X rays for a potential of 58,700 volts supplied by an induction-coil. Since the average potential of the discharge due to an induction-coil is less than the maximum, it is to be expected that the efficiency of the coil would be somewhat higher than for an equal steady voltage supplied by a machine.

Beatty (*loc. cit.*) found the efficiency to be given by

$$E = 2 \cdot 54 \times 10^{-4} A \beta^2.$$

The value of the numerical factor involves the average energy required to produce a pair of ions. This was deduced by a very indirect method by a combination of distinct investigations by Glasson and Whiddington. From the data given by Beatty, it can be calculated that the energy assumed to produce a pair of ions in methyl iodide corresponds to 110 volts. This is undoubtedly more than three times too large, and the value of $E = 7 \cdot 6 \times 10^{-5} A \beta^2$ is nearer the truth, assuming the correctness of the other data involved. The formula gives an efficiency for 96,000 volts of $4 \cdot 1 \times 10^{-3}$, which is about twice as high as that to be expected from our experiments with a Coolidge tube after the probable correction for absorption has been made.*

In these calculations no correction has been made for reflexion or scattering of cathode rays by the tungsten. No doubt the correction for this varies with the speed of the electrons, and must be considerable for very high voltages. In the absence of any definite data on this question, it seems desirable, however, to give the efficiency of the conversion of cathode rays into X rays under actual working conditions.

The relations given by Beatty only apply to the 'general' or 'independent' radiation from an X-ray tube. As pointed out by Beatty, the efficiency rises rapidly when a characteristic radiation is strongly excited. The low percentage value obtained for the efficiency of a Coolidge tube for high voltages is thus an indication that the radiation is mainly of the 'independent' type, and that the 'K' characteristic radiation is not so strongly excited as in the case of metals of lower atomic weight. This is borne out by the difficulty of detecting the presence of the 'K' characteristic of tungsten by absorption experiments, or by reflexion from crystals. It is only in the case of radioactive substances that the characteristic radiations of high frequency are strongly excited. No doubt this is due to the ideal conditions of excitation in this case, for the exciting electrons all come from the nucleus of the atom.

In the preceding paper we have shown that the voltage required to excite

---

* Dr. Beatty has drawn my attention to the recent measurements of Barkla (Phil. Mag. xxv. p. 838, 1913), in which he finds that the total ionization in methyl iodide for cathode rays is $1 \cdot 48$ times that in air. Using the amended data, Beatty's relation becomes $E = 0 \cdot 51 \quad 10^{-4} A \beta^2$, which is in fair accord with our measurements.

the most penetrating rays in tungsten is about twice that to be expected on the quantum theory, indicating that about half of the energy of the exciting electron can be transformed into radiation. From the low value of the efficiency at high voltages, viz. about 1/500, it is clear that, on the average, 1 electron only in 200 is efficient in producing radiation.

University of Manchester
July 1915

Long-range α Particles from Thorium

by SIR ERNEST RUTHERFORD, F.R.S., *and* A. B. WOOD, M.SC.,
Lecturer in Physics, University of Liverpool

From the *Philosophical Magazine* for April 1916, ser. 6, xxxi, pp. 379–86

In the course of an examination of a strong source of the active deposit of thorium by the scintillation method, one of us observed the presence of a small number of bright scintillations which were able to penetrate through a thickness of matter corresponding to $11 \cdot 3$ cm. of air at 760 mm. and 15° C. These scintillations were undoubtedly due to alpha particles and of greater velocity than any previously observed; for the swiftest alpha particles hitherto known, viz. those from thorium C, have a range in air of $8 \cdot 6$ cm. The number of these long-range alpha particles is only a small fraction of the total number emitted by the source. The actual number of long-range particles decreased exponentially with time, falling to half value in $10 \cdot 6$ hours—the normal period of decay of the active deposit of thorium. Owing to the pressure of other work, the experiments were kindly repeated and extended by Mr. A. B. Wood, who examined in detail the variation of the number of scintillations with thickness of matter traversed. There are still a number of points that require further examination, but as neither of the authors is likely to have time to continue the experiments in the near future, it has been thought desirable to give a brief account of the preliminary results.

Experimental arrangements

The end of a brass rod, 1 mm. in diameter, was exposed as negative electrode in a small vessel containing a strongly emanating preparation, either of radio-thorium or meso-thorium. By suitable adjustment of the electrodes, the active deposit was concentrated almost entirely on the end of the rod—a condition essential to the accurate determination of the ranges. After two days' exposure to the electric field, the wire was removed and placed end-on at 4 mm. distance from a small screen of zinc sulphide viewed with a low-power microscope. Care was taken that the axis of the microscope passed through the centre of the rod. The screen was permanently covered with a mica plate whose thickness corresponded to $8 \cdot 6$ cm. of air—the maximum range of the alpha particle from thorium C. All scintillations then observed were due to alpha particles which had passed through the mica plate and 4 mm. of air, *i. e.* a distance corresponding to 9 cm. of air. With the most intense source available, about 20 scintillations per minutes were counted on the microscope. The number fell off rapidly with increase of distance of the source, but an occasional α particle was still observed at a distance of 2 cm. from the source. In order to determine the variation of number of these particles with distance of matter traversed, thin screens of aluminium, each corresponding in thickness to 1·25 mm. of air, were successively interposed between the source and screen. It was found that the number of long-range particles remained constant between 8·6 and 9·3 cm. of air, but decreased in number from 9·3 cm., vanishing at 11·3 cm. The grouped average of all the observations in a large number of experiments is shown in fig. 1.

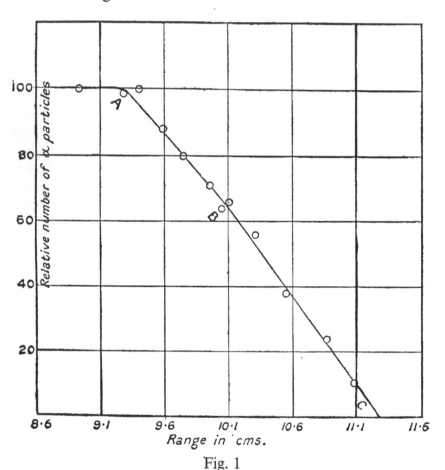

Fig. 1

It is seen that the curve shows evidence of two fairly definite slopes AB, BC, as if there were two sets of alpha particles present of different ranges. This important point was very carefully examined, and the results of a special series of observations are shown in fig. 2. It will be seen that there appears to be a fairly definite change in the slope when the number of scintillations is reduced to about two-thirds of the total.

It will be seen from the curves that the alpha particles start decreasing in number from about 2 cm. of the maximum range. The variation of number with distance is much slower than that to be expected for a single group of alpha particles of corresponding range. This is brought out in fig. 3, which shows the results obtained when the scintillation-distance curve in air was obtained for the two well-known groups of alpha particles emitted from thorium C of ranges 5·0 and 8·6 cm. respectively. In these cases, the scintillations fall off rapidly, beginning at about 1 cm. from the end of the corres-

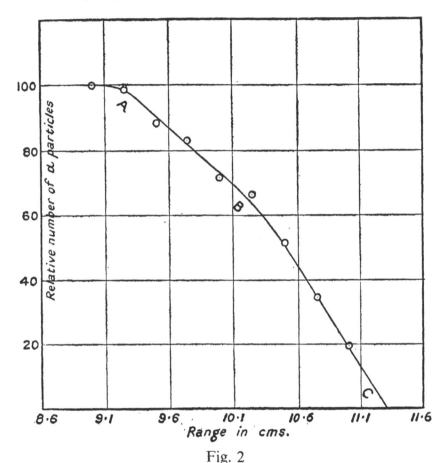

Fig. 2

ponding range. The difference between the slopes of the scintillation-curve for the long-range alpha particles and those from thorium C cannot be explained by the oblique path taken by some of the rays through the mica on account of the nearness of the source and screen. Calculations showed that the influence of obliquity could only account for a small fraction of the difference actually observed.

The results we have so far obtained certainly seem to indicate either that (1) the long-range alpha particles are expelled with variable velocities over a comparatively narrow range, and in this respect differ markedly from alpha particles from ordinary radioactive products which are known to be expelled with identical velocity, or (2) that two homogeneous groups of alpha rays of characteristic ranges are present. In order to distinguish definitely between these two hypotheses, it would be necessary to count many thousands of alpha particles, but other evidence suggests that (2) is the more probable explanation.

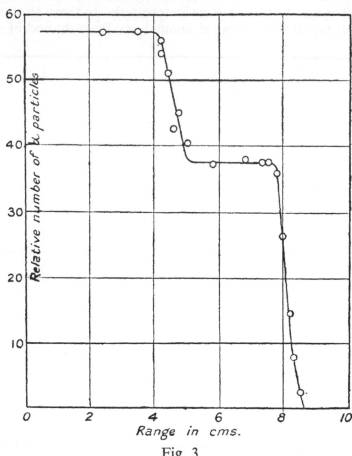

Fig. 3

The slope AB ends at about a range 10·2 cm. and when the number of alpha particles is reduced to about two-third of the total. This suggests that two groups of homogeneous rays are present, one-third of maximum range 10·2 cm. and two-thirds of range 11·3 cm. The slope of the scintillation curve to be expected on this hypothesis agrees within the experimental error with the observed curve. This division of the alpha particles into two homogeneous groups may be compared with the two well-known groups of alpha particles emitted from thorium C, for it is known that one-third have a range 5·0 cm. and two-thirds a range 8·6 cm. This suggests that the new groups of alpha particles have their origin in thorium C, and that one-third of range 10·2 cm. accompany the alpha particles of range 5·0 cm., and the remainder of range 11·3 cm. accompany the alpha particles of range 8·6 cm. While it is very difficult to prove the correctness of such a deduction, the numerical agreement in the divisions of alpha particles of different ranges is certainly striking.

We have not so far examined experimentally whether the new alpha particles are expelled from the alpha ray product thorium C, but this seems very probable. To settle this point, it will be necessary to prepare a strong preparation of thorium C and to determine whether the period of transformation, measured by the new alpha particles, is in agreement with the accepted value for thorium C, viz. half value in 60 minutes.

Number of Long-range Alpha Particles

Since the number of long-range particles decreases at the same rate as the alpha-ray activity of the active deposit of thorium, it is convenient to express their number as a fraction of the total number of alpha particles emitted per second from thorium C. For this purpose, the number of long-range particles per minute was measured with the source fixed at a known distance from the screen. The active deposit was then allowed to decay *in situ* for 32 hours. The absorbing screen was then removed and the number of alpha particles from thorium C measured at distances from 6 to 7 cm., so as to include only the longer range alpha particles (8·6 cm.) from thorium C. In this way it was found that the fraction of long-range alpha particles was 1/6700. Taking into consideration that the alpha particles of range 8·6 cm. from thorium C are two-thirds of the total, the fraction becomes 1/10000.

Preliminary observations by different methods gave a somewhat lower value, but the above number cannot be much in error. We thus see that the long-range alpha particles are expelled in a very small proportion (1/10000) compared with the ordinary alpha particles. Unless a very intense source be employed, it will not be easy to detect the presence of the long-range alpha particles when the ordinary alpha particles are first absorbed by a layer of 8·6 cm. of air.

In the Bragg ionization curves from thorium C given by Marsden and Perkins,* a small residual activity is to be noticed beyond the distance 8·6 cm. which is relatively more marked than for the corresponding curve for radium C. This no doubt is to be ascribed to the effect of the very long-range alpha particles.

Discussion of Results

It is now well established that thorium C is anomalous in breaking up in two distinct ways. One-third of the atoms are transformed with the emission of alpha particles of range 5·0 cm., and the remainder gives alpha particles of range 8·6 cm. These modes of transformation of thorium C have been examined in detail by Marsden and Darwin,† and an ingenious scheme of changes has been suggested to account for the facts observed. It is known

---

* Marsden & Perkins, Phil. Mag. xxvii. p. 691 (April 1914).

† Marsden & Darwin, Proc. Roy. Soc. A. lxxxvii. p. 17 (1912); see also Marsden & Barratt, Proc. Phys. Soc. xxiv. 1, p. 50 (1911); Marsden & Wilson, Phil. Mag. xxvi. p. 354 (1913).

that the products corresponding to thorium C in the radium and actinium series, viz. radium C and actinium C, also have two distinct modes of transformation. Fajans* showed that 1/6000 of the atoms of radium C give rise to a new product of half period $1\cdot38$ minutes, which emits beta rays in its transformation. In a similar way actinium C has been found to emit two sets of alpha particles of range $5\cdot4$ and $6\cdot4$ cm.† This is ascribed to a double mode of transformation, $1\cdot5/1000$ of the atoms breaking up with the emission of alpha particles of range $6\cdot4$ cm.

Assuming that the new alpha particles of thorium can be divided into two homogeneous groups of range $10\cdot2$ and $11\cdot3$ cm., it is seen that thorium C must break up in four distinct ways with the expulsion of alpha particles of ranges $5\cdot0$, $8\cdot6$, $10\cdot2$, and $11\cdot3$ cm. at 15° C.

The possible modes of transformation of thorium C are thus more complicated than was at first supposed, and it is obvious that the suggestions given by Marsden and Darwin as to the modes of transformation of this substance can be only a partial explanation. From the close analogy of the 'C' products of radium, thorium, and actinium, it is probable that further examination will show an analogous complexity in the modes of breaking up of radium C and actinium C. The loss of energy in the form of expelled alpha particles is very different in the four modes of transformation of thorium C, and in consequence it does not seem likely that the resulting products can be the same in all cases. The differences in the energies emitted by the two branch products of thorium C formed a serious difficulty in the original explanation given by Marsden and Darwin of the two main modes of transformation of thorium C, and this difficulty is now further increased. A more detailed discussion on these interesting points will be reserved until further experimental information is available. The relation found by Geiger between the range of the expelled alpha particles and the life of the radioactive product, suggests that the average life of the atoms which expel the long-range alpha particles must be exceedingly short, and of the order of 10^{-13} and 10^{-16} sec. for the products emitting alpha particles of range $10\cdot2$ and $11\cdot3$ cm. respectively.

The following table gives the velocity of the four groups of alpha particles from thorium C, taking as the basis of calculation the measurements of Rutherford and Robinson that the velocity V of the alpha particles from radium C of range $6\cdot94$ cm. at 15° C. is $1\cdot922 \times 10^9$ cm. per second, and assuming Geiger's relation $V^3 = kR$ where R is the range.

	Range at 15° C.	Ratio of velocities	Calculated velocity
Thorium C_1	$4\cdot95$ cm.	1	$1\cdot71 \times 10^9$ cm.
Thorium C_2	$8\cdot6$,,	$1\cdot205$	$2\cdot06$,, ,,
New product (C_3)	$10\cdot2$,,	$1\cdot275$	$2\cdot18$,, ,,
New product (C_4)	$11\cdot3$,,	$1\cdot32$	$2\cdot26$,, ,,

* Fajans, *Phys. Zeit.* xii. p. 369 (1911); xiii. p. 699 (1912).
† Marsden & Perkins, Phil. Mag. xxvii. p. 694 (1914).

Recently one of us* showed by direct measurement that the velocities in two main groups of alpha particles from thorium C were in good agreement with the calculated values if the range of the alpha particles from thorium C_1 is 4·95 cm. instead of 4·80 cm.—the value usually taken. From our measurements, there appears to be no doubt that the higher value is more correct.

Summary

Evidence is given that the active deposit of thorium emits a small number of alpha particles of greater velocity than any previously observed. These alpha particles are believed to have their origin in the transformation of thorium C, and appear to be divided into two homogeneous groups of maximum range 10·2 and 11·3 cm. The number of these alpha particles is about 1/10000 of the total number emitted from thorium C, two-thirds of the number having a range 11·3 cm.

The results indicate that the atoms of thorium C can break up in three and probably four distinct ways with the emission of four characteristic groups of alpha particles of ranges 5·0, 8·6, 10·2, and 11·3 cm.

University of Manchester
February 1916

* A. B. Wood, Phil. Mag. xxx. p. 702 (1915).

Penetrating Power of the X Radiation from a Coolidge Tube

by SIR E. RUTHERFORD, F.R.S.,
Professor of Physics, University of Manchester

From the *Philosophical Magazine* for September 1917, ser. 6, xxxiv, pp. 153–62

THE present paper contains an account of some experiments made to determine the maximum penetrating power of the X rays excited by high voltages in a Coolidge tube, using lead as the absorbing material. Owing to the lack of time at my disposal, the experiments, made a year ago, are incomplete; but they may prove of interest in indicating the penetrating power of the X radiation that can be obtained from this source under practicable conditions, and in throwing light indirectly on the probable frequency of the very penetrating gamma radiation from radioactive bodies.

In these experiments, the absorption of the X radiation by lead has been examined over a very much wider range of intensity and of thickness of absorber than in the original experiments of Rutherford, Barnes, and Richardson.*

To excite the radiation, a large induction-coil of 20-inch spark was used, actuated by a mercury motor-break in an atmosphere of coal-gas. The heating current through the tungsten spiral was adjusted to give a radiation of maximum intensity at the voltage required, which was fixed by an alternative spark-gap between points. The radiation was found to be most constant when a fairly rapid stream of sparks passed between the points during the measurements. The well-insulated Coolidge tube was placed inside a large lead box, and the X rays, issuing through a rectangular opening in the box, passed into the measuring vessel which was placed close to the opening. The ionization current was measured by means of lead electroscopes of the self-contained type used for gamma rays. Three of these electroscopes, of cubical form, respectively 11 cm., 10 cm., and 12 cm. side, were employed in the course of the experiments. For determining the initial absorption, the lead face of the electroscope was cut away, and replaced by thin aluminium-foil. For greater thicknesses of absorber, a lead electroscope with sides 3 mm. thick was used; while for still greater thicknesses, a lead electroscope 8 mm. thick was used in some experiments. In order to avoid disturbances due to stray radiations, the windows of the electroscopes were of thick plate-glass, and still further pro-

* R., B., and R., Phil. Mag. xxx. p. 339 (1915). (*This vol., p. 505.*)

tected by lead extensions. Such precautions are essential when, as in the present experiments, the intensity of the end radiation under measurement was in some cases less than one millionth of its initial value.

In order to make experiments over such a wide range, the heating current through the tungsten spiral was adjusted to give a convenient rate of leak in the electroscope in each experiment.

The voltage corresponding to the alternative spark-gap was determined by comparison with the sparking potential between two large brass spheres 20 cm. in diameter.

The absorbing lead plates, which were of much greater area than the face of the electroscope, were placed close to the electroscope. In such a case, the greater part of the radiation scattered in the absorber in a forward direction enters the electroscope. The average absorption coefficients μ for different thicknesses of the absorber were determined in the usual way. The results for different voltages are given in the following table:—

Max. Voltage	Range of thickness in lead, mm.	Absorption Coefficient, μ. cm.$^{-1}$	Mass Abs. Coef. μ/ρ	Max. Voltage	Range of thickness in lead, mm.	Absorption Coefficient, μ. cm.$^{-1}$	Mass Abs. Coef. μ/ρ
79,000	1·8—2·5	27	2·37	183,000	0·7— 1·3	26	2·28
	2·5—3·1	26	2·28		1·3— 2·0	24	2·11
92,000	1·8—2·5	25	2·19		2·4— 4·0	20·5	1·80
	2·5—3·4	24	2·11		4·0— 4·6	18	1·58
	3·1—3·7	24	2·11		4·6— 5·3	15	1·32
105,000	2·7—3·3	23	2·02		5·3— 6·4	13	1·14
	3·3—3·9	22	1·93		6·4— 7·0	12	1·05
	3·9—4·6	22	1·93	196,000	4·3— 5·5	13	1·14
118,000	2·7—3·3	22	1·93		5·5— 6·4	12	1·05
	3·3—3·9	22	1·93		6·4— 7·8	11	0·96
144,000	2·7—3·3	22	1·93		7·8— 9·2	10	0·88
	3·4—4·6	22	1·93		8·8—10·0	8·5	0·75
170,000	3·1—3·7	18	1·58				
	3·7—4·3	17	1·49				
	4·3—5·5	15	1·32				

It will be seen from the above table that the thickness of lead through which the radiation was measurable increased with the voltage applied. This is a result not only of the increase of the penetrating power of the radiation, but also of the large increase with voltage of the intensity of the radiation. With a voltage of 196,000, the radiation was detected and measured through 10 mm. of lead. In this case, the intensity of the radiation after passing through this thickness of lead was considerably less than one millionth of its initial value. No doubt by the use of still more powerful rays and more sensitive methods of measurement, the radiation could be detected through a still greater thickness. The maximum voltage applied (196,000 volts) was about the limit of capacity of the induction-coil under the working conditions. In addition, I should adjudge this voltage to be about the limit of safety for the bulb itself, so that no attempt was made to examine the penetrating power of the radiation for still higher voltages.

Certain interesting points arise in considering the results given in the table:—

(1) There is not much change in the value of μ for the end radiations between 79,000 and 144,000 volts, and no observable change in μ between 105,000 and 144,000 volts.

(2) Between 105,000 and 144,000 volts the radiation is absorbed nearly exponentially with a value of $\mu = 22$. Above 144,000 volts the absorption is no longer exponential, but the value of μ decreases progressively with increase of thickness of absorber. This is best shown by the results for 183,000 volts, in which the value of μ decreases from 26 to 12 as the thickness of absorber is increased from $0 \cdot 7$ to $7 \cdot 0$ mm.

These results, which are at first sight peculiar and unexpected, can be very readily explained by taking into account the absorption of rays of different frequency by lead. In a recent paper,* Hull and Miss Rice have carefully examined the absorption coefficient of lead for X rays of different wave-lengths, obtained by reflexion from a rock-salt crystal. For wave-lengths greater than $0 \cdot 149$ Å.U., the absorption in lead obeys the law $\mu/\rho = 430\lambda^3 + 0 \cdot 12$, where λ is the wave-length in Ångström units and $0 \cdot 12$ is the assumed mass-scattering coefficient, σ/ρ. The value of μ/ρ suddenly increases for values of λ below $0 \cdot 149$ Å.U. owing to the presence of a characteristic absorption-band in lead. The presence of this sharp absorption-band has been shown also photographically by Hull and Miss Rice and by De Broglie. By plotting the logarithm of λ (fig. 1) against the logarithm μ/ρ for lead, Hull has shown that the curve is nearly a straight line AB. At B, where $\lambda = 0 \cdot 149$ Å.U., the absorption suddenly increases, shown by the nearly horizontal line BC. Assuming that the law of absorption after passing through the absorption-band is similar to that observed before, the line CDE should represent the new portion of the curve. The circles represent values actually found by Hull and Miss Rice. Taking the quantum relation, $\lambda = 0 \cdot 149$ Å.U. corresponds to 83,000 volts, and the minimum corresponding value of μ/ρ for lead found by Hull was $1 \cdot 50$, $i.\ e.\ \mu = 17 \cdot 5$. From the dotted portion of the curve the radiation emitted between $\lambda = 0 \cdot 149$ Å.U. and $\lambda = 0 \cdot 098$ Å.U., $i.\ e.$ between 83,000 and 125,000 volts, should be more absorbed than that emitted for voltages slightly less than 83,000. We should thus expect the value of μ for the end radiation to be sensibly constant for the above range of voltages. Actually we find μ nearly constant between 92,000 and 144,000 volts. This difference is not important, and is to be anticipated from the nature of the measurements. A radiation more penetrating than $\mu = 22$ must be present in some quantity before its presence can be detected by absorption methods.

* Hull and Miss Rice, Phys. Rev. viii. p. 326 (1916).

The minimum value found by Hull, $\mu = 17\cdot5$, is somewhat less than the value, $\mu = 22$, found in these experiments, but the difference is no doubt to be ascribed to the difficulties of accurate measurement of μ in both cases.

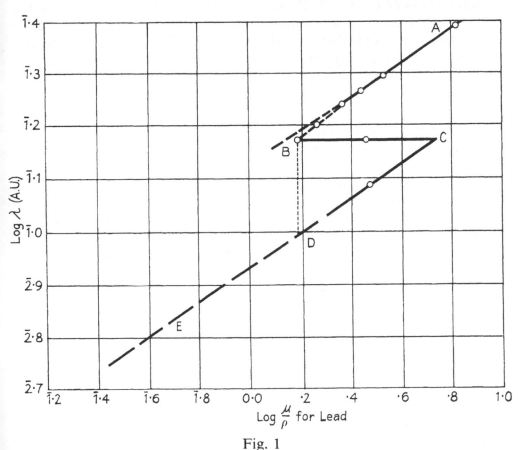

Fig. 1

From the dotted portion of the curve, the minimum value of μ for lead at 196,000 volts ($\lambda = 0\cdot063$ Å.U.) should be about 5. The observed value is $8\cdot5$. Taking into account that the minimum value of μ for 196,000 volts must be somewhat less in any case than $8\cdot5$, and that the actual curve of absorption is probably somewhat steeper than the dotted portion of the curve, there is not a marked divergence between the observed and the calculated results. Taking these factors into consideration, the absorption measurements are not in themselves inconsistent with the view that the maximum frequency of the radiation from a Coolidge tube is given by the quantum relation, $E = h\nu$, over the range of voltage examined. Hull and others have already shown by crystal methods that this relation certain holds up to 100,000 volts and probably up to 150,000 volts.

The peculiarities of the absorption by lead of X rays of different frequencies

* *Loc. cit.*

affords a simple explanation of the results obtained by Rutherford, Barnes, and Richardson.* In their experiments the absorption of the end rays by aluminium was found unchanged between 142,000 and 175,000 volts after the rays had passed through 2·49 mm. of lead as absorber. A reference to the table shows that under these conditions the issuing radiation consisted mainly of the characteristic radiation of lead with a value of $\mu = 22$, and no observable change in the absorption by aluminium is to be expected under the experimental conditions.

Absorption by Aluminium

A few isolated and approximate measurements were made of the absorption of the rays by aluminium under different conditions. In order to avoid complications due to the characteristic radiations of heavy elements like lead, the greater part of the radiation was first absorbed by its passage through an element of low atomic weight like iron. Under such conditions, the absorption results should not be seriously influenced for frequencies much higher than that of the K radiation of iron. The following results were obtained for the absorption by aluminium of the end radiation after passing through iron:—

Volts	μ	μ/ρ
92,000	0·38	0·14
144,000	0·30	0·11
183,000	0·23	0·085

The corresponding values of μ were found to be higher if lead were used as initial absorber instead of iron.

The absorption was measured by placing the aluminium plates close to the electroscope between the latter and the iron plate. Under such conditions the greater part of the forward scattered radiation enters the electroscope, and consequently the absorption coefficient as measured is intermediate between μ and $\mu + \sigma$ (where μ is the true absorption coefficient and σ the scattering coefficient), and probably closer to the former. The value of μ as given by Hull and Miss Rice corresponds to $\mu + \sigma$ in the above notation.

In a recent paper, S. J. Allen and Alexander* have examined the absorption of X rays from a Coolidge tube when different metals are used as filters for the rays. With a tin filter, they found that the absorption coefficient in aluminium for the issuing rays was lower than for any other metal. The value, $\mu/\rho = 0·12$, for aluminium was observed with a steady voltage of about 120,000 volts; with an iron filter $\mu/\rho = 0·134$ under the same conditions. These numbers are in good agreement with those found by the writer.

Application to the wave-lengths of gamma rays

The observations on the absorption of X rays in aluminium and lead throw important light on the difficult question of the probable wave-lengths of the

* Allen and Alexander, Phys. Rev. ix. p. 198 (1917).

penetrating gamma rays from radioactive substances. For convenience, the approximate results so far obtained are collected in the following table. The minimum wave-length is deduced from the voltage or *vice versa* on the assumption that the quantum relation, $E = h\nu$, holds.

The rows with an asterisk give values of μ/ρ obtained by Hull and Miss Rice (*loc. cit.*). In their case, the values of μ/ρ include the effect of scattering as well as absorption, and are consequently not strictly comparable with the values found by the author for aluminium, in which the correction for scattering is less important. The values of μ/ρ for the penetrating gamma rays from radium C are those given in a recent paper by Ishino,[*] where the coefficients of absorption and scattering were separately determined. The values of the mass-scattering coefficients, σ/ρ, for the gamma rays were found by him to be $0\cdot045$ for aluminium and $0\cdot034$ for lead—values much smaller than those previously found for ordinary X rays.

Voltages	Wave-length in Å.U.	Mass Absorption Coefficient μ/ρ	
(Volts)	λ	in Aluminium	in Lead
*84,000	0·147	0·154	1·50
92,000	0·135	0·14	.....
*102,000	0·122	.....	3·00
144,000	0·086	0·11	.....
183,000	0·068	0·085	1·05
196,000	0·063	.....	0·75
Gamma rays from radium C	?	0·026	0·042

It will be observed from the table that the value of μ/ρ in aluminium decreases very slowly between 84,000 and 196,000 volts, even at a slower rate than the first power of the wave-length; while for longer waves it is well known that the value of μ varies approximately as the cube of the wave-length. As we should expect, the variation in μ/ρ with wave-length is much more rapid for lead than for aluminium over the same range. It will be noted that, while the value of μ/ρ for aluminium for X rays generated at 183,000 volts is only 3 times the value for the gamma rays, the corresponding ratio in the case of lead is more than 20. The general results suggest that when the value of μ/ρ becomes of the same order of magnitude as that of σ/ρ, the former coefficient varies slowly with the wave-length, the latter probably remaining constant. In addition, it appears not unlikely that there is a definite connexion between absorption and scattering, and that, for very short waves, the absorption like the scattering may ultimately reach a minimum value independent of wave-length. From some points of view such a connexion between these two quantities is not improbable, but unfortunately no waves of sufficiently short wave-length are available to test the relation experimentally.

* Ishino, Phil. Mag. xxxiii. p. 129, January 1917.

The two shortest wave-lengths of the gamma rays observed in the experiments of Rutherford and Andrade* were 0·072 and 0·099 Å.U., corresponding on the quantum relation to waves excited by 174,000 and 125,000 volts. The values of μ/ρ for aluminium corresponding to X rays excited at these voltages are about 0·09 and 0·12 respectively, while the observed value of μ/ρ for the penetrating gamma rays from radium C is much less, viz. 0·026. Since undoubtedly for such high frequencies, μ/ρ varies very slowly with frequency, it is clear that the wave-length of the more penetrating radiation is considerably smaller than that of the shortest waves observed by Rutherford and Andrade. In other words, the wave-length of the main gamma rays is much shorter than was previously supposed. This conclusion is still more strongly confirmed by the observations on the absorption of the radiation by lead. For a voltage of 196,000 volts, corresponding to a still shorter wave-length than the shortest observed by Rutherford and Andrade, the observed value of μ/ρ in lead was 0·75, while the value of μ/ρ found by Ishino for the penetrating gamma rays was 0·042—a ratio of nearly 20 times. Even allowing that the true value of μ/ρ for waves generated at 196,000 volts is somewhat smaller than the value observed, the largeness of the ratio shows that the gamma rays must be much shorter than those generated at 200,000 volts, *i. e.* much shorter than $\lambda = 0·062$ Å.U.

In our present ignorance of the law of variation of μ/ρ with frequency in this region of the spectrum, it is only possible to estimate the actual wave-length of the most penetrating gamma rays. It is clear, however, that the waves are at least three times and may be ten times shorter than those which correspond to 200,000 volts, *i. e.* they correspond to waves generated by voltages between 600,000 and 2,000,000 volts, and thus lie between 0·02 and 0·007 Å.U. It is thus clear that the gamma rays from radium C consist mainly of waves of about $\frac{1}{100}$ the wave-length of the soft gamma rays from radium B, and are of considerably shorter wave-length than any so far observed in an X-ray tube, with the highest voltages at our disposal.

Another very interesting and important point arises from this discussion. It is well known that the β rays from radium B and radium C when examined in a magnetic field give a veritable spectrum of bright lines corresponding to definite groups of β rays, each group consisting of electrons expelled with a characteristic and definite velocity. The energy of motion of each of these groups of electrons have been measured by Rutherford and Robinson,† and the more intense groups (labelled with letters in the original paper) are given in the following table:—

The column headed 'voltage' gives the potential difference in volts between which the electron must move to acquire the observed energy.

Apart from the low-velocity groups L, M, N, the β rays from radium C consist mainly of groups lying between 500,000 and 2,000,000 volts. This is

* Rutherford and Andrade, Phil. Mag. xxviii. p. 263 (1914). (*This vol., p. 456.*)
† Rutherford and Robinson, Phil. Mag. xxvi. p. 717 (1913). (*This vol., p. 371.*)

about the same range of voltage as we estimated to excite the penetrating gamma rays from consideration of the absorption of X rays and gamma rays by aluminium and lead. It would thus appear probable that the observed

β rays from Radium B				β rays from Radium C			
Group	Intensity	Energy ÷ $10^{13}e$	Voltage (volts)	Group	Intensity	Energy ÷ $10^{13}e$	Voltage (volts)
A	s.	3·332	333,200	A	m.f.	21·02	2,102,000
B	v.s.	2·610	261,000	B	m.f.	17·51	1,751,000
C	v.s.	2·039	203,900	C	m.	16·71	1,671,000
D	v.s.	1·519	151,900	D	m.	14·09	1,409,000
E	s.	0·503	50,300	E	m.s.	13·28	1,328,000
F	v.s.	0·376	37,600	F	m.	11·49	1,149,000
				G	m.s.	10·31	1,031,000
				H	m.s.	5·94	594,000
				K	s.	5·16	516,000
				L	m.	2·96	296,000
				M	m.	2·59	259,000
				N	m.	1·81	181,000

groups of β rays are due to the conversion of the energy, $E = h\nu$, of a wave of frequency ν into electronic form, and that consequently the energy of the β-ray groups may be utilized by the quantum relation to determine the wave-lengths of the penetrating gamma rays.

Such a conclusion is borne out by consideration of the groups of rays from radium B.

H. Richardson* has determined the absorption of these rays by lead, and concluded that they could be analysed approximately into three component groups for which the absorption coefficients, μ, in lead were 45, 6, and 1·5 cm.$^{-1}$ respectively. From the observations with a Coolidge tube, the value, $\mu = 6$, should correspond to waves excited at about 200,000 volts, and it is to be noted in the table that three strong groups, B, C, and D, of β rays from radium B correspond to voltages between 261,000 and 152,000 volts, an average of about 200,000 volts. The value of $\mu = 1·5$ may correspond to group A or a still swifter group, of voltage about 500,000 volts, observed in the spectrum of β rays excited in lead by the gamma rays from radium B and radium C together.†

The results as a whole suggest that the groups of β rays are due to the transformation of the gamma rays in *single* and not *multiple* quanta, according to the relation $E = h\nu$. The multiple relations observed between the energy of some of the groups of β rays‡ must on this view indicate approximate multiple relations between the frequencies of the gamma rays.

* Richardson, Proc. Roy. Soc. xci. p. 396.
† Rutherford, Robinson and Rawlinson, Phil. Mag. xxviii. p. 285 (1914). (*This vol., p. 469.*)
‡ Rutherford, Phil. Mag. xxviii. p. 305 (1914). (*This vol., p. 473.*)

S

With the assistance of Mr. J. West, B.Sc., I have made some experiments to see whether it is possible to detect by the crystal method the presence of waves shorter than those observed in the experiments of Rutherford and Andrade (*loc. cit.*). A narrow pencil of gamma rays and strong sources were employed, but no certain evidence of the existence of such waves was obtained. This may be due either to the overlapping of the numerous lines that should be present, or to the failure of the crystal to resolve waves whose length is very small compared with the grating space.

If the single quantum relation should prove to hold generally for the conversion of γ rays into β rays, the magnetic spectrum of β rays should afford a reliable method of extending the investigation of X-ray spectra into the region of very short waves where the crystal method either breaks down or is practically ineffective, and thus places in our hands a new and powerful method of analysing waves of the highest obtainable frequency. The complexity of the β-ray spectrum for radium B and radium C indicates that the spectrum of the gamma rays, and presumably the very high-frequency spectra of heavy elements in general, are as complicated as the ordinary light spectra of such elements.

University of Manchester
May 12, 1917

Collision of α Particles with Light Atoms

I. HYDROGEN

by PROFESSOR SIR E. RUTHERFORD, F.R.S.

From the *Philosophical Magazine* for June 1919, ser. 6, xxxvii, pp. 537–61

§ 1. ON the nucleus theory of atomic structure, it is to be anticipated that the nuclei of light atoms should be set in swift motion by intimate collisions with α particles. From consideration of impact, it can be simply shown that as a result of a head-on collision, an atom of hydrogen should acquire a velocity $1 \cdot 6$ times that of the α particle before impact, and should possess $0 \cdot 64$ of the energy of the incident α particle. Such high speed 'H' atoms should be readily detected by the scintillation method. This was shown to be the case by Marsden,* who found that the passage of α particles through hydrogen gave rise to numerous faint scintillations on a zinc sulphide screen placed far beyond the range of the α particles. The maximum range of the H particles, set in motion by the α particles from radium C, was over 100 cm. in hydrogen or about four times the range of the colliding α particles in that gas. This range agreed well with the value calculated by Darwin† from Bohr's‡ theory of the absorption of α particles by matter.

In most of the experiments of Marsden, a thin glass α-ray tube, containing purified radium emanation, was used as an intense source of rays. This was placed in a closed vessel at a suitable distance from a zinc sulphide screen, and the space between filled with compressed hydrogen. It was found that the number of H scintillations fell off approximately according to an exponential law when absorbing screens of matter were interposed, and the relative absorption of metal foils was in good accord with the square root law observed by Bragg for α particles.

In a second paper, Marsden§ showed that the α-ray tube itself gave rise to a number of scintillations like those from hydrogen. Similar results were observed with an α-ray tube made from quartz instead of glass, and also with a nickel plate coated with radium C. The number of H scintillations observed in all cases appeared to be too large to be accounted for by the possible presence of hydrogen in the material, and Marsden concluded that there was strong evidence that hydrogen arose from the radioactive matter itself. Further experiments were interrupted by the departure of Mr. Marsden to New Zealand early in 1915 to fill the Professorship of Physics in Victoria

* Marsden, Phil. Mag. xxvii. p. 824 (1914).   † Darwin, Phil. Mag. xxvii. p. 499 (1914).
‡ Bohr, Phil. Mag. xxv. p. 10 (1913).   § Marsden, Phil. Mag. xxx. p. 240 (1915).

College, Wellington. The quantity of radium available there was too small to continue observations, while the possibility of further work was precluded by the return of Professor Marsden to Europe on Active Service.

We have seen that Marsden in his second paper had some indications that the radioactive matter itself gave rise to swift H atoms. This, if correct, was a very important result, for previously the presence of no light element except helium had been observed in radioactive transformations.

It was thought desirable to continue these experiments in more detail, and during the past four years I have made a number of experiments on this point and on other interesting problems that have arisen during the progress of the work. The experiments recorded in this and subsequent papers have been carried out at very irregular intervals, as the pressure of routine and war-work permitted, and in some cases experiments have been entirely dropped for long intervals.

§ 2. *Source of the scintillations from active matter*

Marsden had observed that the number of H scintillations from a nickel plate, coated with radium C, was considerably greater than for a corresponding quantity of emanation—measured by γ rays—from an α-ray tube. It thus seemed possible that H atoms might arise from the disintegration of radium C, for it is well known that this product is transformed in an anomalous manner. In order to test this point, observations were made on the variations of the number of H scintillations from an α-ray tube immediately after it was filled with emanation. It is well known that the amount of radium C in such a tube increases at first very slowly. For example, after filling a tube with emanation, the fraction of the final amount of radium C present after 10 minutes is only 2 per cent., but reaches 9 per cent. after 20 minutes.* Consequently, observations made on the number of scintillations within 10 minutes after filling should decide definitely whether the scintillations arise from radium C alone and not from the other α-ray products present, viz. the emanation and radium A. In the latter case, the number of scintillations after 10 minutes should be only 2 per cent. of the final number reached about three hours later when radium C is in transient equilibrium with the emanation.

A number of α-ray tubes were kindly made and filled for me by Mr. N. Tunstall, B.Sc. The whole process of filling and removal for testing was done as rapidly as possible, and the counting of scintillations was usually begun within four minutes after filling. The α-ray tube was placed between the poles of a strong electromagnet in order to reduce the luminosity due to β rays on the zinc sulphide screen, placed 2 centimetres beyond the range of the α rays. After every precaution had been taken to avoid radioactive contamination, the number of scintillations observed between 4 and 10 minutes was greatly in excess of the number to be expected if they had their origin in the transformation of radium C alone. The actual ratio of the maximum number varied

---

* 'Radioactive Substances and their Radiations,' Rutherford, p. 499.

with the thickness of the α-ray tube, but the fraction observed initially was from 20 to 40 per cent. of the maximum reached three hours later.

These results showed conclusively that, if the H atoms from a glass α-ray tube were a product of radioactive disintegration, they arose not only from radium C but also from radium A or the emanation or both. It is hoped to discuss in a later paper the results of a number of experiments to test whether hydrogen is a product of radioactive change. It is not easy to give a decisive answer to this important problem on account of the numerous factors involved. It will be seen later that the number of scintillations from hydrogen is much greater than is to be expected on the simple theory, and it is difficult to be sure of the absence of hydrogen as a contamination in the source and absorbers of the radiation. In addition, both nitrogen and oxygen atoms are set in such swift motion by collision with α particles that they cause scintillations outside the range of the α particles. It seems probable that the large number of scintillations observed by Marsden (*loc. cit.*) from a nickel plate coated with radium C were mainly due, not to H atoms, but to high-velocity N and O atoms produced from the air between the source and the screen.

§ 3. *Source of radiation*

While the use of α-ray tubes as an intense source of radiation has many advantages, it has the drawback that the α radiation is heterogeneous arising from the three products radium A, radium C, and the emanation. In addition, it is difficult to make α-ray tubes of uniform thickness whose stopping power is less than two centimetres of air. For these reasons, I have discarded the use of α-ray tubes and have conducted the majority of the experiments with a homogeneous source of radiation, consisting of the active deposit of radium. Twenty minutes after removal from the emanation, the α radiation arises entirely from radium C and is homogeneous with a range in air of 7 cm. A brief account will now be given of the method for obtaining an intense source of radiation of convenient dimensions. The source usually consisted of a circular bevelled brass disk which was screwed on the lower end of a glass stopcock. The emanation, after removal from the radium solution, was sparked with oxygen to remove excess of hydrogen until the volume was reduced to about 0·5 c.c. This emanation was introduced by means of a mercury trough into a small transfer pump and the mercury raised until its level was 1 or 2 mm. below the disk to be activated. The disk was connected through the stopcock with the negative pole of the lighting circuit and the mercury with the positive pole, in order to concentrate to some extent the active matter on the surface of the disk. After two hours' exposure, the emanation was pumped out and the active disk removed. Theoretically, in order to obtain the maximum activity, the exposure to the emanation should be more than three hours, but in practice it is found that an exposure of two hours gives more activity, while an exposure of twenty-four hours gives much less than an exposure of two hours. This anomalous effect had been previously observed by Ratner,* and is apparently due to the loss of active matter from the disk through the intermediary of the electric wind.

Using a large quantity of emanation, it is possible to obtain in this way a disk, coated on one side with radium C which has a gamma-ray activity equal to 80 mg. of radium. In most experiments, sources were employed of activity between 5 and 80 mg. of Ra.

The active disk after removal was washed in alcohol and then heated for a minute in an exhausted tube inside an electric furnace at about 300° C. As Ratner (*loc. cit.*) has pointed out, the treatment with alcohol reduces greatly the loss of active matter by so-called volatilization, while the heating tends to remove the surface gases and the emanation occluded in the disk during its exposure. The quantity of active matter on the disk was determined with the aid of a standardized gamma-ray electroscope. The decrease of intensity with time is known from the well-known curve of decay.

§ 4. *Counting scintillations*

As the systematic counting of H scintillations under varied conditions is a rather difficult and trying task, it may be of some value to mention the general arrangements found most suitable and convenient in practice. Using the excellent zinc sulphide screens, specially prepared by Mr. Glew, the scintillation due to a high-speed H atom appears as a fine brilliant star or point of light, very similar in appearance and intensity to that produced by an alpha particle about 3 mm. from the end of its range. Near the end of the range of the H atom, the scintillation becomes very feeble, and can only be observed on a dark background. Consequently, in a heterogeneous beam of H atoms, the actual number counted per minute is to some extent dependent on the luminosity of the background seen in the microscope. It is important to adjust and keep the luminosity of the screen to the right amount throughout the whole interval of an experiment. This is most simply done by means of a small 'pea'-lamp fixed in a metal tube in which the current is varied. While weak scintillations are readily counted on a dark background, it is difficult under such conditions to keep the eye focussed on the microscope image and the eye rapidly becomes fatigued and counting becomes erratic. The microscope employed had a magnification of about 40 and covered a field of 2 mm. diameter. This in practice was found to be a very convenient magnification. In later experiments, special zinc sulphide screens were prepared in which the smaller crystals were sifted through a fine gauze on to a glass plate covered with a thin layer of adhesive material. These fine crystals completely covered the plate several crystals deep. With such a screen, the H scintillations appeared larger and more diffuse, probably due to the scattering of the light in passing through the thick layer of crystals, and were more easily counted, while weak scintillations could be counted on a brighter background than with the ordinary screen. At the same time, the layer of crystals was so uniform, that each incident H atom produced a scintillation.

* Ratner, Phil. Mag. xxxiv. p. 429 (1917); xxxvi. p. 397 (1918).

In these experiments, two workers are required, one to remove the source of radiation and to make experimental adjustments, and the other to do the counting. Before beginning to count, the observer rests his eyes for half an hour in a dark room and should not expose his eyes to any but a weak light during the whole time of counting. The experiments were made in a large darkened room with a small dark chamber attached to which the observer retired when it was necessary to turn on the light for experimental adjustments. It was found convenient in practice to count for 1 minute and then rest for an equal interval, the times and data being recorded by the assistant. As a rule, the eye becomes fatigued after an hour's counting and the results become erratic and unreliable. It is not desirable to count for more than 1 hour per day, and preferably only a few times per week.

Under good conditions, counting experiments are quite reliable from day to day. Those obtained by my assistant Mr. W. Kay and myself were always in excellent accord under the most varied conditions. It was usually arranged that the number of scintillations to be counted varied between 15 and 40 per minute.

§ 5. *Experimental arrangement*

For experiments with hydrogen and other gases, the active disk D (fig. 1) was mounted at a convenient height parallel to the screen on a metal bar B which slid into a rectangular brass box A, 18 cm. long, 6 cm. deep, and 2 cm.

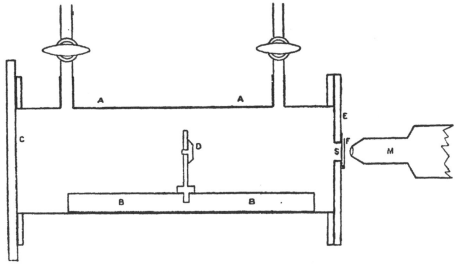

Fig. 1

wide, with metal flanges at both ends fitting between the rectangular poles of a large electromagnet. One end was closed by a ground glass plate C, and the other by a waxed brass plate E, in the centre of which was cut a rectangular opening 1 cm. long and 3 mm. wide. This opening was covered by a thin plate of metals of silver, aluminium or iron, whose stopping power for α particles lay between 4 and 6 cm. of air. The zinc sulphide screen F was fixed opposite the opening and distant 1 or 2 mm. from the metal covering. By means of two stopcocks, the vessel was filled with the gas to be examined either by exhaustion or displacement. It is a great advantage to have the zinc sulphide screen outside the apparatus, in order to avoid contamination due to volatilized active matter, and for the easy introduction of absorbing material between the end plate and the screen.

In practice, the source was introduced into the brass vessel at a convenient distance from the screen, and the air exhausted. The α rays after traversing the end plate fell on the screen, and the marked luminosity due to them was a guide in fixing the microscope M in the centre of the opening. The diameter of the field of view (2 mm.) was less than the width of the opening (3 mm.).

Since the number of H atoms observed under ordinary conditions is less than one in a hundred thousand of the number of α particles, H atoms, projected in the direction of the α particles, can only be detected when the α rays are stopped by the absorbing screens. It was not found possible to bring an intense source closer than 3 cm. from the screen on account of the luminosity excited in it by the γ rays and swift β rays, which prevented counting of weak scintillations. A strong magnetic field was necessary to bend away the β rays which caused a very marked luminosity on the screen. A field of 6000 gauss was generally employed for this purpose.

§ 6. *Scintillations due to source and absorbing screens*

When the containing vessel was exhausted of air, scintillations were always observed on the screen proportional in number to the activity of the source.

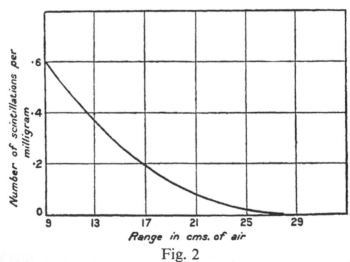

Fig. 2

The number fell off rapidly between 7 and 12 cm. air absorption and then more slowly, but a few could be observed nearly to 28 cm. The variation of number with amount of absorption in terms of cms. of air is shown in fig. 2.

This refers to a heated brass source, $3 \cdot 3$ cm. from the screen, with a heated silver plate of stopping power 6 cm. of air just before the screen.

These scintillations appear to be due mainly to H atoms excited partly in the source and partly in the absorbing screens. Thin foils of aluminium, for example, placed close to the source increase the number of scintillations. This is due to the occlusion of hydrogen, which can be removed by heating the aluminium in an exhausted furnace just below the melting-point. Similar effects were observed with silver but not with gold. In practice, all screens to be used in the path of the α rays were heated to drive off occluded gases as far as possible. This is very necessary when small numbers of scintillations have to be counted. Usually a silver plate was used to absorb the α rays. Gold was found to be very free from hydrogen, but it could not be used in place of silver close to the screen on account of the marked luminosity set up on the screen well beyond the range of the α particles. This peculiarity of gold had been previously noted by Marsden, but I was surprised to observe the magnitude of the effect with strong sources of radiation. A fuller account of the nature and cause of this luminosity will be postponed till a later paper. In a similar way, mica was found to cause a good deal of luminosity, apparently due to gamma rays. In addition, as is to be expected, mica gives rise to numerous H atoms and swift oxygen atoms. For these reasons, mica is unsuitable for an absorbing screen for α particles in this type of experiment.

§ 7. *Theory of Collision of α particles with light atoms*

It will be seen later that the number of H atoms and their distribution with velocity differ markedly from the results to be expected theoretically. It is consequently desirable to consider first with some detail the results to be anticipated on simple theoretical grounds, before discussing the experimental results.

The effect of collision of swift α particles with light atoms has been worked out by C. Darwin.*

α particle: M mass, E charge, v initial velocity, ϕ angle of scattering from original direction.

Light atom: m mass, e charge, u velocity after impact, θ angle of deflexion from original direction of α particle.

From considerations of simple impact, it follows that

$$u = 2v \frac{M}{M + m} \cdot \cos \theta, \quad . \quad . \quad . \quad . \quad . \quad (1)$$

$$\tan \phi = \frac{m \sin 2\theta}{M - m \cos 2\theta} \quad . \quad . \quad . \quad . \quad . \quad . \quad (2)$$

If there is no loss of energy in the impact we should consequently expect $u = \frac{8}{5}v \cos \theta$ for the hydrogen atom, quite independently of any assumption

* C. Darwin, Phil. Mag. (*loc. cit.*).

S*

as to the nature and magnitude of the forces involved in the impact. In order, however, to calculate the number of H atoms scattered within a given angle θ, it is necessary to make special assumptions as to the magnitude and direction of the forces. Assuming that the forces arise from the charges carried by the atomic nuclei which are to be regarded as points, and that the forces vary as the inverse square, Darwin has shown that

$$p = \mu \tan \theta, \qquad \cdots \cdots \cdots \quad (3)$$

where p is the perpendicular distance from the atom on the initial direction

of motion of the α particle and $\mu = \dfrac{Ee}{v^2} \left(\dfrac{1}{m} + \dfrac{1}{M} \right)$.

If Q α particles pass normally through a layer of gas thickness dx, which contains N atoms per c.c. at N.T.P., then the number dn of H atoms projected between the angles 0 and θ is given by

$$dn = QN\pi p^2 dx$$

$$= \pi NQ\mu^2 \tan^2 \theta \,.\, dx.$$

Since the reduction of velocity of the α particle in passing through 1 cm. of hydrogen is small, the number n of H atoms produced per cm. of path is given by

$$n/Q = \pi N\mu^2 \tan^2 \theta. \quad \cdots \cdots \cdots \quad (4)$$

In this case n/Q is the fraction of α particles which give rise to an H atom between 0 and θ.

Taking $e = \tfrac{1}{2}E = 4 \cdot 77 \times 10^{-10}$ e.s. unit, $v = 1 \cdot 922 \times 10^9$ cm. per sec., $N = 5 \cdot 41 \times 10^{19}$, and $e/m = 9570$ for hydrogen,

then $\qquad\qquad\qquad \mu = 9 \cdot 27 \times 10^{-14}$

and $\qquad\qquad\qquad n/Q = 1 \cdot 46 \times 10^{-6} \tan^2 \theta. \quad \cdots \cdots \quad (5)$

It was found experimentally that the swiftest H atoms due to an α particle from radium C had a range corresponding to 28 cm. of air or four times the range of the α particle. Generally it was found that the maximum range of the H atom was four times the range of the α particle producing it. Since the range of α particles varies as the cube of their velocity, it follows that the range of H atoms is proportional, at any rate approximately, to the cube of their velocity. Since the velocity of an H atom projected at an angle θ with the α particle is $u_0 \cos \theta$ where u_0 is the maximum velocity of the H atom, the range R of an H atom projected at angle θ is given by $R/R_0 = \cos^3 \theta$ where R_0 is the maximum range. Since, however, the α particles fall nearly normally on the screen, the H atoms deflected at an angle θ travel a distance R sec θ.

Consequently the range R in the direction of the α particles is given by $R/R_0 = \cos^4 \theta$. Substituting the value of θ in equation (5),

$$n/Q = 1\cdot46 \times 10^{-6}\left(\sqrt{\frac{R_0}{R}} - 1\right).$$

This equation only applies to α particles of velocity v_0 emitted by radium C. Since $p \propto 1/v^2$, it is seen that the number of H atoms varies as $1/v^4$. Remembering that the range of the α particle varies as v^3, it is easily seen that for α particles of range r

$$n/Q = 1\cdot46 \times 10^{-6}(r_0/r)^{\frac{4}{3}}\left(\sqrt{\frac{R_{max.}}{R}} - 1\right),$$

where r_0 is range of α particle from radium C, viz. $7\cdot0$ cm., and $R_{max.}$ is the maximum range of the H atom for a range r, viz. $4r$.

The values of n/Q for different values of R are given in Table I for α particles of range 7, 5, and 3 cm. in air respectively. It should be noted that Q/n represents the number of α particles required to produce on the average in traversing one centimetre of the gas one H atom, which has a range equal to or greater than R cms. of air.

TABLE I

Range of α particles = 7 cms.		Range of α particles = 5 cms.		Range of α particles = 3 cms.	
Range R of H atoms	n/Q	Range R of H atoms	n/Q	Range R of H atoms	n/Q
1 cm.	$6\cdot3 \times 10^{-6}$	1 cm.	$7\cdot9 \times 10^{-6}$	1 cm.	$11\cdot1 \times 10^{-6}$
2 ,,	4·0 ,,	2 ,,	4·9 ,,	2 ,,	6·5 ,,
4 ,,	2·8 ,,	4 ,,	2·8 ,,	3 ,,	4·5 ,,
7 ,,	1·46 ,,	7 ,,	1·53 ,,	4 ,,	3·3 ,,
10 ,,	0·98 ,,	10 ,,	0·95 ,,	5 ,,	2·5 ,,
14 ,,	0·60 ,,	14 ,,	0·45 ,,	6 ,,	1·9 ,,
18 ,,	0·34 ,,	16 ,,	0·27 ,,	8 ,,	1·0 ,,
22 ,,	0·19 ,,	18 ,,	0·12 ,,	10 ,,	0·4 ,,
24 ,,	0·12 ,,	20 ,,	0 ,,	12 ,,	0 ,,
26 ,,	0·05 ,,				
28 ,,	0 ,,				

These results are shown graphically in fig. 3, curves A, B, and C respectively, for ranges of the H atoms from 5 to 28 cm. It is seen that the curve A is approximately exponential between 8 and 18 cm., falling to half value in about $5\cdot3$ cm. This holds equally for curves B and C over corresponding ranges. These curves give the theoretical variation in number of the H atoms with range such as would be observed if the numbers of H atoms were counted for different thicknesses of absorber.

As the value of n/Q is less than 1/100000, it is not feasible with the present arrangement to detect H atoms within the range of the much more numerous α particles. For this reason, it is not possible to compare theory with experiment in the region of short ranges for which some of the values are calculated. It is seen that while the number of H atoms for short ranges increases rapidly

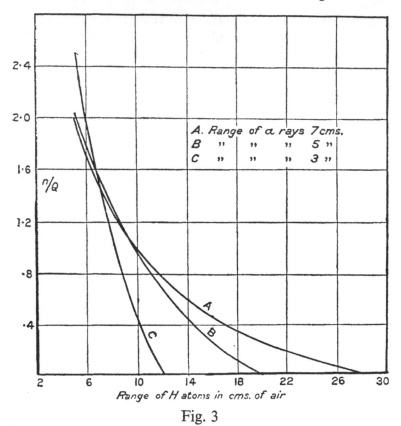

Fig. 3

with the reduction of range of the α particle, the three curves show approximately the same ordinates for 6·5 cm. range of the H atoms.

§ 8. *Absorption of* H *atoms*

The source of α rays was a brass disk coated with radium C only in the central part, in order to reduce the area emitting α rays. The initial γ-ray activity of the disk was equivalent to about 10 mg. Ra. The zinc sulphide screen was mounted parallel to the disk in the apparatus shown in fig. 1 at 3·3 cm. distance from the source. An opening in the end of the box was covered with a heated silver plate, whose absorption for α particles was equivalent to 5·8 cm. of air, and the whole apparatus was filled by exhaustion with dry hydrogen, at atmospheric pressure. Suitable absorbing screens of aluminium foil were interposed between the silver plate and the screen, and the scintillations counted.

The results obtained are shown in fig. 4, curve A, where the ordinates represent number on an arbitrary scale and the abscissæ the thickness of

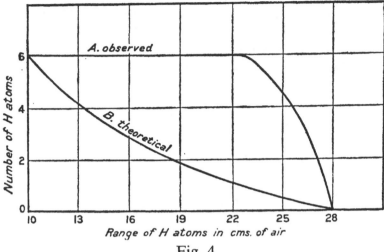

Fig. 4

absorbing material measured in terms of cms. of air for α particles. The equivalent absorption in the silver plate and in the hydrogen is included. The latter was taken as equivalent to 8 mm. of air or one quarter the length of the path of the α particles in hydrogen. The correction due to the natural scintillations from the source and silver plate was small.

It is seen that there is no diminution in the number of scintillations for absorptions between 9 and 19 cm. of air.* After 19 cm., there was a slow decrease followed by a rapid fall near the end of the curve. No scintillations were observed beyond 28 cm., *i. e.* for a range four times that of the α particles from radium C.

The shape of the absorption curve is entirely different from that to be expected theoretically. The latter is shown in curve B, calculated from the data given in § 7, the same ordinate being taken for an absorption of 10 cm. Between 9 and 19 cm. absorption, the number of scintillations according to theory should fall from 100 to 28.

This peculiarity of the absorption curve is only marked for long-range α particles. In fig. 5 the absorption curves for initial ranges of the α particles 7 cm., 6·6, 6·1, 5·7, 4·8, 3·9 cm. are shown. The range was reduced by interposing gold or aluminium screens of known stopping power for the α particles close to the source. Even when the range of the α particles is reduced from 7 to 6 cm., the absorption curve already shows an evident decrease of number with thickness of absorber, and this decrease becomes much more marked for decreasing ranges between 6 and 3 cm.

* It should be remarked that, for the distances employed, the width of the testing vessel (fig. 1) was sufficient to give the correct average distributions of H atoms with velocity, corresponding to a source at the centre of a sphere.

The absorption curves for ranges between 5·7 and 3·9 cm. are very similar in shape to the theoretical curves. For example, in curve F for an initial range of α particles of 3·9 cm. the number of H particles is reduced to $\frac{1}{2}$, $\frac{1}{4}$, $\frac{1}{8}$ for increase of absorption of 2·4, 4·0, 5·5 cm. respectively. The numbers are in good agreement with the calculated values, viz. 2·8, 5·0, 6·4 cm. respectively. The numbers are in still closer agreement if we take the *average* range of the α particles acting on the hydrogen column, viz. 3·9 − 0·4 = 3·5 cm. The corresponding numbers are then 2·3, 4·2, 5·0 cm. respectively.

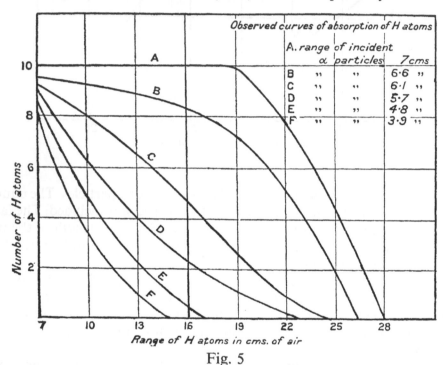

Fig. 5

We shall now consider the interpretation of the anomalous absorption curve for a range of 7 cm. shown in curve A. The curve is very similar to that to be expected if the hydrogen atoms were thrown forward mainly in the direction of the α particles, and all with the same velocity; in fact, the absorption curve for a pencil of H atoms is very similar in shape to that for a pencil of homogeneous α rays from radium C.

It is well known that the number of α particles counted by the scintillation method in a homogeneous pencil of α rays from radium C remains constant from 0 to 6 cm. of the range, and then rapidly falls to zero in the last centimetre of the range. This end effect is usually ascribed to the scattering of the α particles in their passage through the absorbing material. Now if the H particles consist of H atoms carrying unit positive charge e and projected with a velocity $u = 1·6\,v$, where v is the velocity of the α particle, the average angular scattering per cm. should be proportional to e/mu^2 and should thus be 0·78 of that suffered by the α particle for an equal range. Since the H atom has four times the range of the α particle, the average angular scattering of the H atoms before absorption should be approximately $2 \times 0\cdot78 = 1\cdot56$ that of the α particle. It follows, therefore, that the decrease in the number for a homogeneous beam of H atoms should begin about 6 cm. from the end of the maximum range 28 cm.

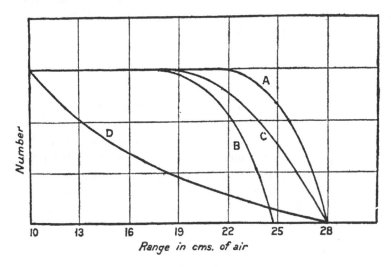

Fig. 6

This theoretical curve is shown in fig. 6, curve A. Remembering that the stopping power of the hydrogen column for α rays corresponds to 8 mm. of air, the absorption curve of H atoms due to α particles of range $6\cdot2$ cm. has a maximum range $24\cdot8$ cm. The corresponding absorption curve is given in curve B. The intermediate curve C shows the distribution to be expected for the hydrogen column, supposing the H atoms are all projected in the direction of the α particles with a velocity proportional to the velocity of the α particle at each point of the hydrogen column.

This theoretical curve C is very similar in all respects to the experimental curve (fig. 4, curve A), showing that the H atoms produced in a thin film of hydrogen are nearly homogeneous in velocity and are thrown forward in the direction of the colliding α particles.

It does not follow that the direction of the H atom coincides with the direction of the α particles, but the average deflexion cannot be much more than 10° or 15°. For an angle of deflexion of θ, the range of the H atom in the direction of the α particles is $R_{max.} \times \cos^4 \theta$. The value of $\cos^4 \theta$ is $0\cdot94$ for 10°, $0\cdot87$ for 15°, and $0\cdot78$ for 20°. An average value of θ of 20° would make the decrease in the number begin about 13 cm. instead of 19 cm.

It is to be anticipated that the average angle of deflexion should increase rapidly with decrease of the velocity of the α particle. The rapid changes in shape of absorption curve with change of velocity of α particle are at any rate partly due to this cause.

It is difficult to determine directly the actual average angle of deflexion of H atoms, since the H atoms are scattered considerably in passing through the minimum 7 cm. of air or other absorbing material required to stop the α particles.

There seems to be little doubt that if a film of hydrogen were exposed to α particles of greater initial velocity than those from radium C, a nearly homogeneous beam of H rays would be obtained, all of which would travel nearly in the direction of the α particles.

§ 9. *Variation of number of* H *particles with velocity of* α *particles*

In order to reduce the velocity of the α particles, the vertical source in the apparatus shown in fig. 1 was completely covered with different thicknesses of gold foil whose stopping power in terms of air was accurately determined. The distance between the source and screen was $3 \cdot 3$ cm., and the apparatus filled with hydrogen at atmospheric pressure. The gold foils were pressed tightly against the source to prevent production of H atoms between the source and foils. The number of H particles was determined after passing through absorbers of known stopping power. The results are given in the table below.

Absorption of gold foil in terms of cms. of air	Issuing range of α rays in cms. of air	Absorption in terms of air between source and screen including the hydrogen	Observed number of H atoms	Calculated number of H atoms
0 cm.	7·0 cm.	8·3 cm.	100	100
1·7 „	5·3 „	8·3 „	77	103
2·5 „	4·5 „	7·5 „	51	119
3·3 „	3·7 „	6·6 „	25	139
4·0 „	3·0 „	6·6 „	5	128

In the last column are given the relative numbers of H atoms to be expected on the simple theory given in § 7, when account is taken of the maximum range of the H atoms and the thickness of absorber traversed. Since the number of H atoms for corresponding ranges varies as the inverse fourth power of the velocity of the α particles, *i. e.*, as the inverse four thirds power of the effective range of the α particles, the number of H atoms should increase with lowering of the velocity of the incident α particles. The observed numbers, however, instead of increasing with reduction of velocity of α particles, fall off slowly at first and then very rapidly for ranges between $3 \cdot 5$ and 3 cm.

In these experiments, the intensity of the H radiation was reduced by this passage through absorbing material equal to $6 \cdot 6$ cm. of air. In order to reduce this absorption, another series of experiments was made in which the silver plate of stopping power $5 \cdot 8$ cm. was replaced by an aluminium plate of stopping power $3 \cdot 7$ cm. The velocity of the α particles was reduced by aluminium foil instead of gold foil, placed close to the source. The aluminium foils used in these experiments were freed as far as possible from hydrogen by heating in a vacuum, and the results obtained with aluminium as absorber were similar to those with gold. The presence of numerous H atoms was observed for α particles of range 2·5 cm., but the number was small and just measurable with certainty for α particles of range 2·0 cm. The actual number in the latter case was small compared with that observed for α particles of range 3 cm. Experiments at low ranges are rendered somewhat difficult by the necessity of taking into account the H scintillations which arise from the absorbing material and source. These are always present, and in number comparable with those produced by the admission of hydrogen. While the general results show that, under the experimental conditions, the number of H atoms becomes relatively small for ranges of α particles between 2 and 3 cm., it is not possible to say with certainty whether the number falls to zero for still smaller velocities of the incident α particles. We are unable to continue observations with this experimental arrangement for absorptions less than 7 cm. of air, so that no information is available of the number of H atoms of range less than 7 cm.

In § 8, the absorption curves for H atoms produced by α particles of different velocities have already been given.

§ 10. *Number of* H *atoms*

We have already mentioned that the number of H atoms is considerably greater than that to be expected on the simple theory. It is important to determine the number as accurately as possible, as it gives us important information on the nature of the collision. The apparatus of fig. 1 was employed. A thick copper plate with a hole 1·02 mm. diameter was placed over the end silver plate of stopping power about 6 cm., and the zinc sulphide screen placed about 1 mm. away. Even allowing for possible scattering, all the H atoms passing through the opening were counted by the microscope, which had a field of view of diameter 2·0 mm. The source was part of a small hemisphere whose outer surface was active, placed 2·85 cm. from the end of the vessel. The space between was filled with hydrogen at atmospheric pressure. The initial γ-ray activity of the source was about 10 mg. Ra.

The zinc sulphide screen was specially made for the purpose and was estimated to have about 90 per cent. efficiency in giving scintillations. As a result of three separate concordant determinations, it was found that the number of H atoms for hydrogen at N. T. P. falling on the screen corresponded to 5·1 per minute per milligram of activity, including an allowance of 10 per cent. for inefficiency of the screen.

If l = length in cms. of path of α particles in hydrogen,
  A = area of opening in sq. cms.,
   n = number of α particles emitted per second by one milligram of radium,
   $ρ$ = fraction of α particles which produce an H atom per centimetre of path in hydrogen at N. T. P.

Then obviously

$$\text{number of H atoms per second on area A} = \frac{5 \cdot 1}{60} = \frac{\rho A l n}{4 \pi l^2}.$$

Taking $l = 2 \cdot 85$ cm., $A = 0 \cdot 84$ sq. mm., $n = 3 \cdot 72 \times 10^7$,

then $\rho = 9 \cdot 7 \times 10^{-6}$,

or in round numbers $\rho = 10^{-5}$.

This number was obtained for a total absorption in path of H atom corresponding to about 15 cm. of air, but we have already seen that the number under conditions of experiment does not vary sensibly between 9 and 19 cm. absorption. We have seen in § 7 that the number of H atoms to be expected on the simple theory is $0 \cdot 98 \times 10^{-6}$ for 10 cm. absorption and $0 \cdot 31 \times 10^{-6}$ for 19 cm. We thus see that for an absorption of 10 cm., the observed number of H atoms is 10 times the theoretical value and for 19 cm. 31 times.

Using the observed result that 1 in 10^5 of the α particles produces one H atom per centimetre of path of hydrogen, it is easy to calculate the maximum distance of the direction of flight of the α particles from the centre of the hydrogen atom in order to produce a high speed atom.

If $p =$ this perpendicular distance,
    $N =$ number of atoms of H per c.c. at N. T. P.

Then $\pi p^2 N = 10^{-5}$.

Taking $N = 2 \times 2 \cdot 705 \times 10^{19}$,

then $p = 2 \cdot 4 \times 10^{-13}$ cm.,

or, on an average, each α particle of radium C of range 7 cm. produces an H atom when the perpendicular distance of its path from the centre of the H atom is equal or less than $2 \cdot 4 \times 10^{-13}$ cm. It should be remembered that this calculation deals with the α particles of range 7 cm. when the H atoms are projected mainly in the direction of the incident α particles and with a range not less than 19 cm. of air, *i. e.*, with a velocity comparable with the maximum velocity of the H atom. As already shown, the distribution of velocity is very different for α particles of shorter range, although the actual number in all cases exceeds considerably the value calculated on the simple theory.

§ 11. *Closeness of approach of α particles to H nucleus*

The experimental results considered show that the number and distribution of H particles are very different from those calculated on the assumption that the α particle and H atom are to be regarded as point nuclei carrying charges $+2e$ and $+e$ respectively, and indicate that the forces involved in a close collision differ considerably in magnitude and probably in direction from those to be expected on the simple theory.

In order to throw light on the magnitudes involved, consider the following case. Assume that for distances greater than D between the centres of the colliding atoms, the forces are given by the simple theory but for decreasing distances the forces between the nuclei augment rapidly according to other laws, and that all the collisions of closer approach than D result in the production of a high-speed H atom which for α particles of range about 7 cm. tends to be projected approximately in the line of flight of the α particles.

Darwin (*loc. cit.*) has shown that the apsidal distance D between an α particle and H atom is given on the simple theory by

$$D = \frac{\mu v_0^2}{v^2}(1 + \sec \theta),$$

where $\mu = \frac{Ee}{v_0^2}\left(\frac{1}{m} + \frac{1}{M}\right) = 9 \cdot 27 \times 10^{-14}$ for α particles of maximum

range 7 cm., where θ is the angle of deflexion of H atom and v_0 the velocity of α particles from radium C. In the same notation (§ 7)

$$p = \mu \frac{v_0^2}{v^2} \tan \theta.$$

Eliminating θ from these two equations, $p^2 = D\left(D - 2\mu \frac{v_0^2}{v^2}\right)$.

We have seen (§ 9) that for α particles of range about 7 cm. the value of $p = 2 \cdot 4 \times 10^{-13}$. Substituting this value of p and putting $v = v_0$ we find the corresponding value of $D = 3 \cdot 5 \times 10^{-13}$ cm. It will be seen later that all collisions for which D on the simple theory is greater than this value, should give rise to H atoms of velocity too small for detection. We may consequently conclude that all collisions for which D is equal or less than $3 \cdot 5 \times 10^{-13}$ cm. give rise to a high-speed H atom.

It is of interest to consider, on these assumptions, how the number of H atoms should vary with the velocity of the incident α particles. From the above equation, it is seen that $p = 0$ when $D = 2\mu \frac{v_0^2}{v^2}$. Substituting the value $D = 3 \cdot 5 \times 10^{-13}$, $\mu = 9 \cdot 27 \times 10^{-14}$, we find $v_0/v = 1 \cdot 89$. The range of α particles of this velocity v is $2 \cdot 7$ cm. This result means that α particles of range less than $2 \cdot 7$ cm., acted on by the forces given by the simple theory, are unable to approach within the critical distance D of the nucleus of the hydrogen atom.

Since the number of H atoms produced is proportional to the value of p^2 given in equation

$$\frac{\text{Number of H atoms of velocity } v}{\text{,, \quad ,, \quad ,, \quad ,, \quad } v_0} = \frac{D - 2\mu \frac{v_0^2}{v^2}}{D - 2\mu}.$$

Substituting the values of D, μ, v_0/v, the relative number of H atoms to be expected for different values of v are given below:—

Range of incident α particles in cms.	7	6	5	4	3·5	3·0	2·7
Relative number of H atoms	100	88	72	50	35	15	0

It has been previously pointed out that the observed number of H atoms shows a rapid decrease for ranges between 3 and 2 cm., a result in general accord with these calculations. It is, however, not to be expected that there would be any close agreement between theory and experiment, for the theory supposes that there is an abrupt variation in the magnitude and direction of the forces for an apsidal distance D, a condition which is physically improbable. We may, however, conclude that the variations of number of H atoms with velocity is not inconsistent with the view that the forces between colliding atoms augment rapidly for values of $D < 3\cdot5 \times 10^{-13}$ cm.

From the known values of D and μ, we are able to calculate the value of θ, i. e., the angle of deflexion of the H atom for a collision of apsidal distance $D = 3\cdot5 \times 10^{-13}$ cm. For α rays of range 7 cm., $\theta = 69°$; the corresponding effective range of the H atom is $28 \cos^4 \theta$ or 4·6 mm. It is thus clear that, on the assumptions made, no atoms for which $D > 3\cdot5 \times 10^{-13}$ should be detected under the experimental conditions.

The general results are consistent with the view that the field of force between the α particle and hydrogen nucleus undergoes rapid changes in magnitude and probably also in direction when the nuclei approach within $3\cdot5 \times 10^{-13}$ cm. of each other.

§ 12. *Summary*

1. The production of high-speed hydrogen atoms due to close collisions between α particles and atoms of hydrogen has been studied using the α particles from radium C as a homogeneous source of radiation. In such close collisions, where the nuclei approach within a distance of about 3×10^{-13} cm., the number and distribution of the H atoms are entirely different from those calculated on the assumption that the nuclei are to be regarded as point charges repelling each other according to the law of inverse squares.

2. The H atoms produced by swift α particles of range 7 cm. are shot forward mainly in the direction of the α particles and are nearly uniform in velocity.

3. The distribution with velocity of H atoms becomes more and more heterogeneous with decrease of velocity of the α particles. For α particles of range less than 4 cm. of air, the distribution and absorption of H atoms are in fair accord with the simple theory although the observed numbers are greater than those calculated on the theory.

4. The number of swift H atoms produced by α particles of range 7 cm. is 30 times greater than the theoretical number. The number falls off rapidly for ranges of α particles between 3 and 2 cm. On an average 10^5 α particles give rise to one swift hydrogen atom in traversing one centimetre of hydrogen.

5. It has been calculated that all α particles of range 7 cm. projected within a perpendicular distance $p = 2 \cdot 4 \times 10^{-13}$ cm. of the centre of the hydrogen nucleus give rise to swift H atoms. The corresponding apsidal distance is about $3 \cdot 5 \times 10^{-13}$ cm.

6. As observed by Marsden, hydrogen atoms are emitted by the radioactive source. The number observed is small, and it is difficult to decide whether these H atoms arise from the radioactive transformation or from occluded hydrogen in the source.

Discussion of results

On the nucleus theory of the atom, the charged nucleus is supposed to be of such small dimensions that it may be regarded as a point charge for distances of the order of 10^{-11} cm. The correctness of this point of view in the case of hydrogen is strongly supported by the remarkable success of Bohr and those who have followed him in explaining by its aid the finer points of the hydrogen spectrum. The experiments of Geiger and Marsden* on the large angle scattering of heavy atoms like those of gold showed that the nucleus of the gold atom could be regarded as a point charge for distances of the order of 3×10^{-12} cm., and that the law of inverse squares held up to that distance within the limits of experimental error. In the present experiments on the collision of particles with hydrogen atoms, the atomic nuclei approach still closer, viz. to a distance of the order of 3×10^{-13} cm. It is to be anticipated that for such small distances of the order of the diameter of the electron, the structure of the helium nucleus can no longer be regarded as a point, and this is borne out by experiment. Such a conclusion in no way invalidates the nucleus theory as ordinarily understood; but a study of the forces close to the nucleus is of great importance in throwing light on its actual dimensions.

It is clear from the results given in this paper that a close collision between an α particle and a hydrogen nucleus is an exceedingly rare occurrence. Only 1 in 100000 of the α particles passing through 1 cm. of hydrogen at normal pressure and temperature gives rise to a high-speed H atom, while in the same distance each α particle on an average passes through the sphere of action of about 10000 hydrogen molecules. Thus for every 10^9 collisions with the molecules, in only one case does the α particle pass close enough to the nucleus to give rise to a swift H atom. No doubt a much greater number of H atoms are set into comparatively swift motion by less direct collisions, but these do not give rise to H atoms which can be detected beyond the range of the α particle.

It is clear that for such close collisions, each hydrogen atom in any complex molecule acts as an independent unit, so that swift H atoms should be liberated by α particles from every substance containing free or combined hydrogen. This is fully borne out by experiment.

* Geiger and Marsden, Phil. Mag. xxv. p. 604 (1913).

In seeking for an explanation of these anomalous results, there are two salient facts to bear in mind, viz., that (1) the H atoms produced by α particles of range greater than 6 cm. are projected mainly in the direction of the α particles and over a narrow range of velocity, and (2) the number of such swift H atoms is far in excess of the number on the simple theory of point charges.

If we consider the nuclei of the atoms in collision to act as point charges, no advantage in explanation is gained by supposing that the free charges carried by the nuclei are greater than those usually supposed; for while such an assumption gives an increased number of H atoms of all velocities, it fails to account for (1) above.

If we suppose the central forces fall off more rapidly than the inverse square law, the proportion of swift atoms increases relatively. This can be deduced from consideration of the calculations given by Darwin* for the case of the inverse cube law, and it is not difficult to see that this relative increase of high speed particles becomes more marked the more rapid the law of variation of the central force. In all cases, however, the pencil of H atoms should be widely heterogeneous for all velocities of the colliding α particle. It thus seems clear that no theory of single central forces can account for the experimental facts.

This is not unexpected, for we have every reason to believe that the α particle has a complex structure consisting probably of four hydrogen nuclei and two negative electrons.† If we assume, for simplicity, that the hydrogen nucleus acts as a point charge for the distances under consideration, we still have a complicated system of forces near the nucleus of the α particle.

Now we have seen that the anomalous effects in hydrogen manifest themselves when the two nuclei approach within about 3×10^{-13} cm. of each other. Geiger and Marsden have shown that the scattering of α particles in passing through atoms of a heavy element like gold, is consistent within experimental error with an inverse square law of repulsion, and in the case of a head-on collision, the closest distance of approach is about 3×10^{-12} cm. or about 10 times the distance in the case of a close collision between the α particle and the hydrogen atom. It appears significant that, in the latter case, the closest distance of approach is about the same as the accepted value of the diameter of the negative electron, viz. $3 \cdot 6 \times 10^{-13}$ cm. The observed effects are similar to those to be expected if the helium nucleus, for example, consisted of a charged disk of radius about 3×10^{-13} cm. with its plane perpendicular to the direction of motion, and it seems clear that the helium nucleus must have dimensions of this order of magnitude.

If the helium nucleus is composed of two electrons and four hydrogen nuclei, we should expect a complicated field of force round the nucleus and rapid variations in direction and magnitude of the forces for distances of the order of the diameter of the electron. In our ignorance of the detailed structure of the nucleus, we can only speculate as to the magnitude and direction

---

* Darwin (*loc. cit.*).   † Rutherford, Phil. Mag. xxvii. p. 488 (1914). (*This vol., p. 423.*)

of the forces close to it. Considering, however, the enormous repulsive force between two positive nuclei in collision at a distance of 3×10^{-13} cm.—about five kilograms weight on the inverse square law,—it is to be anticipated that not only the structure of the complex helium nucleus should be much deformed, but that the electron itself may suffer strong deformation under the intense electric forces. If such deformation of the electron be possible, it is not difficult to see in a general way that the forces between the nuclei in collision may vary exceedingly rapidly close to the nucleus, and may even change rapidly from one of repulsion to one of attraction. It may be possible in this way to explain the experimental effects observed, including both the projection in the direction of the α particle and the increase over the number to be expected on the simple theory.

It is of course possible to suppose that the actual law of force, apart from deformation, does not follow the inverse square for very small distances; but since the inverse square law appears to hold at any rate approximately for positive charges up to a distance 3×10^{-12} cm., it seems simpler to suppose that the rapid alteration in magnitude and direction of the force close to the nucleus is due rather to a deformation of its structure and of its constituent parts. Taking into account the intense forces brought into play in such collisions, it would not be surprising if the helium nucleus were to break up. No evidence of such a disintegration, however, has been observed, indicating that the helium nucleus must be a very stable structure.

It will be shown in a later paper that the anomalous effects observed in hydrogen are shown also by collision of swift α particles with nitrogen and oxygen atoms and for about the same distance between the nuclei.

My thanks are due to Mr. W. Kay for his assistance in counting and in all the experimental work.

University of Manchester
April, 1919

Collision of α Particles with Light Atoms

II. VELOCITY OF THE HYDROGEN ATOMS

by PROFESSOR SIR E. RUTHERFORD, F.R.S.

From the *Philosophical Magazine* for June 1919, ser. 6, xxxvii, pp. 562–71

IN the first paper giving an account of the number of H atoms produced by α particles and their absorption by matter, it has been implicitly assumed that the long-range scintillations observed in hydrogen are due to swift hydrogen atoms set in motion by close collisions with α particles. This is supported by the observations that the range of the atoms is in good accord with the value calculated by Darwin from Bohr's theory of absorption of charged particles.

Taking into account, however, the intense forces developed in such collisions and the possibility of the disruption of the structure of the nuclei involved in the collisions, it was thought desirable to determine experimentally the mass and velocity of these flying atoms, and to compare the values with those deduced from the collision theory. Such a determination was rendered the more necessary by certain apparent anomalies observed in connexion with the brightness and distribution of H atoms, an account of which will be given later in this paper.

To determine the mass and velocity of the H atom, it was necessary to measure the deflexions of a stream of H atoms both in a magnetic and in an electric field. The experiments were somewhat tedious and difficult on account of the small number of H scintillations present under the experimental restrictions.

Magnetic deflexion of H atoms

In these experiments it was necessary to produce the H atoms at a definite point, and for this purpose a film of paraffin wax of convenient thickness, exposed to an intense beam of α rays, was used. The method finally adopted was to compare directly the deflexion of a pencil of H atoms produced from the film of paraffin wax, with the deflexion of a pencil of α rays using the same source of α rays in both cases.

The experimental arrangement is shown in fig. 1.

The horizontal slits A and B, about 1 cm. broad and 1 mm. wide, were mounted on a rectangular brass bar C. The source R, consisting of a circular brass disk coated on one side with radium C, was mounted on a vertical block D, close to the slit A and making a small angle with the horizontal. This carrier was then introduced into the rectangular brass vessel shown in fig. 1 of the previous paper, and the whole apparatus was placed between the poles of a strong electromagnet. A vertical slit 1·5 cm. long and 3 mm. wide, cut

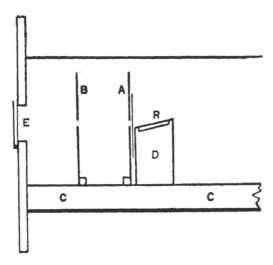

Fig. 1

in the end plate of the box, was covered with a thin sheet of iron E whose stopping-power for α articles corresponded to 4 cm. of air. Close to this was placed the zinc sulphide screen S. The distance between the slits A and B was 2·85 cm., and between B and the iron screen 3·25 cm. On exhausting the apparatus of air a well-defined band of α rays about 2 mm. broad was observed on the screen. The distance measured by the microscope between the centres of the two bands on reversing the field of about 9000 gauss was about 3·9 mm. A film of paraffin wax about 30 μ thick mounted on a frame was then placed close to the slit A between it and the source. This gave a band of H scintillations on the screen, which was of about the same width as the beam of α rays. Aluminium screens were introduced between the iron plate and zinc sulphide screen, so that the total absorption between the source and screen was equivalent to 14·4 cm. of air. Under these conditions the two bands of H scintillations obtained by reversal of the field were carefully determined by the microscope, and the centres of the bands were found to be about 6 mm. apart.

As a result of three concordant determinations, it was found that under the experimental conditions the average deflexion of the pencil of H atoms was 1·45 times the deflexion of the pencil of α rays from radium C. Since the value of $\dfrac{Mv}{E}$ for α particles from radium C has been accurately determined in other experiments and found to be $3\cdot98 \times 10^5$,* the average value of $\frac{mu}{e}$ for the pencil of H atoms is $2\cdot74 \times 10^5$. Now, on passing the slit A, the maximum range of the H atom is $28 - 3\cdot4 = 24\cdot6$ cm., while the minimum range for H atoms to be observed on the screen is $28 - 14\cdot4 = 13\cdot6$ cm. Since the range of the H atom is proportional to the cube of its velocity, the velocity of the H atoms observed which passed through the magnetic field lies between $0\cdot96u_0$ and $0\cdot79u_0$, where u_0 is the maximum velocity of the H atom due to an α particle from radium C. Since the relation between number and velocity was approximately linear over this range, the average velocity of the H atoms was $0\cdot87u_0$. Consequently the value of $\frac{mu_0}{e}$ for the swiftest H rays produced by α particles from radium C is $1/0\cdot87 \times 2\cdot74 \times 10^5 = 3\cdot15 \times 10^5$. On the collision theory, the velocity u of the H atoms of mass m is given by

$$u = \frac{2M}{M + m} \cdot v \cos \theta,$$

and the maximum value

$$u_0 = 1\cdot6\, v_0.$$

The value of $\frac{mu_0}{e}$ for H atoms carrying unit charge should consequently be

$3\cdot2 \times 10^5$. The agreement between theory and experiment is closer than we should expect considering the difficulty of the measurements. In these experiments it was found that all the H atoms carried a positive charge, and no sign of scintillations was observed indicating the presence of negatively charged or neutral atoms.

Electrostatic deflexion of H atoms

The determination of the deflexion of H atoms in an electric field was a much more difficult and lengthy process. The experimental arrangement finally adopted is shown in fig. 2. The α rays from a slanting source R passed through a film of paraffin wax about 30 μ thick placed at the end of two parallel insulated brass plates A and B, $6\cdot02$ cm. long, and $0\cdot155$ cm. apart. These were mounted on an ebonite frame DD with circular ebonite ends which slipped into a glass tube T. A brass plate with a slit 1 cm. long and 3 mm. wide covered with a thin silver plate of stopping power 6 cm. of air, was fixed at the end of the tube T. The zinc sulphide screen was mounted outside close to the silver plate. The electric connexions with the plates A and B were made through ground-glass stoppers shown in the figure.

As radium C was employed as a source, it was necessary to arrange for rapid exhaustion of the apparatus to stand 5000 volts between the plates after

---

* Rutherford and Robinson, Phil. Mag. xxviii. p. 552 (1914). (*This vol., p. 383.*)

a short interval. After preliminary evacuation by a Fleuss and Gaede mercury-pump a Langmuir pump was used, and the process was so rapid that the necessary vacuum was reached and held within two minutes of introducing the apparatus into the glass tube.

In order to deflect completely the H atoms in passing between the parallel plates, it was calculated that about 30,000 volts would be required. Apart from the difficulty of obtaining rapidly a vacuum sufficient to support and maintain such a voltage, a steady supply of not more than 7000 volts was

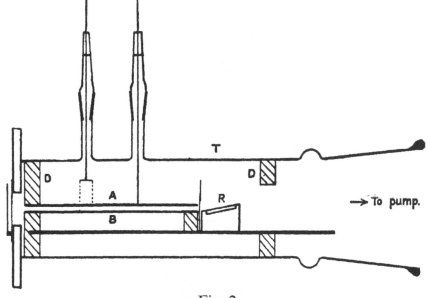

Fig. 2

available in the Laboratory. To overcome this difficulty, it was arranged to compare the deflexions of H atoms due to a magnetic field with that due to a combined magnetic and electric field. The glass tube carrying the source and parallel plates was placed between the poles of a strong electromagnet, the plane of the plates being parallel to the direction of the magnetic field. The microscope was fixed in the centre line of the plates A and B so as to count the scintillations emerging from the plates, and the variation of the number with strength of the magnetic field was determined. The reduction of the number with increase of the magnetic field depended on two causes:—(1) the removal of H atoms bent to the sides of the plates, and (2) the bending of the H atoms emerging from the plates in the magnetic field in the short distance between the end of the plates and the zinc sulphide screen. These two effects were difficult to separate, but (2) was made as small as possible by reducing to a minimum the distance between the end of the plates and the screen. The relation between the number of scintillations and strength of field with no electric field acting, is shown diagrammatically in fig. 3. Suppose the magnetic field to be of a strength H corresponding to a point P on the curve. If a voltage be now applied so as to bend the H atoms in the same direction as the magnetic field, the number of scintillations on the screen decreases corresponding to a point Q on the curve of field H_2. On reversing the voltage the two fields oppose each other, and the number of scintillations correspond to the point R of field H_1. Suppose for simplicity that the number of H atoms

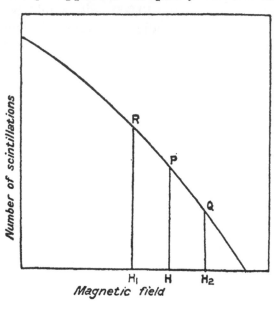

Fig. 3

are counted as they emerge from the plates A and B. Let H be the steady magnetic field and X the electric field applied between A and B. Then it is clearly seen that, if u be the velocity of the H atom,

for assisting fields, $He\,u + Xe = H_2 eu,$

for opposing fields, $He\,u - Xe = H_1 eu.$

   Subtracting $2eX = (H_2 - H_1)eu,$

and $$u = \frac{2X}{H_2 - H_1},$$

so that the velocity of the H atoms can be determined directly. In practice, it was found that the curve PQR over the experimental range was nearly a straight line. The initial field H was varied in different experiments, but was usually about 4000 gauss. The steady voltage employed was about ± 4500 volts. The ratio of the number of scintillations observed on reversal of the electric field varied in different experiments according to the magnetic field H and the voltage, but lay between 1·8 and 3 in the various experiments. Each experiment was complete in itself, for not only were the scintillations counted on reversal of the electric field, but also the number for two magnetic fields on either side of the fixed field which give nearly the same ratio of number of scintillations as that obtained by reversal of the electric field.

The paraffin film had a stopping power of 3·2 cm. of air, and the stopping power of the silver plate together with the aluminium screens was 11·4 cm. The range of H atom, which passed between the parallel plates and produced scintillations on the screen, thus lay between $28 - 3·2 = 24·8$ cm. and $28 - 14·6 = 13·4$ cm.

The corresponding velocities are $0·96u_0$ and $0·78u_0$ where u_0 is the maximum velocity of an H atom due to an α particle from radium C. As in the experiments on the magnetic deflexion the average velocity was found to be $0·87u_0$. In calculating the relative effect due to a magnetic and electric field, a small correction is necessary to allow for the fact that the electric field was only effective the length of the plates, while the magnetic field acted on the H rays from the slit A to the zinc sulphide screen. Making this correction, estimated to be about 12 per cent., the deflexion due to 1000 volts between the plates was found to correspond in five different experiments to 235, 227, 260, 235, 221 gauss respectively, with an average value of 238 gauss.

Thus,

$$u = \frac{X}{H} = \frac{10^{11}}{0·155} \times \frac{1}{238} = 2·71 \times 10^9 \text{ cm. per sec.}$$

The maximum value

$$u_0 = \frac{1}{0·87} \times 2·71 \times 10^9 = 3·12 \times 10^9 \text{ cm. per sec.}$$

The calculated value of $u_0 = 1·6 \times 1·92 \times 10^9$

$$= 3·07 \times 10^9 \text{ cm. per second.}$$

The experimental and calculated values agree well within the probable error of experiments. From the magnetic deflexion we found that

$$\frac{mu_0}{e} = 3·15 \times 10^5,$$

and from the electric deflexion

$$u_0 = 3·12 \times 10^9,$$

consequently, $e/m = 10^4$ e.m. units.

The value of e/m for the hydrogen atom in the electrolysis of water is 9570. The agreement is sufficiently close to show that the long-range scintillations produced by α particles in hydrogen are due to hydrogen atoms carrying a unit positive charge. The agreement between the calculated and observed velocities shows that, within the margin of experimental error, the conservation of momentum and energy hold for close collisions between the atomic nuclei and that there is no sensible loss of energy due to radiation.

Brightness of scintillations

The maximum energy communicated to an H atom is $0\cdot64$ of the energy of the colliding α particle. After passing through 12 cm. of air, for example, the energy of the H atom is reduced to $0\cdot44$ of the energy of the particle. Supposing that the H atoms are produced by α particles of radium C of range 7 cm., the energy of the H atom after passing through 12 cm. of air, corresponds to an α particle of range about 2 cm. In practice, the brightness of the corresponding H scintillations is much less than is to be expected from its energy, and is not greater than that produced by an α particle of range 5 mm. This relative lack of brightness of H scintillations compared with α particles of corresponding energy holds for all velocities of the H atoms. Since we have seen that we can rely on the calculations of the energy of the H atom, it seems clear that the H atom is less effective in producing light on a zinc sulphide screen than an α particle of equal energy. This may be a consequence of the much weaker ionization along the path of the H atom, for since its range is four times that of the α particle and energy $0\cdot64$, the energy spent per unit path is only 1/6 of that due to an α particle.

In this connexion it is of interest to note, that nitrogen atoms set in motion by α particles from radium C have a range in air of about 9 cm. Although the energy of the nitrogen atoms after traversing 7 cm. of air is less than that of the H atoms after traversing 12 cm., the nitrogen atom gives a much brighter scintillation than the H atom.

Probability distribution of H scintillations

In the course of counting H scintillations, it was often noted that a number of the scintillations appeared as *instantaneous* doubles, *i. e.* two points of light of about equal brightness appeared in the field of view at the same instant. Some preliminary experiments seemed to show that the number of these doubles was greater than was to be expected from probability considerations. For example, in counting bright scintillations due to the active deposit of thorium, on an average, about $1\cdot5$ doubles per minute were counted for an average of 30 scintillations per minute, while for a similar number of H scintillations the number of doubles was about 5. If these 'doubles' from hydrogen were instantaneous doubles, it was obviously a matter of great importance, possibly indicating the disruption by collision of one of the nuclei into two parts.

A large number of experiments were made to test this question, using both hydrogen and paraffin wax as a source of H atoms, but very similar results were obtained under all conditions of experiment. The most favourable theoretical conditions were chosen to increase the number of such doubles if they existed. For example, the H atoms were liberated in a thin film of paraffin covering an opening 1 mm. in diameter, placed near the zinc sulphide screen. The distance of the screen from the paraffin film and the nature of the absorber between was adjusted, so that even if two atoms were shot forward nearly in the same direction, the scattering would separate them on an average a convenient distance in the field of view of the microscope, which included a field of view 2 mm. diameter. No apparent advantage as regards the number of doubles was gained by this arrangement.

I was fortunate in January of this year, to obtain for a short time the skilled assistance of Professor E. Marsden before his return to New Zealand. Systematic observations were undertaken to record electromagnetically the time of appearance of each scintillation on a chronograph tape while the number of doubles was separately recorded. Mr. Marsden and Mr. Kay counted alternately for a minute interval, and the counts of each observer were separately analysed by the former at leisure. On the probability theory, the number of intervals between t_1 and t_2 seconds is given by $Ne^{-\mu(t_1-t_2)}$ where $1/\mu$ is the average interval between each scintillation and N the total number of intervals. Marsden and Barratt* had previously verified the correctness of this theory, which shows that short intervals are more probable than long ones. If the average number of scintillations is 30 per minute, $\mu = 1/2$, and if the eye fails to distinguish an interval less than 1/10 of a second, the average number of doubles to be expected is 1·5 per minute. In practice, under favourable conditions, the eye is just able to detect 1/10 second intervals for bright α ray scintillations.

Comparisons were made to test the probability distribution of α particles from polonium, whose range was adjusted to give a scintillation of about the same average brightness as the H atom.

The results of a typical series of counts both for α rays and H atoms are included in the following table. The theoretical and observed number of intervals <1/10, <1/2, and <1 second are given in the table below:—

Observer	Average number scintillations per min.	Total number of scintillations	Number of doubles	Calculated number of intervals <1/10 sec.	Observed number of intervals < 1/2 sec.	Calculated number of intervals <1/2 sec.	Observed number of intervals <1 sec.	Calculated number of intervals <1 sec.
α particles from polonium								
M	28·0	280	13	12·9	53	60	106	105
K	25·4	229	10	9·6	45	45	83	79
Hydrogen atoms								
M	24·3	243	15	9·6	50·5	46	84	81
K	22·7	250	25	9·2	58·5	45	92	79
M	31·0	216	24	11	60·5	50	95	88
K	29·6	148	18	7	33	35	59	58

* Marsden and Barratt, Proc. Phys. Soc. xxiii. p. 367 (1911); xxiv. p. 50 (1913).

The calculated numbers are the sum of each observation worked out separately.

It will be seen that while there is a very satisfactory agreement between theory and experiment for the α rays from polonium, the agreement is not so good for the H atoms. In the case of the α rays, the number of doubles shows that the eye cannot distinguish an interval less than 1/10 second; while in the case of H atoms the number of doubles is nearly twice the theoretical number calculated on this power of distinction. Whether this difference is apparent or real is difficult to decide, for it must be remembered that counting such weak scintillations and at the same time distinguishing time intervals make a difficult task.

It is clear that under the experimental conditions, only a small fraction of the number of scintillations can be regarded as possible instantaneous doubles, and the effect is too small and uncertain to draw any very definite conclusions. It may be urged that a question of this kind could be settled more definitely by arranging that a small number of scintillations fell on the screen per minute when the probability of short intervals becomes very small. On the other hand, it takes a long time to count a sufficient number to compare theory with experiment, and it is very fatiguing to the eye and unreliable to count for long under such conditions.

I am much indebted to Professor Marsden for his valuable help in obtaining and analysing data for me on this important point.

University of Manchester
April, 1919

Collision of α Particles with Light Atoms

III. NITROGEN AND OXYGEN ATOMS

by PROFESSOR SIR E. RUTHERFORD, F.R.S.

From the *Philosophical Magazine* for June 1919, ser. 6, xxxvii, pp. 571–80

BOHR* has worked out a general theory of the absorption of electrified atoms in passing through matter, and has verified his conclusions by consideration of the absorption of α particles. On this theory, Darwin† has shown that the range of a swift hydrogen atom in hydrogen can be calculated, and the value so found is in good accord with experiment. It is not difficult to deduce by the same method that the range x in hydrogen of an electrified atom of charge e and mass m moving with a speed equal to an α particle of range R in hydrogen is given by

$$x/\mathrm{R} = \frac{m}{\mathrm{M}} \cdot \frac{\mathrm{E}^2}{e^2}, \qquad \cdots \qquad \cdots \qquad (1)$$

where M is the mass and E the charge on the α particle.

It is to be expected that this relation would hold approximately for the passage of electrified atoms through light substances like air and aluminium. Since $\mathrm{M} = 4$ and $\mathrm{E} = 2e$ where e is the unit charge, the range x of a particle carrying a single charge is obviously $x = m\mathrm{R}$. The velocity u acquired by an atom of mass m due to a close collision with an α particle of velocity v is given by

$$u = \frac{2\mathrm{M}}{\mathrm{M} + m} \cdot v \cos \theta,$$

w ere θ is the angle of deflexion of the atom after the collision. Assuming that the range of electrified atoms in general like the range of α particles varies as the cube of the velocity, the range x after collision of an atom carrying unit charge is given by

$$x = m\mathrm{R} \left(\frac{2\mathrm{M}}{\mathrm{M}+m} \right)^3 \cos^3\theta.$$

Applying this result to H atoms, the maximum velocity should be $(8/5)^3\mathrm{R} = 4\cdot1\mathrm{R}$, while the observed value is about 4R. As a further test of this relation, consider the range to be expected for the recoil atom of radium B of mass m resulting from the expulsion of an α particle of range $4\cdot75$ cm. from radium A. By the principle of momentum $\mathrm{M}v = mu$ and the velocity of

* Bohr, Phil. Mag. xxv. p. 10 (1913). † Darwin, Phil. Mag. xxvii. p. 499 (1914).

T

recoil $u = \dfrac{M}{m}v$ where $m = 214$. Consequently the range in air x

$$= 214 \cdot \left(\frac{4}{214}\right)^3 \times 4\cdot75 = 0\cdot067 \text{ cm.}$$

The value found by Wertenstein[*] is $0\cdot12$ mm., but, considering the very wide range of velocity, the agreement is fairly satisfactory. If it be assumed that the range is proportional to the power $2\cdot85$ instead of 3, this is a good agreement both for the hydrogen and recoil atoms.

If the atom after collision with an α particle carries a charge of two units, its range from (1) should be only about 1/4 of the same atom carrying a single charge. For example, in a collision of an α particle with the helium nucleus of equal mass, the range of the helium atom should be the same as the α particle before the collision if it carries two charges, but four times this range if it carries one charge.

We have collected in the following table data connected with the collision of α particles with the lighter atoms of matter. The maximum velocity, momentum, and energy of the atom after collision are given as fractions of that of the incident α particle. The range is calculated on the assumption that it is proportional to the power $2\cdot9$ of the velocity and that the atom carries unit charge.

TABLE I

Element	Atomic weight	Ratio of velocity to that of α particle	Ratio of momentum to that of α particle	Ratio of energy to that of α particle	Ratio of range to that of incident α particle
Hydrogen	1	1·6	0·4	0·64	3·91
?	2	1·33	0·66	0·89	4·6
?	3	1·14	0·85	0·98	5·05
Helium	4	1·00	1·00	1·00	4·00
Lithium	7	0·727	1·27	0·925	2·78
Beryllium	9	0·615	1·38	0·85	2·20
Boron	11	0·533	1·46	0·78	1·77
Carbon	12	0·500	1·50	0·75	1·61
Nitrogen	14	0·444	1·55	0·69	1·33
Oxygen	16	0·400	1·60	0·64	1·12
Fluorine	19	0·348	1·65	0·575	0·89
Neon	20	0·333	1·67	0·55	0·82
Sodium	23	0·296	1·70	0·50	0·67
Magnesium	24	0·286	1·71	0·49	0·64
Aluminium	27	0·258	1·74	0·45	0·53
Iron	56	0·133	1·86	0·25	0·19
Silver	108	0·071	1·92	0·136	0·05
Gold	197	0·040	1·92	0·079	0·017

[*] Wertenstein, *C. R.* cl. p. 869 (1910); cli. p. 469 (1910).

For convenience, the data for hypothetical atoms of mass 2 and 3 times that of hydrogen are included.

It is seen that, on the assumption of unit charge, all the atoms of atomic weight up to oxygen should be detected beyond the range of the α particle. Supposing that α particles of range 7 cm. are used, the maximum range to be expected for unit charge are for He 28·0, Li 19·6, Be 15·4, B 12·4, C 11·2, N 9·3, O 7·8 cm.

Some preliminary experiments have been made with helium, using the apparatus similar to that employed for hydrogen and described in paper I but on a smaller scale. The results show that if the collisions of α particles with helium atoms give any long-range scintillations of the order of 28 cm. range, their number is very small compared with that produced in hydrogen under similar conditions. We may consequently conclude that the swift helium atoms produced by collision carry a double charge like the α particle.

A few experiments have been made to test whether the atoms of lithium, boron, or beryllium have the range to be expected if they carry a single charge. The salts Li₂CO₃, B₂O₃, BeO were spread in a thin layer over the active source which was inclined at a small angle with the horizontal, and determinations made of the number of scintillations beyond the range of the α particle. The air was exhausted and the α particles absorbed in aluminium and silver foils. No certain evidence was obtained of the presence of appreciable numbers of scintillations at the ranges to be expected if the atoms carry a single charge. Experiments of this kind are not easy on account of the difficulty of obtaining thin uniform films of the salts or metal under examination, and of the necessity of getting rid of all traces of hydrogen and water vapour, which give rise to numerous H atoms. It is intended later to make a systematic examination of these elements to determine the range of the atoms produced by close collisions with α particles.

Experiments in Air and Oxygen

Experiments on the range of swift atoms become much easier and more certain when the elements are in the gaseous state, for there is then no uncertainty with regard to the uniformity of the absorbing column and usually no difficulty in ensuring absence of hydrogen and water vapour. Thin films of rolled metals like aluminium, silver, or gold are usually very irregular in thickness. This irregularity comes out very obviously when intense sources of radiation are employed under conditions when one in a million of the incident particles can be detected. It is not unusual in these cases to find that α particles can be detected at a distance 10 per cent. beyond the average range of the α particles as determined by ordinary methods. Mica films are very uniform and show none of these irregularities, but unfortunately mica contains both hydrogen and oxygen and gives rise to numerous H and O atoms beyond the range of the bombarding α particles.

We have seen that both N and O atoms carrying a single charge should be detected beyond the range of the α particles, and this is borne out by experiment. In the case of air, the active disk coated with radium C of γ-ray activity about 30 mg. Ra, was mounted with its plane vertical at a distance of about 7 cm. from a zinc sulphide screen in the open air. Both the source and screen were placed between the poles of a large electromagnet to deflect the β rays. The vertical convection currents due to the heated electromagnet prevented any contamination of the screen by active matter escaping from the source.

The end of the range of the α particles was sharply defined, but numerous bright scintillations were observed for distances nearly 2 cm. beyond the range of the α particles. There was a steady decrease both in number and brightness up to 9 cm. of air, and beyond that distance the small number of scintillations observed, due to H atoms from the source and from the water vapour in the air, fell off slowly.

The range of these atoms was best determined by placing the screen just outside the range of the α particles (7·1 cm. at 15° C.) and then adding thin screens of aluminium foil close to the zinc sulphide screen. The variation of number with absorption in terms of cms. of air is shown in fig. 1.

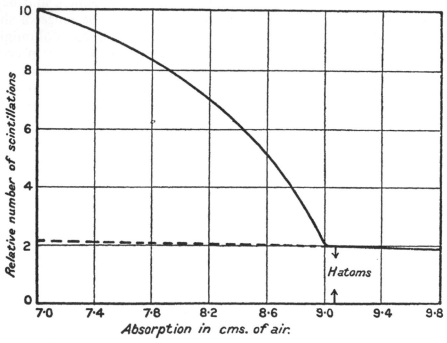

Fig. 1

It will be seen that the scintillations fall off at first slowly with increase of absorption and more rapidly near the end of their range, which was equal to 9·0 cm. of air at normal pressure and 15° C. The scintillations, presumably due to swift N and O atoms, are bright and easily counted for a total absorption corresponding to about 7·5 cm. of air. At this stage they appear equal in brightness to those given by an α particle of range about 1 cm.

In other experiments with air, nitrogen, and oxygen, and carbon dioxide, the screen and source were placed in a rectangular box and a slow current of the dried gas passed through during the experiment. This prevented contamination of the screen by diffusion of active matter from the source, and the range was determined by altering the distance between source and screen.

The scintillations in pure oxygen and carbon dioxide were about the same brightness for corresponding ranges, and had nearly the same equivalent ranges in air as those due presumably to N atoms from the air.

This was rather surprising, as we should expect the O atoms to have considerably less range than the lighter N atom. The calculated ranges (see table above) are $7 \cdot 8$ and $9 \cdot 3$ cm. respectively. This suggested the possibility that the scintillations might be due not to N or O atoms but to actual α particles of range 9 cm. which were expelled from the radioactive source. If this were the case, the total range of the α particles should not be altered by placing an absorbing screen of aluminium or gold of known stopping power close to the source in the path of the α rays. On the other hand, if the scintillations were due to swift N or O atoms from the air, the range should be diminished. For example, if a screen of stopping power equal to $3 \cdot 5$ cm. of air were placed in the path of the α rays of range $7 \cdot 0$ cm., the resulting range of the α particles acting on the gas is $3 \cdot 5$ cm., and the total range of the N or O atoms measured from the source should be $3 \cdot 5 + \frac{2}{7} \times 3 \cdot 5 = 8 \cdot 0$ cm. instead of $9 \cdot 0$ cm. Experiments of this kind were made with an aluminium and a gold screen of stopping powers $3 \cdot 7$ and $4 \cdot 2$ cm. respectively, but were not altogether satisfactory on account of the inequalities of the films already referred to. They showed, however, that no appreciable number of scintillations could be detected beyond 8 cm. The results indicated that the scintillations were due to atoms of N and O and not to α particles from the source. This was further confirmed by experiments with mica screens of stopping power $7 \cdot 0$ cm. The number of bright scintillations which resembled α particles were less than half the number observed in air or oxygen gas under similar conditions, but the presence of numerous H atoms from the mica interfered with an accurate determination. Since mica contains oxygen as well as hydrogen we should obtain swift O atoms, and the number of scintillations observed was about that to be expected from the amount of oxygen present, but was less than the number observed in air. There appears to be no doubt that the scintillations observed in air between the ranges 7 and 9 cm. arise from collision of α particles with N and O atoms. The observation that the range of the swift atoms, produced by α particles in their passage through carbon dioxide, is equivalent to the range of O atoms, indicates that there are no carbon atoms carrying a single charge, for in that case bright scintillations should have been observed for ranges up to 12 cm. of air (see Table I.).

It will be remembered that in the beautiful photographs of Mr. C. T. R. Wilson* showing the trails of α particles, an example is given where the α

* C. T. R. Wilson, Proc. Roy. Soc. A. lxxxvii. p. 277 (1912).

particle in air shows a sudden deflexion of 43°, and there is clear evidence of a well-marked spur presumably showing the trail of the N or O recoil atom. It is of interest to compare the length of this spur with the range to be expected for a collision with an O atom. If ϕ be angle of deflexion of the α particle and θ the deflexion of the O atom,

$$\tan \phi = \frac{m \sin 2\theta}{M - m \cos 2\theta},$$

where m = mass of O atom and M = mass of α particle.

Putting M = 4, m = 16, ϕ = 43°, then θ = 63°·55.

If v = velocity of the α particle before the collision, the velocity of the O atom

$$= \frac{2M}{M + m} \cdot v \cos \theta = 0 \cdot 178 \, v,$$

while the velocity of the α particle after the collision is $0 \cdot 934 \, v$.

$$\frac{\text{Range of recoil O atom}}{\text{Range of } \alpha \text{ particle after collision}} = 16 \times \left(\frac{0 \cdot 178}{0 \cdot 934}\right)^{2 \cdot 9} = 0 \cdot 13.$$

This is based on the calculation that the maximum range of O atoms due to α particles from radium C is $7 \cdot 8$ cm., while the observed range is $9 \cdot 0$ cm. Making this correction, the value $0 \cdot 13$ becomes $0 \cdot 15$.

It is possible to compare only roughly the ranges of the α particle and recoil atom in the photograph, but the results are in fair accord with the calculation.

In the same photograph the α particle shows another sudden bend of $10° \cdot 5$. In this case, the range of the recoil O atom should be only about 1/800 of the α particle and could not be distinguished on the photograph.

Number of N atoms

In a previous paper we have calculated the number and distribution of H atoms produced by α particles on the assumption that the nuclei may be regarded as point centres of force repelling according to the law of the inverse square. When these calculations are applied to the collision of α particles with nitrogen or oxygen nuclei, the distribution with velocity of the N and O atoms is very similar to that for H atoms. We should consequently expect on the simple theory that the number of N and O atoms should fall off very rapidly between 7 and 9 cm., and that the number of short-range atoms should greatly preponderate. Quite the contrary is observed in the experiments (fig. 1), where it is seen that the number of scintillations fall off quite gradually with range.

There seems to be no doubt that the effects produced by the collision of α particles with N and O atoms are very similar to those observed in hydrogen. The observations only receive an explanation on the assumption that the N and O atoms like the H atoms are thrown forward mainly in the direction of the α particles and, at any rate for swift α particles, the velocities of the recoil atoms are nearly uniform for a given velocity of the α particles. It should be pointed out that the experiments with air and oxygen differ in one respect from those with hydrogen. In the case of air the α particles are completely absorbed in the column of gas, while in the case of hydrogen the stopping power was usually equivalent to less than 1 cm. of air. Consequently in the air experiments, the scintillations observed are due to N and O atoms which are produced by α particles of all ranges between 7 and 0 cm., and thus have a wide range of velocities.

A number of experiments were made by the use of absorbing-screens of aluminium and gold in order to determine the number of N and O atoms produced by α particles of different range. The result as a whole showed that, for example, the number produced in the first 3·5 cm. of the range of the α particle from radium C was greater than in the last 3·5 cm., but accurate deductions were vitiated by the lack of uniformity in thickness of the metal films.

A number of concordant measurements were made to fix the total number of scintillations observed in air for a known activity of the source. The number of scintillations per minute due to N and O atoms at a distance of 7·5 cm. in air at 15° C. was 2·2 on an area of the zinc sulphide screen equal to 3·14 sq. mm. Referring to curve 1, it is seen that the number corresponding to an absorption of 7 cm. should be 2·6 and the number for 8 cm. absorption 1·5.

All those atoms of range equal to or greater than 8 cm. must be produced in the first 3·5 cm. of the path of the α rays; for the O atoms produced by α particles of range 3·5 cm. cannot travel further than 8 cm. from the source, and probably only a small fraction reach this distance owing to scattering and straggling.

For the purpose of calculation, suppose that the production of swift atoms is uniform over the first 3·5 cm. of the range and that ρ is the ratio of the number of swift atoms produced per cm. of path to the number of α particles passing through the gas.

The number Q of the recoil atoms falling per second on the screen of area A after passing through l cm. of gas is given by

$$Q = \rho \cdot \frac{AlN}{4\pi r^2},$$

where N is the total number of α particles emitted by the source per second ($3·7 \times 10^7$ per second per mg. Ra of activity) and r is the distance of the source from the screen.

Putting

$$Q = \frac{1·5}{60}, \quad A = 0·0314 \text{ sq. cm.}, \quad l = 3·5 \text{ cm.}, \quad r = 7·5 \text{ cm.},$$

then the average value of $\rho = 4·3/10^6$.

When we take into consideration the well-known way in which the α particles fall off near the end of the range in consequence of scattering, it is obvious that the true value of ρ is considerably greater than the above and is probably about $7/10^6$.

In the experiments with hydrogen, it was shown that $\rho = 1/10^5$ about—a value not very different from that observed in these experiments. We may consequently conclude that about the same number of swift atoms are produced per centimetre of path by the passage of α particles through air, oxygen, and hydrogen. As in the case of hydrogen, it can be shown that all α particles, shot within a perpendicular distance $p = 2\cdot4 \times 10^{-13}$ cm. of the atomic nucleus, give rise to swift atoms of nitrogen and oxygen.

It is clear from these results that the nuclei under consideration can no longer be regarded as point charges for distances of approach of the order of the diameter of the electron. As far as experiment has so far gone, it is difficult to fix with certainty the distance at which the forces between the nuclei become abnormal, but a rough estimate can be made. Regarding the nuclei as point charges, the closest distance of approach in a collision is $1\cdot9 \times 10^{-13}$ cm. in the case of a hydrogen atom and $3\cdot8 \times 10^{-13}$ cm. in the case of the oxygen atom. Taking into account the close similarity of the effects produced by α particles in hydrogen and oxygen and the greater repulsive forces between the nuclei in the latter case, it seems probable that the abnormal forces in the case of oxygen manifest themselves at about twice the distance observed in the case of hydrogen. This would mean that the rapid variation in the magnitude and direction of the forces between the nuclei which lead to the recoil of swift atoms mainly in the direction of the α particle should begin at a distance about 7×10^{-13} cm. Such a result is to be anticipated on general grounds, for presumably the oxygen nucleus is more complex and has larger dimensions than that of helium.

In a paper published three years ago Mr. A. B. Wood and the writer* described experiments which showed that the active deposit of thorium gave rise to a few α particles of range $11\cdot3$ cm. in addition to the main group of ranges $5\cdot0$ and $8\cdot6$ cm. In these experiments, the α rays of range $8\cdot6$ cm. were absorbed in mica. In the light of the present experiments, the oxygen present in the mica should give rise to scintillations like α particles of range

$$8\cdot6 \times \frac{9\cdot0}{7\cdot0} = 11\cdot1 \text{ cm.}$$

This range is nearly the same as that observed in the thorium experiment, and raises the question whether these long range α particles are not in reality due to collision of α particles with the oxygen atoms in the mica. A fraction of the scintillations must undoubtedly have been due to this cause, but on the other hand the number of scintillations observed, about $1/10000$ of the number of α particles, is considerably greater than is to be expected from the experiments with radium C. Further experiments to clear up this important point have been undertaken by Professor Marsden in New Zealand.

University of Manchester
April, 1919

* Rutherford and Wood. Phil. Mag. xxxi p. 379 (1916). (*This vol., p. 531.*)

It is clear here ~~~ that chemical
nitrogen gives long range of particles
which ~~~ ~~~ at least as
bright as H. ~ have about the
same range (to be tested ~~~)

$L\alpha = \cdot 64$ for ~~~ N ~~~ length 8 ~~

$Nv \quad \cdot 51$ for ~~ $p \; 21$ $\Big\}$ $\cdot 47$
$\quad \cdot 42$ $\; - -$ $p \; 42$

2° ~~~ due to N above ~~~
$= \frac{4}{7} \times \cdot 64 = \cdot 51$ ~~~ agreement
with ~~~ numbers.

~~~ ~~~ of CO_2 gives very small
effect, it is clear that the ~~~ ~~~
~~~ these ~~~ ~~~ from C ~ O_2
"important"

To settle whether these scintillations are
N, He, H ~ ~~ ?

A page from Rutherford's Laboratory Notebook; recorded 9 November, 1917

Collision of α Particles with Light Atoms

IV. AN ANOMALOUS EFFECT IN NITROGEN

by PROFESSOR SIR E. RUTHERFORD, F.R.S.

From the *Philosophical Magazine* for June 1919, ser. 6, xxxvii, pp. 581–87

It has been shown in paper I that a metal source, coated with a deposit of radium C, always gives rise to a number of scintillations on a zinc sulphide screen far beyond the range of the α particles. The swift atoms causing these scintillations carry a positive charge and are deflected by a magnetic field, and have about the same range and energy as the swift H atoms produced by the passage of α particles through hydrogen. These 'natural' scintillations are believed to be due mainly to swift H atoms from the radioactive source, but it is difficult to decide whether they are expelled from the radioactive source itself or are due to the action of α particles on occluded hydrogen.

The apparatus employed to study these 'natural' scintillations is the same as that described in paper I. The intense source of radium C was placed inside a metal box about 3 cm. from the end, and an opening in the end of the box was covered with a silver plate of stopping power equal to about 6 cm. of air. The zinc sulphide screen was mounted outside, about 1 mm. distant from the silver plate, to admit of the introduction of absorbing foils between them. The whole apparatus was placed in a strong magnetic field to deflect the β rays. The variation in the number of these 'natural' scintillations with absorption in terms of cms. of air is shown in fig. 1, curve A. In this case, the air in the box was exhausted and absorbing foils of aluminium were used. When dried oxygen or carbon dioxide was admitted into the vessel, the number of scintillations diminished to about the amount to be expected from the stopping power of the column of gas.

A surprising effect was noticed, however, when dried air was introduced. Instead of diminishing, the number of scintillations was increased, and for an absorption corresponding to about 19 cm. of air the number was about twice that observed when the air was exhausted. It was clear from this experiment that the α particles in their passage through air gave rise to long-range scintillations which appeared to the eye to be about equal in brightness to H scintillations. A systematic series of observations was undertaken to account for the origin of these scintillations. In the first place we have seen that the passage of α particles through nitrogen and oxygen gives rise to numerous bright scintillations which have a range of about 9 cm. in air. These scintillations have about the range to be expected if they are due to swift N or O atoms, carrying unit charge, produced by collision with α particles. All experiments have consequently been made with an absorption greater than 9 cm. of air, so that these atoms are completely stopped before reaching the zinc sulphide screen.

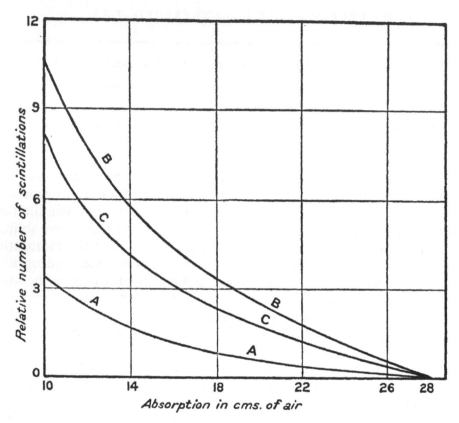

Fig. 1

It was found that these long-range scintillations could not be due to the presence of water vapour in the air; for the number was only slightly reduced by thoroughly drying the air. This is to be expected, since on the average the number of the additional scintillations due to air was equivalent to the number of H atoms produced by the mixture of hydrogen at 6 cm. pressure with oxygen. Since on the average the vapour pressure of water in air was not more than 1 cm., the effects of complete drying would not reduce the number by more than one sixth. Even when oxygen and carbon dioxide saturated with water vapour at 20° C. were introduced in place of dry air, the number of scintillations was much less than with dry air.

It is well known that the amount of hydrogen or gases containing hydrogen is normally very small in atmospheric air. No difference was observed whether the air was taken directly from the room or from outside the laboratory or was stored for some days over water.

There was the possibility that the effect in air might be due to liberation of H atoms from the dust nuclei in the air. No appreciable difference, however, was observed when the dried air was filtered through long plugs of cotton-wool, or by storage over water for some days to remove dust nuclei.

Since the anomalous effect was observed in air, but not in oxygen, or carbon dioxide, it must be due either to nitrogen or to one of the other gases present in atmospheric air. The latter possibility was excluded by comparing the effects produced in air and in chemically prepared nitrogen. The nitrogen was obtained by the well-known method of adding ammonium chloride to sodium nitrite, and stored over water. It was carefully dried before admission to the apparatus. With pure nitrogen, the number of long-range scintillations under similar conditions was greater than in air. As a result of careful experiments, the ratio was found to be 1·25, the value to be expected if the scintillations are due to nitrogen.

The results so far obtained show that the long-range scintillations obtained from air must be ascribed to nitrogen, but it is important, in addition, to show that they are due to collision of α particles with atoms of nitrogen through the volume of the gas. In the first place, it was found that the number of the scintillations varied with the pressure of the air in the way to be expected if they resulted from collision of α particles along the column of gas. In addition, when an absorbing screen of gold or aluminium was placed close to the source, the range of the scintillations was found to be reduced by the amount to be expected if the range of the expelled atom was proportional to the range of the colliding α particles. These results show that the scintillations arise from the volume of the gas and are not due to some surface effect in the radioactive source.

In fig. 1 curve A the results of a typical experiment are given showing the variation in the number of natural scintillations with the amount of absorbing matter in their path measured in terms of centimetres of air for α particles. In these experiments carbon dioxide was introduced at a pressure calculated to give the same absorption of the α rays as ordinary air. In curve B the corresponding curve is given when air at N.T.P. is introduced in place of carbon dioxide. The difference curve C shows the corresponding variation of the number of scintillations arising from the nitrogen in the air. It was generally observed that the ratio of the nitrogen effect to the natural effect was somewhat greater for 19 cm. than for 12 cm. absorption.

In order to estimate the magnitude of the effect, the space between the source and screen was filled with carbon dioxide at diminished pressure and a known pressure of hydrogen was added. The pressure of the carbon dioxide and of hydrogen were adjusted so that the total absorption of α particles in the mixed gas should be equal to that of the air. In this way it was found that the curve of absorption of H atoms produced under these conditions was somewhat steeper than curve C of fig. 1. As a consequence, the amount of hydrogen mixed with carbon dioxide required to produce a number of scintillations equal to that of air, increased with the increase of absorption. For example, the effect in air was equal to about 4 cm. of hydrogen at 12 cm. absorption, and about 8 cm. at 19 cm. absorption. For a mean value of the absorption, the effect was equal to about 6 cm. of hydrogen. This increased absorption of H atoms under similar conditions indicated either that (1) the swift atoms from air had a somewhat greater range than the H atoms, or (2) that the atoms from air were projected more in the line of flight of the α particles.

While the maximum range of the scintillations from air using radium C as a source of α rays appeared to be about the same, viz. 28 cm., as for H atoms produced from hydrogen, it was difficult to fix the end of the range with certainty on account of the smallness of the number and the weakness of the scintillations. Some special experiments were made to test whether, under favourable conditions, any scintillations due to nitrogen could be observed beyond 28 cm. of air absorption. For this purpose a strong source (about 60 mg. Ra activity) was brought within $2 \cdot 5$ cm. of the zinc sulphide screen, the space between containing dry air. On still further reducing the distance, the screen became too bright to detect very feeble scintillations. No certain evidence of scintillations was found beyond a range of 28 cm. It would therefore appear that (2) above is the more probable explanation.

In a previous paper (III) we have seen that the number of swift atoms of nitrogen or oxygen produced per unit path by collision with α particles is about the same as the corresponding number of H atoms in hydrogen. Since the number of long-range scintillations in air is equivalent to that produced under similar conditions in a column of hydrogen at 6 cm. pressure, we may consequently conclude that only one long-range atom is produced for every 12 close collisions giving rise to a swift nitrogen atom of maximum range 9 cm.

It is of interest to give data showing the number of long-range scintillations produced in nitrogen at atmospheric pressure under definite conditions. For a column of nitrogen $3 \cdot 3$ cm. long, and for a total absorption of 19 cm. of air from the source, the number due to nitrogen per milligram of activity is $0 \cdot 6$ per minute on a screen of $3 \cdot 14$. sq.mm. area.

Both as regards range and brightness of scintillations, the long-range atoms from nitrogen closely resemble H atoms, and in all probability are hydrogen atoms. In order, however, to settle this important point definitely, it is necessary to determine the deflexion of these atoms in a magnetic field. Some preliminary experiments have been made by a method similar to that employed in measuring the velocity of the H atom (see paper II). The main difficulty is to obtain a sufficiently large deflexion of the stream of atoms and yet have a sufficient number of scintillations per minute for counting. The α rays from a strong source passed through dry air between two parallel horizontal plates 3 cm. long and $1 \cdot 6$ mm. apart, and the number of scintillations on the screen placed near the end of the plates was observed for different strengths of the magnetic field. Under these conditions, when the scintillations arise from the whole length of the column of air between the plates, the strongest magnetic field available reduced the number of scintillations by only 30 per cent. When the air was replaced by a mixture of carbon dioxide and hydrogen of the same stopping power for α rays, about an equal reduction was noted. As far as the experiment goes, this is an indication that the scintillations are due to H atoms; but the actual number of scintillations and the amount of reduction was too small to place much reliance on the result. In order to settle this question definitely, it will probably prove necessary to employ a solid nitrogen compound, free from hydrogen, as a source, and to use much stronger sources of α rays. In such experiments, it will be of importance to discriminate between the deflexions due to H atoms and possible atoms of atomic weight 2. From the calculations given in paper III, it is seen that a collision of an α particle with a free atom of mass 2 should give rise to an atom of range about 32 cm. in air, and of initial energy about $0 \cdot 89$ of that of the H atom produced under similar conditions. The deflexion of the pencil of these rays in a magnetic field should be about $0 \cdot 6$ of that shown by a corresponding pencil of H atoms.

Discussion of results

From the results so far obtained it is difficult to avoid the conclusion that the long-range atoms arising from collision of α particles with nitrogen are not nitrogen atoms but probably atoms of hydrogen, or atoms of mass 2. If this be the case, we must conclude that the nitrogen atom is disintegrated under the intense forces developed in a close collision with a swift α particle, and that the hydrogen atom which is liberated formed a constituent part of the nitrogen nucleus. We have drawn attention in paper III to the rather surprising observation that the range of the nitrogen atoms in air is about the same as the oxygen atoms, although we should expect a difference of about 19 per cent. If in collisions which give rise to swift nitrogen atoms, the hydrogen is at the same time disrupted, such a difference might be accounted for, for the energy is then shared between two systems.

It is of interest to note, that while the majority of the light atoms, as is well known, have atomic weights represented by $4n$ or $4n + 3$ where n is a whole number, nitrogen is the only atom which is expressed by $4n + 2$. We should anticipate from radioactive data that the nitrogen nucleus consists of three helium nuclei each of atomic mass 4 and either two hydrogen nuclei or one of mass 2. If the H nuclei were outriders of the main system of mass 12, the number of close collisions with the bound H nuclei would be less than if the latter were free, for the α particle in a collision comes under the combined field of the H nucleus and of the central mass. Under such conditions, it is to be expected that the α particle would only occasionally approach close enough to the H nucleus to give it the maximum velocity, although in many cases it may give it sufficient energy to break its bond with the central mass. Such a point of view would explain why the number of swift H atoms from nitrogen is less than the corresponding number in free hydrogen and less also than the number of swift nitrogen atoms. The general results indicate that the H nuclei, which are released, are distant about twice the diameter of the electron (7×10^{-13} cm.) from the centre of the main atom. Without a knowledge of the laws of force at such small distances, it is difficult to estimate the energy required to free the H nucleus or to calculate the maximum velocity that can be given to the escaping H atom. It is not to be expected, *a priori*, that the velocity or range of the H atom released from the nitrogen atom should be identical with that due to a collision in free hydrogen.

Taking into account the great energy of motion of the α particle expelled from radium C, the close collision of such an α particle with a light atom seems to be the most likely agency to promote the disruption of the latter; for the forces on the nuclei arising from such collisions appear to be greater than can be produced by any other agency at present available. Considering the enormous intensity of the forces brought into play, it is not so much a matter of surprise that the nitrogen atom should suffer disintegration as that the α particle itself escapes disruption into its constituents. The results as a whole suggest that, if α particles—or similar projectiles—of still greater energy were available for experiment, we might expect to break down the nucleus structure of many of the lighter atoms.

I desire to express my thanks to Mr. William Kay for his invaluable assistance in counting scintillations.

University of Manchester
April 1919

THE END